Handbook of Animal Lectins

PROPERTIES AND BIOMEDICAL APPLICATIONS

Handbook of Animal Lectins

PROPERTIES AND BIOMEDICAL APPLICATIONS

A compendium of galectins, collectins, selectins, pentraxins and other carbohydrate-binding proteins from throughout the animal kingdom

DAVID C. KILPATRICK

Scottish National Blood Transfusion Service, Edinburgh, UK

JOHN WILEY & SONS, LTD
Chichester · New York · Weinheim · Brisbane · Singapore · Toronto

Baffins Lane, Chichester,
West Sussex PO19 1UD, England

National 01243 779777
International (+44) 1243 779777
e-mail (for orders and customer service enquiries): cs-books@wiley.co.uk
Visit our Home Page on: http://www.wiley.co.uk
or http://www.wiley.com

Other Wiley Editorial Offices

John Wiley & Sons, Inc., 605 Third Avenue,
New York, NY 10158-0012, USA

WILEY-VCH Verlag GmbH, Pappelallee 3,
D-69469 Weinheim, Germany

Jacaranda Wiley Ltd, 33 Park Road, Milton,
Queensland 4064, Australia

John Wiley & Sons (Asia) Pte Ltd, 2 Clementi Loop #02-01,
Jin Xing Distripark, Singapore 129809

John Wiley & Sons (Canada) Ltd, 22 Worcester Road,
Rexdale, Ontario M9W 1L1, Canada

Library of Congress Cataloging-in-Publication Data

Kilpatrick, David C.
Handbook of animal lectins : properties and biomedical applications : a compendium of galectins, collectins, selectins, pentraxins and other carbohydrate-binding proteins from throughout the animal kingdom / David C. Kilpatrick.
p.cm.
Includes bibliographical references and index.
ISBN 0-471-89981-X (alk. paper)
1. Lectins – Handbooks, manuals, etc. I. Title.

QP552.L42 K556 2000
572′.68–dc21 00 032469

British Library Cataloguing in Publication Data

A catalogue record for this book is available from the British Library

ISBN 0 471 89981 X

Typeset in 10/12pt Palatino from the author's disk by Dobbie Typesetting Limited, Devon
Printed and bound in Great Britain by Biddles Ltd, Guildford, Surrey
This book is printed on acid-free paper responsibly manufactured from sustaintable forestry, in which at least two trees are planted for each one used for paper production

Contents

Preface

Around 1977, when I was a post-doctoral fellow in the University of Edinburgh, I was asked to give a seminar on the relatively new and rapidly expanding field of plant lectins. I additionally volunteered to do a companion lecture on animal lectins. At that time, a few haemagglutinins had been isolated from animal sources including snakes, snails, crustaceans, frogs and fish, but entirely new lectin-like molecules were starting to be discovered, such as the rabbit hepatic lectin and the β-galactoside binding protein (electrolectin) from the electric eel. After preparing and delivering these talks, I wrote to the editor of *Nature* suggesting a review article on the topic of animal lectins. This offer was politely declined on the grounds (as I recall) that it would not interest a wide enough readership.

In the intervening quarter-century, lectins from animal sources have been studied intensively. While knowledge of invertebrate lectins continued to accumulate, the major advances and the greatest interest were focused on the various families of carbohydrate-binding proteins described in mammals and other 'higher' animals. The literature had become diffuse and much less amenable to reviewing and summarising. When my old friend Arpad Pusztai phoned to tell me that John Wiley and Sons were looking for someone to write an animal lectin handbook and suggested I undertake the task, I soon warmed to the idea despite having reservations about the enormous scope of the subject matter.

I would therefore like to thank Arpad, whose kindness and encouragement over the many years I have known him have been much appreciated, for bringing author and publisher together. I am also indebted to Lisa Tickner, my initial contact at John Wiley, whose efforts ensured our proposal was accepted. I would like to thank Dr Shunju Natori for his critical reading of the chapter on invertebrate lectins at the pre-submission stage. Thanks are also due to the staff at the Erskine Medical Library, University of Edinburgh, and the authors who sent me offprints/preprints of their papers. Special thanks go to my wife, Alison, for helping me with the computing and for everything else; without her assistance, this book could not have been written. Finally, I am grateful to the Scottish National Blood Transfusion Service for tolerating my academic interest in lectins over the years.

D.C. Kilpatrick
December, 1999

Abbreviations

CHO (cells)	Chinese hamster ovary (cells)
CRP	C-reactive protein
CSF	Cerebrospinal fluid
Da	Daltons
DEAE	Diethylaminoethyl
DNA	Deoxyribonucleic acid
Gal	Galactose
GalNAc	*N*-acetylgalactosamine
HPLC	High performance liquid chromatography
IFN	Interferon
Ig(G,A,M)	Immunoglobulin(G,A,M)
IL	Interleukin
Mol. wt	Molecular weight (=Mr)
Mr	Relative molecular mass (=Mol. Wt)
NeuAc	*N*-acetylneuraminic acid
NK	Natural killer
PAGE	Polyacrylamide gel electrophoresis
PCR	Polymerase chain reaction
RGD	Arginine-glycine-aspartic acid (peptide)
RNA	Ribonucleic acid
RT-PCR	Reverse transcriptase-polymerase chain reaction
SDS	Sodium dodecyl sulphate

Safety Notice

Neither the author nor John Wiley & Sons Ltd. accepts any responsibility or liability for loss or damage occasioned to any person or property through using the materials, instructions, methods or ideas contained herein, or acting or refraining from acting as a result of such use. While the author and publisher believe that the data, practical procedures and other information, as set forth in this book, are in accord with current recommendations and practice at the time of publication, they accept no legal responsibility for any errors or omissions, and make no warranty, express or implied, with respect to material contained herein. Attention to safety aspects is an integral part of all laboratory procedures and national legislations impose legal requirements on those persons planning or carrying out such procedures. It remains the responsibility of the reader to ensure that any procedures that are followed are carried out in a safe manner and that all necessary safety instructions and national regulations are implemented.

In view of ongoing research, equipment modifications and changes in governmental regulations, the reader is urged to review and evaluate the information provided by the manufacturer for each lectin, piece of equipment or device, for any changes in the instructions or usage, and for added warnings and precautions.

1 Introduction to Animal Lectins

DEFINITION OF A LECTIN

The term lectin, from the Latin *legere* meaning to pick out, choose or select, was coined by W.C. Boyd [1]. It was chosen to have a precise meaning: those plant agglutinins, recognised only as activities in crude tissue extracts (usually of seeds), which displayed blood antigen specificity. Lectins agglutinated the erythrocytes of some species more readily than the erythrocytes of other species, and/or displayed blood-group specificity within the human species. A few years later [2], it was demonstrated that human blood group antigenicity depends on the terminal sugars of complex glycoconjugates, and increasingly thereafter it became apparent that the agglutinating preferences of lectins reflected their recognition of saccharide structures. By the time Sharon and Lis published their influential review of lectins in 1972 [3], they defined their subjects as 'cell agglutinating and sugar-specific proteins'. By that time, the term lectin had come into common use but had clearly departed from its original meaning. The Sharon–Lis definition also took account of the more recently acquired knowledge that lectins were proteins with the common characteristic of carbohydrate-binding ability, and that this property enabled lectins to agglutinate cells other than erythrocytes and indeed to precipitate glycoconjugates from solution.

According to George Bird [4], Boyd was relaxed about the loose usage of the term he had created, opining that language altered with time and people were welcome to use 'lectin' to convey whatever they wanted it to mean. Others were more concerned. Goldstein and others [5] in 1980 proposed that the term 'lectin' should only be used for 'a sugar-binding protein or glycoprotein of non-immune origin which agglutinates cells and/or precipitates glycoconjugates'.

The main problem with the above definition was that it excluded monovalent carbohydrate-binding proteins such as the toxins abrin and ricin, and therefore it provoked considerable opposition among researchers working in the field [6]. The term therefore continued to be used for carbohydrate-binding proteins irrespective of whether they could agglutinate cells or precipitate polysaccharides, provided they were non-immunoglobulin in nature and reversibly bound carbohydrate moieties without altering the covalent structure of the latter. Therefore carbohydrate-specific immunoglobulins and carbohydrate-specific enzymes were not to be classified as lectins [6]. By this time, it was almost universally accepted that a lectin could be of animal or indeed protist origin and not necessarily just from plants. Barondes [7] proposed that a lectin should be 'any carbohydrate-binding protein other than an enzyme or an antibody'.

Barondes felt his definition might be criticised for being so inclusive. However, I take the opposite view. Although few others seem to object, I myself am troubled by the inclusion of a phrase like 'non-immune origin' or its equivalent in any definition. It is my view that lectins are very much part of the immune systems of animals, either functioning directly in first-line defence or indirectly by promoting appropriate protective mechanisms via leukocyte traffic management or by transmitting signals between cells involved in the immune response. Most (but not all) animal lectins are clearly distinct from antibodies in structure, but there are immunoglobulins with exclusively carbohydrate specificity. Several 'natural' antibodies display specificity for simple sugars [8,9], including the so-called anti-Gal immunoglobulins responsible for preventing clinical xenotransplantation [10]. In what way are they fundamentally different from other animal lectins? I submit that such immunoglobulins should be considered as constituting a separate lectin family; after all, I-type lectins are accepted as such and are structurally defined by their immunoglobulin domain. It is perverse to include I-type adhesion molecules but not antibodies. Nevertheless, sugar-specific antibodies will not be included in this book.

Moreover, I am not convinced all enzymes should be excluded either. Obviously, most glycosidases and glycosyltransferases should not be regarded as lectins, but should some exceptions be permitted? Ricin and some other toxins have enzymic activity towards RNA, yet are widely, if not universally, regarded as plant lectins. Therefore, should those frog RNAases with carbohydrate-binding activity [11], for example, not be considered animal lectins? Additionally, some lectins consist of a carbohydrate-recognition domain coupled to a structurally unrelated domain. Should the latter invalidate the former if it has enzymic activity?

I am afraid a precise, concise and universally acceptable definition of lectin eludes us. For the purpose of inclusion in this text, a lectin shall be considered a carbohydrate-binding protein not obviously involved in carbohydrate metabolism and which does not belong to any of the main classes of immunoglobulins. Lectin-like molecules, which possess a primary structure homologous to one of the recognised animal lectin domains, will not be regarded as lectins on that basis alone and will only be included in the directory if they have been shown to have biological activity that can be inhibited by saccharides. For example, convulxin [12], a snake venom glycoprotein which structurally belongs to the C-type lectin family, will not be included since it does not appear to bind carbohydrate.

For completeness I must also define 'animal', a task superficially easier and less controversial. The difference between a squirrel and a cabbage, after all, seems self-evident even to a layperson. However, not all organisms are as readily distinguished. I am inclined to regard plants as organisms made up of single cells or tissues containing rigid cell walls protecting the plasma membranes, and animals as organisms the cells of which are bound by plasma membranes alone. By this definition cellular slime moulds like *Dictyostelium* or traditional slime moulds like *Physarum* are typically animal at one stage of their life cycles yet typically plant at the fruiting body stage. Nevertheless, the slime mould amoebae are typical animal cells, and therefore any lectins synthesised by such amoebae shall be deemed as falling within the scope of this book. Perhaps I should stress that an animal lectin is a lectin found in animals and not a lectin peculiar to animals (e.g. galectins are animal lectins, even if they occur in fungi too).

HISTORICAL OVERVIEW OF ANIMAL LECTINS

From the preceding account, it is clear that plant lectins were known as such long before the term was readily used to include carbohydrate-binding proteins of animal origin. Interestingly, Sharon and Lis [3], reviewing the literature in 1972, listed only two animal sources out of eighteen from which lectins had been isolated and studied in purified form. Nevertheless, what we now call animal lectins have been known about for as long or longer than plant lectins even though they were not recognised as such for many years after being first discovered. Indeed, I have argued elsewhere [13] that S. Weir Mitchell was the first person known to have observed lectin activity and as such he, rather than Herman Stillmark, deserves to be regarded as the father of lectin science. There is certainly no doubt that Weir Mitchell was the first to observe snake venom-mediated haemagglutination during his extensive studies on the rattlesnake *Crotalus durissus*. Fletcher and Noguchi [14] in 1902 reported the characteristics of snake venom agglutinins in more detail and were the first to liken the active principle to the antibodies of vertebrates.

Throughout the 20th century, haemagglutination activities were steadily reported in molluscs, crustaceans, reptiles, amphibians and fish. These were, oxymoronically speaking, plant lectins from animals. That is to say, they were discovered in the same way as those from plant sources, by investigators detecting their ability to agglutinate red blood cells in crude homogenates or other extracts. Some of their other properties were also plant lectin-like: blood-group specificity and mitogenic activities were not uncommon.

However, other animal lectins were discovered in other contexts, and as a result were not considered as lectins for many years afterwards. Examples include bovine conglutinin (first discovered as early as 1906) and C-reactive protein (first reported in 1930). The authors of a case report in 1968 of a sick child whose apparent immune deficiency was associated with defective opsonisation of baker's yeast [15] could not have suspected that this functional activity was due to mannan-binding lectin (MBL). Yet the latter, itself discovered independently from different interests in the early 1970s, was found to be responsible for the opsonic activity in 1989 [16], and MBL is now one of the most studied and thoroughly characterised animal lectins. To my knowledge, however, MBL has not been found to agglutinate red blood cells from any species.

Lest I be thought to overstate my case, I should emphasise that recent research into major animal lectin families arose precisely because plant agglutinin-like, haemagglutinating activity was initially observed. Electrolectin from the electric

Table 1.1. Some Milestones in the History of Animal Lectins

1860–86	Weir Mitchell's studies on rattlesnake venom [40,41]
1902	Fletcher and Noguchi's explicit description of snake venom lectin activity [14]
1906	Discovery of bovine conglutinin: first association of an animal lectin with the immune system and the first mammalian lectin [42]
1935–46	Discovery of eel lectins and their identification as potential blood typing reagents [43]
1964	Anti-A lectin in butter clam (*Saxidomus*) and beginning of gastropod lectin research [44]
1974	Discovery of the rabbit hepatic lectin/asialoglycoprotein receptor [20]
1975	Discovery of electrolectin and beginning of galectin research [17]
1989	Demonstration of the identity between mannan-binding lectin and the factor responsible for opsonisation of yeasts etc. [16]
1989	Publication of primary sequences of three independently studied adhesion molecules permitting recognition of a new family of C-type lectins, the selectins [45]

eel was reported in 1975 [17] and similar lactose-extractable activity from various solid vertebrate tissues was found soon after [18]. Another lectin initially called erythroid developmental agglutinin [19] exhibited a most interesting developmental regulation. Changes in specific activity with embryonic age were just one of many unusual properties members of the family we now call galectins had in common, and its unusual features stimulated much research. Yet another example was the discovery of a protein in rabbit liver homogenate able to agglutinate untreated human or rabbit erythrocytes [20]. In 1974, the idea of a mammalian lectin was quite strange, and for that reason, the hepatic lectin was mistakenly claimed as the first from a mammalian source.

An alternative screening approach has been to use labelled glycoconjugates or neoglycoconjugates with cells or tissue sections. For example, by probing human placental tissue sections with labelled bovine albumin to which *N*-acetylneuraminic acid (NeuAc) had been covalently attached, a sialic acid-binding lectin distinct from others in placenta could be inferred. This observation subsequently led to the isolation and characterisation of the lectin responsible [21], which was purified conventionally on immobilised bovine submaxillary mucin and monitored conventionally by haemagglutination of protease-treated rabbit red blood cells.

In the past two decades, some lectins have been identified and cloned using molecular biological techniques. By this means several lectins have been characterised without having been extracted from a tissue source. Characterisation and classification have been vastly aided by the use of computer homology searches exploiting knowledge of the structural aspects of carbohydrate-recognition domains.

Knowledge of primary structures enabled Drickamer [22] to propose that most of the animal lectins known up to the late 1980s fell into two categories: the calcium-dependent (or C-type lectins) and the sulphydryl-dependent (or S-type) lectins. Since then it has become clear that many animal lectins do not fall into these categories, and around ten structural classes are now recognised. Moreover, not all of the calcium-dependent lectins have the C-type structure, and not all of the S-type lectins (galectins) are thiol dependent.

Some discoveries that might be considered milestones in the history of animal lectin science are listed in Table 1.1.

COMPARISON OF PLANT AND ANIMAL LECTINS

Some of the differences between animal and plant lectins have been touched on in the preceding section. These are not fundamental differences, but circumstantial. Animal lectins are less likely to have been identified as haemagglutinins. Animal lectins are more likely to have been studied as endogenous receptors with potentially important roles in cell recognition and cell communication, and less likely to have been viewed largely as potential reagents.

A consequence of this emphasis on the study of animal lectins in its own right is that knowledge of the function of lectins from animal sources is much more advanced. Suggestions being made today for the possible physiological roles of plant lectins are not so different from those proposed by Boyd [23] half a century ago and repeated regularly by reviewers ever since. Although knowledge of plant lectin function has increased over the years, there has been no dramatic breakthrough, and the accumulated evidence remains circumstantial and none too persuasive. It is noteworthy, however, that one of the suggestions that has withstood the passage of time, and is currently rather popular, is that some plant lectins function in host defence ('plant antibodies'). Some animal lectins certainly do function in host defence, and, generally speaking, the evidence for putative animal lectin functions is much more compelling. Selectins are clearly involved in leukocyte homing and the initiation of inflammation; collectins are probably defence molecules, ante-antibodies with a broad recognition capacity for non-self; pentraxins may exist primarily to control and regulate inflammatory responses. There are abundant and ever-increasing data implicating animal lectins in such diverse processes as cell adhesion, cell growth regulation, embryogenesis, morphogenesis and cancer.

Another difference of major relevance when writing a book such as this is that the 'one-species/one-lectin' scenario of the plant kingdom does not apply. The entries in the *Handbook of Plant Lectins* [24] are listed alphabetically by species. While there may be two or even three distinct lectins obtained from a single plant, this is a feasible approach. It would be absurd, however, to refer to 'the mouse lectin' or 'the *Homo sapiens* lectin'. Indeed, it is no longer practicable to define a lectin simply from the tissue and species of origin. When Hirabayashi and Kasai first described what we now call galectin-1 from human placenta they called it 'human placental lectin', and that term was used for some time [25]. Yet the human placenta appears to contain numerous lectins: the sialic acid-binding lectin referred to earlier [21], the macrophage mannose receptor [26], an HIV glycoprotein receptor [27], MBL [28], and a group of plant lectin analogues detected serologically [29]. The same is true to some extent of other tissues.

Therefore, animal lectins are better described as members of families of homologues, not as unique molecules. This classification system is itself not without problems. For example, consider the galectins, a family defined by a highly conserved carbohydrate-binding domain. Galectins-1 to 10 (subfamilies) differ from each other in structure, including primary structure, as do members of the same galectin subfamily from different species. For those with a similar domain structure, like galectins-1 and -2, primary structure is definitive. Although the galectins-1 from mice and rats may resemble each other more than galectins-1 and 2 from either species, the same would not necessarily be true if a larger phylogenetic difference applied. It therefore becomes impossible to classify accurately all galectins into one of those subfamilies.

The differences between animal and plant lectins, however, are insignificant compared to the surprising resemblances. Some similarities (haemagglutination, blood-group specificity, mitogenic activity etc.) are not surprising, but the structural similarities across a vast phylogenetic divide are truly remarkable. Again, the human placenta provides us with the first example. Polyclonal antisera raised in rabbits against plant lectins were found to display characteristic cross-reactivity with placental antigens by immunohistochemistry [29]. Antibodies to wheat germ agglutinin (WGA) bound to cytotrophoblasts and fibroblasts; antibodies to Con A additionally bound to syncytiotrophoblast and amniotic epithelium; antibodies to phytohaemagglutinin (PHA) bound to endothelial cells as well as most of the structures recognised by the anti-Con A antisera. This does not necessarily mean that placenta contains plant lectins or plant lectin homologues; the primary structure of the epitopes concerned may have no sequence similarity to that of the plant protein used for immunisation, and the proteins bearing those epitopes may not have lectin (carbohydrate-binding) activity. Instead, it may be that plant and animal proteins share conformational epitopes formed by a three-dimensional arrangement of similar shape and charge.

Indeed, in very general terms, conformational similarities exist. Lectins, whether plant or animal, differ from other carbohydrate-binding proteins in having a shallow depression as the sugar binding site [30]. This topographical feature is shared with antibodies. In contrast, enzymes and transport proteins which bind carbohydrate have a deep binding site in which the ligand is engulfed. Another resemblance is the requirement for divalent cations for lectin activity, in some plant and some animal lectins. In the case of legume lectins, calcium ions act indirectly to modify the shape of the carbohydrate binding site, while in C-type animal lectins the calcium ions bind carbohydrate directly.

Yet the most remarkable finding has come from the analysis of three-dimensional structures by X-ray crystallography. The galectin-1 monomer was found to share a striking structural homology with legume lectin monomers despite the absence of primary structural homology [31]. Similar polypeptide folding was subsequently observed with human serum amyloid P (SAP) component [32] and human C-reactive protein [33]. Legume lectins, galectins and pentraxins share a characteristic β-sheet topography, the so-called jelly-roll structural motif. Although differences are apparent in the positions of the binding sites, the existence of a common fold in phylogenetically unrelated lectins appears to be a remarkable example of convergent evolution.

OUTLINE OF THIS BOOK

Gauss [34], introducing another book on lectins, indicated the extent of the vast and expanding literature involving lectins evidenced by computer searches. He concluded that there is a considerable knowledge base in the lectin field coupled with a notable lack of knowledge. This is particularly true of the animal lectin field, which has been expanding at a particularly high rate in the past decade or so. Interestingly, Gauss made the claim that 'really good knowledge retrieval requires a lectin facts database rather than a literature database'. This provides the primary motive for this book, which is an attempt to reduce the disparate, unweildy, interdisciplinary animal lectin literature to a coherent summary.

The past decade has also seen a long overdue recognition that innate immunity is of fundamental importance and not just some vestigial ancestor of the acquired immune system [35–37]. The innate immune system consists of a network of cells and proteins providing essential first-line defence, members of which interact in complex ways with each other and with agents of acquired immunity. This provides a secondary motive for this book, which tries to place lectins in their rightful context as important components of the innate immune system.

This volume is intended to be of use to the widest possible readership, from established research workers to newcomers seeking an introduction to the field. The first four chapters provide introductions, overviews and generalisations. The aim has been to be brief but comprehensive, condensing large areas of literature into digestible summaries of essential information. Inevitably this selection requires personal judgement, and inevitably there will be some readers who will not share that judgement. Undoubtedly there will be some areas neglected from poor judgement or just plain ignorance; hopefully these are in the minority and their absence will not spoil the book. My conscious biases have been towards human lectins and towards medically significant activities and applications. This reflects my personal preferences but I believe they can be objectively justified.

Chapters 2 to 4 also provide explicit indications, either from simple mention in the text or by listing in tables, of which lectins are to be found in the alphabetical directory (Chapter 5). It is also envisaged that the reader will transfer from an individual entry in the directory to the appropriate passage(s) in the introductory chapters in order to place the individual lectin of interest in its wider context.

Chapter 5 is the main part of the book and its *raison d'être*. Sometimes the entry consists of a single molecule from a particular species (e.g. *Helix pomatia* agglutinin), and such entries are arranged in alphabetical order of their most popular names, which, as in the example just given, may be simply (species) lectin/agglutinin. Lectins most intensively studied in vertebrates are more likely to be entered as family names (e.g. galectin-3) where a substantial number of homologous proteins has been reported. Sometimes an individual family member merits a separate entry. For example, electrolectin (for both phylogenetic and historical reasons), *Caenorhabditis elegans* galectin and *Geodia cydonium* galectin (because they are sufficiently different from vertebrate galectins) are given separate entries from galectins-1 to -10. (Incidentally, it should be apparent that any reader wishing to consult, for example, the directory entry for the sponge galectin but who is ignorant of the scientific name of the species would be wise to browse through the appropriate introductory section in Chapter 2 to elicit the necessary information.)

At the top of each entry in the directory chapter, below the principal lectin name, is a summary of the information which distinguishes that lectin from others and establishes its identity. The essential information for each lectin is listed in a reader-friendly way under the headings of **isolation, structure, biological activities, tissue and/or subcellular distribution** and **possible function(s)**. A final **comments** (=miscellaneous) section differs with each lectin and permits flexibility. Much is included under that heading for some lectins, less for others, and the nature of this information varies. As far as possible, only positive information is included; negative statements (like 'no applications have been described') are usually eschewed.

The following three chapters attempt to overview the entire field of animal lectins. The reader might also like to refer to other comprehensive reviews by Gabius [38] and by Gabius and Romero [39]. The former places emphasis on structural aspects, while the latter is briefer but contains a most helpful table classifying animal lectins on a functional basis.

REFERENCES

1. Boyd, W.C. & Shapleigh, E. (1954). Specific precipitating activity of plant agglutinins (lectins). *Science* **119**, 419.
2. Watkins, W.M. (1955). Blood group substances. *Science* **152**, 172–181.
3. Sharon, N. & Lis, H. (1972). Lectins: cell agglutinating and sugar-specific proteins. *Science* **177**, 949–958.
4. Bird, G.W.G. (1989). Lectins in immunohematology. *Transfusion Medicine Reviews* **3**, 55–62.
5. Goldstein, I.J., Hughes, R.C., Monsigny, M., Osawa, T. & Sharon, N. (1980). What should be called a lectin? *Nature* **285**, 66.
6. Kocourek, J. & Horejsi, V. (1983). A note on the recent discussion on definition of the term 'lectin'. In: *Lectins: Biology, Biochemistry and Clinical Biochemistry*, Volume 3 (eds T.C. Bog-Hansen & G.A. Spengler) pp. 3–6. Walter de Gruyter, Berlin.
7. Barondes, S.H. (1988). Bifunctional properties of lectins: lectins redefined. *Trends in Biochemical Sciences* **13**, 480–482.
8. Bird, G.W.G. & Roy, T.C.F. (1980). Human antibodies to melibiose and other carbohydrates. *Vox Sanguinis* **38**, 169–171.
9. Lalezari, P. & Jiang, A.F. (1984). Carbohydrate-specific antibodies in normal human sera. II Structural definition of antigenic determinants. *Vox Sanguinis* **47**, 146–156.
10. Galili, U. (1993). Interaction of the natural anti-Gal antibody with alpha-galactosyl epitopes: a major obstacle for xenotransplantation in humans. *Immunology Today* **14**, 480–482.
11. Irie, M., Nitta, K. & Nonaka, T. (1998). Biochemistry of frog ribonucleases. *Cellular and Molecular Life Sciences* **54**, 775–784.

12. Leduc, M. & Bon, C. (1998). Cloning of subunits of convulxin, a collagen-like platelet-aggregating protein from *Crotalus durissus terrificus* venom. *Biochemical Journal* **333**, 389–393.
13. Kilpatrick, D.C. & Green, C. (1992). Lectins as blood typing reagents. In: *Advances in Lectin Research* Volume 5 (ed. H. Franz) pp. 51–94. Ullstein Mosby, Berlin.
14. Fletchner, S. & Noguchi, H. (1902). Snake venom in relation to haemolysis, bacteriolysis, and toxicity. *Journal of Experimental Medicine* **6**, 277–301.
15. Miller, M.E., Seals, J., Kaye, R. & Levitsky, L. C. (1968). A familial, plasma-associated defect of phagocytosis. *Lancet* **2**, 60–63.
16. Super, M., Thiel, S., Lu, J., Levinsky, R.J. & Turner, M.W. (1989). Association of low levels of mannan-binding protein with a common defect of opsonisation. *Lancet* **2**, 1236–1239.
17. Teichberg, V.I., Silman, I., Beitsch, D.D. & Resheff, G. (1975). A beta-D-galactoside binding protein from the electric organ tissue of *Electrophorus electricus*. *Proceedings of the National Academy of Sciences USA* **72**, 1383–1387.
18. Harrison, F.L. & Chesterton, C.J. (1980). Factors mediating cell–cell recognition and adhesion. Galaptins, a recently discovered class of bridging molecules. *FEBS Letters* **122**, 157–165.
19. Harrison, F.L. & Chesterton, C.J. (1980). Erythroid developmental agglutinin is a protein lectin mediating specific cell–cell adhesion between differentiating rabbit erythroblasts. *Nature* **286**, 502–504.
20. Stockert, R.J., Morell, A.G. & Scheinberg, I.J. (1974). Mammalian hepatic lectin. *Science* **186**, 365–366.
21. Ahmed, H. & Gabius, H.-J. (1989). Purification and properties of a Ca-independent sialic acid-binding lectin from human placenta with preferential affinity to *O*-acetylsialic acids. *Journal of Biological Chemistry* **264**, 18673–18678.
22. Drickamer, K. (1988). Two distinct classes of carbohydrate-recognition domains in animal lectins. *Journal of Biological Chemistry* **263**, 9557–9560.
23. Boyd, W.C. (1963). The lectins: their present status. *Vox Sanguinis* **3**, 1–32.
24. Van Damme, E.J.M., Peumans, W.J., Pusztai, A. & Bardocz, S. (1998). *Handbook of Plant Lectins: Properties and Biomedical Applications*. John Wiley, Chichester.
25. Hirabayashi, J. & Kasai, K.-I. (1984). Human placenta beta-galactoside binding lectin. Purification and some properties. *Biochemical and Biophysical Research Communications* **122**, 938–944.
26. Lennartz, M.R., Cole, F.S., Shepherd, V.L., Wileman, T.E. & Stahl, P.D. (1987). Isolation and characterization of a mannose-specific endocytosis receptor from human placenta. *Journal of Biological Chemistry* **262**, 9942–9944.
27. Curtis, B.M., Scharnowske, S. & Watson, A.J. (1992). Sequence and expression of a membrane-associated C-type lectin that exhibits CD4-independent binding of human immunodeficiency virus envelope glycoprotein gp120. *Proceedings of the National Academy of Sciences USA* **89**, 8356–8360.
28. Kilpatrick, D.C., Bevan, B.H. & Liston, W.A. (1995). Association between mannan binding protein deficiency and recurrent miscarriage. *Molecular Human Reproduction* **1** *Human Reproduction* **10**, 2501–2505.
29. Debbage, P.L., Hanisch, U.-K., Reisinger, P.M.W. & Lange, W. (1993). Visualization of lectin-like proteins in human placenta by means of anti-plant lectin antibodies. *Anatomy and Embryology* **187**, 465–473.
30. Elgavish, S. & Shaanan, B. (1997). Lectin–carbohydrate interactions: different folds, common recognition principles. *Trends in Biochemical Sciences* **22**, 462–467.
31. Rini, J.M. (1995). Lectin structure. *Annual Review of Biophysical and Biomolecular Structure* **24**, 551–577.
32. Emsley, J., White, H.E., O'Hara, B.P., Srinivason, N., Tickle, I.J., Blundell, T.L., Pepys, M.B. & Wood, S.P. (1994). Structure of pentameric human serum amyloid P component. *Nature* **387**, 338–345.

33. Shrive, A.K., Cheetham, G.M.T., Holden, D., Miles, D.A.A., Turnell, W.G., Pepys, M.B., Bloomer, A.C. & Greenhough, T.J. (1996). Three dimensional structure of human C-reactive protein. *Nature Structural Biology* **3**, 346–354.
34. Gauss, D.H. (1993). Lectins: insights into the state of knowledge by literature searches. In: *Lectins and Glycobiology* (eds H.-J. Gabius & S. Gabius) pp. 3–5. Springer-Verlag, Berlin.
35. Fearon, D.T. (1997). Seeking wisdom in innate immunity. *Nature* **388**, 323–324.
36. Janeway, C.A. (1998). The road less travelled by: the role of innate immunity in the adaptive immune response. *Journal of Immunology* **161**, 539–544.
37. Magor, K.E. & Vasta, G.R. (1998). Ancestral immunity comes of age. *Immunology Today* **19**, 54–56.
38. Gabius, H.-J. (1997). Animal lectins. *European Journal of Biochemistry* **243**, 543–576.
39. Gabius, H.-J. & Romero, A. (1998). Structure and function of animal lectins. *Carbohydrates in Europe* **23**, 24–29.
40. Mitchell, S.W. (1860). Researches upon the venom of the rattlesnake. *Smithsonian Contributions to Knowledge* **XII**, Article VI, 89–90.
41. Mitchell, S.W. & Reichert, E.T. (1886). Researches upon the venoms of poisonous serpents. *Smithsonian Contributions to Knowledge* **XXVI**, Article I, 155.
42. Bordet, J. & Gay, F.P. (1906). Sur les relations des sensibilisatrices avec l'alexine. *Annals Institute Pasteur* **20**, 467–498.
43. Kocourek, J. (1986). In: *The Lectins* (eds I.E. Liener, N. Sharon & I.J. Goldstein) pp. 1–32. Academic Press, Orlando, FL.
44. Johnson, H.M. (1964). Human blood group A_1 specific agglutinin of the butter clam *Saxidomus giganteus*. *Science* **146**, 548–549.
45. Bevilacqua, M.P. & Nelson, R.M. (1993). Selectins. *Journal of Clinical Investigation* **91**, 379–387.

2 Invertebrate Lectins

LECTINS FROM SLIME MOULDS, PROTOZOA AND SPONGES

Lectins undoubtedly exist in all animal phyla, and probably in most tissues of all species. However, most of the animal lectins studied, either as activities in crude extracts or as purified proteins, have come from a minority of phyla. This chapter gives an overview of invertebrate lectins on a phylogenetic basis, to give some idea of the taxonomic variety of known sources and also because it is a convenient way of classifying lectins in order to introduce them to the reader. In Chapter 3, vertebrate lectins have been grouped on a structural basis, and in principle this method could be used for the invertebrate lectins also. Although invertebrate analogues of well-characterised lectins from mammals exist, direct comparison is rather difficult. For example, lectins structurally homologous to vertebrate galectins have been described in sponges and nematodes but the differences are so great they could not be placed within any one of the mammalian galectin subfamilies. Similarly, lectins from the horseshoe crab *Limulus* are homologous to mammalian pentraxins, but is limulin the equivalent of human serum amyloid P (SAP) component or C-reactive protein?

Lectins have been isolated from the simplest protozoa and metazoa, but first let us consider the two unrelated kinds of slime mould. The cellular slime moulds (Acrasiales) differ from the true or traditional slime moulds (Myxomycetes) most obviously in not forming a syncytium when individual cells aggregate. Both classes have species which secrete slime and which undergo complex life cycles from spores that germinate into unicellular amoebae and, after an interesting series of developmental changes, differentiate into fungus-like fruiting bodies.

There are about two dozen species of cellular slime moulds, of which one in particular, *Dictyostelium discoideum*, has become popular as a model for the study of eukaryotic development and differentiation [1,2]. Its life cycle may be considered to start with the germination of spores to form free-living myxamoebae (diameter 10–12 μm) feeding on bacteria and dividing by binary fission. This is referred to as the vegetative stage of the life cycle. When food becomes scarce, growth ceases. After a latent period of a few hours when the cells become more cohesive (interphase), the cells stream together forming specific end-to-end contacts in a process (aggregation) ultimately visible to the naked eye as a wave pattern. These aggregates (1–2 mm long) then topple over and migrate across the supporting surface within a sheath of secreted slime (slug stage). Finally, the cell mass rounds up again (culmination) and a sorocap is raised into the air on a narrow cylinder of dead cells. The whole differentiation process can be completed in 24 hours. Many biochemical and cell biological changes have been correlated with each of those developmental stages [2].

Rosen and co-workers first reported a carbohydrate-binding protein with the ability to agglutinate formalin-treated sheep erythrocytes in aggregating, but not vegetative, *Dictyostelium* amoebae [3]. This agglutinating activity was inhibited by galactose and related sugars, and exhibited a 400-fold increase in specific activity as the cells became more cohesive. The lectin responsible was affinity purified on Sepharose 4B and named 'discoidin' [4]. Discoidin was subsequently shown to have a cell-surface location [5] consistent with its presumed function as an adhesion molecule. Before long it was discovered that 'discoidin' consisted of major and minor haemagglutinins with distinct molecular weights and amino acid compositions [6]. So discoidin was renamed 'discoidin I' and its less abundant relative was called 'discoidin II'.

Evidence subsequently accumulated that discoidin I was more likely to function in cell–substratum adhesion than in cell–cell adhesion, and that this function was mediated by a second, non-lectin binding site [7]. The carbohydrate binding site may function to package discoidin I into multilamellar bodies for externalisation; externalised lectin was more abundant in the extracellular matrix than on the cell surface [8].

The existence in these primitive slime moulds of a galactoside-binding lectin which exhibits a dramatic developmental regulation, possesses two distinct binding sites (one for carbohydrate, the other not), and may be found intracellularly, at the cell surface or extracellularly, is remarkable and intriguing, especially as the same combination of unusual characteristics is shared with the family of (typically vertebrate) galectins. Yet discoidin I is a Ca^{2+}-dependent lectin [9] and has some primary sequence homology to the plant lectin, ricin, from *Ricinus communis* [10].

A number of other lectins have been found in other species of cellular slime moulds and have been well characterised. These include purpurin from *Dictyostelium purpureum*, mucoroidin from *Dictyostelium mucoroides* and pallidin from *Polysphondylium pallidin* (Table 2.1), as well as others detected in crude extracts from various species.

The phylogenetically unrelated slime mould *Physarum polycephalum* also contains a haemagglutinin, physarumin, associated with the multinucleate plasmodial stage of its life cycle [11]. Although not as well characterised as the discoidins, there appear to be two separate physarumins as well [12]. Additionally, *P. polycephalum* possesses two other lectin-like proteins on the plasmodial surface [13]. Tectonins I and II have been cloned; the smaller tectonin I (Mr=25 000) is homologous to the C-terminal part of tectonin II (Mr=39 000). Both molecules share with the horseshoe crab lectin L6 (tachylectin-1) six repeats (33–37 amino acids each) of a novel consensus sequence, while tectonin II has an additional N-terminal domain exhibiting homology to the B-chain of ricin. It is therefore likely that the tectonins have carbohydrate-binding ability but this has not yet been demonstrated.

Collectively, the traditional and the cellular slime moulds constitute the most primitive eukaryotic cells and it is significant that they possess a variety of

Table 2.1. Lectins from Primitive Species Listed in Chapter 5

Aaptos papillata agglutinins I, II and III (sponge)
Axinella polypoides agglutinins I and II (sponge)
Discoidin I (from *Dictyostelium discoideum*)
Discoidin II (from *Dictyostelium discoideum*)
Entamoeba histolytica lectin
Geodia cydonium galectin
Halichondria okadai lectin I
Halichondria okadai lectin II
Pallidin (from *Polysphondylium pallidum*)
Physarumin (from *Physarum polycephalum*)
Purpurin (from *Dictyostelium purpureum*)
Taglin (from *Giardia lamblia*)
Tritrichomonas foetus lectin (protozoan)
Tritrichomonas mobilensis lectin (protozoan)

carbohydrate-binding proteins, some of which have resemblances to both plant and animal lectins. The obvious inference is that protein–carbohydrate interactions originated very early in evolution and continued to be exploited as more complex life forms developed.

The protozoa are the most primitive of the true animal phyla and the majority, like slime moulds, seldom inconvenience human beings. The minority which do cause problems, however, are of enormous clinical significance [14]. Such protozoa include *Plasmodium falciparum*, the most harmful of the four malaria-causing parasites, responsible for a million human deaths a year, mostly of children. The rapid spread of resistance to the most effective prophylactic drug, chloroquine, coupled with the lack of an effective vaccine means that malaria will continue to be a major global health problem for some time to come. Other disease-causing protozoa include *Entamoeba histolytica*, which infects 10% of the world's population, causing 50 million cases of colitis and perhaps 100 000 deaths per annum. Even *Giardia lamblia* infection, which is rarely fatal, causes substantial morbidity and has become more common in developed countries with the advent of the acquired immune deficiency syndrome. Infection of their respective host cells by all of those organisms appears to be mediated by protozoal surface lectins binding to host glycoconjugates.

The presumed *P. falciparum* lectin is the most recently and least well described. Erythrocytes infected with *P. falciparum* bind to chondroitin-4-sulphate [15]. This interaction appears to be of considerable importance during pregnancy, when plasmodia can bind to chondroitin-4-sulphate on the syncytiotrophoblast. It seems that the deleterious effects of malaria on fetal development are more common in primigravid compared to multigravid women because the latter produce anti-adhesion antibodies which inhibit sequestration of plasmodia in the placenta [16]. Since these antibodies are strain-independent, immunisation with an appropriate antigen or some other means of inhibiting this putative lectin–carbohydrate interaction may ultimately provide a globally effective prevention strategy.

The surface lectin of *E. histolytica* is perhaps the most fully characterised lectin of protozoal origin. It is a 260 kDa heterodimer consisting of a 170 kDa polypeptide (with specificity for galactose and *N*-acetylgalactosamine and a modest degree of sequence homology to wheat germ agglutinin (WGA)) linked by disulphide bonds to 35 kDa subunits reported to function as receptors for fibronectin [17]. Another surface lectin has been found in *E. histolytica* [18] which may also be implicated in adhesion and virulence, while a mitogenic lectin of much smaller molecule size has been prepared from the soluble fraction of *E. histolytica* sonicates [19,20]. The relationship of these additional lectins, all of which can be inhibited by galactose or *N*-acetylgalactosamine, to the original adhesion factor is not clear.

Two more protozoal lectins are of considerable interest and importance. The flagellate *G. lamblia* possesses a unique mannose-6-phosphate-specific lectin, named taglin, which is activated by limited trypsin-mediated proteolysis [21]. It is thought that the cell-surface prolectin is converted by host trypsin in the proximal small intestine to the active form, which promotes adherence and infection [22]. The rhizopod *Hartmannella vermiformis* possesses a surface galactose/GalNAc lectin related to the corresponding molecule from *E. histolytica* [23]. This lectin permits the uptake of the legionnaires' disease bacterium, *Legionella pneumophila*, into the hartmannella amoebae. Although hartmannellae do not themselves cause disease in humans, these protozoa play a significant role in maintaining *L. pneumophila* in the environment as well as acting as vectors. It is noteworthy that *L. pneumophila* is the only known example of an intracellular pathogen which can replicate both within protozoal and mammalian cells, and it does so via protein–carbohydrate interaction.

The sponges (Porifera) constitute the most primitive metazoan animals. It has long been recognised from investigations into the species-specific reaggregation of mechanically or enzymatically dissociated sponge cells that aggregation factors are responsible for the integrity of these multicellular masses [24]. Lectins were first described in sponges by Dodds and co-workers [25] in crude extracts from *Cliona celata* and *Axinella* species. Although both *Axinella* species and *Cliona* species preparations agglutinated erythrocytes from many species, the relative activities were very different. For example, *Cliona* species lectin(s) showed equal activity towards goose and mouse red blood cells, whereas *Axinella* species lectin(s) only agglutinated goose erythrocytes. Shortly afterwards it was found that these agglutinins could be inhibited by galactose and related saccharides [26]. Later, two distinct lectins from *A. polypoides* were purified by affinity chromatography [27]. Even at this time it was thought that aggregation of sponge cells required at least two proteins acting cooperatively. The aggregation factor was known to be Ca^{2+}-dependent while these purified lectins were not, and therefore it was thought that the aggregation factor might act to bridge lectins attached to adjacent cell surfaces.

In fact, good evidence for a similar concept came from studies with another sponge, *Geodia cydonium* [28]. By binding to the 'anti-aggregation receptors', the lectin converts reversibly aggregation-deficient cells into aggregation-susceptible cells, which can be brought together in the presence of the Ca^{2+}-dependent aggregation factor. The lectin concerned has now been cloned and found to be related to the vertebrate galectins [29]. The workers responsible provided compelling evidence that both the galectin and the heterogeneous Ca^{2+}-dependent aggregation factor are essential components of the aggregation process, and concluded that the aggregation factor binds to aggregation receptors on cell surfaces via galectin bridges.

These and related studies indicate that during the evolution of unicellular eukaryotes to multicellular animals, lectins (adhesion molecules) and their receptors played a fundamental role. We may tentatively conclude that lectins began their evolutionary journey as adhesion molecules promoting cell recognition and specific adhesion. It is likely that lectins have played a fundamental role in the development of all animals as diversification of function took place within evolutionary niches.

MOLLUSC LECTINS

The phylum Mollusca includes animals varying in size from tiny clams to the largest invertebrate, the giant squid. The best-known species are found in three classes: the gastropods (snails and slugs), the pelecypods (bivalves) and the cephalopods (squids,etc). Lectins have been found in representatives of all three of those classes.

The gastropods comprise a very numerous and widely distributed class of substantial importance to humans, since some species are edible, some are garden/agricultural pests, and some serve as intermediate hosts of parasites causing major health problems (e.g. schistosomiasis). Shortly after the brief report of an anti-blood group A lectin in the butter clam [30], another lectin of similar blood-group specificity was discovered in the snail *Helix pomatia* [31]. The *H. pomatia* agglutinin has become one of the most studied and utilised of all the animal lectins, with applications as a reagent both in blood grouping and in cancer diagnostics (see Chapter 4). Several other species of snail possess similar GalNAc-binding lectins. Notably different, however, is the lectin from the haemolymph of *Achatina fulica*, or achatinin$_H$. This lectin not only recognises sialic acid, rather than GalNAc, but displays a narrow specificity for 9-*O*-acetylneuraminic acid [32]. Other gastropod lectins with a more typical sialic acid specificity include those from the snail *Cepaea hortensis* [33] and the slug *Limax flavus* [34]. Another land slug, *Incilaria fruhstorferi*, is the source of another lectin activity towards human erythrocytes. Three polypeptides of 15 kDa, named incilarin A, B and C, were isolated from surface mucus, but although the authors claim the first sequence report of a molluscan C-type lectin, the isolated polypeptides lost their haemagglutinating activity (presumably due to denaturation) and characterisation was therefore restricted [35]. Nevertheless, it is interesting that all three exhibited some sequence homology to chicken hepatic lectin. The *L. flavus* lectin has also been cloned and its amino acid sequence found to have significant similarity to sequences in human tenascin-C, P35 (human L-ficolin) and porcine ficolin [36].

Table 2.2. Mollusc Lectins Listed in Chapter 5

Achatinin$_H$ (from *Achatina fulica*)
Anadarin MS (from *Anadara granosa*)
Anadarin P (from *Anadara granosa*)
Aplysia kurodai skin lectin
Biomphalaria glabrata lectin-I
Cepaea hortensis sialic acid-binding lectin
Helix pomatia agglutinin
Limax flavis (slug) lectin
Modiolus modiolus lectins H and E (modiolin E/modiolin H)
Pinctada fucata (pearl oyster) lectin
Saxidomus purpuratus isolectins
Tridacna derasa lectin
Tridacna maxima lectin (tridacnin)

Although the best-known mollusc lectins occur in gastropods, numerous lectin activities have been reported in bivalves. In the haemolymph of several of the clams and oysters investigated, the lectin is the major component. These agglutinins are typically Ca^{2+}-dependent proteins with specificity for galactose and related structures; they generally possess a high molecular weight in native form and consist of many much smaller subunits. Several have been well characterised (Table 2.2), and are entered in the directory in Chapter 5. Quite a variety of lectins with other specificities, however, have been found in this class of mollusc. These include the Ca^{2+}-independent galactose-binding lectin from *Aplysia kurodai* skin [37], the sialic acid-binding lectin from horse mussel (*Modiolus modiolus*) haemolymph [38], the GlcNAc-binding lectin from *Saxidomus purpuratus* [39], the heparin-binding lectin from the clam *Anadara granosa* [40] and the galacturonic acid-binding lectin from the sea hare *Aplysia depilans* [41].

Lectins have been found in gonadal tissue of several species of sea hares of the genus *Aplysia*. The *A. depilans* lectin is a glycoprotein with an apparent molecular weight of 55–60 kDa determined by gel filtration which is mitogenic for human lymphocytes and strongly agglutinates human erythrocytes [41]. The presence of lectins in the reproductive organs of species throughout the phylum Mollusca may suggest a specific functional role in reproduction, but it may equally be speculated that they serve a defensive purpose, protecting the tissues that ensure the continuation of the species. Certainly it is arguable that mollusc haemolymph lectins function in defence. Several have been found to agglutinate bacteria [42,43] and to act as opsonins [42,44]. Although strong evidence is lacking, it can be said that the known properties of mollusc lectins are consistent with a primary role as defence molecules.

ANNELID LECTINS

Although lectin activities have been found when sought in several species of worms belonging to the phylum Annelida, none has been really well characterised in purified form. Among the best characterised is 'amphitritin' from *Amphitrite ornata*, a Ca^{2+}-independent agglutinin with limited specificity for human blood group A erythrocytes and the ability to inhibit the growth of Ehrlich ascites tumours in mice [45]. Lectins from earthworms have also been the subject of several studies [46–49]. No annelid lectins appear in Chapter 5, and the main purpose of this brief section is to refer the interested reader to the publications already cited and to the recent discussion of earthworm lectins and haemolysins by Eue and colleagues [50].

ARTHROPOD LECTINS

It is no exaggeration to state that most animal species belong to the phylum Arthropoda. Its members, which include horseshoe crabs, scorpions, spiders, ticks, crustaceans, insects, centipedes and millipedes, have undergone extreme adaptive radiation, exploiting marine, freshwater and terrestrial habitats to the full. General reviews of arthropod lectins and their putative functions in recognition and defence include those by Vasta and Marchalonis [51] and Mandal and Mandal [52].

The horseshoe crabs are chelicerates and as such are rather distantly related to the crustacean crabs familiar to Europeans. Nevertheless, horseshoe crabs were one of the first sources of animal lectins to be discovered; interest in *Limulus polyphemus* lectins was revived by workers such as Cohen [53] in the 1960s, and studies on that and other species have continued up to the present time. The first attempts at isolating 'limulin' (as the major lectin came to be known) from *L. polyphemus* haemolymph by starch gel electrophoresis/gel filtration [54] or by ion-exchange chromatography/gel filtration [55] resulted in a 20–30-fold purification of an apparently homogeneous protein of 335–400 kDa made up of non-covalently associated subunits of 19–22.5 kDa. However, other purification schemes incorporating an affinity chromatography step [56,57] yielded a similar apparently homogeneous protein but with a purification ratio of 1700–3000-fold. Subsequently, Roche *et al.* [58] devised an affinity chromatographic procedure achieving the separation of two Mn^{2+}-dependent agglutinins of low specific activities ('limulin-I' and 'limulin-II') from the Ca^{2+}-dependent, sialic acid-binding principal lectin, 'limulin-III'. Even the affinity purified limulin [56] was found to consist mainly ($>90\%$) of limulin-II with less than 10% present as limulin-III. Highly purified limulin was found to bind phosphorylcholine and cross-reacted with an ovine antiserum to the C-reactive protein (CRP) in rabbits [59]; therefore limulin was thought to be an invertebrate homologue of mammalian CRP. This deduction was consistent with the pentraxin-like ring structure observed by electron microscopy many years previously in what was mistakenly thought to be pure limulin [60]. However, it is now clear that there are at least two pentraxins that could be regarded as CRP equivalents in what have previously been regarded as the most highly purified limulin preparations [61]. These two pentraxins are closely related, but the abundant form has no lectin activity and may be considered the true *Limulus* CRP, while the sialic acid-binding lectin is a minority component. It therefore follows that virtually all the information accumulated about 'limulin' summarised in Chapter 5 was unwittingly obtained from partially purified material, which explains some of the minor discrepancies regarding its characteristics reported in the literature. Moreover, a further complication is the fact that *L. polyphemus* haemolymph undoubtedly contains other lectins, including other sialic acid-binding lectins [58,61,62].

Limulin, albeit not in genuinely homogeneous form, is undoubtedly the best known and most studied of the horseshoe crab lectins and arguably the best known of all the arthropod lectins. Polyphemin, a teichoic acid-binding protein [63] and the 18 kDa agglutinin-aggregation factor [64] are among the other lectins found in *L. polyphemus*. Two other species of chelicerates have been found to possess lectins: the Japanese horseshoe crab (*Tachypleus tridentatus*) and the Indian horseshoe crab (*Carcinoscorpius rotunda cauda*). The haemolymph of the former appears to contain a bewildering complexity of lectins/isolectins, some of which are antigenically distinct and exhibit differing species specificity for erythrocytes [65,66]. Shimizu *et al.* [65] obtained four fractions with lectin activity towards human erythrocytes but concentrated on the major lectin, which was calcium-independent, had subunits of apparent Mr 23 000 and a strong preference for human erythrocytes relative to those of other species. Curiously, although 90% of lectin activity bound to their bovine submaxillary mucin–Sepharose affinity column, less than 23% was recovered in total, implying that most of the lectin activity in the haemolymph remained attached to the column even after exposure to three separate eluants. This loss of activity is in marked contrast to the work of Shishikura and Sekiguchi [66], who obtained a quantitative recovery from a similar column and appeared to achieve a massive increase in total activity (loss of inhibitor?) during the course of further purification. These workers concentrated on a lectin which agglutinated equine, but not human, erythrocytes. They obtained an apparently homogeneous protein with a single subunit of apparent Mr 42 000. The best-characterised lectins from this species are four haemocyte lectins ('tachylectins') and one sialic acid-specific lectin from haemolymph plasma (Tables 2.3 and 2.4). Very recently, two other homologous lectins (tachylectins-5) have been purified to homogeneity [67]. The Indian horseshoe crab *Carcinoscorpius rotunda cauda* appears to contain one major haemolymph lectin which has been well characterised (Table 2.3).

Lectins occur widely among species of the Arachnida, the spiders and scorpions [51,52]. These are often sialic acid-binding agglutinins from haemolymph, which have been studied in crude extracts from the standpoint of serological relatedness and specificity for erythrocytes from different species. An exception is the Indian scorpion *Heterometrus granulomanus* [68], which has been studied in purified form.

Multiple lectins have been found in several species of crustaceans, including the prawn *Macrobrachium rosenbergii* [69], the lobster *Homarus americanus* [70] and the acorn barnacle *Megabalanus rosa* [71]. In each instance, at least one of the lectins is specific for sialic acid. For example, of the two major *H. americanus* lectins differing in molecule size, electrophoretic mobility and erythrocyte species specificity, the larger lectin (which aggutinates both human and murine

Table 2.3. Arthropod Lectins Listed in Chapter 5

Acorn barnacle isolectins (from *Megabalanus rosa*)
Allomyrina dichotoma isolectins
Blaberus discoidalis lectin-1
Blaberus discoidalis lectin-2
Blaberus discoidalis β-1,3-glucan-specific lectin
Bombyx mori β-1,3-glucan-binding protein
Cancer antennarius lectin
Carcinoscorpin (from *Carsinoscorpius rotunda*)
Charybdis japonica blood type B lectin
Diogenes affinis agglutinin
Drosophila melanogaster lectin
Galleria mellonella β-1,3-glucan-binding lectin
Granulocytin (from *Sarcophoga peregrina*)
Heterometrus granulomanus lectin (scorpin)
Homarus americanus (lobster) agglutinins I and II
Limulin (from *Limulus polyphemus*)
Liocarcinus depurator lectin
Melanoplus sanguinipes (grasshopper) lectin
Pacifasticus leniusculus (crayfish) β-1,3-glucan-binding protein
Periplaneta americana (original) lectin
Periplaneta americana lipopolysaccharide-binding protein
Periplaneta americana 26 kDa lectin
Phalera flavescens lectin
Pieris brassicae (butterfly) lectin
Regenectin (from *Periplaneta americana*)
Sarcophaga peregrina humoral lectin
Scylla serrata lectin
Scyllin (from *Scylla serrata*)
Spodoptera exigua lectin
Tachylectin-1 (limulus lectin L-6)
Tachylectin-2 isolectins (from *Tachypleus tridentatus*)
Tachylectin-3 (from *Tachypleus tridentatus*)
Tachylectin-4 (from *Tachypleus tridentatus*)
Tachylectins-5A and -5B (from *Tachypleus tridentatus*)
Tachypleus tridentatus sialic acid-binding lectin
Teleogryllus commodus (cricket) lectin

cells) is sialic acid-specific, whereas the smaller lectin (which agglutinates murine but not human cells) is specific for GalNAc [70].

Insect lectins, and in particular their role in immunity and in relation to parasite–vector interactions, were thoroughly reviewed a decade ago by Ingram and Molyneux [72]. It seems that biosynthesis of insect lectins occurs mainly in the fat body, although some lectins may be produced by sex organs or haemocytes. Many of the known insect lectins are specific for galactose, but insect agglutinins are heterogeneous with regard to sugar specificity, and examples of lectin activities specific for sialic acid, GlcNAc, glucuronic acid/heparin, rhamnose/fucose, sucrose/stachyose, GalNAc/fucose, glucose and sulphated polysaccharides are cited [72].

Table 2.4. Comparison of Well-Characterised Lectins from *Tachypleus tridentatus*

	TL-1	TL-2	TL-3	TL-4	SABL
Mr (native)	24 000	27 000	28 000	470 000	530 000
Mr (subunit)	24 000	27 000	14 000	30 000 31 500	30 000 32 000
Glycoprotein?	No	No	No	Yes	No
LPS-binding?	Yes	No	No	Yes	?
Ca^{2+}-dependent?	Yes	No	No	Yes	Yes
pI	9.7	?	9.1	6.1	?
RBC specificity	None	hA	hA	hA	Equine
Sugar specificity	gal	GlcNAc	GAPS	Fuc/SA	SA

Abbreviations: GAPS, group A pentasaccharide; hA, human blood group A; gal, galactose; Fuc, fucose; SA, sialic acid; GlcNAc, *N*-acetylglucosamine.

Among the best-known insect lectins are those from the flesh fly *Sarcophaga peregrina* [73] and the beetle *Allomyrina dichotoma* [74], both with galactose-type specificity. The *Sarcophaga* lectin has been particularly well studied, and is of particular interest because it appears to have at least two independent functions. First of all, like many insect haemolymph agglutinins, it is believed to function in innate immunity. This assumption is supported by its acute-phase protein-like synthesis in response to injury and its immunological properties, which include the ability to recognise and opsonise foreign material [75]. Secondly, the same lectin appears to have a crucial role in development. There is experimental evidence that *Sarcophaga* lectin influences the normal differentiation of imaginal discs into adult structures [76].

Two other groups of insect lectins are worth a brief mention. Since the original discovery of haemagglutinating activity in the American cockroach in 1971 [77,78], four separate lectins have been characterised from an extended family [79]. Another cockroach, *Blaberus discoidalis*, as well as possessing multiple galactose-specific lectins [80], contains a β-1,3-glucan-specific lectin [81]. Lectins of this specificity, first reported in haemocytes of *Galleria mellonella* [82], also occur in a silkworm [83] as well as in various cockroach species from several genera [81]. These and other endogenous insect lectins can activate the prophenoloxidase system, an important complement-like component of the insect innate immune system [81,84]. Indeed, it has been shown for *B. discoidalis* that multiple endogenous lectins are capable of enhancing the phagocytosis of a wide variety of micro-organisms [84].

Most insect agglutinins have not been isolated and characterised, but, in addition to those mentioned above, there are a few exceptions which have been described in sufficient detail to be included in Chapter 5 (Table 2.3).

Arthropod lectins, like invertebrate lectins generally, were initially investigated as possible components of invertebrate immune systems, and early studies on structure were conducted to ascertain the degree of resemblance to vertebrate immunoglobulins. It has long since become apparent that invertebrate lectins have little in common with the domains of immunoglobulins, but the original hypothesis that lectins fulfil an immunoglobulin-like role remains plausible. It is clear that lectins with multiple specificities exist, not only between species but within the same species. Lectins are usually found in the haemolymph and often two or more lectins of different specificity co-exist within that tissue. Many have been shown to possess anti-bacterial properties and some have been found to be opsonic. Many arthropod lectins have been found to have specificity for sialic acid in some linkages, a sugar commoner in bacteria than on cells from arthropod species. Sialic acid-binding lectins are rare in the plant kingdom, with only seven listed in the companion volume to this text [85], and most of those come from species of the same genus. Recently, isolectins from mulberry (*Morus alba*) leaves were found to be highly specific for *N*-glycolylneuraminic acid [86], a specificity otherwise only reported for the crab (*Scylla serrata*) lectin [87]. In contrast to plants, sialic acid-binding lectins occur frequently in animals, vertebrates as well as invertebrates. The arthropods and molluscs together provide a rich variety of such lectins. A list of the lectins recovered from the phyla overviewed so far summarising their different erythrocyte specificities is provided for easy reference (Table 2.5).

Table 2.5. Comparison of Invertebrate Sialic Acid-Binding Lectins

Name/source	Erythrocyte specificity					
	Human	Horse	Mouse	Sheep	Rabbit	Chicken
Achatinin$_H$	-	-	-	-	+++	-
B. glabrata-I	+++(A)	NR	-	NR	-	-
C. antennarius	-	++	+++	-	+	-
Carcinoscorpin	NR	+++	NR	NR	+	NR
C. hortensis	++	-	++	+++	-	-
LAg-1	+++	+	+++	++	+	++
L. flavus	++	++	++	++	+++	++
L. depurator	+	++	+++	+	+	NR
Limulin	++	+++	++	+	+++	+
Modiolin	+	+++	++	+	+	NR
Scorpin	+++	NR	NR	NR	NR	NR
S. serrata	+++(B)	+	+++	-	+++	-
Tachylectin-4	+++(A)	-	NR	-	-	NR
T. tridentatus	+	+++	NR	NR	NR	NR
TFL	+++	+++	NR	NR	++	NR
TML	+++	+++	+++	NR	NR	NR
Walker insect	+++	-	NR	+	-	+++

Relative titres towards the cells of different species are arbitrarily denoted as +, ++ and +++; NR, not reported.

ECHINODERM LECTINS

The Echinodermata is by far the best known of the four deuterostome phyla containing exclusively invertebrate animals. The division of higher multicellular species into proterostomes and deuterostomes, based definitively on the embryological origins of the mouth and anus, is considered to be of major taxonomic importance. The echinoderms may be regarded as among the most advanced animals in evolutionary terms, rather close phylogenetically to tunicates and vertebrates.

Echinoderms are mainly small, but never microscopic, marine animals belonging to several classes, including starfish (Asteroidea), sea urchins (Echinoidea) and sea cucumbers (Holothuroidea). Characteristic features of these animals include an internal skeleton and an enlarged coelomic cavity. The coelomic fluid bathes and presumably protects the internal tissues and organs.

Lectins have long been known to occur in numerous echinoderms [88]. Multiple lectins have been purified and characterised from at least three species: the sea urchin, *Anthocidaris crassispina* [89,90] and the sea cucumbers *Cucumaria echinata* [91] and *Stichopus japonicus* [92,93]. Of particular interest is the haemolytic lectin CEL-III from *C. echinata* [94]; it appears to be a pore-forming protein, which, in a single molecule, combines the functional equivalent of vertebrate antibody and complement system.

Echinoderm lectins appearing as separate entries in Chapter 5 are listed in Table 2.6.

Table 2.6. Echinoderm Lectins Listed in Chapter 5

Anthocidaris crassispina egg lectin (SUEL)
Cucumaria echinata lectins I and II
Cucumaria echinata lectin III (CEL-III)
Echinoidin (from *Anthocidaris crassispina*)
Echinonectin (from *Lytechinus variegatus*)
Stichopus japonicus coelomic plasma lectin-1 (SPL-1)
Stichopus japonicus coelomic plasma lectin-2 (SPL-2)
Stichopus japonicus lectin I (SJL-I)
Stichopus japonicus lectin II (SJL-II)

PROTOCHORDATE (TUNICATE) LECTINS

The most advanced organisms in evolutionary terms belong to the phylum Cordata. Most cordates are vertebrates so their lectins are overviewed in the next chapter. That leaves the invertebrate cordates (protochordates), and in practice only the urochordates, or tunicates as they are more often known. All tunicates exhibit a dorsal hollow nerve cord, a notocord and pharyneal clefts at some stage in their life cycles (only the last is present in adults), indicating their phylogenetic closeness to the vertebrates, yet it may be difficult to appreciate that these sea squirts and other small marine animals are the closest relatives vertebrates have. The highly specialised present-day tunicates are probably some way off the main evolutionary stem of the Cordata, but are of unique interest as a means to search for ancestral characteristics of vertebrates that were established before their divergence from invertebrates. In this respect, the discoveries of an ascidian mannan binding lectin-associated serine protease [95] and an ascidian homologue of mammalian complement component C3 [96] are of major significance.

Lectins occur widely in the plasma of ascidian tunicates [97]. The first tunicate lectins to be purified were the D-galactose-specific lectins from *Ascidia malaca* [98], the sialic acid-specific lectin from *Halocynthia pyriformis* [99], the D-galactose-specific lectins from *Didemnum candidum* [99], and the α-lactose-specific lectins from *Phallusia mamillata* [100], but these were not thoroughly characterised at the time. Conversely, the two lactose-specific lectins (HA-1 and HA-2) from *Botrylloides leachii* were perhaps the first tunicate lectins to be described in considerable detail but only in partially purified form [101]. HA-1 is a Ca^{2+}-dependent, D-galactose-specific haemagglutinin of Mr 200 000 with a narrow species specificity for guinea-pig erythrocytes. HA-2 is smaller (Mr 63 000), agglutinates erythrocytes from a wide variety of species (but with a marked preference for pigeon cells), does not require divalent cations for activity and is most readily inhibited by lactose. HA-2 is also active as an opsonin [102].

The galactose-specific lectin from *Halocynthia roretzi* was arguably the first to be well characterised in purified form [103,104], although by the time of the second report, an even more detailed characterisation of what had been revealed as two isolectin families (DCL-1 and DCL-2) from *D. candidum* was appearing in the literature [105,106]. The *Polyandrocarpa misakiensis* (TC-14) lectin was the first tunicate lectin to be completely sequenced [107]. While the *P. misakiensis* lectin is a 'primitive' C-type lectin (see Chapter 3), the four fucosyl-binding lectins from *Clavelina picta* are related to the pentraxins, C-reactive protein and serum amyloid P component [108]. More recently, a putative C-type lectin has been cloned from *Botryllus schlosseri* which possesses an immunoglobulin domain as well as a carbohydrate-recognition domain [109].

Tunicate lectins listed in the Directory (Chapter 5) are given in Table 2.7. A review of humoral factors in tunicates (including lectins) and their possible role in host defence is given by Saito [110].

Table 2.7. Tunicate Lectins Listed in Chapter 5

Botryllus schlosseri galectin-like lectin
Didemnum candidum I and II isolectin families
Didemnum ternatanum lectin
Halocynthia roretzi lectin
Polyandrocarpa misakiensis lectin
Styela clava opsonin

REFERENCES

1. Bonner, J.T. (1967). *The Cellular Slime Moulds*, 2nd Edn. Princeton University Press, Princeton, NJ.
2. Loomis, W.F. (1975). Dictyostelium discoideum. *A Developmental System*. Academic Press, New York.

3. Rosen, S.D., Kafka, J.A., Simpson, D.L. & Barondes, S.H. (1973). Developmentally regulated carbohydrate binding protein in *Dictyostelium discoideum*. *Proceedings of the National Academy of Sciences USA* **70**, 2554–2557.
4. Simpson, D.L., Rosen, S.D. & Barondes, S.H. (1974). Discoidin, a developmentally regulated carbohydrate binding protein in *Dictyostelium discoideum*. Purification and characterisation. *Biochemistry* **13**, 3487–3493.
5. Chang, C.-M., Reitherman, R.W., Rosen, S.D. & Barondes, S.H. (1975). Cell surface location of discoidin, a developmentally regulated carbohydrate-binding protein from *Dictyostelium discoideum*. *Experimental Cell Research* **95**, 136–142.
6. Frazier, W.A., Rosen, S.D., Reitherman, R.W. & Barondes, S.H. (1975). Purification and comparison of two developmentally regulated lectins from *Dictyostelium discoideum* – discoidin I and II. *Journal of Biological Chemistry* **250**, 7714–7721.
7. Gabius, H.-J., Springer, W.R. & Barondes, S.H. (1985). Receptor for the cell binding site of discoidin I. *Cell* **42**, 449–456.
8. Barondes, S.H. (1986). Lectins in cellular slime molds. In: *The Lectins* (eds I.E. Liener, N. Sharon & I.J. Goldstein), pp. 467–491. Academic Press, Orlando.
9. Alexander, S., Cebulsky, A.M. & Lerner, R.A. (1983). Ion dependence of the discoidin I lectin from *Dictyostelium discoideum*. *Differentiation* **24**, 209–212.
10. Robertus, J.D. & Ready, M.P. (1984). Ricin B chain and discoidin I share a common primitive protein fold. *Journal of Biological Chemistry* **259**, 13953–13956.
11. Yokota, M. & Nitta, K. (1996). Purification and some properties of haemagglutinin from the Myxomycete, *Physarum polycephalum*. *Experientia* **52**, 544–548.
12. Morita, M. & Nishi, A. (1992). Isolation and partial characterisation of haemagglutinins from plasmodia of *Physarum polycephalum*. *Journal of General Microbiology* **138**, 619–625.
13. Huh, C.-G., Aldrich, J., Mottahedeh, J., Kwon, H., Johnson, C. & March, R. (1998). Cloning and characterisation of *Physarum polycephalum* tectonins. *Journal of Biological Chemistry* **273**, 6565–6574.
14. Plorde, J.J. (1990). Introduction to pathogenic parasites: pathogenesis and chemotherapy of parasitic diseases. In: *Medical Microbiology. An Introduction to Infectious Diseases*, 2nd edn (ed. J.C. Sherris) pp. 683–700. Elsevier, New York.
15. Rogerson, S.J., Chaiyaroj, S.C., Ng, K., Reeder, J.C. & Brown, G.V. (1995). Chondroitin sulphate A is a cell surface receptor for *Plasmodium falciparum*-infected erythrocytes. *Journal of Experimental Medicine* **182**, 15–20.
16. Fried, M., Nosten, F., Brockman, A., Brabin, B.J. & Duffy, P.E. (1998). Maternal antibodies block malaria. *Nature* **395**, 851–852.
17. Tannich, E., Ebert, F. & Horstmann, R.D. (1991). Primary structure of the 170 kDa surface lectin of pathogenic *Entamoeba histolytica*. *Proceedings of the National Academy of Sciences USA* **88**, 1849–1853.
18. Cheng, X.-J., Tsukamoto, H., Kaneda, Y. & Tachibana, H. (1998). Identification of the 150 kDa surface antigen of *Entamoeba histolytica* as a galactose- and *N*-acetyl-D-galactosamine-inhibitable lectin. *Parasitology Research* **84**, 632–639.
19. Ravdin, J.I., Murphy, C.F., Salata, R.A., Guerrant, R.L. & Hewlett, E.L. (1985). *N*-Acetyl-D-galactosamine-inhibitable adherence lectin of *Entamoeba histolytica*. I. Partial purification and relation to amoebic virulence in vitro. *Journal of Infectious Diseases* **151**, 804–815.
20. Salata, R.A. & Ravdin, J.I. (1985). *N*-Acetyl-D-galactosamine-inhibitable adherence lectin of *Entamoeba histolytica*. II. Mitogenic activity for human lymphocytes. *Journal of Infectious Diseases* **151**, 816–822.
21. Ward, H.D., Lev, B.I., Kane, A.V., Keusch, G.T. & Pereira, M.E.A. (1987). Identification and characterisation of taglin, a mannose-6-phosphate binding, trypsin-activated lectin from *Giarda lamblia*. *Biochemistry* **26**, 8669–8675.
22. Ward, H.D., Keusch, G.T. & Pereira, M.E.A. (1990). Induction of a phosphomannosyl binding lectin activity in *Giardia*. *Bioessays* **12**, 211–215.

23. Venkataraman, C., Haack, B.J., Bondada, S. & Kwaik, Y.A. (1997). Identification of a Gal/GalNAc lectin in the protozoan *Hartmannella vermiformis* as a potential receptor for attachment and invasion by the Legionnaires' disease bacterium. *Journal of Experimental Medicine* **186**, 537–547.
24. Winzler, R.J. (1970). Carbohydrates in cell surfaces. In: *International Review of Cytology*, Volume 29 (eds G.H. Bourne & J.F. Daniello), pp. 77–125. Academic Press, New York.
25. Dodds, R.Y., MacLennan, A.P. & Hawkins, D.C. (1968). Haemagglutinins from marine sponges. *Vox Sanguinis* **15**, 386–391.
26. Khalap, S., Thompson, T.E. & Gold, E.R. (1970). Haemagglutination and haemagglutination-inhibition reactions of extracts from snails and sponges. I. Agglutination of human and various animal red cells: its inhibition by sugars and amino sugars. *Vox Sanguinis* **18**, 501–526.
27. Bretting, H. & Kabat, E.A. (1976). Purification and characterization of the aggglutinins from the sponge *Axinella polypoides* and a study of their combining sites. *Biochemistry* **15**, 3228–3236.
28. Muller, W.E.G., Kurelec, B., Zahn, R.K., Muller, I., Vraith, P. & Uhlenbruck, G. (1979). Aggregation of sponge cells. Function of a lectin in its homologous biological system. *Journal of Biological Chemistry* **254**, 7479–7481.
29. Wagner-Hulsmann, C., Bachinski, N., Diehl-Seifert, B., Blumbach, B., Steffen, R., Pancer, Z. & Muller, W.E.G. (1996). A galectin links the aggregation factor to cells in the sponge (*Geodia cydonium*) system. *Glycobiology* **6**, 785–793.
30. Johnson, H.M. (1964). Human blood group A_1, specific agglutinin of the butter clam, *Saxidomus giganteus*. *Science* **146**, 548–549.
31. Prokop, O., Rackwitz, A. & Schlesinger, D. (1965). A 'new' human blood group receptor A_{hel}. *Journal of Forensic Medicine* **12**, 108–110.
32. Mandal, C. & Basu, S. (1987). A unique specificity of a sialic acid binding lectin achatinin$_H$, from the hemolymph of *Achatina fulica* snail. *Biochemical and Biophysical Research Communications* **148**, 795–801.
33. Holm, S.E., Bergholm, A.-M., Wagner, B. & Wagner, M. (1985). A sialic acid-specific lectin from *Cepaea hortensis* that promotes phagocytosis of a group-B, type Ia, streptococcal strain. *Journal of Medical Microbiology* **19**, 317–323.
34. Miller, R.L., Collawn, J.F. & Fish, W.W. (1982). Purification and macromolecular properties of a sialic acid-specific lectin from the slug *Limax flavus*. *Journal of Biological Chemistry* **257**, 7574–7580.
35. Yuasa, H.J., Furuta, E., Nakamura, A. & Takagi, T. (1998). Cloning and sequencing of three C-type lectins from body surface mucus of the land slug, *Incilaria fruhstorferi*. *Comparative Biochemistry and Physiology Part B* **119**, 479–484.
36. Kurachi, S., Song, Z., Takagaki, M., Yang, Q., Winter, H.C., Kurachi, K. & Goldstein, I.J. (1998). Sialic acid-binding lectin from the slug *Limax flavus*. Cloning, expression of the polypeptide, and tissue localisation. *European Journal of Biochemistry* **254**, 217–222.
37. Ozeki, Y. (1998). Purification and cell attachment activity of a D-galactose-binding lectin from the skin of sea hare, *Aplysia kurodai*. *Biochemistry and Molecular Biology International* **45**, 989–995.
38. Tunkijjanukij, S., Mikkelsen, H.V. & Olafsen, J.A. (1997). A heterogeneous sialic acid-binding lectin with affinity for bacterial LPS from horse mussel (*Modiolus modiolis*) hemolymph. *Comparative Biochemistry and Physiology* **117B**, 273–286.
39. Tatsumi, M., Arai, Y. & Itoh, T. (1982). Purification and characterization of a lectin from the shellfish, *Saxidomus purpuratus*. *Journal of Biochemistry* **91**, 1139–1146.
40. Dam, T.K., Bandyopadhyay, P., Sarkar, M., Ghosal, J., Bhattacharya, A. & Choudhury, A. (1994). Purification and partial characterisation of a heparin-binding lectin from the marine clam, *Anadara granosa*. *Biochemical and Biophysical Research Communications* **203**, 36–45.
41. Gilboa-Garber, N., Susswein, A.J., Mizrahi, L. & Avichezer, D. (1985). Purification and characterization of the gonad lectin of *Aplysia depilans*. FEBS Letters **181**, 267–270.

42. Olafsen, J.A., Fletcher, T.C. & Grant, P.T. (1992). Agglutinin activity in pacific oyster (*Crassostrea gigas*) hemolymph following in vivo *Vibrio anguillarum* challenge. *Developmental and Comparative Immunology* **16**, 123–138.
43. Tunkijjanukij, S. & Olafsen, J.A. (1998). Sialic acid-binding lectin with antibacterial activity from the horse mussel: further characterization and immunolocalization. *Developmental and Comparative Immunology* **22**, 139–150.
44. Renwrantz, L. (1983). Involvement of agglutinins (lectins) in invertebrate defense reactions: the immuno-biological importance of carbohydrate-specific binding molecules. *Developmental and Comparative Immunology* **7**, 603–608.
45. Garte, S.J. & Russell, C.S. (1976). Isolation and characterization of a hemagglutinin from *Amphritrite ornata*, a polychaetous annelid. *Biochimica et Biophysica Acta* **439**, 368–379.
46. Stein, E.A., Wojdani, A. & Cooper, E.L. (1982). Agglutinins in the earthworm *Lumbricus terrestris*: naturally occurring and induced. *Developmental and Comparative Immunology* **6**, 407–421.
47. Wojdani, A., Stein, E.A., Cooper, E.L. & Alfred, L.J. (1982). Agglutinins and proteins in the earthworm *Lumbricus terrestris,* before and after injection of erythrocytes, carbohydrates and other materials. *Developmental and Comparative Immunology* **6**, 613–624.
48. Stein, E.A. & Cooper, E.L. (1988). In vitro agglutinin production by earthworm leukocytes. *Developmental and Comparative Immunology* **12**, 531–547.
49. Eue, I., Kauschke, E. & Mohrig, W. (1991). Comparative characterization of lectins in the coelomic fluid of lumbricides. *Developmental and Comparative Immunology* **15**, 575.
50. Eue, I., Kauschke, E., Mohrig, W. & Cooper, E.L. (1998). Isolation and characterization of earthworm hemolysis and agglutinins. *Developmental and Comparative Immunology* **22**, 13–25.
51. Vasta, G.R. & Marchalonis, J.J. (1983). Humoral recognition factors in the Arthropoda. The specificity of chelicerata serum lectins. *American Zoology* **23**, 157–171.
52. Mandal, C. & Mandal, C. (1990). Sialic acid binding lectins. *Experimentia* **46**, 433–441.
53. Cohen, E., Rowe, A.W. & Wissler, F.C. (1965). Heteroagglutinins of the horseshoe crab, *Limulus polyphemus*. *Life Sciences* **4**, 2009–2016.
54. Marchalonis, J.J. & Edelman, G.M. (1968). Isolation and characterization of a hemagglutinin from *Limulus polyphemus*. *Journal of Molecular Biology* **32**, 453–465.
55. Roche, A.-C. & Monsigny, M. (1974). Purification and properties of limulin: a lectin (agglutinin) from hemolymph of *Limulus polyphemus*. *Biochimica et Biophysica Acta* **371**, 242–254.
56. Oppenheim, J.D., Nachbar, M.S., Salton, M.R.J. & Aull, F. (1974). Purification of a hemagglutinin from *Limulus polyphemus* by affinity chromatography. *Biochemical and Biophysical Research Communications* **58**, 1127–1134.
57. Nowak, T.P. & Barondes, S.H. (1975). Agglutinin from *Limulus polyphemus*. Purification with formalinized horse erythrocytes as the affinity adsorbent. *Biochimica et Biophysica Acta* **393**, 115–123.
58. Roche, A.-C., Schauer, R. & Monsigny, M. (1975). Protein–sugar interactions. Purification by affinity chromatography of limulin, an *N*-acyl-neuraminidyl-binding protein. *FEBS Letters* **57**, 245–249.
59. Robey, F.A. & Liu, T.-Y. (1981). Limulin: a C-reactive protein from *Limulus polyphemus*. *Journal of Biological Chemistry* **256**, 969–975.
60. Fernandez-Moran, H., Marchalonis, J.J. & Edelman, G.M. (1968). Electron microscopy of a hemagglutinin from *Limulus polyphemus*. *Journal of Molecular Biology* **32**, 467–469.
61. Armstrong, P.B., Swarnaker, S., Srimal, S., Misquith, S., Hahn, E.A., Aimes, R.T. & Quigley, J.P. (1996). A cytolytic function for a sialic acid-binding lectin that is a member of the pentraxin family of proteins. *Journal of Biological Chemistry* **271**, 14717–14721.
62. Saito, T., Kawabata, S., Hirata, M. & Iwanaga, S. (1995). A novel type of limulus lectin-L6. *Journal of Biological Chemistry* **270**, 14493–14499.

63. Brandin, E.R. & Postole, T.G. (1983). Polyphemin: a teichoic-acid binding lectin from the horseshoe crab, *Limulus polyphemus*. *Biochemical and Biophysical Research Communications* **113**, 611–617.
64. Fuju, N., Minetti, C.A.S.A., Nakhasi, H.L., Chen, S.-W., Barhehenn, E., Nunes, P.H. & Nguyen, N.Y. (1992). Isolation, cDNA cloning and characterization of an 18-kDa hemagglutinin and amebocyte aggregation factor from *Limulus polyphemus*. *Journal of Biological Chemistry* **267**, 22452–22459.
65. Shimizu, S., Ito, M. & Niwa, H. (1977). Lectins in the hemolymph of the Japanese horseshoe crab, *Tachypleus tridentatus*. *Biochemica et Biophysica Acta* **500**, 71–79.
66. Shishikura, F. & Sekiguchi, K. (1983). Agglutinins in the horseshoe crab hemolymph: purification of a potent agglutinin of horse erythrocytes from the hemolymph of *Tachypleus tridentatus*. *Journal of Biochemistry* **93**, 1539–1546.
67. Gokudian, S., Muta, T., Tsuda, R., Koori, K., Kawahara, T., Seki, N., Mizunoe, Y., Wai, S.N., Iwanago, S. & Kawabata, S. (1999). Horseshoe crab acetyl group-recognizing lectins involved in innate immunity are structurally related to fibrinogen. *Proceedings of the National Academy of Sciences USA* **96**, 10086–10091.
68. Ahmed, H., Chatterjee, B.P., Kelm, S. & Schauer, R. (1986). Purification of a sialic acid-specific lectin from the Indian scorpion, *Heterometrus granulomanus*. *Biological Chemistry Hoppe-Seyler* **367**, 501–506.
69. Vasta, G.R., Warr, G.W. & Marchalonis, J.J. (1983). Serological characterization of humoral lectins from the freshwater prawn *Macrobrachium rosen bergii*. *Developmental and Comparative Immunology* **7**, 13–20.
70. Hall, J.L. & Rowlands, D.T. (1974). Heterogeneity of lobster agglutinins I. Purification and physiochemical characterisation. *Biochemistry* **13**, 821–827.
71. Muramoto, K., Ogata, K. & Kamiya, H. (1985). Comparison of the multiple agglutinins of the acorn barnacle, *Megabalanus rosa*. *Agricultural and Biological Chemistry* **49**, 85–93.
72. Ingram, G.A. & Molyneux, D.H. (1991). Insect lectins: role in parasite–vector interactions. In: *Lectin Reviews*, Volume 1 (eds D.C. Kilpatrick, E. Van Driessche & T.-C. Bøg-Hansen) pp. 103–127. Sigma Chemical Company, St Louis, MO.
73. Komano, H., Mizuno, D. & Natori, S. (1980). Purification of lectin induced in the hemolymph of *Sarcophaga peregrina* larvae on injury. *Journal of Biological Chemistry* **256**, 7087–7089.
74. Umetsu, K., Kosaka, S. & Suzuki, T. (1984). Purification and characterization of a lectin from the beetle, *Allomyrina dichotoma*. *Journal of Biochemistry* **95**, 239–245.
75. Komano, H. & Natori, S. (1985). Participation of *Sarcophaga peregrina* humoral lectin in the lysis of sheep red blood cells injected into the abdominal cavity of larvae. *Developmental and Comparative Immunology* **9**, 31–40.
76. Kawaguchi, N., Komano, H. & Natori, S. (1991). Involvement of *Sarcophaga* lectin in the development of imaginal discs of *Sarcophaga peregrina* in an autocrine manner. *Developmental Biology* **144**, 86–93.
77. Scott, M.T. (1971). A naturally occurring hemagglutinin in the hemolymph of the American cockroach. *Archives of Zoology and Experimental Genetics* **112**, 73–80.
78. Scott, M.T. (1972). Partial characterization of the hemagglutinating activity in hemolymph of the American cockroach (*Periplaneta americana*). *Journal of Invertebrate Pathology* **19**, 66–71.
79. Arai, T., Kawasaki, K., Kubo, T. & Natori, S. (1998). Cloning of cDNA for regenectin, a humoral C-type lectin of *Periplaneta americana*, and expression of the regenectin gene during leg regeneration. *Insect Biochemistry and Molecular Biology* **28**, 987–994.
80. Chen, C., Ratcliffe, N.A. & Rowley, A.F. (1993). Detection, isolation and characterization of multiple lectins from the haemolymph of the cockroach *Blaberus discoidalis*. *Biochemical Journal* **294**, 181–190.
81. Chen, C., Rowley, A.F. & Ratcliffe, N.A. (1998). Detection, purification and immunoaffinity chromatography and properties of β-1,3-glucan-specific lectins from the sera of several insect species. *Insect Biochemistry and Molecular Biology* **28**, 721–731.

82. Matha, V., Grubhopper, L. & Söderhäll, K. (1990). New glucose and zymosan specific lectin from *Galleria mellonella* hemocyte lysate. In *Lectins: Biology, Biochemistry, Clinical Biochemistry*, Volume 7 (eds J. Kocourek & D.L.J. Freed), pp. 91–95. Sigma Chemical Company, St Louis, MO.
83. Ochiai, M. & Ashida, M. (1988). Purification of a β-1,3-glucan recognition protein in the prophenoloxidase activating system from hemolymph of the silkworm, *Bombyx mori*. *Journal of Biological Chemistry* **263**, 12056–12062.
84. Wilson, R., Chen, C. & Ratcliffe, N.A. (1999). Innate immunity in insects: the role of multiple, endogenous serum lectins in the recognition of foreign invaders in the cockroach, *Blaberus discoidalis*. *Journal of Immunology* **162**, 1590–1596.
85. Van Damme, E.J.M., Peumans, W.J., Pusztai, A. & Bardocz, S. (1998). *Handbook of Plant Lectins: Properties and Biomedical Applications*. John Wiley, Chichester.
86. Ratanapo, S., Wayakorn, N., Ngamjunyaporn, W. & Chulavatnatol, M. (1998). Sialic acid binding lectins from leaf of mulberry (*Morus alba). Plant Science* **139**, 141–148.
87. Mercy, P.D. & Ravindranath, M.H. (1993). Purification and characterisation of *N*-glycolylneuraminic acid specific lectin from *Scylla serrata*. *European Journal of Biochemistry* **215**, 697–704.
88. Yeaton, R.W. (1981). Invertebrate lectins. I. Occurrence. *Developmental and Comparative Immunology* **5**, 391–402.
89. Sasaki, H. & Aketa, K. (1981). Purification and distribution of a lectin in sea urchin (*Anthocidaris crassispina*) egg before and after fertilisation. *Experimental Cell Research* **135**,15–19.
90. Giga, Y., Sutoh, K. & Ikai, A. (1985). A new multimeric hemagglutinin from the coelomic fluid of the sea urchin *Anthocidaris crassispina*. *Biochemistry* **24**, 4461–4467.
91. Hatakeyama, T., Kohzaki, H., Nagatomo, H. & Yamasaki, N. (1994). Purification and characterization of four Ca^{2+}-dependent lectins from the murine invertebrate, *Cucumaria echinata*. *Journal of Biochemistry* **116**, 209–214.
92. Hatakeyama, T., Himeshima, T., Komatsu, A. & Yamasaki, N. (1993). Purification and characterization of two lectins from the sea cucumber *Stichopus japonicus*. *Bioscience, Biotechnology and Biochemistry* **57**, 1736–1739.
93. Matsui, T., Ozeki, Y., Suzuki, M., Hino, A. & Titani, K. (1994). Purification and characterization of two Ca^{2+}-dependent lectins from coelomic plasma of sea cucumber *Stichopus japonicus*. *Journal of Biochemistry* **116**, 1127–1133.
94. Hatkeyama, T., Nagatomo, H. & Yamasaki, N. (1995). Interaction of the hemolytic lectin CEL-III from the marine invertebrate *Cucumaria echinata* with the erythrocyte membrane. *Journal of Biological Chemistry* **270**, 3560–3564.
95. Ji, X., Azumi, K., Sasaki, M. & Nonaka, M. (1997). Ancient origin of the complement lectin pathway revealed by molecular cloning of mannan binding protein-associated serine protease from a urochordate, the Japanese ascidian, *Halocynthia roretzi*. *Proceedings of the National Academy of Sciences USA* **94**, 6340–6345.
96. Nonaka, M., Azumi, K., Ji, X., Namikawa-Yamida, C., Sasaki, M., Saiga, H., Dodds, A.W., Sekine, H., Homma, M.K., Matsushita, M., Endo, Y. and Fujita, T. (1999). Opsonic complement component C3 in the solitary ascidian, *Halocynthia roretzi*. *Journal of Immunology* **162**, 387–391.
97. Vasta, G.R., Warr, G.W. & Marchalonis, J.J. (1982). Tunicate lectins: distribution and specificity. *Comparative Biochemistry and Physiology* **73B**, 887–900.
98. Parrinello, N. & Canicatti, C. (1982). Carbohydrate binding specificity and purification by biospecific affinity chromatography of *Ascidia malaca* traust hemagglutinins. *Developmental and Comparative Immunology* **6**, 53–64.
99. Vasta, G.R. & Marchalonis, J.J. (1983). Lectins from tunicates and cyclostomes. A biochemical characterization. In: *Lectins – Biology, Biochemistry, Clinical Chemistry*, Volume 3 (eds T.C. Bøg-Hansen & G.A. Spengler). Walter de Gruyter, Berlin.
100. Parrinello, N. & Canicatti, C. (1983). α-Lactose binding hemagglutinins from the ascidian, *Phallusia mamillata* (Cuv.). *Biological Bulletin* **164**, 124–135.

101. Schluter, S.F., Ey, P.L., Keough, D.R. & Jenkin, C.R. (1982). Identification of two carbohydrate-specific erythrocyte agglutinins in the haemolymph of the protochordate *Botrylloides leachii*. *Immunology* **42**, 241–250.
102. Coombe, D.R., Ey, P.L. & Jenkin, C.R. (1984). Particle recognition by hemocytes from the colonial ascidian *Botrylloides leachii*. Evidence that *B. leachii* HA-2 agglutinin is opsonic. *Journal of Comparative Physiology* **154**B, 509–521.
103. Yokosawa, H., Sawada, H., Abe, Y., Numakunai, T. & Ishii, S. (1982). Galactose-specific lectin in the hemolymph of solitary ascidian, *Halocynthia roretzi*: isolation and characterisation. *Biochemical and Biophysical Research Communications* **107**, 451–457.
104. Yokosawa, H., Harada, K., Igarashi, K., Abe, Y., Takahashi, K. & Ishii, S. (1986). Galactose-specific lectin in the hemolymph of solitary ascidian, *Halocynthia roretzi*. Molecular, binding and functional properties. *Biochimica et Biophysica Acta* **870**, 242–247.
105. Vasta, G.R., Hunt, J.C., Marchalonis, J.J. & Fish, W.W. (1986). Galactosyl-binding lectins from the tunicate *Didemnum candidum*. Purification and physiochemical characterization. *Journal of Biological Chemistry* **261**, 9174–9181.
106. Vasta, G.R. & Marchalonis, J.J. (1986). Galactosyl-binding lectins from the tunicate *Didemnum candidum*. Carbohydrate specificity and characterization of the combining site. *Journal of Biological Chemistry* **261**, 9182–9186.
107. Suzuki, T., Takagi, T., Furukahri, T., Kawamura, K. & Nakauchi, M. (1990). A calcium-dependent galactose-binding lectin from the tunicate *Polyandrocarpa misakiensis*. *Journal of Biological Chemistry* **265**, 1274–1281.
108. Elola, M.T. & Vasta, G.R. (1994). Lectins from the colonial tunicate *Clavelina picta* are structurally related to acute-phase reactants from vertebrates. *Annals of the New York Academy of Sciences* **712**, 321–323.
109. Pancer, Z., Seifert, B.D., Rinkevich, B. & Muller, W.E.G. (1997). A novel tunicate (*Botryllus schlosseri*) putative C-type lectin features an immunoglobulin domain. *DNA and Cell Biology* **16**, 801–806.
110. Saito, Y. (1996). Humoral factors in tunicates. In: *Invertebrate Immunology* (eds B. Rinkevich & W.E.G. Muller) pp. 218–234. Springer-Verlag, Berlin.

3 Vertebrate Lectins

S-TYPE (GALECTINS)

After the discovery of electrolectin, it soon became apparent that β-galactoside binding proteins were very widely distributed in vertebrate tissues. Typically these lectins required reducing conditions for activity, could be inhibited by lactose, and usually had subunit molecular weights of around 14 000 [1]. These were sometimes referred to vaguely as soluble lectins, soluble tissue lectins or soluble β-galactoside-binding lectins [2] or alternatively as lactose lectins, or S-Lac lectins [2]. The term S-type (for sulphydryl-dependent) had more to recommend it, and was adopted by Drickamer when he proposed that the structurally homologous group of Ca^{2+}-independent lectins extracted with lactose solutions constituted the second major family of animal lectins after the C-type [3]. The term S-type is not entirely suitable, since not all of its members are sulphydryl-dependent, but it is still widely used and sounds suitably antithetic to C-type. The term 'galaptins' [4] also seemed very appropriate but never gained widespread usage. Therefore, there was a deliberate decision to invent a new term, 'galectin', defined on primary structural grounds [5]. Fortunately, this term and the nomenclature derived from it have obtained wide acceptance. Several reviews of galectins have been published in recent years [1,6–9].

It should be stressed that some lectins inhibited by lactose either are definitely not galectins or cannot be assigned as yet (because the primary structure has not been determined). Galectins are defined by the presence of a conserved carbohydrate-recognition domain which is distinct from the conserved carbohydrate-recognition domain of C-type lectins. This domain is composed of about 130 amino acid residues, and, remarkably, has a similar tertiary structure (three-dimensional shape) to legume lectins by folding as a 'jelly roll' or sandwich composed of two β-sheets [10]. The amino acid residues of galectins most closely involved in sugar binding are those most highly conserved during evolution, and this accounts for two properties shown by all members of the family: a lectin activity that is Ca^{2+}-independent, and a specificity for β-galactosides, which, though it might exhibit minor variations from galectin to galectin, always includes *N*-acetyllactosamine, Gal-β1,4-GlcNAc. Members of this intriguing family also have several other characteristics in common: a blocked (acetylated) N-terminus, lack of a signal sequence, lack of attached carbohydrate, a predominantly cytoplasmic location, and a vulnerability to extraction with lactose-containing but detergent-free buffers. Additionally, galectins are often developmentally regulated, have cysteine residues existing as free thiols, and are often unstable under oxidising conditions, requiring (for example) 2-mercaptoethanol for storage.

Although galectins have been found in amphibians [11], fish [12], birds [13] and occasionally in phylogenetically distant animals like nematodes [14] and sponges [15], the literature predominantly describes galectins from mammalian sources. Mammalian galectins have been classified into ten families based on amino acid sequence, but a more fundamental structural classification of subunits, which incorporates non-mammalian galectins, is the division into proto-type, chimera-type and tandem-repeat-type galectins (Figure 3.1). Proto-type subunits (galectins-1, -2, -5, -7 and -10) consist of a single domain (Mr about 14 000) and usually form homodimers with haemagglutinating activity. The chimera-type subunit (galectin-3) has, in addition to the carbohydrate-recognition domain on the C-terminal half, an unrelated N-terminal half consisting of a short domain followed by an intervening proline/glycine/tyrosine-rich domain with short repetitive sequences; it is essentially monomeric, and therefore non-haemagglutinating, but can form multimers on binding to surfaces containing appropriate carbohydrate ligands. Tandem-repeat-type subunits (galectins-4, -6, -8 and -9) consist of two homologous carbohydrate-recognition domains side-by-side in a single polypeptide joined by a short glycine-rich region.

Galectins have been found in a very wide variety of organs and tissues. Galectins-1 and -3 have been the most extensively studied. Galectin-1 is found in spleen, liver, kidney, lung, skin, placenta and neurons, as well as in all kinds of muscular tissue; it is also expressed in tumour cells. Galectin-3 is particularly associated with certain leukocytes and various epithelial cells as well as numerous malignancies. Within the cell, galectins are most abundantly expressed in the cytoplasm, but have also been found within the nucleus, on the cell surface

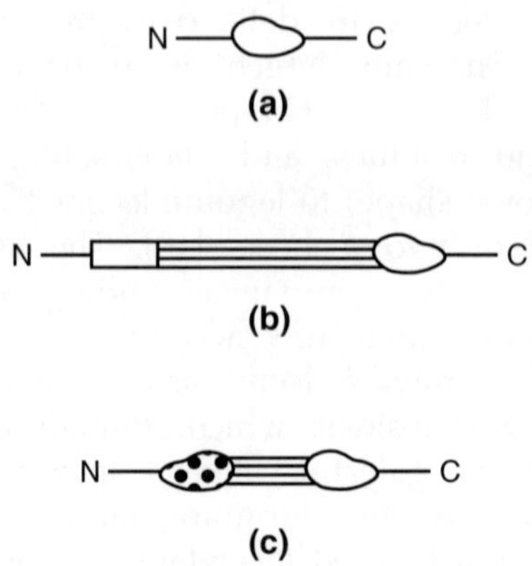

Figure 3.1. Gross structural division of the galectins: (a) proto-type, (b) chimera-type and (c) tandem-repeat-type

and extracellularly. Knock-out mice lacking galectin-1 [16] or galectins-1 and -3 [17] have apparently normal phenotypes. The superficial first impression is that galectins are everywhere but do nothing!

Nevertheless, there is considerable circumstantial evidence that galectins are involved in cell adhesion and the related function of regulation of cellular proliferation and differentiation. Galectin-1 certainly has growth modulatory qualities. Galectin-1 from bovine lung promotes the growth of arterial smooth muscle and endothelial cells [18], while experiments with a recombinant galectin-1 provided evidence for its involvement in the attachment, spreading and migration of human vascular smooth muscle cells [19]. The corresponding protein from murine embryonic fibroblasts acted as a growth inhibitor for autologous cells; most importantly, this cytostatic activity was unrelated to its lectin activity [20].

Indeed, there is primary structural homology between galectins and the transforming growth factor (TGF) family, and galectin-1 has TGF activity (designated TGFγ2) separate from its lectin activity [21]. Members of the TGF family have been shown to stimulate or inhibit growth depending on the circumstances. Similarly, overexpression of rat galectin-1 by transfecting its cDNA into fibroblasts induces malignant transformation in rodent model systems [21,22], while recombinant galectin-1 has been shown to induce apoptosis in human T cells [23], thymocytes [24] and B lymphocytes [25]. Rat galectin-1/TGFγ2 is mitogenic for fibroblasts in an oxidising environment, but is without carbohydrate-binding activity; conversely, under reducing conditions, the mitogenic activity is greatly diminished but lectin activity is manifested. It has been suggested that intramolecular disulphide bonds are essential for the conformation conferring TGF activity, while the loss of such bonds in the presence of reducing agents alters the conformation to one conferring lectin activity [26].

Preliminary experiments with natural human placental galectin-1 and human mononuclear leukocytes yielded puzzling but interesting results [27]. Under reducing conditions that permitted lectin activity, the galectin was mitogenic, but under lectin-labile conditions, the galectin was anti-mitogenic. The essence of these results was confirmed and extended independently in a different system [28]. Adams and co-workers found that recombinant human galectin-1 was mitogenic towards fibroblasts at relatively low concentrations and that this activity could be inhibited by lactose; at higher concentrations, however, the galectin behaved as a cell-growth inhibitor, and the latter activity was not inhibited by lactose. Finally, as discussed in Chapter 4, galectin-1 has immunomodulatory properties *in vivo* [29].

Although the literature is not entirely consistent, there is little doubt that galectin-1 has both lectin and TGF-like activities and that the latter may be stimulative or inhibitory depending on the context. The physiological influence of this protein may depend on the aggregate effect of several factors, including concentration, the availability of particular ligands and the redox state of the micro-environment. A key question is whether the lectin activity can regulate the TGF activity and, if so, how.

The presence of free SH groups is not essential for sugar binding, but their oxidation renders lectin activity labile by promoting a conformational change [30]. Protection from loss of activity can be obtained by alkylation with iodoacetamide, or, as demonstrated with the recombinant protein, by substituting cysteine with serine [32]. As similar modifications have not arisen through evolutionary processes, it seems the lability of lectin activity may have a purpose. As discussed in the context of human pregnancy [33], it is possible that the lectin activity enables the galectin to reach and accumulate at its physiological location where the growth modulatory activity can be used appropriately. The labile carbohydrate-binding activity would limit the TGF activity, since its rapid loss in an oxidising environment would lead to a decrease in local concentration by diffusion away from the site. This could be a crucial regulatory mechanism governing the organisation of connective tissue and cell-to-matrix signalling.

Although much of the earlier galectin research concerned galectin-1, the literature on galectin-3 has been steadily catching up. Galectin-3 is the only known example of a chimera-type galectin. It was originally reported independently under various circumstances and given names like CBP-35, RL-29, Mac-2 and IgE-binding protein. Galectin-3 can also influence cell adhesion, but (unlike galectin-1) seems to function in inflammatory processes (via effects on basophils/mast cells, neutrophils and monocytes/macrophages) and in pre-messenger RNA nuclear splicing. Notably, a reciprocal or cross-regulatory relationship between galectins-1 and -3 may exist [9]. An extensive and at times contradictory literature concerning galectin-3 and cancer has also accumulated.

The associations of galectins with cancer are reviewed in Chapter 4. Galectins are also discussed in Chapter 4 under the headings of blood grouping reagents, mitogenic agents, lectins and immunity, and even under drugs and drug targets. The large literatures on galectin-1 and galectin-3 are summarised in the Directory; separate entries are also found for the other mammalian galectins (Table 3.1) as well as those from *Geodia cydonium* (sponge), *Caenorhabditis elegans* (nematode) and electric eel (electrolectin). Additionally, amphibian and chicken galectins are given group entries.

Table 3.1. Comparative Summary of Mammalian Galectins

No.	Distribution	Structural type	Native Mr
1	General	Proto	28 000 (dimer)
2	Intestinal epithelium	Proto	28 000 (dimer)
3	General	Chimera	35 000 (monomer)
4	Alimentary epithelium	Tandem-repeat	36 000 (monomer)
5	Erythroid cells	Proto	15 000 (monomer)
6	Alimentary epithelium	Tandem-repeat	34 000 (monomer)
7	Epidermis	Proto	14 000 (monomer)
8	General	Tandem-repeat	35 000 (monomer)
9	General	Tandem-repeat	36 000 (monomer)
10	Eosinophils	Proto	17 000 (monomer)

Kasai and Hirabayashi [7] list seven great mysteries concerning galectins, one of which is: why are they found and appear to function outside the cell despite apparently being designed as intracellular proteins? Perhaps I could be permitted a guess at answering the question? I suggest it is because the original function of galectins (in the cytoplasm) has become obsolete, and therefore galectins have only survived where they have evolved additional (secondary) functions. Despite all the hype about being widely distributed, with nematodes and sponges being cited, the fact is that galectins have been found extremely rarely in species other than vertebrates so far. That may change, but until it does, galectins should be considered vertebrate lectins, even though homologous molecules occur (rarely) in other parts of the animal kingdom.

C-TYPE (GENERAL)

Although the term C-type derives from the calcium-dependence of the activity common to all C-type lectins, there are some Ca^{2+}-dependent lectins that are not C-type and others that cannot yet be classified. This is because the term C-type is strictly reserved for members of a family of proteins showing a homologous carbohydrate-binding domain. This defining domain contains up to 130 amino acid residues, including about 30 highly conserved (14 invariate) residues directly involved in carbohydrate binding [34,35]. In some C-type lectins (e.g. the collectins) two calcium-binding sites are highly conserved; in the selectins, however, site 1 has not been conserved, and in the NK group of membrane receptors both sites have changed considerably. The C-type domain is completely unrelated to the S-type domain of galectins and is associated with a very different collection of characteristics (Table 3.2).

In contrast to the galectins, which are all rather similar with a very limited subunit repertoire and a common β-galactoside saccharide specificity, the C-type lectins are enormously varied. Native molecular weights may be rather small or very large; likewise the subunits, which may combine in numerous different ways; and entirely different saccharide specificities have been found. Unlike galectins, which have been been identified mainly in vertebrates, C-type lectins have been found copiously throughout the animal kingdom. Examples mentioned in Chapter 2 include the fly *Sarcophaga peregrina*, the barnacle *Megabalanus rosa* and the tunicate *Polyandrocarpa misakiensis*.

Table 3.2. Gross Comparison of C-type and S-type Lectins

	C-Type	S-Type
Ca^{2+} requirement	Yes	No
Specificity	Various	β-Galactosides
Location	Mainly extracellular	Mainly intracellular
Cysteines	Disulphides	Free thiols
Function	Defence	Unknown

The C-type lectin family is sometimes divided into seven subgroups [35], reflecting different gene and protein structures (Table 3.3). Some C-type lectins will be introduced in Chapter 4: for example, some snake C-type lectins are mentioned under 'Lectins and haemostasis' while tetranectin is overviewed under 'Lectins and cancer'. Others, which might be called complex or compound C-type lectins (because they also possess domains other than the lectin domain), are introduced in the following pages. Hyalectans, collectins and selectins each get a separate subheading, while the miscellany remaining are grouped together as membrane-bound receptor lectins.

Table 3.3. Classification of C-type Lectins

	Additional features
1. Proteoglycans	Hyaluronate and GMG-binding regions
2. Endocytic type-II receptors	Transmembrane anchor (type II)
3. Collectins	Collagen-like domain
4. Selectins	EGF-like and short consensus repeat
5. Leukocyte receptors	Transmembrane anchor (type II)
6. Macrophage mannose receptor	Fibronectin-type repeats
7. Simple (primitive) lectins	None

C-TYPE (HYALECTANS)

Molecular cloning has permitted the identification of a family of proteoglycans with core proteins that are structurally homologous. These proteoglycans, also known as hyalectans or lecticans, were recently reviewed by Iozzo [36]. The primary structure concerned is rather complex but may be considered to have three regions in all family members: an N-terminal region that binds to hyaluronic acid, a central region carrying glycosaminoglycan side chains and a C-terminal region that includes the C-type lectin domain. There are currently four members of this family: aggrecan [37], versican [38], neurocan [39] and brevican [40].

Aggrecan is the major proteoglycan of cartilage. Versican, the largest member of the family, is expressed by fibroblasts. Neurocan and brevican are smaller molecules associated with neurological tissue. All of them can bind to tenascin-R, but this property is mediated by protein–protein (non-lectin) interactions [41]. Nevertheless, the C-type lectin domain of the hyalectans is capable of binding immobilised sugars in a calcium-dependent manner. Most notably, brevican has been found to bind to sulphated glycolipids on cell surfaces [42].

C-TYPE (COLLECTINS)

The collectins are proteins that combine a C-type lectin domain with a collagen-like region (reviewed in [43–48]). The collagen-like region is preceded by a short cysteine-rich N-terminal sequence and is connected to the carbohydrate recognition domain by a short link or neck sequence. The collectins have a considerable resemblance to the complement component, C1q, in overall three-dimensional structure (but not in amino acid sequence). The subunits consisting of a globular lectin head at the carboxyl terminal end and a long collagen-like tail associate into trimers, then combine to form either cruciform or 'bunch-of-tulips' structures (Figure 3.2). There are five distinct, well-characterised members of this group: conglutinin, CL-43, surfactant protein A (SP-A), surfactant protein D (SP-D) and mannan-binding lectin (MBL). All are believed to function primarily in defence, distinguishing self from non-self.

Conglutinin was one of the first lectins to be discovered [49]. This bovine GlcNAc-binding protein binds to immune complexes (via iC3b) [50]; its anti-microbial activity may depend on bridging complement-opsonised bacteria and effector leukocytes. Although a human analogue of conglutinin has been described [51], it is by no means certain that conglutinin really exists in humans or indeed in any species outside the Bovidae [52]. Another bovine collectin is best known by its abbreviation CL-43 (for collectin-43); CL-43 is closely related to both conglutinin and SP-D, but has a distinct saccharide specificity [53].

The lung surfactant proteins A and D have been described in several mammalian species and have received much attention in recent years. Both SP-A

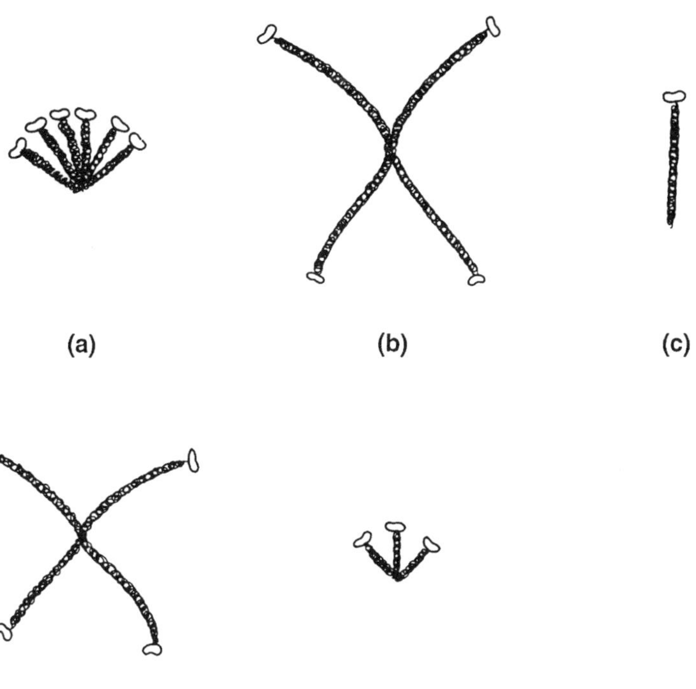

(a) (b) (c)

(d) (e)

Figure 3.2. The collectins: (a) SP-A, (b) SP-D, (c) CL-43, (d) conglutinin and (e) MBL. These drawings are based on appearance under the electron microscope [176]

[54–57] and SP-D [56–59] have been the subject of several recent reviews. Both are synthesised and secreted by epithelial cells into lung surfactant, and both can bind to bacterial, viral and fungal glycoconjugates as well as pollen and dust mite allergens [57]. SP-A, with a relatively short collagenous region (subunit Mr approximately 30 000), is the product of two genes in humans, each with several alleles, but is encoded by a single gene in other species. SP-D, the more water-soluble of the two, is encoded by a single allelic gene, and has a relatively long collagenous region (subunit Mr 43 000) as a result of which it resembles conglutinin in overall structure (Figure 3.2). It seems plausible that the surfactant proteins function to protect lung tissue from inhaled pathogens, and are part of the innate immune system's equivalent of the mucosal immune system. The latter is an independent and complementary component of adaptive immunity. Viewed thus, SP-A and SP-D are equivalent to IgA, protectors of mucosal surfaces, while conglutinin and MBL are similarly related to IgG and IgM as protectors of internal tissues.

MBL (originally called mannan-binding *protein* and sometimes unsuitably called *mannose*-binding protein or lectin) has been intensively studied, particularly during the past decade. Two excellent reviews by Turner [60,61] are recommended for overviews, as is a brief journalistic account by Thompson [62].

This collectin was first noticed as a plasma factor that opsonised baker's yeast, the absence of which was implicated in a child with severe immune deficiency in association with phagocytic dysfunction [63]. The association of this opsonic defect with recurrent infections in infancy was supported by several subsequent reports, including a convincing prospective study by Richardson *et al.* [64] which linked the deficiency both to infectious and allergic disorders. In the meantime, Kawasaki *et al.* [65] had isolated MBL from rabbit liver and later from human serum [66], while Wild *et al.* [67] had independently isolated similar 'mannose-binding proteins' from both rat and human livers. In due course, both the baker's yeast opsonin and the mannose-binding lectins were independently linked to complement activation. Finally, the identity of MBL to the yeast opsonin was demonstrated by the correlation between the opsonic defect and low serum MBL concentrations [68], and their co-inheritance within three separate families with probands experiencing severe, recurrent infections [69]. Another complement-activating, anti-bacterial protein, the Ra-reactive factor, of murine serum [70], was subsequently found to be a rodent equivalent of MBL, and indeed a highly conserved group of mannan-binding lectin homologues are now known to occur within vertebrates, including chickens [71].

Rodents have two functional MBL genes encoding two homologous but distinct proteins, MBL-A (serum MBL) and MBL-C (liver MBL). Humans have only one MBL gene, and the MBLs isolated from liver and serum have identical amino acid sequences [72]; the observed differences between the two, which include an inability of the liver MBL to activate complement, must therefore result from post-translational modification [72].

The single human gene nevertheless exhibits considerable allelic polymorphism. Sumiya *et al.* [69] first identified a genetic basis of low circulating levels of MBL activity as a single point mutation in codon 54 (Figure 3.3). This mutant

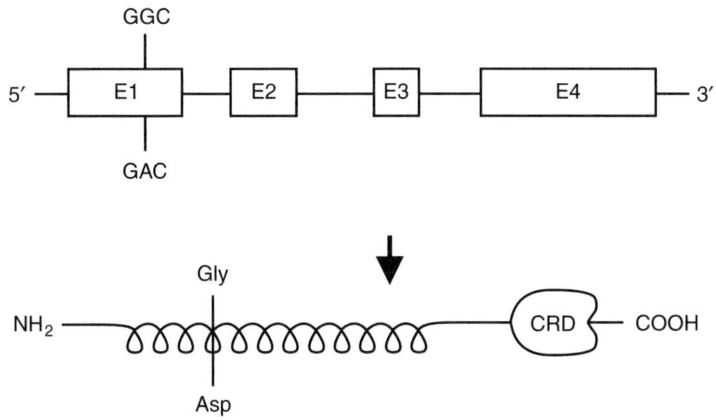

Figure 3.3. Major mutant allele of MBL. The structure of the human MBL gene on chromosome 10 consists of four exons as shown. The commonest basis for low MBL concentrations in Europeans arises from a G to A mutation in codon 54 in the first exon (E1), leading to the replacement of a normal glycine by an aspartic acid residue within the collagen-like helix. CRD, carbohydrate recognition domain

allele does indeed seem to be the main cause of relative MBL deficiency in Europeans. However, two other point mutations in the first exon have since been discovered [60,61]: at codon 57 (relatively common in African populations) and at codon 52 (occurring in both Europeans and Africans, but at a lower frequency than the other mutations). The situation is further complicated by the influence of two additional dimorphisms in the promoter region of the gene [73]. The genetic influence on MBL concentration therefore is inherited in the form of haplotypes (Table 3.4). As can be seen from the examples given, inheritance of two haplotypes each possessing the wild-type structural gene (irrespective of promoter region inheritance) is usually, but not always, associated with a higher circulating MBL level than haplotypes that are heterogeneous for the structural gene. It is equally clear, however, that simply genotyping for the presence of a structural gene mutant allele gives a rather imprecise indication of the MBL activity to be expected. It is interesting that the number of haplotypes detected so far is much smaller than the number theoretically possible [73,74].

The dominant inheritance of low MBL concentrations associated with all of the structural gene mutations is readily explained by a structural alteration in the collagen-like region resulting in an abnormal and unstable triple helix [75]. The abnormal triplet formation leads to defective oligomerisation [76], inability to bind MASP-1 [77], inefficient complement activation [78,79] and (at least in mice) a shorter half-life in plasma [80]. Although the physiological distribution pattern is not known for certain, it seems that in subjects homozygous for the wild-type structural gene, MBL exists predominantly as dimers, trimers and tetramers [76,81], while in heterozygotes and mutant allele homozygotes, lower forms predominate. The less dramatic influence of the promoter region variants may be related to ability to respond to environmental stimuli as in acute phase responses [82,83].

Table 3.4. MBL Haplotypes Related to MBL Concentrations

Genotype	MBL (μg/ml) Median	Range
HYA/HYA	2.28	1.00–4.13
HYA/LYA	1.90	0.95–4.83
LYA/LYA	1.90	0.84–2.63
HYA/LXA	1.40	0.67–3.67
LYA/LXA	0.99	0.40–1.40
HYA/O	0.40	0.19–1.2
LXA/LXA	0.35	0.10–0.8
LYA/O	0.25	0.13–0.62
LXA/O	0.03	0.00–0.28

The letter A represents the normal or wild-type structural gene, while O is used to represent any of the three mutant alleles. The promoter combinations HY, LY and LX complete the haplotypes found on each chromosome. These data have been extracted from Madsen *et al.* [73] to illustrate the relationship between allelic variants in the structural gene and MBL concentration and the influence of promoter genetic variants on the structural gene. Individuals who are homozygous for structural gene mutants (O/O) have virtually undetectable levels of MBL irrespective of the promoter region genetic component.

All individuals who lack the wild-type gene (who are 'homozygous', with the same or two different mutant genes) and many who are heterozygotes (wild-type/mutant) have very low MBL levels and may be considered to be deficient. Some 4–10% of the general population could be so described, depending on the exact level at which the normal/deficient boundary is drawn. Nevertheless, evidence is accumulating that such individuals are more susceptible to infectious diseases in general, particularly in childhood but also throughout adult life [64, 84–86]. The MBL-deficient minority is also over-represented in cohorts with a variety of specific infectious diseases (Table 3.5). MBL deficiency is commoner in patients with recurrent miscarriage than in blood donors [87]; this could be a consequence of uterine infection. Several independent studies confirm a weak

Table 3.5. Specific Infections Associated with MBL Insufficiency

Atherosclerosis (*Chlamydia pneumoniae*?)
Burkholderia cepacia (in cystic fibrosis)
Cryptosporidium parvum (co-infected with HIV)
Hepatitis B
Hepatitis C (interferon-resistant)
Human immune deficiency virus (HIV)
Meningococcal disease
Plasmodium falciparum malaria

but undoubtedly real association with systemic lupus erythematosus (SLE) (Table 3.6); in this instance, the association may be secondary to dysfunctional immune complex formation, via inadequate complement activation, in the same way that deficiencies of C2 and other complement components predispose to SLE.

The relatively high frequency of mutant genes in all populations studied suggests some selective advantage for the heterogeneous state and the diminished MBL concentrations that accompany it. High levels of MBL, for example, may sometimes have undesirable consequences, like actually promoting tuberculous meningitis [87]. A particularly elegant hypothesis concerns the inflammation of rheumatoid arthritis [88]. It has been suggested that in binding to the agalactotosyl IgG (G_0 isoforms) characteristic of rheumatic disease, MBL causes excess complement activation and release of inflammatory mediators. MBL deficiency should limit the tissue damage and associated clinical symptoms. Although this hypothesis is partially supported by one study which identified a small subgroup of patients with high MBL levels and a severe disease course, overall a lower proportion of rheumatoid arthritis patients had detectable MBL, and below-median MBL concentrations were associated with early onset and more severe disease [89]. Moreover, two other independent studies found no difference in the frequency of MBL deficiency between rheumatoid arthritis patients and controls [90,91].

While there are some inconsistencies in the literature, particularly in relation to HIV disease, the bulk of published work in the area is consistent with an important role for MBL in the innate immune system. Substitution therapy has already been attempted and was concluded to be 'safe, practical and possibly efficacious' [92].

Very recently, a new collectin-like molecule has been cloned from human liver [93] which may prove to be the exception that contradicts all the generalisations previously made. This 277 amino acid molecule has all the required collectin structural features, but is intracellular and seems to be expressed in most tissues. Moreover, the gene for CL-L1 (as it has been named) has been localised to chromosome 8 (not chromosome 10, like other human collectins). Although there is some evidence that CL-L1 has weak carbohydrate-binding activity, its putative lectin nature has not yet been unequivocally demonstrated. Therefore CL-L1 is

Table 3.6. MBL Genotype and Systemic Lupus Erythematosus

	Percentage with mutant gene(s)			
Population	Patients	Controls	OR	CI
English [169]	41	30	1.6	1.0–2.8
Chinese [170]	33	23	1.7	0.95–3.0
Black American [171]	46	23	2.0	1.1–4.6
Spanish [172]	52	31	2.4	1.5–5.6

OR, Odds ratio; CI, confidence interval.

not included in Chapter 5. At present, it seems reasonable to suppose that CL-L1 shares a common ancestor with the collectins, but has evolved separately to serve different biological functions.

C-TYPE (SELECTINS)

The selectins comprise a small subfamily (CD62) defined by having a C-type lectin domain combined with an epidermal growth factor-like domain and a variable number of complement regulatory protein-like repeat domains. In contrast to the collectins, the C-type lectin domain is found at the N-terminus, while the last repeat domain is connected to a short cytoplasmic tail by a hydrophobic transmembrane domain. The basic structure therefore resembles the C-terminal region of the hyalectans.

There are only three selectins, but this small group has been the exclusive subject of several extensive review articles [94–98] and even a recent monograph [99], since the independent cloning of several apparently unrelated molecules revealed their common membership of the same C-type lectin group.

Leukocyte-selectin (L-selectin), endothelial-selectin (E-selectin) and platelet-selectin (P-selectin) from the same species possess about 50% sequence identity in the C-type lectin domain, but the most obvious difference between the three members of the family is in the number of complement regulatory protein-type (sometimes called short consensus repeat) domains and hence molecular size (Figure 3.4). L-selectin (CD62L) is by far the smallest with only two such domains, but is also the exception in being expressed constitutively on most leukocytes. E-selectin (CD62E), the middle-sized member, is synthesised *de novo* in endothelial cells on activation. P-selectin (CD62P) which occurs on endothelial cells as well as platelets, is stored intracellularly in granules and moves to the cell surface [94–99].

That selectins function in cell traffic, promoting extravasation via the capture and rolling of leukocytes, is well established from arguably the best available evidence for the function of any group of lectins [94–99]. This general function explains the seminal observation made originally in 1964 [100] that sugar recognition mediated the 'homing' of rat lymphocytes, an observation reinforced by subsequent studies of lymphocyte interactions with endothelial cells of postcapillary venules [101]. It also provides the basis for using saccharide structures as anti-inflammatory agents with selectins as drug targets [102], discussed in Chapter 4.

Selectins can bind to a number of naturally occurring ligands. The three family members appear to have similar, but not identical, saccharide specificities. Sialic acid-containing structures are recognised, and most particularly sialyl-Lewis^x and closely related structures appear to be a common minimal ligand (Figure 3.4). With such similar structures and similar specificities, it is unsurprising that there seems to be considerable redundancy of function between the selectins. For example, although L-selectin binds high endothelial venules more efficiently than E-selectin, both L- and E-selectin could inhibit *in vivo* trafficking equally well

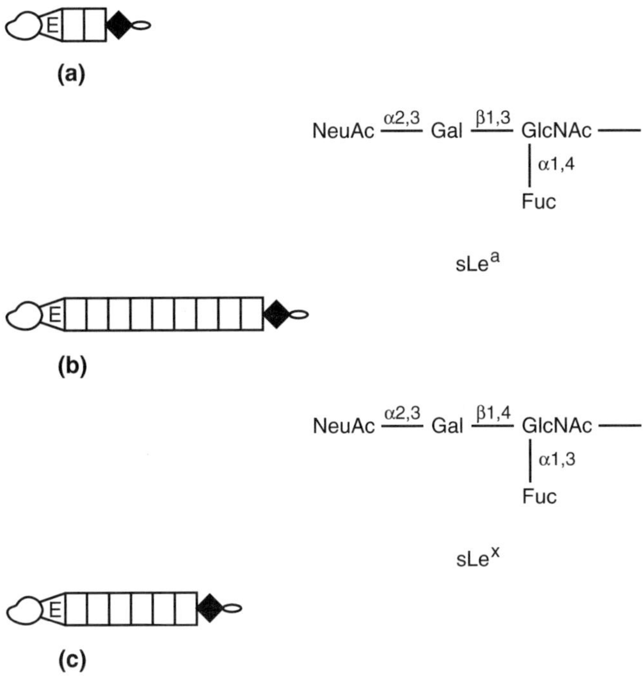

Figure 3.4. Selectins and their ligands: (a) L-selectin, (b) P-selectin and (c) E-selectin

[103]. Similarly, neutrophil accumulation after an inflammatory stimulus was barely reduced in mice genetically deficient in either E- or P-selectin, but was dramatically impaired in mice deficient in both selectins [104]. The subtle functional differences that probably exist between the selectins may be extended by the nature of the stimulus involved, the adherant leukocyte type or subset, and/or the nature of the vascular bed concerned. For each physiological situation, a panel of different knock-out mice is required to analyse the changes attributable to any one selectin. Jung and Dey [105] investigated leukocyte rolling after trauma and after inflammatory (TNFα) stimulus. By this means, they were able to demonstrate that any one of the selectins can support some neutrophil recruitment, although the presence of P-selectin was more influential than that of the others.

No selectin deficiency disease has been reported in humans, but a congenital absence of the potential selectin ligand, sialyl-Lewis^x, has been described [106]. This 'leukocyte adhesion deficiency-II syndrome', a disorder of fucose metabolism reported in a total of three patients, was associated with neutrophilia/neutrophil adhesion deficiency and severe, recurrent infections. Physical stigmata and psychomotor retardation were also present [106]. (There is also a report of a girl who presented aged ten with recurrent bacterial infections and an absence of immunohistochemical staining for E-selectin on inflammed tissue. However, E-selectin message was present in endothelium, soluble

Table 3.7. Selectins in Disease States

Selectin	Disease	Observation
L-selectin	Acute myeloid leukaemia	Increased in blood
	Acute lymphocytic leukaemia	Increased in blood
	Hairy cell leukaemia	Decreased on cell surface
	Systemic sclerosis	Decreased in blood
	Vasculitis	Decreased in blood
	Meningoencephalitis	Increased in CSF
	Meningeal leukaemia	Increased in CSF
E-selectin	Asthma	Increased in lavage fluid
	Atopic dermatitis	Increased in blood
	Pre-eclampsia	Increased in blood
	Rheumatoid arthritis	Increased in synovial fluid
	Septic shock	Increased in blood
	Sepsis	Increased on cell surface
	Vasculitis	Increased on cell surface
	Inflammatory bowel diseases	Increased on cell surface
	Hepatitis	Increased on cell surface
	Organ transplant rejection	Increased on cell surface
	Graft versus host disease	Increased on cell surface
P-selectin	Cardiovascular disease	Increased in blood
	Insulin-dependent diabetes	Increased in blood
	Pre-eclampsia/eclampsia	Increased in blood
	Rheumatoid arthritis	Increased in blood

E-selectin was present in blood, and other immune-related abnormalities were found [107].)

The realisation of the importance of selectin-mediated neutrophil–endothelial cell interactions in lung inflammation [108] and myocardial reperfusion injury [109] has encouraged the investigation of saccharide selectin inhibitors as pharmaceutical agents. This application is reviewed further in the next chapter, as is the possible role of selectins in tumour metastasis [110,111].

Increased E- and/or P-selectin expression is upregulated in many inflammatory disorders, while L-selectin levels may be increased or decreased in disease states [98]. Selectins may therefore be regarded as disease markers, either measured cell-bound or in soluble form in blood or other body fluids. Some examples are given in Table 3.7, and are discussed further in Chapter 4.

Finally, it is of major interest that selectins appear to mediate pain control [112]. Selectins mediate the initial migration of opioid-containing immunocytes into sites of inflammation, where the latter release β-endorphin and thus promote peripheral analgesia. This finding could be exploited in new approaches to pain control based on analgesics that target immunocytes.

C-TYPE (MEMBRANE RECEPTORS)

As well as selectins, three other groups of C-type lectins exist as plasma membrane receptor molecules. Two of these groups are comprised of type-II

(N-terminus inside the cell) transmembrane receptors: the asialoglycoprotein receptor group and the NK cell group. The former group has an extra module, but otherwise a similar molecular architecture to the latter group. A third group is represented by the macrophage mannose receptor (MMR), a multi-lectin domain protein with an extracellularly orientated N-terminus.

Until recently, the MMR was the only known member of its family. It appears to function in first-line defence, promoting phagocytosis and antigen presentation [113–115]. A transmembrane glycoprotein with a C-terminal cytoplasmic tail, it is unusual in having eight homologous C-type lectin domains on the same polypeptide. The same structural characteristics are shared with phospholipase A2 receptors from rabbit skeletal muscle [116] and bovine corpus luteum [117], and with another receptor that was cloned from a mouse heart cDNA library after surfing a database using E-selectin as a probe [118]. In addition, a dendritic cell and thymic epithelial cell receptor, DEC-205, implicated in antigen processing, differs only in having ten instead of eight C-type carbohydrate-recognition domains [119,120].

Despite the structural homology to MMR, there is no compelling evidence that any of the other three family members is a genuine lectin. The phospholipase A2 receptor binds to phospholipase enzymes by protein–protein interaction(s) although this interaction can be inhibited by the neoglycoprotein, BSA-mannose, 'proving' that the receptor can recognise carbohydrate [116]. This interaction cannot be inhibited by other MMR inhibitors like mannose, fucose, invertase or mannan, and it is susceptible to BSA-galactose, which does not inhibit MMR [116]. The effect of BSA-mannose is therefore not satisfactorily explained, and it is possible that inhibition is due to the influence of altered BSA rather than by a specific recognition of mannose.

The natural killer (NK) gene complex encodes two structurally distinct groups of proteins, some of which are inhibitory receptors that recognise major histocompatibility complex (MHC) class-1 antigens and recruit the tyrosine phosphatase SHP-1 to block activation [121–123]. One group of these receptors belongs to the immunoglobulin superfamily, the other to the C-type lectin superfamily. The best known members of the latter group include the Ly-49 family, a polymorphic series of murine homodimers, which switch off NK cytotoxicity on recognising self-MHC on target cells; the functionally equivalent CD94/NKG2 molecules of humans, heterodimers composed of a monomorphic CD94 chain and a variable NKG2 chain; and the NKR-P1 series, homodimers that are polymorphic in rodents but with only one known homologue in humans [121–123]. Despite the sequence similarities to C-type lectins, the corresponding domain in these NK cell receptors is sufficiently different to be considered a distinct variant [124,125]. Most significantly, the highly conserved calcium- and carbohydrate-binding residues found in other C-type lectins are absent, leading to the prediction that these receptors would not be able to bind carbohydrate in a manner analogous to the true C-type lectins. Nevertheless, although Ly-49 binds to its natural ligand, H-2D, by a carbohydrate-independent mechanism [126], Ly-49 is able to bind certain saccharides [127,128]. NKR-P1 also appears to have lectin activity [129–131], as do other related receptors such as CD69 [132] and

NKCL [133]. Some NK gene complex-associated lectin-like molecules are listed in Table 3.8.

Although NKCL [133] and DCIR (dendritic cell immunoreceptor) [134] are encoded by genes mapping close to the NK gene region (human chromosome 12p13; and syntenic region on mouse chromosome 6), structurally at the protein level these molecules perhaps have more in common with the group II transmembrane receptors, the asialoglycoprotein receptor family. The discovery of the latter group arose from the seminal observation of Morell and co-workers [135] that asialocaeruloplasmin (but not the normal, sialylated protein) was rapidly cleared from the circulation after injection into rabbits. The first purified glycoprotein known to possess binding activity towards asialoglycoproteins, and which agglutinated human and rabbit erythrocytes, was described at the time as 'the first lectin of mammalian origin' [136]. Although other mammalian lectins had been described (if not recognised as such), this work was highly influential in stimulating research into vertebrate lectins. Several mammalian hepatic lectins with the capacity to bind asialoglycoproteins were discovered: three of them (plus an avian homologue) belong to a family including the originally described member, and are listed in Chapter 5 as 'asialoglycoprotein receptors' [137]; a fucose-specific receptor found only in Kupffer cells [138]; and a human placental lectin that binds the gp120 envelope glycoprotein of the AIDS virus [139]. The B lymphocyte signalling molecules, CD23 and CD72, are closely related structurally to these receptors, but are less clearly established as lectins [140].

Table 3.8. Some Lectin-like Molecules Encoded in the NK Gene Region

Molecule(s)	Distribution	Lectin activity?
CD69 [132]	Platelets; activated leukocytes	Yes
CD94/NKG2 [121,122]	Mainly NK cells	?
DCIR [134]	Myelo-monocytic and B cells	?
LOX-1 [173,174]	Endothelium, smooth muscle, macrophages	?
Ly-49 [126–128]	NK subsets (murine)	Yes
MAFA/2F1-Ag [175]	Mast (rat), NK (mouse) cells	Yes
NKCL [133]	Myelo-monocytic cells	Yes
NKR-P1 [129]	Mainly NK cells	Yes

The five gene products with demonstrated carbohydrate-binding activity have been given individual entries in the Directory (Chapter 5).

PENTRAXINS

The pentraxins constitute a small family with a characteristic ring structure found in the two well-known human examples, C-reactive protein (CRP) and serum amyloid P (SAP) component [141,142]. The human pentraxins have five identical subunits cyclically arranged around a hole 2 nm in diameter, and possess a 'jelly roll' tertiary fold not dissimilar to that of legume lectins [143]. Homologues of human CRP and SAP appear to be universal among vertebrate species, and related molecules have also been found in some invertebrates, including tunicates [144] and horseshoe crabs [145]. *Limulus polyphemus* haemolymph has at least two separate pentraxins; the arguably more SAP-like molecule has been studied by X-ray crystallography and has the pentraxin fold, but is a hexadecamer (not a pentamer) with a doubly stacked octameric ring [145].

Human CRP and SAP are both monomorphic proteins for which no deficiency state is known; each is encoded by a single gene on chromosome 1 and secreted by the liver. There is approximately 50% primary sequence identity between the two. Both are calcium-dependent, complement-activating lectins, but are emphatically not C-type lectins in the accepted sense, since no primary sequence homology exists between the binding sites characteristic of the two families. Moreover, despite binding carbohydrates, the pentraxins have catholic tastes in ligands, and could be thought of as readily as DNA-binding proteins or as phospholipid-binding proteins. SAP is always a glycoprotein, but in most species CRP is not glycosylated. Both are acute-phase reactants, but in any one species, only CRP or SAP will be massively induced in response to infection or trauma [141–143].

CRP is the major pentraxin acute-phase reactant in humans, and indeed is the most widely used biochemical marker of inflammation. This is because its concentration can increase enormously (from around 1 μg/ml to up to 1 mg/ml) in less than a day, and because it has a short biological half-life. Sometimes CRP measurement is the only objective indication of an inflammatory event and may expose covert infection; usually it is used for assessing disease activity and response to treatment; and sometimes it is a helpful means of distinguishing viral (relatively low levels) from bacterial infections [146]. CRP was given its name for its ability to bind the C-polysaccharide of *Streptococcus pneumoniae*; phosphocholine binding distinguishes it from SAP [147].

SAP, unlike CRP, recognises phosphoethanolamine [147] and (in humans) shows little concentration rise in response to inflammatory stimuli. It is the major DNA/chromatin-binding protein of plasma, a universal component of amyloid deposits and a normal component of extracellular matrix [142].

Both CRP and SAP may be important components of innate immunity, acting at the initiation and also the resolution of an inflammatory response. CRP recognises foreign pathogens, activates complement and induces pro-inflammatory cytokines, yet its net effect is anti-inflammatory in transgenic mice expressing large amounts [148]. SAP may opsonise immune complexes [149], but SAP knock-out mice develop a disease with clinical and serological features resembling SLE, perhaps indicating that SAP is required for the properly regulated clearance of tissue debris resulting from inflammatory tissue damage [150].

A related molecule named PTX3 expressed on several cell types may be considered the third human member of the pentraxin family; with a molecular weight of 42 000 it is twice the size of the others, and only the C-terminal half shows homology with CRP and SAP [151].

I-TYPE LECTINS

The I-type lectins are carbohydrate-binding proteins (other than sugar-specific antibodies) that belong to the immunoglobulin superfamily [152–154]. The latter is a vast collection of cell-surface proteins with domains homologous to those of immunoglobulins. All of the I-type lectins are integral membrane proteins, preferentially expressed on the plasma membrane, with extracellular domain(s) possessing an immunoglobulin motif with carbohydrate-recognition ability. These modules of around 100 amino acids have been divided into four groups: V, C1, C2 and I. All have two β-sheets packed together, but other features show variation in hydrogen bonding patterns. The I-type lectins differ in the number and nature of these modules (Table 3.9).

The I-type lectins may be considered to fall into two main groups: the sialoadhesin group of sialic acid-specific lectins (siglecs) and the miscellany of lectin-like adhesion molecules of variable specificity (Table 3.9).

Table 3.9. I-Type Lectins listed in Chapter 5

	Modules	Specificity
CD22	$V(C2)_6$	NeuAcα2, 6-X
CD33	V C2	NeuAcα2, 3-X
Myelin-associated glycoprotein	$V(C2)_4$	NeuAcα2, 3-X
Sialoadhesin	$V(C2)_{16}$	NeuAcα2, 3-X
CD31 (PECAM-1)	$(C2)_6$	Heparin
CD54 (ICAM-1)	$(C2)_5$	Hyaluronic acid
N-CAM	$(C2)_5$	Oligomannose

P-TYPE LECTINS

Following early work indicating that lysosomal hydrolases were specifically recognised and taken up by fibroblasts in culture [155,156], a receptor binding to phosphorylated carbohydrates was implicated [157]. In fact, there are at least two such receptors which function to separate lysosomal enzymes from those destined for secretion after their co-synthesis in the Golgi apparatus [158]. These type-I integral membrane glycoproteins are sometimes known as P-type lectins.

Phosphomannosyl receptor I has a subunit Mr of $>275\,000$ and binds ligands containing mannose-6-phosphate in a calcium-independent manner; it is identical to the receptor for insulin-like growth factor-II. Phosphomannosyl receptor II has a subunit Mr of only 46 000. In some species this receptor requires divalent cations for optimal ligand binding, but as the human lectin has little or no such requirement, the expression 'cation-dependent mannose phosphate receptor' that is sometimes used is clearly a misnomer [158].

FICOLINS

The ficolins are a small family characterised by having a fibrinogen-like domain combined with a collagen-like domain [159,160]. In quaternary structure, the ficolins resemble the collectins, but do not have a C-type carbohydrate-recognition domain and instead depend on their fibrinogen-like domain for lectin activity.

Ficolins were first identified as two transforming growth factor-β1-binding proteins in porcine uterus. However, it was the discovery of P35 [161] that indicated their lectin nature; P35 (or L[iver]-ficolin) was isolated as an opsonic, GlcNAc-specific protein apparently with much in common with MBL. A second human ficolin (M-ficolin or ficolin-1) corresponds to the other porcine molecule, and equivalent molecules have been found in rodents. While human P35/L-ficolin and ficolin-1/M-ficolin have 80% primary sequence identity, a protein with less than 50% identity to the other ficolins has been identified as a third member of this family. The Hakata antigen had previously been identified by autoantibodies present in the serum of patients with SLE and was subsequently found to be a normal plasma protein and a D-fucose/GlcNAc-specific, calcium-independent lectin [162].

CYTOKINE LECTINS

A series of papers by Muchmore and colleagues in the late 1980s were the first indication that proteins formerly known as cytokines might have carbohydrate-binding activity (reviewed by Zanetta *et al.* [163]). These early observations were based on the binding of Tamm–Horsfall glycoprotein to interleukin-1 (IL-1), interleukin-2 (IL-2) or tumour necrosis factor (TNF) immobilised on ELISA plates. The significance of these observations was questioned, however, by claims that IL-1 and TNF require a conformational change to bind to the Tamm–Horsfall glycoprotein and that the latter binds non-specifically to proteins under acid conditions [164]. Indeed, Lambert and colleagues [165] have confirmed that TNF does not bind the Tamm–Horsfall glycoprotein under physiological conditions. Nevertheless, the trypanocidal activity of TNF is potently inhibited by N,N′-diacetylchitobiose but not via its cytokine receptors [166]. Moreover, recent independent evidence confirms the lectin nature of IL-1 [167] and IL-2 [168]. All three have therefore been included in Chapter 5.

REFERENCES

1. Harrison, L. (1991). Soluble β-galactoside binding lectins in vertebrates. In: *Lectin Reviews*, Volume 1 (eds D.C. Kilpatrick, E. Van Driessche & T.-C. Bøg-Hansen) pp. 17–39, Sigma Chemical Company, St Louis, MO.
2. Barondes, S.H. (1984). Soluble lectins: a new class of extracellular proteins. *Science* **223**, 1259–1264.
3. Drickamer, K. (1988). Two distinct classes of carbohydrate-recognition domains in animal lectins. *Journal of Biological Chemistry* **263**, 9557–9560.
4. Harrison, F.L. & Chesterton, C.J. (1980). Factors mediating cell–cell recognition and adhesion. Galaptins, a recently discovered class of bridging molecules. *FEBS Letters* **122**, 157–165.

5. Barondes, S.H., Castronovo, V., Cooper, D.N.W., Cummings, R.D., Drickamer, K., Feizi, T., Gitt, M.A., Hirabayashi, J., Hughes, R.C., Kasai, K., Leffler, H., Liu, F.-T., Lotan, R., Mercurio, A.M., Monsigny, M., Pillai, S., Poirier, F., Raz, A., Rigby, P., Rini, J.M. & Wang, J.L. (1994). Galectins: a family of animal β-galactoside binding lectins. *Cell* **76**, 597–598.
6. Barondes, S.H., Cooper, D.N.W., Gitt, M.A. & Leffler, H. (1994). Galectins. *Journal of Biological Chemistry* **269**, 20807–20810.
7. Kasai, K. & Hirabayashi, J. (1996). Galectins: a family of animal lectins that decipher glycocodes. *Journal of Biochemistry* **119**, 1–8.
8. Hughes, R.C. (1997). The galectin family of mammalian carbohydrate-binding molecules. *Biochemical Society Transactions* **25**, 1194–1198.
9. Perillo, N.L., Marcus, M.E. & Baum, L.G. (1998). Galectins: versatile modulators of cell adhesion, cell proliferation and cell death. *Journal of Molecular Medicine* **76**, 402–412.
10. Rini, J.M. (1995). Lectin structure. *Annual Review of Biophysical and Biomolecular Structure* **24**, 551–577.
11. Ahmed, H., Pohl, J., Fink, N.E., Strobel, F. & Vasta, G.R. (1996). The primary structure and carbohydrate specificity of a β-galactosyl binding lectin from toad (*Bufo arenarum* Hensel) ovary reveal closer similarities to the mammalian galectin-1 than to the galectin from the clawed frog, *Xenopus laevis*. *Journal of Biological Chemistry* **271**, 33083–33094.
12. Levi, G. & Teichberg, V.I. (1981). Isolation and physiochemical characterization of electrolectin, a β-galactoside binding lectin from the electric organ of *Electrophorus electricus*. *Journal of Biological Chemistry* **256**, 5735–5740.
13. Den, H. & Malinzak, D.A. (1977). Isolation and properties of β-D-galactoside-specific lectin from chick embryo thigh muscle. *Journal of Biological Chemistry* **252**, 5444–5448.
14. Greenhalgh, C.J., Beckham, S.A. & Newton, S.E. (1999). Galectins from sheep gastrointestinal nematode parasites are highly conserved. *Molecular and Biochemical Parasitology* **98**, 285–289.
15. Wagner-Hülsmann, C., Bachinski, N., Diehl-Seifert, B., Blumbach, B., Steffen, R., Pancer, Z. & Muller, W.E.G. (1996). A galectin links the aggregation factor to cells in the sponge (*Geodia cydonium*) system. *Glycobiology* **6**, 785–793.
16. Poirier, F. & Robertson, E.J. (1993). Normal development of mice carrying a null mutation in the gene encoding the L14 S-type lectin. *Development* **119**, 1229–1236.
17. Colnot, C., Ripoche, M.-A., Scaeron, F., Fowlis, D. & Poirier, F. (1996). Galectins in mouse embryogenesis. *Biochemical Society Transactions* **24**, 1–599.
18. Sanford, G.L. & Harris-Hooker, S. (1990). Stimulation of vascular cell proliferation by β-galactoside specific lectins. *FASEB Journal* **4**, 2912–2918.
19. Moiseeva, E.P., Spring, E.L., Baron, J.H. & de Bono, D.P. (1999). Galectin-1 modulates attachment, spreading and migration of cultured vascular smooth muscle cells via interactions with cellular receptors and components of extracellular matrix. *Journal of Vascular Research* **36**, 47–58.
20. Wells, V. & Mallucci, L. (1991). Identification of an autocrine negative growth factor: mouse β-galactoside-binding protein is a cytostatic factor and cell growth regulator. *Cell* **64**, 91–97.
21. Yamaoka, K., Ohno, S., Kawasaki, H. & Suzuki, K. (1993). Mammalian lectin as transforming growth factor. In: *Lectins and Glycobiology* (eds H.-J. Gabius & S. Gabius) pp. 492–499. Springer-Verlag, Berlin.
22. Yamaoka, K., Ohno, S., Kawasaki, H. & Suzuki, K. (1991). Overexpression of a β-galactoside binding protein causes transformation of BALB 3T3 fibroblast cells. *Biochemical and Biophysical Research Communications* **179**, 272–279.
23. Perillo, N.L., Pace, K.E., Seilhamer, J.J. & Baum, L.G. (1995). Apoptosis of T cells mediated by galectin-1. *Nature* **378**, 736–739.
24. Perillo, N.L., Uittenbogaart, C.H., Nguyen, J.T. & Baum, L.G. (1997). Galectin-1, an endogenous lectin produced by thymic epithelial cells, induces apoptosis of human thymocytes. *Journal of Experimental Medicine* **185**, 1851–1858.

25. Murphy, J.J., Stevanin, T., Schoendorf, D.S., Wells, V. & Mallucci, L. (1997). β-Galactoside binding protein inhibits B cell growth and induces apoptosis. *Biochemical Society Transactions* **25**, 248S.
26. Yamaoka, K., Ingendoh, A., Tsubuki, S., Nagai, Y. & Sanai, Y. (1996). Structural and functional characterization of a novel tumor-derived rat galectin-1 having transforming growth factor (TGF) activity: the relationship between intramolecular disulphide bridges and TGF activity. *Journal of Biochemistry* **119**, 878–886.
27. Kilpatrick, D.C. (1993). Is human placental lectin an immunoregulatory molecule? In: *Lectins – Biology, Biochemistry, Clinical Chemistry,* Volume 9 (eds J. Basu, M. Kundu & P. Chakrabarti) pp. 100–104. Wiley Eastern, New Delhi.
28. Adams, L., Scott, G.K. & Weinberg, C.S. (1996). Biphasic modulation of cell growth by recombinant human galectin-1. *Biochimica et Biophysica Acta* **1312**, 137–144.
29. Offner, H., Celnik, B., Bringman, T.S., Casentini-Borocz, D., Nedwin, G.E. & Vanderbark, A.A. (1990). Recombinant human β-galactoside-binding lectin suppresses clinical and histological signs of experimental autoimmune encephalomyelitis. *Journal of Neuroimmunology* **28**, 177–184.
30. Tracey, B.M., Feizi, T., Abbott, W.M., Carruthers, R.A., Green, B.N. & Lawson, A.M. (1992). Subunit molecular mass assignment of 14 654 kDa to the soluble β-galactoside-binding lectins from bovine heart muscle and demonstration of intramolecular disulphide bonding associated with oxidative inactivation. *Journal of Biological Chemistry* **267**, 10342–10347.
31. Whitney, P.L., Powell, J.T. & Sanford, G.L. (1986). Oxidation and chemical modification of lung β-galactoside-binding-specific lectin. *Biochemical Journal* **238**, 683–689.
32. Hirabayashi, J. & Kasai, K. (1991). Effect of amino acid substitution by site-directed mutagenesis on the carbohydrate recognition and stability of human 14 KDa β-galactoside binding lectin. *Journal of Biological Chemistry* **266**, 23648–23653.
33. Kilpatrick, D.C., Bevan, B.H., Liston, W.A., Hirabayashi, J., Kasai, K. & Cooper, B.L. (1996). Lectins in the human feto-placental unit: a review. In: *Lectins – Biology, Biochemistry, Clinical Biochemistry,* Volume 11 (eds E. Van Driessche, P. Rougé, S. Beeckmans & T.-C. Bøg-Hansen) pp. 161–167. Textop, Hellerup.
34. Drickamer, K. (1993). Recognition of complex carbohydrates by Ca^{2+}-dependent animal lectins. *Biochemical Society Transactions* **21**, 456–459.
35. Day, A.J. (1994). The C-type carbohydrate recognition domain (CRD) superfamily. *Biochemical Society Transactions* **22**, 83–88.
36. Iozzo, R.V. (1998). Matrix proteoglycans: from molecular design to cellular function. *Annual Review of Biochemistry* **67**, 609–652.
37. Doege, K., Sasaki, M., Horigan, E., Hassell, J.R. & Yamada, Y. (1987). Complete primary structure of the rat cartilage proteoglycan core protein deduced from cDNA clones. *Journal of Biological Chemistry* **262**, 17757–17767.
38. Zimmerman, D.R. & Ruoslahti, E. (1989). Multiple domains of the large fibroblast proteoglycon, versican. *EMBO Journal* **8**, 2975–2981.
39. Rauch, U., Karthikeyan, L., Maurel, P., Margolis, R.U. & Margolis, R.K. (1992). Cloning and primary structure of neurocan, a developmentally regulated, aggregating chondroitin sulphate proteoglycan of brain. *Journal of Biological Chemistry* **267**, 19536–19547.
40. Yamada, H., Watanbe, K., Shimonaka, M. & Yamaguchi, Y. (1994). Molecular cloning of brevican, a novel brain proteoglycan of the aggrecan/versican family. *Journal of Biological Chemistry* **269**, 10119–10126.
41. Aspberg, A., Muira, R., Bourdoulous, S., Shimonika, M., Heinegard, D., Schachner, M., Ruoslahti, E. & Yamaguchi, Y. (1997). The C-type lectin domains of lecticans, a family of aggregating chondroitin sulphate proteoglycans, bind tenascin-R by protein–protein interactions independent of carbohydrate moiety. *Proceedings of the National Academy of Sciences USA* **94**, 10116–10121.
42. Muira, R., Aspberg, A., Ethell, I.M., Hagihara, K., Schnaar, R.L., Ruoslahti, E. & Yamaguchi, Y. (1999). The proteoglycan lectin domain binds sulphated cell surface

glycolipids and promotes cell adhesion. *Journal of Biological Chemistry* **274**, 11431–11438.
43. Thiel, S. & Reid, K.B.M. (1989). Structures and functions associated with the group of mammalian lectins containing collagen-like sequences. *FEBS Letters* **250**, 78–84.
44. Reid, K.B.M. (1993). Structure/function relationships in the collectins (mammalian lectins containing collagen-like regions). *Biochemical Society Transactions* **21**, 464–468.
45. Holmskov, U., Malhotra, R., Sim, R.B. & Jensenius, J.C. (1994). Collectins: collagenous C-type lectins of the innate immune defense system. *Immunology Today* **15**, 67–74.
46. Epstein, J., Eichbaum, Q., Sheriff, S. & Ezekowitz, R.A.B. (1996). The collectins in innate immunity. *Current Opinion in Immunology* **8**, 29–35.
47. Lu, J. (1997). Collectins: collectors of microorganisms for the innate immune system. *BioEssays* **19**, 509–518.
48. Crouch, E.C. (1998). Collectins and pulmonary host defense. *American Journal of Respiratory Cell and Molecular Biology* **19**, 177–201.
49. Bordet, J. & Gay, F.P. (1906). Sur les relations des sensibilisatrices avec l'alexine. *Annals Institut Pasteur* **20**, 467–498.
50. Hirani, S., Lambris, J.D. & Muller-Eberhard, H.J. (1985). Localization of the conglutinin binding site on the third component of human complement. *Journal of Immunology* **134**, 1105–1109.
51. Thiel, S., Baatrup, G., Friis-Christiansen, P., Svehag, S.-E. & Jensenius, J.C. (1987). Characterization of a lectin in human plasma analogous to bovine conglutinin. *Scandinavian Journal of Immunology* **26**, 461–468.
52. Lu, J., Laursen, S.B., Thiel, S., Jensenius, J.C. & Reid, K.B.M. (1993). The cDNA cloning of conglutinin and identification of liver as a primary site of synthesis of conglutinin in members of the Bovidae. *Biochemical Journal* **292**, 157–162.
53. Holmskov, U., Laursen, S.B., Malhotra, R., Weidemann, H., Timpl, R., Stuart, G.R., Tornøe, I., Madsen, P.S., Reid, K.B.M. & Jensenius, J.C. (1995). Comparative study of the structural and functional properties of a bovine plasma C-type lectin, collectin-43, with other collectins. *Biochemical Journal* **305**, 889–896.
54. McCormack, F.X. (1997). The structure and function of surfactant protein A. *Chest* **111**, 114S–119S.
55. McCormack, F.X. (1998). Structure, processing and properties of surfactant protein A. *Biochimica et Biophysica Acta* **1408**, 109–131.
56. Mason, R.J., Greene, K. & Voelker, D.R. (1998). Surfactant protein A and surfactant protein D in health and disease. *American Journal of Respiratory Cell and Molecular Biology* **275**, L1–L13.
57. Crouch, E.C. (1998). Collectins and pulmonary host defense. *American Journal of Respiratory Cell and Molecular Biology* **19**, 177–201.
58. Crouch, E.C. (1998). Structure, biologic properties and expression of surfactant protein D (SP-D). *Biochimica et Biophysica Acta* **1408**, 278–289.
59. Reid, K.B.M. (1998). Interactions of surfactant protein D with pathogens, allergens and phagocytes. *Biochimica et Biophysica Acta* **1408**, 290–295.
60. Turner, M.W. (1996). Mannose-binding lectin: the pluripotent molecule of the innate immune system. *Immunology Today* **17**, 532–540.
61. Turner, M.W. (1998). Mannose-binding lectin (MBL) in health and disease. *Immunobiology* **199**, 327–339.
62. Thompson, C. (1995). Protein proves to be a key link in innate immunity. *Science* **269**, 301–302.
63. Miller, M.E., Seals, J., Kaye, R. & Levitsky, L.C. (1968). A familial plasma-associated defect of phagocytosis. *Lancet* **2**, 60–63.
64. Richardson, V.F., Larcher, V.F. & Price, J.F. (1983). A common congenital immunodeficiency predisposing to infection and atopy in infancy. *Archives of Disease in Childhood* **58**, 799–802.

65. Kawasaki, T., Etoh, R. & Yamashina, I. (1978). Isolation and characterization of a mannan-binding protein from rabbit liver. *Biochemical and Biophysical Research Communications* **81**, 1018–1024.
66. Kawasaki, N., Kawasaki, T. & Yamashina, I. (1983). Isolation and characterization of a mannan-binding protein from human serum. *Journal of Biochemistry* **94**, 937–947.
67. Wild, J., Robinson, D. & Winchester, B. (1983). Isolation of mannose-binding proteins from human and rat liver. *Biochemical Journal* **210**, 167–174.
68. Super, M., Lu, J., Thiel, S., Levinsky, R.J. & Turner, M.W. (1989). Association of low levels of mannan-binding protein with a common defect of opsonisation. *Lancet* **2**, 1236–1239.
69. Sumiya, M., Super, M., Tabona, P., Levinsky, R.J., Arai, T., Turner, M.W. & Summerfield, J.A. (1991). Molecular basis of opsonic defect in immunodeficient children. *Lancet* **337**, 1569–1570.
70. Kawakami, M., Ihara, I., Ihara, S., Suzuki, A. & Fukui, K. (1984). A group of bactericidal factors conserved by vertebrates for more than 300 million years. *Journal of Immunology* **132**, 2578–2581.
71. Laursen, S.B., Dalgaard, T.S., Thiel, S., Lim, B.-L., Jensen, T.V., Madsen-Juul, H.R., Takahashi, A., Hamana, T., Kawakami, M. & Jensenius, J.C. (1998). Cloning and sequencing of a cDNA encoding chicken mannan-binding lectin (MBL) and comparison with mammalian analogues. *Immunology* **93**, 421–430.
72. Kurata, H., Sannoh, T., Kozutsumi, Y., Yokota, Y. & Kawasaki, T. (1994). Structure and function of mannan-binding proteins isolated from human liver and serum. *Journal of Biochemistry* **115**, 1148–1154.
73. Madsen, H.O., Garred, P., Thiel, S., Kurtzhals, J.A.L., Lamm, L.U., Ryder, L.P. & Svejgaard, A. (1995). Interplay between promotor and structural gene variants control basal serum level of mannan-binding protein. *Journal of Immunology* **155**, 3013–3020.
74. Matsushita, M., Hijikata, M., Matsushita, M., Ohta, Y. & Mishiro, S. (1998). Association of mannose-binding lectin gene haplotype LXPA and LYPB with interferon-resistant hepatitis C virus infection in Japanese patients. *Journal of Hepatology* **29**, 695–700.
75. Bella, J., Eaton, M., Brodsky, B. & Berman, H.M. (1994). Crystal and molecular structure of a collagen-like peptide at 1.9 Å resolution. *Science* **266**, 75–81.
76. Lipscombe, R.J., Sumiya, M., Summerfield, J.A. & Turner, M.W. (1995). Distinct physiochemical characteristics of human mannose binding protein expressed by individuals of differing genotype. *Immunology* **85**, 660–667.
77. Matsushita, M., Ezekowitz, R.A.B. & Fujita, T. (1995). The Gly-54 → Asp allelic form of human mannose-binding protein (MBP) fails to bind MBP-associated serine protease. *Biochemical Journal* **311**,1021–1023.
78. Super, M., Gillies, S.D., Foley, S., Sastry, K., Schweinle, J.-E., Silverman, V.J. & Ezekowitz, R.A.B. (1992). Distinct and overlapping functions of allelic forms of human mannose binding protein. *Nature Genetics* **2**, 50–55.
79. Yokata, Y., Arai, T. & Kawasaki, T. (1995). Oligomeric structures required for complement activation of serum mannan-binding proteins. *Journal of Biochemistry* **117**, 414–419.
80. Naito, H., Ma, Y., Uemura, K., Arano, Y. & Kawasaki, T. (1999). Metabolic properties of normal and mutant mannan-binding proteins in mouse plasma. *Biochemical and Biophysical Research Communications* **256**, 231–234.
81. Lu, J., Thiel, S., Niedemann, H., Timpl, R. & Reid, K.B.M. (1990). Binding of the pentamer/hexamer forms of mannan-binding protein to zymosan activates the proenzyme $C1r_2C1s_2$ complex, of the classical pathway of complement, without involvement of C1q. *Journal of Immunology* **144**, 2287–2294.
82. Thiel, S., Holmskov, U., Hviid, L., Laursen, S.B. & Jensenius, J.C. (1992). The concentration of the C-type lectin, mannan-binding protein, in human plasma increases during an acute phase response. *Clinical and Experimental Immunology* **90**, 31–35.

83. Arai, T., Tabona, P. & Summerfield, J.A. (1993). Human mannose-binding protein gene is regulated by interleukins, dexamethasone and heat shock. *Quarterly Journal of Medicine* **86**, 575–582.
84. Garred, P., Madsen, H.O., Hofman, B. & Svejgaard, A. (1995). Increased frequency of homozygosity of abnormal mannan-binding protein alleles in patients with suspected immunodeficiency. *Lancet* **346**, 941–943.
85. Summerfield, J.A., Sumiya, M., Levin, M., Turner, M.W. (1997). Association of mutations in mannose binding protein gene with childhood infection in consecutive hospital series. *British Medical Journal* **314**, 1229–1232.
86. Kakkanaiah, V., Shen, G.Q., Ojo-Amaize, E.A. & Peter, J.B. (1998). Association of low concentrations of serum mannose-binding protein with recurrent infection in adults. *Clinical and Diagnostic Laboratory Immunology* **5**, 319–321.
87. Hoal-Van Helden, E.G., Epstein, J., Victor, T.C., Hon, D., Lewis, L.-A., Beyers, N., Zurakowski, D., Ezekowitz, R.A.B., Van Helden, P.D. (1999). Mannose-binding protein B allele confers protection against tuberculosis meningitis. *Paediatric Research* **45**, 459–464.
88. Malhotra, R., Wormald, M.R., Rudd, P.M., Fischer, P.B., Dwek, R.A. & Sim, R.B. (1995). Glycosylation changes of IgG associated with rheumatoid arthritis can activate complement via the mannose-binding protein. *Nature Medicine* **1**, 237–243.
89. Graudal, N.A., Homann, C., Madsen, H.O., Svejgaard, A., Jurik, A.G., Gravdal, H.K. & Garred, P. (1997). Mannan binding lectin in rheumatoid arthritis. A longitudinal study. *Journal of Rheumatology* **25**, 629–635.
90. Kilpatrick, D.C. (1997). Mannan binding protein in sera positive for rheumatoid factor. *British Journal of Rheumatology* **36**, 207–209.
91. Stanworth, S.J., Donn, R.P., Hassall, A., Dawes, P., Ollier, W. & Snowden, N. (1998). Absence of an association between mannose-binding lectin polymorphism and rheumatoid arthritis. *British Journal of Rheumatology* **37**, 186–188.
92. Turner, M.W. (1998). 90 years on: a therapy to 'stimulate the phagocytes'. *Scandinavian Journal of Immmunology* **48**, 124–126.
93. Ohtani, K., Suzuki, Y., Edu, S., Kawai, T., Kase, T., Yamazaki, H., Shimoda, T., Keshi, H., Sakai, Y., Fukuoh, A., Sakamoto, T. & Wakamiya, N. (1999). Molecular cloning of a novel human collectin from liver (CL-L1). *Journal of Biological Chemistry* **274**, 13681–13689.
94. Bevilacqua, M.P. & Nelson, R.M. (1993). Selectins. *Journal of Clinical Investigation* **91**, 379–387.
95. Lowe, J.B. (1994). Carbohydrate recognition in cell–cell interaction. In: *Molecular Glycobiology* (eds M. Fukuda & O. Hindsgaul) pp. 163–205. Oxford University Press, New York.
96. Lasky, L.A. (1995). Selectin–carbohydrate interactions and the initiation of the inflammatory response. *Annual Review of Biochemistry* **64**, 113–139.
97. Kansas, G.S. (1996). Selectins and their ligands: current concepts and controversies. *Blood* **88**, 3259–3287.
98. Bird, M.I., Foster, M.R., Priest, R. & Malhotra, R. (1997). Selectins: physiological and pathophysiological roles. *Biochemical Society Transactions* **25**, 1199–1206.
99. Vestweber, D. (ed.) (1997). *The Selectins. The Initiators of Endothelial Adhesion.* Harwood, Amsterdam.
100. Gesner, B.M. & Ginsburg, V. (1964). Effect of glycosidases on the fate of transfused lymphocytes. *Proceedings of the National Academy of Sciences USA* **52**, 750–755.
101. Stoolman, L.M. & Rosen, S.D. (1983). Possible role for cell-surface carbohydrate-binding molecules in lymphocyte recirculation. *Journal of Cell Biology* **96**, 722–729.
102. Mulligan, M.S., Lowe, J.B., Larsen, R.D., Paulson, J., Zheng, Z., DeFrees, S., Maemura, K., Fukada, M. & Ward, P.A. (1993). Protective effects of sialylated oligosaccharides in immune complex-induced acute lung injury. *Journal of Experimental Medicine* **178**, 623–631.

103. Mebius, R.E. & Watson, S.R. (1993). L- and E-selectin can recognise the same naturally occurring ligands on high endothelial venules. *Journal of Immunology* **151**, 3252–3260.
104. Homeister, J.W., Zhang, M., Frenette, P.S., Hynes, R.O., Wagner, D.D., Lowe, J.B. & Marks, R.M. (1998). Overlapping functions of E- and P-selectin in neutrophil recruitment during acute imflammation. *Blood* **92**, 2345–2352.
105. Jung, U. & Dey, K. (1999). Mice lacking two or all three selectins demonstrate overlapping and distinct functions for each selectin. *Journal of Immunology* **162**, 6755–6762.
106. Marquart, T., Brune, T., Luhn, K., Zimmer, K.-P., Korner, C., Fabritz, L., Van Der Werft, N., Vormoor, J., Freeze, H.H., Louwen, F., Biermann, B., Harms, E., Von Figura, K., Vestweber, D. & Koch, H.G. (1999). Leukocyte adhesion deficiency II syndrome, a generalized defect in fucose metabolism. *Journal of Pediatrics* **134**, 681–688.
107. DeLisser, H.M., Christofidou-Solomidou, M., Sun, J., Nakada, M.T. & Sullivan, K.E. (1999). Loss of endothelial surface expression of E-selectin in a patient with recurrent infections. *Blood* **94**, 884–894.
108. Mulligan, M.S., Polley, M.J., Bayer, R.J., Numm, M.F., Paulson, J.C. & Ward, P.A. (1992). Neutrophil-dependent acute lung injury. Requirements for P-selectin (GMP-140). *Journal of Clinical Investigation* **90**, 1600–1607.
109. Weyrich, A.S., Ma, X., Lefer, D.J., Albertine, K.H. & Lefer, A.M. (1993). In vivo neutralisation of P-selectin protects feline heart and endothelium in myocardial ischemia and reperfusion injury. *Journal of Clinical Investigation* **91**, 2620–2629.
110. Dejana, E., Martin-Padura, I., Lauri, D., Bernasconi, S., Bani, M.R., Garofalo, A., Giavazzi, R., Magnoni, J., Mantovani, A. & Menard, S. (1992). Endothelial leukocyte adhesion molecule-1-dependent adhesion of colon carcinoma cells to vascular endothelium is inhibited by an antibody to Lewis fucosylated type I carbohydrate chain. *Laboratory Investigation* **66**, 324–330.
111. Kim, Y.L., Borsig, L., Varki, N.M. & Varki, A. (1998). P-selectin deficiency attenuates tumour growth and metastasis. *Proceedings of the National Academy of Sciences USA* **95**, 9325–9330.
112. Machelska, H., Cabot, P.J., Mousa, S.A., Zhang, Q. & Stein, C. (1998). Pain control in inflammation governed by selectins. *Nature Medicine* **4**, 1425–1428.
113. Anderson, O., Laursen, S.B., Svehag, S.E., Holmskov, U. & Thiel, S. (1991). Mammalian lectin in defense mechanisms against microorganisms. In: *Lectin Reviews*, Volume 1 (eds D.C. Kilpatrick, E. Van Driessche & T.-C. Bög-Hansen) pp. 41–51. Sigma Chemical Company, St Louis, MO.
114. Taylor, M.E. (1993). Recognition of complex carbohydrates by the macrophage mannose receptor. *Biochemical Society Transactions* **21**, 468–473.
115. Stahl, P.D. & Ezekowitz, R.A.B. (1998). The mannose receptor is a pattern recognition receptor involved in host defense. *Current Opinion in Immunology* **10**, 50–55.
116. Lambeau, G., Ancian, P., Barhanin, J. & Lazdunski, M. (1994). Cloning and expression of a membrane receptor for secretary phospholipases A_2. *Journal of Biological Chemistry* **269**, 1575–1578.
117. Ishizaki, J., Hanasaki, K., Higashino, K., Kishino, J., Kikuchi, N., Ohara, O. & Arita, H. (1994). Molecular cloning of pancreatic group I phospholipase A_2 receptor. *Journal of Biological Chemistry* **269**, 5897–5904.
118. Wu, K., Yuan, J. & Lasky, L.A. (1996). Characterization of a novel member of the macrophage mannose receptor type C lectin family. *Journal of Biological Chemistry* **271**, 21323–21330.
119. Jiang, W., Swiggard, N.J., Heufler, C., Peng, M., Mirza, A., Steinman, R.M. & Nussenzweig, M.C. (1995). The receptor DEC-205 expressed by dendritic cells and thymic epithelial cells is involved in antigen processing. *Nature* **375**, 151–155.
120. Swiggard, W.J., Mirza, A., Nussenzweig, M.C. & Steinman, R.M. (1995). DEC-205, a 205 kDa protein abundant on mouse dendritic cells and thymic epithelium that is

detected by the mouse monoclonal antibody NLDC-145: purification, characterization and N-terminal amino acid sequence. *Cellular Immunology* **165**, 302–311.
121. Brown, M.G., Scalzo, A.A., Matsumoto, K. & Yokoyama, W.M. (1997). The natural killer gene complex: a genetic basis for understanding natural killer cell function and innate immunity. *Immunological Reviews* **155**, 53–65.
122. Lanier, L.L. (1998). NK cell receptors. *Annual Review of Immunology* **16**, 359–393.
123. Long, E.O. (1999). Regulation of immune responses through inhibitory receptors. *Annual Review of Immunology* **17,** 875–904.
124. Adamkiewicz, T.V., McSherry, C., Bach, F.H. & Houchins, J.P. (1994). Natural killer lectin-like receptors have divergent carboxy-termini, distinct from C-type lectins. *Immunogenetics* **39**, 218.
125. Weis, W.I., Taylor, M.E. & Drickamer, K. (1998). The C-type lectin superfamily in the immune system. *Immunological Reviews* **163**, 19–34.
126. Matsumoto, N., Ribaudo, R.K., Abastado, J.P., Margulies, D.H. & Yokayama, W.M. (1998). The lectin-like NK cell receptor Ly-49A recognises a carbohydrate-independent epitope on its MHC class I ligand. *Immunity* **8**, 245–254.
127. Daniels, B.F., Nakamura, M.C., Rosen, S.D., Yokoyama, W.M. & Seaman, W.E. (1994). Ly-49A, a receptor for H-2D^{d}, has a functional carbohydrate recognition domain. *Immunity* **1**, 785–792.
128. Brennan, J., Takei, F., Wong, S. & Mager, D.L. (1995). Carbohydrate recognition by a natural killer cell receptor, Ly-49C. *Journal of Biological Chemistry* **270**, 9691–9694.
129. Bezouska, K., Vlahas, G., Horvath, O., Jinochova, G., Fiserova, A., Giorda, R., Chambers, W.H., Feizi, T. & Popíšil, M. (1994). Rat natural killer cell antigen, NKR-P1, related to C-type animal lectins is a carbohydrate-binding protein. *Journal of Biological Chemistry* **269**, 16945–16952.
130. Bezouska, K., Yuen, C.-T., O'Brien, J., Childs, R.A., Chai, W., Lawson, A.M., Drbal, K., Fiserova, A., Popíšil, M. & Feizi, T. (1994). Oligosaccharide ligands for NKR-P1 protein activate NK cells and cytotoxicity. *Nature* **372**, 150–157.
131. Bezouska, K., Yuen, C.-T., O'Brien, J., Childs, R.A., Chai, W., Lawson, A.M., Drbal, K., Fiserova, A., Popíšil, M. & Feizi, T. (1996). Correction: oligosaccharide ligands for NKR-P1 protein activate NK cells and cytotoxicity. *Nature* **380,** 559.
132. Bezouska, K., Nepovim, A., Horvath, O., Popíšil, M., Hamann, J. & Feizi, T. (1995). CD69 antigen of human lymphocytes is a calcium-dependent carbohydrate-binding protein. *Biochemical and Biophysical Research Communications* **208**, 68–74.
133. Fernandes, M.J., Finnegan, A.A., Siracusa, L.D., Brenner, C., Iscove, N.N. & Calabretta, B. (1999). Characterization of a novel receptor that maps near the natural killer gene complex: demonstration of carbohydrate binding and expression in hematopoietic cells. *Cancer Research* **59**, 2709–2717.
134. Bates, E.E.M., Fournier, N., Garcia, E., Valladeau, J., Durand, I., Pin, J.-J., Zurawski, S.M., Patel, S., Abrams, J.S., Lebecque, S., Garrone, P. & Saeland, S. (1999). APCs express DCIR, a novel C-type lectin surface receptor containing an immunoreceptor tyrosine-based inhibitory motif. *Journal of Immunology* **163**, 1973–1983.
135. Morell, A.G., Irvine, R.A., Sternlieb, I., Scheinberg, I.M. & Ashwell, G. (1968). Physical and chemical studies on ceruloplasmin. V Metabolic studies on sialic acid-free ceruloplasmin in vivo. *Journal of Biological Chemistry* **243**, 155–159.
136. Stockert, R.J., Morell, A.G. & Scheinberg, I.H. (1974). Mammalian hepatic lectin. *Science* **186**, 365–366.
137. Sato, M., Kawakomi, K., Osawa, T. & Toyoshima, S. (1992). Molecular cloning and expression of cDNA encoding a galactose/*N*-acetylgalactosamine-specific lectin on mouse tumoricidal macrophages. *Journal of Biochemistry* **111**, 331–336.
138. Haltiwanger, R.S., Lehrmann, M.A., Eckhardt, A.E. & Hill, R.L. (1986). The distribution and localization of the fucose-binding lectin in rat tissues and the identification of a high affinity form of the mannose/*N*-acetylglucosamine-binding lectin in rat liver. *Journal of Biological Chemistry* **261**, 7433–7439.

139. Curtis, B.M., Scharnowske, S. & Watson, A.J. (1992). Sequence and expression of a membrane-associated C-type lectin that exhibits CD4-independent binding of human immunodeficiency virus envelope glycoprotein gp120. *Proceedings of the National Academy of Sciences USA* **89**, 8356–8360.
140. Gordon, J. (1994). B-cell signalling via the C-type lectins CD23 and CD72. *Immunology Today* **15**, 411–417.
141. Steele, D.M. & Whitehead, A.S. (1994). The major acute phase reactants: C-reactive protein, serum amyloid P component and serum amyloid A protein. *Immunology Today* **15**, 81–88.
142. Tennent, G.A. & Pepys, M.B. (1994). Glycobiology of the pentraxins. *Biochemical Society Transactions* **22**, 74–79.
143. Gewurz, H., Zhang, X.-H. & Lint, T.F. (1995). Structure and function of the pentraxins. *Current Opinion in Immunology* **7**, 54–64.
144. Elola, M.T. & Vasta, G.R. (1994). Lectins from the colonial tunicate *Clavelina picta* are structurally related to acute phase reactants from vertebrates. *Annals of the New York Academy of Sciences* **712**, 321–323.
145. Shrive, A.K., Metcalfe, A.M., Cartwright, J.R. & Greenhough, T.J. (1999). C-reactive protein and SAP-like pentraxin are both present in *Limulus polyphemus* haemolymph: crystal structure of *Limulus* SAP. *Journal of Molecular Biology* **290**, 997–1008.
146. Whicker, J.T. & Dieppe, P.A. (1985). Acute phase proteins. *Clinics in Immunology and Allergy* **5**, 425–446.
147. Schwalbe, R.A., Dahlback, B., Coe, J.E. & Nelsestuen, G.L. (1992). Pentraxin family of proteins interact specifically with phosphorylcholine and/or phosphorylethanolamine. *Biochemistry* **31**, 4907–4915.
148. Gabay, C. & Kushner, I. (1999). Acute-phase proteins and other systemic responses to inflammation. *New England Journal of Medicine* **340**, 448–454.
149. Brown, M.R. & Anderson, B.E. (1993). Receptor–ligand interactions between serum amyloid P component and model soluble immune complexes. *Journal of Immunology* **151**, 2087–2095.
150. Bickerstaff, M.C.M., Botto, M., Hutchinson, W.L., Herbert, J., Tennent, G.A., Bybee, A., Mitchell, D.A., Cook, H.T., Butler, P.J.G., Walport, M.J. & Pepys, M.B. (1999). Serum amyloid P component controls chromatin degradation and prevents antinuclear autoimmunity. *Nature Medicine* **5**, 694–697.
151. Alles, V.V., Bottazzi, B., Peri, G., Golay, J., Introna, M. & Mantovani, A. (1994). Inducible expression of PTX3, a new member of the pentraxin family, in humans mononuclear phagocytes. *Blood* **84**, 3483–3493.
152. Powell, L.D. & Varki, A. (1995). I-type lectins. *Journal of Biological Chemistry* **270**, 14243–14246.
153. Kelm, S., Schauer, R. & Crocker, P.R. (1996). The sialoadhesins – a family of sialic acid-dependent cellular recognition molecules within the immunoglobulin-superfamily. *Glycoconjugate Journal* **13**, 913–926.
154. Gabius, H.-J. (1997). Animal lectins. *European Journal of Biochemistry* **243**, 543–576.
155. Hickman, S. & Neufeld, E.F. (1972). A hypothesis for I-cell disease: defective hydrolases that do not enter lysosomes. *Biochemical and Biophysical Research Communications* **49**, 992–999.
156. Von Figura, K. & Kress, H. (1974). Inhibition of pinocytosis by cytochalasin B. *European Journal of Biochemistry* **48**, 357–363.
157. Ullrich, K., Mersmann, G., Weber, E. & Von Figura, K. (1978). Evidence for lysosomal enzyme recognition by human fibroblasts via a phosporylated carbohydrate moiety. *Biochemical Journal* **170**, 643–650.
158. Kornfeld, S. (1992). Structure and function of the mannose-6-phosphate/insulin like growth factor II receptors. *Annual Review of Biochemistry* **61**, 307–330.
159. Lu, J. & Le, Y. (1998). Ficolins and the fibrinogen-like domain. *Immunobiology* **199**, 190–199.

160. Ohashi, T. & Erickson, H.P. (1999). Oligomeric structure and tissue distribution of ficolins from mouse, pig and human. *Archives of Biochemistry and Biophysics* **360**, 223–232.
161. Matsushita, M., Endo, Y., Taira, S., Sato, Y., Fujita, T., Ichikawa, N., Nakata, M. & Mizuochi, T. (1996). A novel human serum lectin with collagen- and fibrinogen-like domains that functions as an opsonin. *Journal of Biological Chemistry* **271**, 2448–2454.
162. Sugimoto, R., Yae, Y., Akaiwa, M., Kitijima, S., Shibata, Y., Sato, H., Hirata, J., Okochi, K., Izuhara, K. & Hamasaki, N. (1998). Cloning and characterization of the Hakata antigen, a member of the ficolin/opsonin P35 lectin family. *Journal of Biological Chemistry* **273**, 20721–20727.
163. Zanetta, J.-P., Scior, T., Wantyghem, J., Wermuth, C., Aubery, M., Strecker, G. & Michalski, J.-C. (1996). Lectin activities of cytokines and growth factors: function and implications for pathology. *Histology and Histopathology* **11**, 1101–1108.
164. Moonen, P., Gaffner, R. & Wingfield, P. (1988). Native cytokines do not bind to uromodulin (Tamm–Horsfall glycoprotein). *FEBS Letters* **226**, 314–318.
165. Lambert, C., Brealey, R., Steele, J. & Rook, G.A.W. (1993). The interaction of Tamm–Horsfall protein with the extracellular matrix. *Immunology* **79**, 203–210.
166. Lucas, R., Magez, S., De Leys, R., Fransen, L., Scheerlinck, J.-P., Rampelberg, M., Sablon, E. & De Baetselier, P. (1994). Mapping the lectin-like activity of tumour necrosis factor. *Science* **263**, 814–817.
167. Tandai-Hiruma, M., Endo, T. & Kobata, A. (1999). Detection of novel carbohydrate binding activity of interleukin-1. *Journal of Biological Chemistry* **274**, 4459–4466.
168. Zanetta, J.-P., Alonso, C. & Michalski, J.-C. (1996). Interleukin 2 is a lectin that associates its receptor with the T cell receptor complex. *Biochemical Journal* **318**, 49–53.
169. Davies, E.J., Snowden, N., Hillarby, M.C., Carthy, D., Grennan, D.M., Thomson, W. & Ollier, W.E.R. (1995). Mannose-binding protein gene polymorphism in systemic lupus erythematosus. *Arthritis and Rheumatism* **38**, 110–114.
170. Lau, Y.L., Lau, C.S., Chan, S.Y., Karlberg, J. & Turner, M.W. (1996). Mannose-binding protein in Chinese patients with systemic lupus erythematosus. *Arthritis and Rheumatism* **39**, 706–708.
171. Sullivan, K.E., Wooten, C., Goldman, D. & Petri, M. (1996). Mannose-binding protein genetic polymorphisms in black patients with systemic lupus erythematosus. *Arthritis and Rheumatism* **12**, 2046–2051.
172. Davies, E.J., Teh, L.S., Ordi-Ros, J., Snowden, N., Hillarby, M.C., Hajeer, A., Donn, R., Perez-Pemen, P., Vilardell-Tarres, M. & Ollier, W.E.R. (1997). A dysfunctional allele of the mannose binding protein gene associates with systemic lupus erythematosus in a Spanish population. *Journal of Rheumatology* **24**, 485–488.
173. Yamanaka, S., Zhang, X.-Y., Miura, K., Kim, S. & Iwao, H. (1998). The human gene encoding the lectin-type oxidized LDL receptor (OLRI) is a novel member of the natural killer gene complex with a unique expression profile. *Genomics* **54**, 191–199.
174. Draude, G., Hrboticky, N. & Lorenz, R.L. (1999). The expression of the lectin-like oxidized low density lipoprotein receptor (LOX-1) on human vascular smooth muscle cells and monocytes and its down-regulation by lovastatin. *Biochemical Pharmacology* **57**, 383–386.
175. Hanke, T., Corral, L., Vance, R.E. & Raulet, D.H. (1998). 2F1 antigen, the mouse homolog of the rat 'mast cell function-associated antigen', is a lectin-like type II transmembrane receptor expressed by natural killer cells. *European Journal of Immunology* **28**, 4409–4417.
176. Hansen, S. & Holmskov, U. (1998). Structural aspects of collectins and receptors for collectins. *Immunobiology* **199**, 165–189.

4 Biomedical Applications of Animal Lectins

BLOOD GROUPING REAGENTS

The term 'lectin' was invented to describe blood group-specific agglutinins, and lectins have been extensively used in immunohaematology as blood typing reagents [1–4]. The original lectins were almost exclusively of plant origin and initially were used in the form of crude extracts. Most plant extracts were ABO-group non-specific, and relatively few of the remainder were reliable enough for routine use. Nevertheless, several plant lectins became widely used as anti-A or anti-O(H) reagents, although excellent anti-B lectins proved to be more elusive. In addition, several anti-N lectins were found in the plant kingdom, while lectins specific for the T antigen and several rarer cryptantigens were found to be uniquely valuable for the detection and classification of polyagglutinability. The animal kingdom has also yielded anti-A, anti-B and anti-O lectins, but not (to my knowledge) any with anti-M, anti-N, anti-T, anti-Tk, anti-Tn* or anti-Cad specificity. However, the *Helix pomatia* agglutinin reacts with VA but not NOR erythrocytes, and so can be of use in distinguishing polyagglutination caused by these extremely rare cryptantigens.

H. pomatia agglutinin is also unquestionably an excellent anti-A reagent, as are the lectins from the related snails, *H. aspera* and *H. hortensis* [5,6]. These snail lectins are among the best reagents available and have been used successfully in blood banks for routine typing, including use in automated grouping machines [7]. Other animal sources of anti-A lectins include newt [8] and frog [9] eggs.

No group B-specific lectin has ever found a place in routine blood typing, nor become widely used for special investigations. Unlike some anti-A lectins, no readily available and stable anti-B lectin source with equal or greater potency and specificity to alloantisera has been found. Nevertheless, quite a few lectins with a strong preference for group B over other types have been reported. Among the first discovered in animals were those from species of the salmon family [10–12]; these crude extracts cross-reacted with blood group P determinants. Some time later a lectin was purified from a member of that family, the Chinook salmon [13]. Among the first to be obtained in purified form from other families, those from powan roe (qv) and from the Ayu fish, *Plecoglossus altivelis* (qv), seemed to have a

---

*An asialoglycoprotein receptor cloned from cultured human monocytes appears to recognise the Tn antigen, but was not evaluated as a typing reagent [123].

high degree of specificity. More recently, lectins from the crabs *Scylla serrata* (qv) and *Charybdis japonica* (qv) seem to be highly group-B specific, while another lectin, *Didemnum candidum* II lectin (qv), is ABO non-specific but becomes highly B-specific when the cells are treated with pronase. Although the *D. candidum* lectin is specific for galactose, the immunodominant sugar of blood group B, it shows a slight preference for group O cells before pronase treatment. The *S. serrata* lectin is even more puzzling when trying to relate saccharide specificity to ABO specificity; inexplicably, it has a narrow specificity for *N*-glycolylneuraminic acid.

Eel agglutinins (qv) are historically important in blood group serology, contributing to our early understanding of the carbohydrate nature of human erythrocyte polymorphism. The lectin from *Anguilla anguilla* perhaps remains the best animal source of a specific group O lectin.

Finally, special mention should be made of the *Aplysia depilans* gonad lectin. This lectin has been evaluated and found to be a reliable anti-I reagent; moreover, it can be used at room temperature and is independent of the presence of ABH or P system antigens [14].

A list of lectins with marked blood group specificity listed in Chapter 5 is given in Table 4.1. However, there are several other lectins with more limited blood group preferences to be found in Chapter 5. Moreover, galectins, which are not blood group-specific in the obvious sense, sometimes differ in sensitivity to an extended blood group A structure: galectin-3 (qv) has a high affinity for this oligosaccharide, a characteristic that can be used to distinguish amphibian galectins (qv) also.

Table 4.1. Blood Group-Specific Lectins Listed in Chapter 5

Lectin	Specificity
Biomphalaria glabrata lectin-I	A_1
Cepaea hortensis sialic acid-binding lectin	A
Charybdis japonica blood type B lectin	B
Didemnum candidum II isolectins	B (pronase-treated)
Eel agglutinins	O(H)
Helix pomatia agglutinin	A
Perch roe lectin	O(H)
Periplaneta americana (original) lectin	A
Plecoglossus altivelis egg lectin	B
Powan roe lectin	B
Rana catesbeiana egg lectin	A
Saxidomus purpuratus isolectins	A
Scylla serrata lectin	B
Tachylectins-2, -3 and -4	A
Trout egg lectins	B

MITOGENIC AGENTS

Lectins (agglutinins) were originally detected by their ability to agglutinate red blood cells, and are principally thought of as anti-erythrocyte reagents.

Nevertheless, leuko-agglutination has been known for almost as long as erytho-agglutination, indeed since the agglutinins in snake venom extracts were shown to be active towards leukocytes [15]. Undoubtedly, it was the seminal discovery of Nowell [16] that a lectin (PHA) could activate as well as separate lymphocytes that was most influential in this context. That discovery did much to stimulate scientific interest in lectins, and was also of great value to immunology by providing a reagent that could stimulate whole populations of resting lymphocytes in an antigen non-specific manner for biochemical investigations. The existence of mitogenic lectins (mostly of plant origin) incidentally demonstrated the importance of cell-surface glycoconjugates for lymphocyte activation and its regulation.

I have reviewed the interactions of lectins with leukocytes and the molecular mechanisms involved in mitogenesis on several occasions elsewhere [17–20]. In summary, it would seem that a mitogenic lectin is one which binds to the T cell receptor and/or to one or more of its associated molecules in an appropriate manner without concomitantly inducing an opposite effect by binding to inhibitory receptors. A large number of lymphocyte membrane receptors can influence the mitogenic process, some of which are associated with each other to some extent on the cell surface. Further complexity is provided by cellular interactions between leukocytes and accessory cells, and the influence of soluble cytokines. A particular lectin in a particular *in vitro* context may be mitogenic, non-mitogenic, anti-mitogenic or co-mitogenic [17–20].

Unsurprisingly, many animal lectins are mitogenic (Table 4.2). At one time it was common to test newly isolated animal lectins for mitogenic activity. In the past two decades, as more animal lectins have been identified by cloning, fewer lectins are being investigated in this way. Most mitogenic lectins (animal or plant) are polyclonal activators of T lymphocytes, although some stimulate B cells in a T-cell-dependent manner; mitogens specific for B cells are rare. In this regard, it is noteworthy that chicken galectins (qv) and purpurin (qv) exclusively stimulate non-T lymphocytes from murine spleen cell suspensions [21], and the *Liocarcinus depurator* crab lectin (qv) is weakly mitogenic for a human B cell enriched preparation [22]. Of course, testing with lymphocytes is non-physiological; it is unlikely that any natural function would include perturbing

Table 4.2. Mitogenic Lectins Listed in Chapter 5

Achatinin$_H$
Allomyrina dichotoma isolectins
Asialoglycoprotein receptors
Chicken galectins C-16 and C-14
Crotalus atrox venom lectin
Electrolectin
Entamoeba histolytica lectin
Galectin-1 and galectin-3
Geodia cydonium galectin
Interleukin-2
Limulin
Liocarcinus depurator lectin
Purpurin

the immune system of foreign species. Lectins may nevertheless promote growth and differentiation of autologous tissues, but in some known examples (eg. IL-2, galectin-1), the carbohydrate-binding site does not seem to be directly involved. By analogy with plant lectins, it is to be expected that some animal lectins might be anti-mitogenic in some circumstances. In fact, anti-mitogenic activities have been demonstrated for calreticulin (qv), *Rana* sialic acid-binding lectin (qv), serum amyloid P component (qv) and the surfactant proteins A and D (qv).

LECTINS AND HAEMOSTASIS

Haemostasis is a carefully controlled process involving sequential activation of a series of plasma proteins on the membranes of activated platelets and endothelial cells. In outline, when anticoagulant endothelium is damaged, platelets adhere to exposed subendothelium via von Willebrand factor (vWF), become activated and aggregate. The contents of platelet granules are released and receptors for fibrinogen appear on the platelet surface. Bound fibrinogen is subsequently converted to a fibrin clot by a cascade of proteolytic conversions of zymogens to active enzymes culminating in the formation of active thrombin by activated factor X. Control is exerted by the protease inhibitor, anti-thrombin III, and/or by the binding of thrombin to thrombomodulin on endothelial cells and consequent activation of protein C, a potent anticoagulant [23].

Lectins are relevant to the above process in two ways. First, there are endogenous lectins associated with platelets and clotting factors. Second, exogenous lectins can be used as reagents to activate or inhibit platelet function and to investigate mechanisms involving platelet receptors [24].

Plasma factor VIII and vWF exist as a complex in the circulation; this complex was reported to agglutinate trypsin-treated, fixed ovine erythrocytes in a calcium-independent manner, and was susceptible to inhibition by mannose, hexosamines and basic amino acids, particularly arginine [25]. vWF has a heparin binding site which is distinct from its C-terminal domains containing an RGD sequence [26]. Factor VIII and factor V also have two C-terminal domains; each domain shows considerable homology to its adjacent C domain, to the corresponding C domains of the other clotting factor and to a sequence found in discoidin I from *Dictyostelium discoideum* [27]. Factors Va and VIIIa can be inactivated (indirectly) by thrombomodulin, a vascular endothelial cell receptor the N-terminal region of which has a module with approximately 20% primary sequence identity to various C-type lectins from both invertebrate and vertebrate sources [28,29]. It is not clear, however, whether protein–carbohydrate interactions are involved in coagulation or anticoagulation.

Platelets possess several endogenous lectins, including thrombospondin (which may have been contaminating the above-mentioned vWF/factor VIII complex) and P-selectin. Thrombospondin (qv) is an adhesion molecule synthesised by many cells other than platelets; P-selectin (qv) is also an adhesion molecule, but is largely confined to platelets/megakaryocytes and endothelial cells, and might be necessary for optimal haemostasis [30].

Two related groups of snake venom proteins interact with platelets and therefore can be used as reagents for research purposes. The thrombolectin (qv) group of calcium-dependent, lactose-specific haemagglutinins includes *Agkistrodon piscivorus* venom lectin (qv), *Crotalus atrox* venom lectin (qv), *Lachesis muta* venom lectin (qv) and *Trimeresurus stejnegeri* venom lectin (qv). The lectins in this group have a very similar primary structure, yet often differ in detailed saccharide specificity, isoelectric points and (more than might be expected) biological activities. *Crotalus atrox* (rattlesnake), *Agkistrodon piscivorus* (Western cottonmouth moccasin) and *Lachesis muta* (bushmaster) lectins consistently induced platelet aggregation apparently via their carbohydrate-binding sites, while thrombolectin gave inconsistent results [31]. Another group of venom proteins having a significant but relatively modest primary sequence homology with the thrombolectin group, but which appear to lack carbohydrate-binding ability, are C-type lectin-like proteins. Members of this group include convulxin [32], a potent platelet-aggregating factor from the rattlesnake *Crotalus durissus*; the alboaggregins from *Trimeresurus albolabris*, which induce platelet aggregation after binding to platelet glycoprotein Ib (GPIb) [33]; botrocetin from *Bothrops jararaca* [34] and bitiscetin from *Bitis arietans* [35] which bind vWF, promoting vWF–GPIb interaction and subsequent platelet activation; and the various GPIb-binding proteins that inhibit platelet activation, such as agkicetin from *Agkistrodon acutus* [36], echicetin from *Echis carinatus* [37], flavocetins A and B from *Trimeresurus flavoviridis* [38] and tokaracetin from *Trimeresurus tokarensis* [39].

These lectin-like reagents are outstandingly useful for studying platelet receptor function because the natural ligands they replace often bind to more than one receptor [24].

Finally, plasma tetranectin is an endogenous lectin which binds plasminogen and (probably) fibrin [40]. Despite its involvement in haemostasis and fibrinolysis, tetranectin has not found application in this area, but rather as an inverse cancer marker. This is in contrast to P-selectin, which is a reliable marker of platelet activation; both P-selectin and CD31 are elevated in pre-eclampsia, as noted in the following section about lectins and reproduction.

LECTINS AND REPRODUCTION

Gonadal tissues are well-established sources of lectins. Mention has already been made of the blood group A-specific lectins in the albumin glands of snails [5–7], and eggs of amphibians [8,9]; also the group B-specific lectins found in the roe of several fish species [4,10–13], and the I-specific gonad lectin from *Aplysia depilans* [41]. One could add the sperm lectins from the blue mussel *Mytilus edulis* [42] and the varied selection of lectins present in the human placenta [43]. Rat uterine tissue also contains at least two lectins: a calcium-dependent sialic acid-binding lectin [44] and one inhibited by mannosides and glucosides [45]. The former was found at all stages of the oestrous cycle; the latter is present only during pregnancy.

Protein–carbohydrate interactions appear to be common in reproductive mechanisms [46], particularly during fertilisation [47–52], but also during implantation [53] and throughout the maintenance of viable pregnancy. Often the individual lectins involved are not well characterised, but there are a few exceptions. For example, CD31 (qv) is expressed by a subpopulation of human trophoblasts and may mediate essential trophoblast–endothelial cell interactions during placentation [54]. Properly balanced cytokine networks are thought to be crucial for viable feto-maternal cellular communication [55], and these would include the lectin cytokines interleukin-1 (qv), interleukin-2 (qv) and tumour necrosis factor (qv). It is widely acknowledged that decidual leukocytes (mainly NK cells) play a role in successful mammalian pregnancy [56], and the NK group of lectins (see Chapter 3) could be involved.

Mannan-binding lectin (MBL) (qv) may function in successful pregnancy maintenance, since relative deficiency of MBL is associated with recurrent miscarriage [57–59]. MBL has been detected in vascular endothelial cells, Hofbauer cells, cytotrophoblasts and mesenchymal cells of first trimester placenta [57]. Since a low MBL concentration in either parent seems to predispose to spontaneous abortion, it is likely that the level in the fetus/placental tissue (inherited in a dominant fashion) would be relevant rather than the MBL concentration in the maternal circulation. The mechanism by which an adequate level of MBL protects a pregnancy is not known, but protection from intrauterine infections is a plausible and appealing possibility.

Another relatively common disease of pregnancy is pre-eclampsia (proteinuric gestational hypertension). This mysterious syndrome can have very serious consequences for both mothers and babies. Some authors believe the fundamental defect in pre-eclampsia is inadequate trophoblast invasion of the maternal spiral arteries, leading to reduced placental perfusion; this in turn is in some way responsible for a generalised endothelial cell dysfunction resulting in multi-organ-derived maternal symptoms [60]. Several endogenous lectins, including P-selectin and CD31, are elevated in the circulation of pre-eclamptic women, reflecting the common occurrence of platelet activation in the disease; circulating E-selectin (qv) and CD54 (qv) concentrations are also elevated, presumably reflecting endothelial cell dysfunction, and the latter is detectable before the onset of clinical symptoms [61]. Maternal plasma tumour necrosis factor is also increased in pre-eclampsia [62]. This finding is consistent with the hypothesis that oxidative stress is of major significance in the pathogenesis of the disease [63], and possibly can account for some apparently inconsistent immunogenetic findings [64].

Finally, it should be mentioned that the C-type lectin-like protein, thrombomodulin, is a sensitive and specific marker of acute placental abruption [65].

LECTINS AND IMMUNITY

The involvement of lectins in the immune system has been reviewed by this author [18] and others [66]. It is arguable that the main function of animal lectins in general is to mediate recognition phenomena within the immune system, and numerous examples have already been given in the preceding chapters. The purpose of this section is simply to recapitulate (with examples) that there are four aspects of immunity in which lectins play a central role: as antibody-like defence molecules (or ante-antibodies); as promoters of cellular communication between, and trafficking of, immune function cells; as regulators of inflammation; and in the prevention of autoimmunity. These four aspects are not entirely separate, and indeed there is much overlap between the last three.

Information reported in numerous papers has implicated endogenous lectins in the regulation of the immune response, often as suppressor factors [67–75]. The situation is complicated by the finding that antigen-specific (but not mitogen-induced) T-cell activation *in vitro* can be inhibited by certain monosaccharides, and that this inhibition depends critically on the minute proportion of such sugars that exist in the open chain configuration and therefore as Schiff base-forming aldehydes [76]. Caution is therefore necessary in the interpretation of observations concerning carbohydrate–receptor interactions involving cells of the immune system. Nevertheless, there are many clearly defined lectin–carbohydrate interactions relevant to immunity and some of these are reviewed below.

Lectins which appear to function as broad-specificity antibodies or 'protectins' are abundant among invertebrates. Most of the arthropod lectins included in this book probably function in this way, especially those found in insect or crab haemolymphs, and there are strong suspicions that many mollusc and tunicate lectins exist primarily for the same purpose. Vertebrate examples include the collectins like MBL (qv), conglutinin (qv) or surfactant protein A (qv), the macrophage mannose receptor (qv), CD11b/CD18 (qv), and presumably the ficolins, like P35 (qv). All of these lectins distinguish self from non-self, and in many instances have been shown to opsonise microbial invaders for engulfment by phagocytic cells. It is interesting that in an insect species, *Blaberus discoidalis* [77], and a horseshoe crab, *Tachypleus tridentatus* [78], it has been shown that antimicrobial protection is provided by a battery of multiple lectins with differing specificities not so dissimilar to the range of activity provided by polyclonal immunoglobulins in vertebrates.

Lectins also function as cell recognition and cell communication molecules within the immune system. The most obvious examples include the selectins, L-selectin (qv), E-selectin (qv) and P-selectin (qv), which have been shown to be adhesion molecules that direct extravasation of leukocytes. CD44 (qv) and CD11b/CD18 (qv) appear to have rather similar and complementary functions to those of the selectins. Other adhesion molecules possibly implicated in immune cell cross-talk include the 'siglecs', CD22 (qv) and sialoadhesin (qv), while many cytokines, including the lectin examples interleukin-1 (qv), interleukin-2 (qv) and tumour necrosis factor (qv), obviously participate in inter-immune cell communication.

Galectins-1(qv), -3 (qv) and -9 (qv) have also been implicated in cell–cell communication within the immune system, as well as in inflammation, autoimmunity and various roles in non-immune tissues.

Lectins can serve as markers of inflammation: C-reactive protein (qv), like the selectins, is an obvious example, and others include CD31 (qv) and CD54 (qv).

Although CRP and other pentraxins are acute-phase reactants with the ability to participate in first-line defence like the above-mentioned antibody-like lectins, there are grounds for believing that CRP and other pentraxins also function to limit tissue damage and prevent autoimmune responses after successful resolution of an immune response. The surprising complexity of the pentraxin composition recently described in a horseshoe crab, *Tachypleus tridentatus* (22 different pentraxins, although falling into three main groups of CRP-like proteins) [79], is not readily explained and may point to a more sophisticated biological role than is currently envisaged. It would seem plausible that in general terms pentraxins function to regulate inflammation, a function that might to some extent be shared with the apoptosis-inducing galectin-1 [80].

Certainly, the other major mammalian pentraxin, serum amyloid P component (qv), is necessary to prevent the development of autoimmunity [81], another property which might be shared to some extent by galectin-1. It is noteworthy in this context that galectin-1 is itself a potential autoantigen in humans [82], as are some other lectins, including the Hakata antigen (qv) and calreticulin (qv).

Lectins may also be relevant to the immune system as therapeutic agents or their targets in the context of malignant and non-malignant diseases. These applications are surveyed in the following two sections.

LECTINS AND CANCER

Malignant diseases, particularly carcinomas, collectively constitute the greatest health care problem in developed countries at the present time. Although primary tumours can usually be removed or effectively destroyed, metastatic disease is essentially incurable. (The only real exception is in the context of bone marrow transplantation for certain haematological malignancies.)

The discovery by Aub, Burger and others [83,84] that a plant lectin, wheat germ agglutinin, selectively agglutinated murine tumour cells did much to stimulate interest in lectins generally. These and other investigations established that neoplastic cells often differ from their normal counterparts in surface glycoconjugate composition. Several (mainly plant) lectins were found to distinguish malignant from normal tissues to some degree, often by histochemical techniques [85]. Some animal lectins also show an enhanced affinity for neoplastic cells, including achatinin$_H$ (qv), the acorn barnacle isolectins (qv), *Cancer antennarius* lectin (qv), CD33 (qv), *Plecoglossus altivelis* egg lectin (qv), the *Rana* sialic acid-binding lectin (qv) and the *Sarcophaga peregrina* humoral lectin (qv).

One of the most useful lectins in this regard is the *Helix pomatia* agglutinin (HPA) [86]. As a histochemical reagent, HPA is of value in prognoses of breast, colorectal, gastric and oesophageal cancers. Breast tumours have been most thoroughly studied, and the correlation between HPA-binding sites and poor prognosis has been confirmed in several laboratories. HPA binding to the primary tumour correlates with metastatic potential, lymph node involvement and (inversely) with patient survival. Unfortunately, results obtained vary with tissue processing (fixation etc.) and detection methodology; in consequence, these non-standardised procedures have not gained acceptance as a routine investigation in oncology [85].

As well as serving as reagents for the detection of tumour-associated glycoconjugates, animal lectins are of major interest as endogenous tumour-related proteins. Endogenous lectins seem to be a feature of all tumours, and malignant transformation is often associated with an increase in lectin content [87]. Endogenous lectins have been studied as cancer markers (for diagnostic/prognostic purposes), or simply for insights into the biochemistry of neoplastic disease. Examples of the former include achatinin$_H$, which can be used to detect and monitor lymphoblasts in acute lymphoblastic leukaemia [88], and CD33, which has diagnostic and prognostic value in acute myeloid leukaemia [89]. Examples of the latter include surfactant protein A (qv), thrombospondin (qv), sarcolectin (qv), the selectins and, above all, the galectins.

Before leaving the topic of diagnostic/prognostic cancer markers, however, special mention should be made of tetranectin. This plasminogen-binding protein has a C-type lectin structure, although it is not clear whether the defining carbohydrate-recognition site actually binds carbohydrate. However, tetranectin can bind sulphated polysaccharides and therefore qualifies as a lectin, although this activity may be calcium-independent and mediated by a second binding site. Unlike most cancer markers, the concentration of tetranectin in the circulation decreases in association with malignant diseases [90]. Measurement of plasma tetranectin appears to be of some value as a prognostic indicator in ovarian, breast and colorectal cancers, and could profitably be used to select and monitor therapy. For example, pre-operative serum tetranectin inversely correlated with residual tumour and survival in ovarian carcinoma patients undergoing second- or third-look surgery, and possibly could be used to decide if surgery is necessary. Also in the context of ovarian cancer, a decrease in serum tetranectin in patients receiving chemotherapy often indicates recurrence of the disease and may warrant a change of treatment. Similarly, in metastatic breast cancer, where serum tetranectin is a strong prognostic indicator of shortened patient survival, low tetranectin levels may suggest poor response to treatment and the desirability of a change of therapeutic regimen [90].

Galectins, as previously mentioned, are generally regarded as growth regulatory adhesion molecules with a role in cell–cell and cell–extracellular matrix interactions. It is therefore unsurprising that galectins have been widely investigated in the context of malignant and metastatic diseases. Some relationships seem straightforward. Galectin-4 (qv) and galectin-7 (qv) are down-regulated on malignant change, while high levels of galectin-8 (qv) and galectin-9 (qv) are associated with cancers. Relatively little is known about these galectins, however, and these apparently simple relationships may simply be a reflection of our relative ignorance.

Galectins-1 and 3 have been much more intensively investigated, and the emerging picture is much less clear. As a crude generalisation, there is a reciprocal relationship between the two in the context of cancer: high expression of galectin-1 by tumour cells correlates with metastatic potential and poor prognosis, while increased expression of galectin-3 is inversely related to metastatic potential and is positively associated with a more favourable outcome [91,92]. However, there are exceptions, and the findings in colorectal cancer are frankly contradictory [91]. The apparently opposite influences of galectin-1 and -3 in tumour cells can be partly explained by their antagonistic effects on apoptosis: galectin-1 induces lymphocyte apoptosis, while galectin-3 acts to prevent cell death [92,93]. Galectin-1 may therefore protect tumour cells from destruction by tumour infiltrating lymphocytes, while galectin-3 might confer resistance to lymphocyte apoptotic mechanisms. This is unlikely to be the full explanation, however, and the influence of galectin-1 and -3 may also depend on other factors, including relative local concentrations, the nature of the extra-cellular environment and, above all, the presence of other relevant molecules (perhaps including other galectins). A consistent finding with immunohisto-chemistry has been that galectin-3 localisation alters from a nuclear to a cytoplasmic predominance with colon cancer progression, but the meaning of this observation is unclear. The role of galectin-3 specifically in cancer biology was reviewed in depth a few years ago [94].

The observations of Castronovo and colleagues [95,96] that decreased galectin-3 expression (as compared to normal tissue) is a feature of carcinomas generally, including those of breast and colon, are hard to reconcile with the induction of tumorigenicity in both colonic cancer [97] and breast cancer [98] cell lines by increasing the galectin-3 expression by molecular biological techniques. A decrease in expression of galectin-3 (and, to a lesser extent, galectin-1) was also reported for squamous cell carcinomas of the head and neck [99].

A list of cancers with which elevated galectin-1 and/or galectin-3 levels are associated is given in Table 4.3. It should perhaps be emphasised that these galectins are particularly interesting in relation to cancers of the thyroid gland. Overexpression of galectin-1 is selectively associated with papillary carcinoma and anaplastic carcinoma [100], while detection of galectin-3 in fine needle aspirates can be used to differentiate thyroid malignancy from benign thyroid adenomas or normal tissue [101,102]. Detection of galectin-3 could therefore become a useful pre-surgical diagnostic test, which could lead to a reduction in

Table 4.3. Human Cancers with Overexpression of Galectin-1 or Galectin-3

Cancer	Galectin-1	Galectin-3
Anaplastic large cell lymphoma		+
Brain tumours (glioblastomas)		+
Breast carcinoma		+
Colon carcinoma	+	+
Gastric carcinomas		+
Hepatocellular carcinoma		+
Leiomyosarcoma (uterus)		+
Melanoma	+	+
Ovarian carcinoma	+	
Prostate adenocarcinoma	+	
Thyroid carcinoma	+	+

unnecessary surgery, since it is sometimes difficult to distinguish benign from malignant thyroid neoplasms in the absence of metastases.

The selectins, particularly E-selectin, have been implicated in the haematogeneous spread of cancer cells [103–105], and decreased expression of L-selectin on progenitor cells may be relevant to the biology of chronic myeloid leukaemia [106]. Other lectins that could be applied as drugs or drug targets in oncology include calreticulin (qv), CD33 (qv), CD44 (qv), C-reactive protein (qv), tumour necrosis factor (qv) and possibly the *Rana* sialic acid-binding lectin (qv), the *Sarcophaga peregrina* humoral lectin (qv) and mannan-binding lectin (qv).

ENDOGENOUS LECTINS AS DRUGS AND DRUG TARGETS

Few animal lectins have been used as drugs, either clinically or in animal model systems. Electrolectin has been used to prevent experimental myasthenia gravis in rabbits and can even alleviate symptoms if administered after disease induction [107]. Electrolectin was found to be weakly mitogenic for lapine lymphocytes, and the authors suggested that the galectin had immunomodulatory properties; a more popular explanation today would be that mammalian galectins-1 and their homologues regulate immune responses by inducing apoptosis of T cells. Recombinant human galectin-1 (injected intravenously into rats) can prevent the signs and symptoms of experimental allergic encephalomyelitis [108], and a single injection of syngeneic fibroblasts engineered to secrete murine galectin-1 was able to abrogate the manifestations of collagen-induced arthritis when administered to mice on the day of disease onset [109]. A shift towards a Th-2 response was noted which correlated with an increased susceptibility to apoptosis in lymph node cells [109]. Human recombinant galectin-1 likewise suppressed the production of TNF-α and IFN-γ from IL-2 activated human T cells [110]. These findings are of major potential importance, since recombinant human galectin-1 should not be immunogenic in humans (unlike plant or non-human animal proteins), as previously discussed [18]. However, at the present time, galectin-1 does not seem to have been used clinically.

MBL, on the other hand, was first used clinically in the form of whole plasma many years ago, and more recently has been administered in highly purified form to patients [111]. It appears to be safe, and is effective in correcting the laboratory-defined opsonic defect. The number of potential indications for MBL replacement therapy is large, but clinical trials are required to identify the clinical settings (if any) in which replacement therapy might be beneficial and cost-effective. The relative merits of plasma-derived and recombinant MBL products also require to be evaluated. The finding that MBL has anti-tumour activity *in vivo* when administered in the form of a recombinant vaccinia virus carrying the MBL gene and which expresses a molecular form lacking opsonic (complement activating) activity is of considerable interest [112]. It is possible that plasma-derived and recombinant MBLs may each be optimal for separate clinical indications.

Serum amyloid P (SAP) component differs radically from the above examples since it has been administered for diagnostic, not therapeutic, purposes.

Radiolabelled SAP can be detected by scintigraphy, and has been used to locate amyloid deposits [113]. The success of this diagnostic procedure implies that SAP could be used for drug delivery: by coupling SAP to a drug that degrades amyloid, the SAP would target the effective agent to the fibrils. Conversely, since SAP and amyloid fibrils may protect and stabilise each other [114,115], saccharide inhibitors of SAP might permit degradation of amyloid by physiological mechanisms.

The other major pentraxin in humans, C-reactive protein (qv), acts as an anti-tumour agent in an animal model, as does the *Rana* sialic acid-binding lectin (qv). Interleukin-2 (qv) and tumour necrosis factor (qv) have been used clinically as drugs but in both cases the cytokine activities (mediated by cytokine receptors), and not the lectin activities, were being utilised.

SAP, as noted above, could find application either as a drug or a drug target. Some other lectins are more likely to be considered as the latter. The selectins in particular have received much attention as the recipient of anti-inflammatory agents. It has been shown in animal models that selectins mediate the neutrophil–endothelial cell interactions that form the first step in inflammatory events like reperfusion injury and complement-promoted inflammatory lung injury. These processes can be inhibited *in vivo* not only by anti-selectin antibodies but also by oligosaccharides. This work has stimulated a search for ultra-potent saccharide selectin inhibitors with anti-inflammatory activity [116,118]. The pharmaceutical industry is pursuing the ideal synthetic oligosaccharide which is inexpensive to produce and which can alleviate selectin-mediated symptoms without side effects. Clinical trials of compounds with such potential began some years ago [119]. A cautionary note is sounded, however, by the authors of a study of leukocyte rolling and adhesion in post-ischaemic venules; a high degree of inhibition of neutrophil rolling was required to achieve a modest attenuation of adhesion [120].

Although selectins have also been implicated in the metastasis of tumour cells, it was hepatic asialoglycoprotein receptors that were the targets for the therapeutic use of galactose in oncology. Neuraminidase-treated murine sarcoma cells were found to settle in the liver and lungs; pre-injection of mice with galactose or arabinogalactan (but not mannan or dextran) completely prevented colonisation of the liver but not of the lung [121]. These experimental findings provided the rationale for a prospective, randomised trial in colorectal cancer patients, who were given infusions of either galactose or glucose perioperatively. Although the results did not achieve statistical significance, there was a most encouraging trend towards delayed events (hepatic metastasis, death) and increased survival after galactose treatment [122].

Other animal lectins that have been used clinically or in animal models as drug targets include CD33 (qv), targeted both for *ex vivo* marrow purging and for *in vivo* leukaemic blast reduction; CD44 (qv), for differentiation therapy in acute myeloid leukaemia; the macrophage mannose receptor (qv), as a selective drug delivery system; interleukin-1 (qv) for prevention of septic shock; tumour necrosis factor (qv) for amelioration of rheumatoid arthritis symptoms; and *Entamoeba histolytica* lectin (qv) for prevention of amoebic liver abscesses.

REFERENCES

1. Judd, W.J. (1979). The use of purified lectins in immunohematology. *Transfusion* **19**, 768–772.
2. Judd, W.J (1980). The role of lectins in blood group serology. *CRC Critical Reviews in Clinical Laboratory Sciences* **12**, 171–214.
3. Bird, G.W.G. (1989). Lectins in immunohaematology. *Transfusion Medicine Reviews* **3**, 55–62.
4. Kilpatrick, D.C. & Green, C. (1992). Lectins as blood typing reagents. In: *Advances in Lectin Research*, Volume 5 (ed. H. Franz). Ullstein Mosby, Berlin.
5. Prokop, O., Rackwitz, A. & Schlesinger, D. (1965). A 'new' human blood group receptor A_{hel} tested with saline extracts from *Helix hortensis* (garden snail). *Journal of Forensic Medicine* **12**, 108–110.
6. Gold, E.R. & Thompson, T.E. (1969). Serological differences between related species of snails. *Vox Sanguinis* **16**, 63–66.
7. Cann, R.J. (1975). A brief evaluation of snail anti-A in manual blood-grouping procedures. *Medical Laboratory Technology* **32**, 69–71.
8. Schenkel-Brunner, H., Kothbauer, H. & Luttenberger, F. (1975). Hemagglutinins in newt (*Triturus*) eggs: comparative studies on different species, subspecies and a hybrid. *Comparative Biochemistry and Physiology* **50A**, 31–33.
9. Sakakibara, F., Takayanagi, G., Kawauchi, H., Watanabe, K. & Hakomori, S. (1976). An anti-A-like lectin of *Rana catesbiana* eggs showing unusual reactivity. *Biochimica et Biophysica Acta* **444**, 386–395.
10. Todd, G.M. (1971). Blood group antibodies in Salmonidae roe. *Vox Sanguinis* **21**, 451–454.
11. Anstee, D.J., Holt, P.D.J. & Pardoe, G.I. (1973). Agglutinins from fish ova defining blood groups B and P. *Vox Sanguinis* **25**, 347–360.
12. Downie, D.M., Madin, D.F. & Voak, D. (1977). An evaluation of salmon anti-B reagent in manual and automated blood grouping. *Medical Laboratory Science* **34**, 319–324.
13. Voss, E.W., Fryer, J.L. & Banowetz, G.M. (1978). Isolation, purification and partial characterization of a lectin from Chinook salmon ova. *Archives of Biochemistry and Biophysics* **186**, 25–34.
14. Gilboa-Garber, N., Sudakewitz, D. & Levene, C. (1999). A comparison of the *Aplysia* lectin anti-I specificity with human anti-I and several other I-detecting lectins. *Transfusion* **39**, 1060–1064.
15. Flexner, S. & Noguchi, H. (1902). Snake venom in relation to haemolysis, bacteriolysis, and toxicity. *Journal of Experimental Medicine* **6**, 277–301.
16. Nowell, P. (1960). Phytohemagglutinin: an initiator of mitosis in cultures of normal human leukocytes. *Cancer Research* **20**, 462–464.
17. Kilpatrick, D.C. (1991). Lectin interactions with human leukocytes – mitogenicity, cell separation, clinical applications. In: *Lectin Reviews* Volume 1 (eds D.C. Kilpatrick, E. Van Driessche & T.-C. Bøg-Hansen) pp. 69–80. Sigma Chemical Company, St Louis, MO.
18. Kilpatrick, D.C. (1995). Lectins in immunology. In: *Lectins–Biomedical Perspectives* (eds A. Pusztai & S. Bardocz) pp. 155–182. Taylor & Francis, London.
19. Kilpatrick, D.C. (1997). Mechanisms and assessment of mitogenesis – an overview. In: *Methods in Molecular Medicine*, Volume 9: *Lectin Methods and Protocols* (eds J.M. Rhodes & J.D. Milton) pp. 365–378. Humana Press, Totowa.
20. Kilpatrick, D.C. (1999). Mechanisms and assessment of lectin-mediated mitogenesis. *Molecular Biotechnology* **11**, 55–65.
21. Lipsick, J.S., Beyer, E.C., Barondes, S.H. & Kaplan, N.O. (1980). Lectins from chicken tissues are mitogenic for Thy-1 negative murine spleen cells. *Biochemical and Biophysical Research Communications* **97**, 56–61.
22. Fragkiadakis, G.A. & Stratakis, E.K. (1997). The lectin from the crustacean *Liocarcinus depurator* recognises *O*-acetylsialic acids. *Comparative Biochemistry and Physiology* **117B**, 545–552.

23. Meisenberg, G. & Simmons, W.H. (1998). *Principles of Medical Biochemistry*, pp. 534–545. Mosby, St Louis, MO.
24. Clemetson, K.J., Polgar, J. & Clemetson, J.M. (1998). Snake venom C-type lectins as tools in platelet research. *Platelets* **9**, 165–169.
25. Booth, W.J., Furby, F.H., Berndt, M.C. & Castaldi, P.A. (1984). Factor VIII/Von Willebrand factor has potent lectin activity. *Biochemical and Biophysical Research Communications* **118**, 495–501.
26. Girma, J.-P., Meyer, D., Verweij, C.L., Pannekoek, H. & Sixma, J.J. (1987). Structure-function relationship of human Von Willebrand factor. *Blood* **70**, 605–611.
27. Kane, W.H. & Davie, E.W. (1986). Cloning of a cDNA coding for human factor V, a blood coagulation factor homologous to factor VIII and ceruloplasmin. *Proceedings of the National Academy of Sciences USA* **83**, 6800–6804.
28. Petersen, T.E. (1988). The amino-terminal domain of thromomodulin and pancreatic stone protein are homologous with lectins. *FEBS Letters* **231**, 51–53.
29. Patthy, L. (1988). Detecting distant homologies of mosaic proteins. *Journal of Molecular Biology* **202**, 689–696.
30. Subramaniam, M., Frenette, P.S., Saffaripour, S., Johnson, R.C., Hynes, R.O. & Wagner, D.D. (1996). Defects in hemostasis in P-selectin deficient mice. *Blood* **87**, 1238–1242.
31. Ogilvie, M.L., Byl, J.W. & Gartner, T.K. (1989). Platelet aggregation is stimulated by lactose-inhibitable snake venom lectins. *Thrombosis and Haemostasis* **62**, 704–707.
32. Leduc, M. & Bon, C. (1998). Cloning subunits of convulxin, a collagen-like platelet-aggregating protein from *Crotalus durissus terrificus* venom. *Biochemical Journal* **333**, 389–393.
33. Peng, M., Lu, W. & Kirby, E.P. (1992). Characterization of three alboaggregins purified from *Trimeresurus albolabris* venom. *Thrombosis and Haemostasis* **67**, 702–707.
34. Usami, Y., Fujimura, Y., Suzuki, M., Ozeki, Y., Nishio, K., Fukui, H. & Titani, K. (1993). Primary structure of two-chain botrocetin, a von Willebrand factor modulator purified from the venom of *Bothrops jararaca*. *Proceedings of the National Academy of Sciences USA* **92**, 928–932.
35. Hamako, J., Matsui, T., Suzuki, M., Ito, M., Makita, K., Fujimura, Y., Ozeki, Y. & Titani, K. (1996). Purification and characterization of bitiscetin, a novel von Willebrand factor modulator protein from *Bitis arietans* snake venom. *Biochemical and Biophysical Research Communications* **226**, 273–279.
36. Chen, Y.-L. & Tsai, I.-H. (1995). Functional and sequence characterization of agkicetin, a new glycoprotein 1B antagonist isolated from *Agkistrodon acutus* venom. *Biochemical and Biophysical Research Communications* **210**, 472–477.
37. Peng, M., Holt, JC. & Niewiarowski, S. (1994). Isolation, characterization and amino acid sequence of echicetin β subunit, a specific inhibitor of von Willebrand factor, and thrombin interaction with glycoprotein 1b. *Biochemical and Biophysical Research Communications* **205**, 68–72.
38. Taniuchi, Y., Kawasaki, T., Fujimura, Y., Suzuki, M., Titani, K., Sakai, Y., Kaku, S., Hisamichi, N., Satoh, N., Takenaka, T., Handa, M. & Sawai, Y. (1995). Flavocetin-A and -B, two high molecular mass glycoprotein Ib binding proteins with high affinity purified from *Trimeresurus flavoviridis* venom, inhibit platelet aggregation at high shear stress. *Biochemica et Biophysica Acta* **1244**, 331–338.
39. Kawasaki, T., Taniuchi, Y., Hisamichi, N., Fujimura, Y., Suzuki, M., Titani, K., Sakai, Y., Kaku, S., Satoh, N., Takenaka, T., Handa, M. & Sawai, Y. (1995). Tokaracetin, a new platelet antagonist that binds to platelet glycoprotein Ib and inhibits von Willebrand factor-dependent shear-induced platelet aggregation. *Biochemical Journal* **308**, 947–953.
40. Høgdall, C.K. (1998). Human tetranectin: methodological and clinical studies. *APMIS* **106**, Suppl. 86, 7–31.
41. Gilboa-Garber, N., Susswein, A.J., Mizraki, L. & Avichezer, D. (1985). Purification and characterization of the gonad lectin of *Aplysia depilans*. *FEBS Letters* **181**, 267–270.
42. Takagi, T., Nakamura, A., Deguchi, R. & Kyozuka, K.-I. (1994). Isolation, characterization, and primary structure of three major proteins obtained from *Mytilus edulis* sperm. *Journal of Biochemistry* **116**, 598–605.

43. Kilpatrick, D.C., Bevan, B.H., Liston, W.A., Hirabayashi, J., Kasai, K. & Cooper, B.L. (1996). Lectins in the feto-placental unit: a review. In: *Lectins – Biology, Biochemistry, Clinical Biochemistry*, Volume 11 (eds E. Van Driessche, P. Rougé, S. Beeckmans & T.-C. Bøg-Hansen) pp. 161–167. Textop, Hellerup.
44. Chrowdhury M, Sarkar M & Mandal C (1985). Identification and isolation of an agglutinin from uterus of rats. *Biochemical and Biophysical Research Communications* **130**, 1301–1307.
45. Chrowdhury, M. (1986). Isolation of a keto sugar specific agglutinin from rat uterus. *Biochemical and Biophysical Research Communications* **136**, 116–121.
46. Ivell, R. (1999). The role of carbohydrates in reproduction. *Human Reproduction Update* **5**, 277–279.
47. Calvete, J.-J., Toepfer-Petersen, E. & Sanz, L. (1996). Spermadhesions: a novel class of mammalian lectins? In: *Lectins – Biology, Biochemistry, Clinical Biochemistry*, Volume 11 (eds E. Van Driessche, P. Rougé, S Beeckmans & T.-C. Bøg-Hansen) pp. 185–192. Textop, Hellerup.
48. Benoff, S. (1997). Carbohydrates and fertilization: an overview. *Molecular Human Reproduction* **3**, 599–637.
49. Suarez, S.S., Revah, I., Lo, M. & Kölle, S. (1998). Bull sperm binding to oviductal epithelium is mediated by a Ca^{2+}-dependent lectin on sperm that recognises Lewis-a trisaccharide. *Biology of Reproduction* **59**, 39–44.
50. Sinowatz, F., Plendl, J. & Kölle, S. (1998). Protein–carbohydrate interactions during fertilization. *Acta Anatomica* **161**, 196–205.
51. Töpfer-Petersen, E. (1999). Carbohydrate-based interaction on the route of a spermatozoon to fertilization. *Human Reproduction Update* **5**, 314–329.
52. Dell, A., Morris, H.R., Easton, R.L., Patankar, M. & Clark, G.F. (1999). The glycobiology of gametes and fertilisation. *Biochimica et Biophysica Acta* **1473**, 196–205.
53. Biermann, L., Gabius, H.-J. & Denker, H.-W. (1997). Neoglycoprotein-binding sites (endogenous lectins) in the fallopian tube, uterus and blastocyst of the rabbit during the preimplantation phase and implantation. *Acta Anatomica* **160**, 159–171.
54. Coukos, G., Makrigiannakis, A., Amin, K., Albelda, S.M. & Coutifaris, C. (1998). Platelet–endothelial cell adhesion molecule-1 is expressed by a subpopulation of human trophoblasts: a possible mechanism for trophoblast–endothelial interaction during haemochorial placentation. *Molecular Human Reproduction* **4**, 357–367.
55. Robertson, S.A., Seamark, R.F., Guilbert, L.J. & Wegmann, T.G. (1994). The role of cytokines in gestation. *Critical Reviews in Immunology* **14**, 239–292.
56. King, A., Loke, Y.W. & Chaouat, G. (1997). NK cells and reproduction. *Immunology Today* **18**, 64–66.
57. Kilpatrick, D.C., Bevan, B.H. & Liston, W.A. (1995). Association between mannan binding protein deficiency and recurrent miscarriage. *Molecular Human Reproduction* **1** *Human Reproduction* **10**, 2501–2505.
58. Christiansen, O.B., Kilpatrick, D.C., Souter, V., Varming, K., Thiel, S. & Jensenius, J.C. (1999). Mannan-binding lectin deficiency is associated with unexplained recurrent miscarriage. *Scandinavian Journal of Immunology* **49**, 193–196.
59. Kilpatrick, D.C., Starrs, L., Moore, S., Souter, V. & Liston, W.A. (1999). Mannan binding lectin concentration and risk of miscarriage. *Human Reproduction* **14**, 2379–2380.
60. Ness, R.B. & Roberts, J.M. (1996). Heterogeneous causes constituting the single syndrome of pre-eclampsia: a hypothesis and its implications. *American Journal of Obstetrics and Gynecology* **175**, 1365–1370.
61. Krauss, T., Kuhn, W., Lakoma, C. & Augustin, H.G. (1997). Circulating endothelial cell adhesion molecules as diagnostic markers for the early identification of pregnant women at risk for development of preeclampsia. *American Journal of Obstetrics and Gynecology* **177**, 443–449.

62. Kilpatrick, D.C. (1999). Influence of human leukocyte antigen and tumour necrosis factor genes on the development of pre-eclampsia. *Human Reproduction Update* **5**, 94–102.
63. Stark, J.M. (1993) Pre-eclampsia and cytokine induced oxidative stress. *British Journal of Obstetrics and Gynaecology* **100**, 105–109.
64. Kilpatrick, D.C. (1996). HLA-dependent TNF secretory response may provide an immunogenetic link between pre-eclampsia and type 1 diabetes mellitus. *Disease Markers* **13**, 43–47.
65. Magriples, U., Chan, D.W., Bruzek, D., Copel, J.A. & Hsu, C.-D. (1999). Thrombomodulin: a new marker for placental abruption. *Thrombosis and Haemostasis* **81**, 32–34.
66. Ni, Y. & Tizard, I. (1996). Lectin–carbohydrate interaction in the immune system. *Veterinary Immunology and Immunopathology* **55**, 205–223.
67. Greene, W.C., Fleisher, T.A. & Waldmann, T.A. (1981). Soluble suppressor supernatants elaborated by concanavalin A-activated human mononuclear cells. I Characterization of a soluble suppressor of T cell proliferation. *Journal of Immunology* **126**, 1185–1191.
68. Fleisher, T.A., Greene, W.C., Blaese, R.M. & Waldmann, T.A. (1981). Soluble suppressor supernatants elaborated by concanavalin A-activated human mononuclear cells. II Characterization of a soluble suppressor of B cell immunoglobulin production. *Journal of Immunology* **126**, 1192–1197.
69. Parish, C.R., Rylatt, D.B. & Snowden, J.M. (1984). Demonstration of lymphocyte surface lectins that recognise sulphated polysaccharides. *Journal of Cell Science* **67**, 145–158.
70. Kamo, I., Furukawa, S., Akazawa, S., Fujisawa, K., Tada-Kikuchi, A., Nonaka, I. & Satoyoshi, E. (1988). Mitogenic heparin-binding lectin-like protein from cloned thymic myoid cells. *Cellular Immunology* **103**, 183–190.
71. Licastro, M., Tabacchi, P. & Franceschi, C. (1987). Simple sugars inhibit proliferation of human T-lymphocytes in autologous and allogeneic mixed lymphocyte reactions. *Cellular Immmunology* **107**, 15–23.
72. Parish, C.R., Low, L., Warren, H.S. & Cunningham, A.L. (1990). A polyanion binding site on the CD4 molecule. Proximity to the HIV-gp120 binding region. *Journal of Immunology* **145**, 1188–1195.
73. Grillon, C., Monsigny, M. & Kieda, C. (1991). Soluble human lymphocyte sugar binding proteins with immunosuppressive activity. *Immunology Letters* **28**, 47–56.
74. Duan, X., Ackerly, M., Vivier, E. & Anderson, P. (1994). Evidence for involvement of β-glucan-binding cell surface lectins in human natural killer cell function. *Cellular Immunology* **157**, 393–402.
75. Lachgar, A., Le Buanec, H., Burny, A., Bizzini, B. & Zagury, D. (1999). Lectin production by human T-lymphocytes. *Biomedicine and Pharmacotherapy* **53**, 288–289.
76. Rhodes, J., Zheng, B. & Lifely, M.R. (1992). Inhibition of specific T-cell activation by monosaccharides is through their reactivity as aldehydes. *Immunology* **75**, 626–631.
77. Wilson, R., Chen, C. & Ratcliffe, N.A. (1999). Innate immunity in insects: the role of multiple, endogenous serum lectins in the recognition of foreign invaders in the cockroach, *Blaberus discoidalis*. *Journal of Immunology* **162**, 1590–1596.
78. Kawabata, S. & Iwanaga, S. (1999). Role of lectins in the innate immunity of horseshoe crab. *Developmental and Comparative Immunology* **23**, 391–400.
79. Iwaki, D., Osaki, T., Mizunoe, Y., Wai, S.N., Iwanaga, S. & Kawabata, S. (1999). Functional and structural diversities of C-reactive proteins present in horseshoe crab hemolymph plasma. *European Journal of Biochemistry* **264**, 314–326.
80. Perillo, N.L., Pace, K.E., Seilhamer, J.J. & Baum, L.G. (1995). Apoptosis of T cells mediated by galectin-1. *Nature* **378**, 736–739.
81. Bickerstaff, M.C.M., Botto, M., Hutchinson, W.L., Herbert, J., Tennent, G.A., Bybee, A., Mitchell, D.A., Cook, H.T., Butler, P.J.G., Walport, M.J. & Pepys, M.B. (1999). Serum amyloid P component controls chromatin degradation and prevents antinuclear autoimmunity. *Nature Medicine* **5**, 694–697.

82. Lutomski, D., Joubert-Caron, R., Lepebure, C., Salama, J., Belin, C., Bladier, D. & Caron, M. (1977). Anti-galectin-1 autoantibodies in serum of patients with neurological diseases. *Clinica Chimica Acta* **262**, 131–138.
83. Aub, J.C., Tieslau, C. & Lankester, A. (1963). Reactions of normal and tumour cell surfaces to enzymes. I Wheat germ lipase and associated mucopolysaccharide. *Proceedings of the National Academy of Science USA* **50**, 613–619.
84. Burger, M.M. & Goldberg, A.R. (1967). Identification of a tumour-specific determinant on neoplastic cell surfaces. *Proceedings of the National Academy of Sciences USA* **57**, 359–366.
85. Schumacher, U. (1995). Lectins and cancer – an old field revisited. In: *Lectins–Biomedical Perspectives* (eds A. Pusztai & S. Bardocz) pp. 211–224. Taylor & Francis, London.
86. Schumacher, U., Adam, E., Brooks, S.A. & Leathem, A.J. (1996). The use of *Helix pomatia* agglutinin in metastasis research. In: *Lectins: Biology, Biochemistry, Clinical Chemistry*, Volume 11 (eds E. Van Driessche, P. Rougé, S. Beeckman & T.-C. Bøg-Hansen) pp. 307–310. Textop, Hellerup.
87. Lotan, R. & Raz, A. (1988). Lectins in cancer cells. *Annals of the New York Academy of Sciences* **551**, 385–398.
88. Sinha, D., Mandal, C. & Bhattacharya, D.K. (1999). Identification of 9-0-acetyl sialoglycoconjugates (9-0AcSGs) as biomarkers in childhood acute lymphoblastic leukaemia using a lectin, achatin$_H$, as a probe. *Leukaemia* **13**, 119–125.
89. Kristensen, J.S. & Hokland, P. (1991). Monoclonal antibodies in myeloid diseases: prognostic use in acute myeloid leukaemia. *Leukaemia Research* **15**, 693–700.
90. Høgdall, C.K. (1998). Human tetranectin: methodological and clinical studies. *APMIS Supplement 86* **106**, 7–31.
91. Itzkowitz, S.H. (1997). Galectins: multipurpose carbohydrate-binding proteins implicated in tumor biology. *Gastroenterology* **113**, 2003–2005.
92. Perillo, N.L., Marcus, M.E. & Baum, L.G. (1998). Galectins: versatile modulators of cell adhesion, cell proliferation and cell death. *Journal of Molecular Medicine* **76**, 402–412.
93. Rabinovich, G.A., Riera, C.M., Landa, C.A. & Sotomayor, C.E. (1999). Galectins: a key intersection between glycobiology and immunology. *Brazilian Journal of Medical and Biological Research* **32**, 383–393.
94. Akahani, S., Hidenori, I., Nangia-Makker, P. & Raz, A. (1997). Galectin-3 in tumour metastasis. *Trends in Glycoscience and Glycotechnology* **9**, 69–75.
95. Castronovo, V., Van Den Brule, F.A., Jockers, P., Clausse, N., Liu, F.T., Gillet, C. & Sobel, M.E. (1996). Decreased expression of galectin-3 is associated with progression of human breast cancer. *Journal of Pathology* **179**, 43–48.
96. Castronovo, V., Liu, F.T. & Van Den Brule, F.A. (1999). Decreased expression of galectin-3 in basal cell carcinoma of the skin. *International Journal of Oncology* **15**, 67–70.
97. Bresalier, R.S., Mazurek, N., Sternberg, L.R., Byrd, J.C., Yunker, C.K., Nangia-Makker, P. & Raz, A. (1998). Metastasis of human colon cancer is altered by modifying expression of the β-galactoside-binding protein galectin 3. *Gastroenterology* **115**, 287–296.
98. Nangia-Makker, P., Thompson, E., Hogan, C., Ochieng, J. & Raz, A. (1995). Induction of tumorgenicity by galectin-3 in a non-tumorigenic human breast carcinoma cell line. *International Journal of Oncology* **7**, 1079–1087.
99. Choufani, G., Nagy, N., Saussez, S., Marchant, H., Bisschop, P., Burchert, M., Danguy, A., Louryan, S., Salmon, I., Gabius, H.-J., Kiss, R. & Hassid, S. (1999). The levels of expression of galectin-1, galectin-3, and the Thomsen–Friedenreich antigen and their binding sites decrease as clinical aggressiveness increases in head and neck cancers. *Cancer* **86**, 2353–2363.
100. Chiarotti, L., Berlingieri, M.T., Battaglia, C., Benvenuto, G., Martelli, M.L., Salvatore, P., Chiappetta, G., Bruni, C.B. & Fusco, A. (1995). Expression of galectin-1 in normal

human thyroid gland and in differentiated and poorly differentiated thyroid tumours. *International Journal of Cancer* **64**, 171–175.

101. Orlandi, F., Saggiorato, E., Pivano, G., Puligheddu, B., Termine, A., Cappia, S., De Giuli, P. & Angeli, A. (1998). Galectin-3 is a presurgical marker of human thyroid carcinoma. *Cancer Research* **58**, 3015–3020.
102. Inohara, H., Honjo, Y., Yoshii, T., Akahani, S., Yoshida, J., Hattori, K., Okamoto, S., Sawada, T., Raz, A. & Kubo, T. (1999). Expression of galectin-3 in fine-needle aspirates as a diagnostic marker differentiating benign from malignant thyroid neoplasms. *Cancer* **85**, 2475–2484.
103. Dejana, E., Martin-Podura, I., Lauri, D., Bernasconi, S., Boni, M.R., Garofalo, A., Giavazzi, R., Magnoni, J., Mantovani, A. & Menard, S. (1992). Endothelial-leukocyte adhesion molecule-1-dependent adhesion of colon carcinoma cells to vascular endothelium is inhibited by an antibody to Lewis fucosylated type I carbohydrate chain. *Laboratory Investigation* **66**, 324–330.
104. Sawada, R., Tsuboi, S. & Fukuda, M. (1994). Differential E-selectin dependent adhesion efficiency in sublines of a human colon cancer exhibiting distinct metastatic potentials. *Journal of Biological Chemistry* **269**, 1425–1431.
105. Kim, Y.J., Borsing, L., Varki, N.M. & Varki, A. (1998). P-selectin deficiency attenuates tumor growth and metastasis. *Proceedings of the National Academy of Sciences USA* **95**, 9325–9330.
106. Kimura, A., Kawaishi, K., Sasaki, A., Hyodo, H. & Oguma, N. (1998). L-selectin expression in CD34 positive cells in chronic myelogenous leukaemia. *Leukaemia and Lymphoma* **28**, 399–404.
107. Levi, G., Tarrab-Hazdai, R. & Teichberg, V.I. (1983) Prevention and therapy with electolectin of experimental autoimmune myasthenia gravis in rabbits. *European Journal of Immunology* **13**, 500–507.
108. Offner, H., Celnik, B., Bringman, T.S., Casentini-Borocz, D., Nedwin, C.E. & Vandenbark, A.A. (1990). Recombinant human ß-galactoside binding lectin suppresses clinical and histological signs of experimental autoimmune encephalomyelitis. *Journal of Neuroimmunology* **28**, 177–184.
109. Rabinovich, G.A., Daly, G., Dreja, H., Tailor, H., Riera, C.M., Hirabayashi, J. & Chernajovsky, Y. (1999). Recombinant galectin-1 and its genetic delivery suppress collagen-induced arthritis via T cell apoptosis. *Journal of Experimental Medicine* **190**, 385–397.
110. Rabinovich, G.A., Ariel, A., Hershkoviz, R., Hirabayashi, J., Kasai, K.-I. & Lider, O. (1999). Specific inhibition of T cell adhesion to extracellular matrix and proinflammatory cytokine secretion by human recombinant galectin-1. *Immunology* **97**, 100–106.
111. Valdimarsson, H., Stefansson, M., Vikingsdottir, T., Arason, G.J., Koch, C., Thiel, S. & Jensenius, J.C. (1998). Reconstitution of opsonizing activity by infusion of mannan-binding lectin (MBL) to MBL-deficient humans. *Scandinavian Journal of Immunology* **48**, 116–123.
112. Ma, Y., Uemura, K., Oka, S., Kozutsumi, Y., Kawasaki, N. & Kawasaki, T. (1999). Antitumor activity of mannan-binding protein *in vivo* as revealed by a virus expression system: mannan binding protein-dependent cell mediated cytotoxicity. *Proceedings of the National Academy of Sciences USA* **96**, 371–375.
113. Hawkins, P.N., Lavender, J.P. & Pepys, M.B. (1990). Evaluation of systemic amyloidosis by scintigraphy with ^{123}I-labelled serum amyloid component. *New England Journal of Medicine* **323**, 508–513.
114. Hawkins, P.N., Wooton, R. & Pepys, M.B. (1990). Metabolic studies of radioiodinated serum amyloid P component in normal subjects and patients with systemic amyloidosis. *Journal of Clinical Investigation* **86**, 1862–1869.
115. Botto, M., Hawkins, P.N., Bickerstaff, M.C.M., Herbert, J., Bygrave, A.E., McBride, A., Hutchinson, W.L., Tennent, G.A., Walport, M.J. & Pepys, M.B. (1997). Amyloid

deposition is delayed in mice with targetted deletion of the serum amyloid P component gene. *Nature Medicine* **3**, 855–859.

116. Yuen, C.-T., Bezouska, K., O'Brien, J., Stoll, M., Lemoine, R., Lubineau, A., Kiso, M., Hasegawa, A., Bockovich, N.J., Nicolaou, K.C. & Feizi, T. (1994). Sulphated blood group Lewis[a]. A superior ligand for human E-selectin. *Journal of Biological Chemistry* **269**, 1595–1598.
117. Rao, B.N.N., Anderson, M.B., Musser, J.H., Gilbert, J.H., Schaefer, M.E., Foxall, C. & Brandley, B.K. (1994). Sialyl Lewis X mimics derived from a pharmacophone search are selectin inhibitors with anti-inflammatory activity. *Journal of Biological Chemistry* **269**, 19663–19666.
118. Ikami, T., Tsuruta, N., Inagaki, H., Kakigama, T., Matsumoto, Y., Tomiya, N., Jomori, T., Usui, T., Suzuki, Y., Tanaka, H., Miyamoto, D., Ishida, H., Hasegawa, A. & Kiso, M. (1998). Synthetic studies on selectin ligands/inhibitors. Synthesis and biological evaluation of sulphated and phosphorylated beta-D-galacto- and lactopyranosides containing fatty-alkyl residues of different carbon chain lengths. *Chemical and Pharmaceutical Bulletin* **46**, 797–806.
119. Alper, J. (1993). Carbohydrates surge through clinical trials. *Biotechnology* **11**, 1093.
120. Kubes, P., Jutila, M. & Payne, D. (1995). Therapeutic potential of inhibiting leukocyte rolling in ischemia reperfusion. *Journal of Clinical Investigation* **95**, 2510–2519.
121. Beuth, J., Ko, H.L., Oette, K., Pulverer, G., Roszkowski, K. & Uhlenbruck, G. (1987). Inhibition of liver metastasis in mice by blocking hepatocyte lectins with arabinogalactan infusions and D-galactose. *Journal of Cancer Research and Clinical Oncology* **113**, 51–55.
122. Isenberg, J., Stoffel, B., Stützer, H., Otte, K. & Benth, J. (1997). Liver lectin blocking with D-galactose to prevent hepatic metastases in colorectal carcinoma patients. *Anticancer Research* **17**, 3767–3772.
123. Suzuki, N., Yamamoto, K., Toyoshima, S., Osawa, T. & Irimura, T. (1996). Molecular cloning and expression of cDNA encoding human macrophage C-type lectin. Its unique carbohydrate binding specificity for Tn antigen. *Journal of Immunology* **156**, 128–135.

5 Alphabetical Directory of Animal Lectins

Aaptos papillata agglutinins I, II and III

Other names: *Aaptos* lectins I, II and III
Mol. wt: 12 000 and 21 000 for lectin I; 16 000 for lectin II (subunit)
Specificity: $(GlcNAc)_4$
Source: The sponge *Aaptos papillata*

ISOLATION

A mixture of lectins was purified from saline extracts of sponge homogenate by affinity chromatography on hog gastric mucin blood group (A +H) substance and elution with $MgCl_2$. Lectin I (unretained fraction) was isolated by application to a DEAE-cellulose ion-exchanger; bound lectins were eluted at a lower pH, and were subsequently separated into agglutinins II and III by preparative PAGE [1].

STRUCTURE

All three major lectins are proteins rich in acidic amino acids. *Aaptos* lectins II and III have very similar amino acid compositions, including a high content of half-cysteine residues; lectin I has a distinct composition with an absence of half-cysteine and vastly more tryptophan [1]. Lectins II and III appear to be homologous isolectins with a single subunit of apparent Mr 16 000 determined by SDS-PAGE under reducing or non-reducing conditions. The sedimentation coefficients were similar at 6 S and 5.5 S, respectively, for lectins II and III. In contrast, lectin I had a sedimentation coefficient of 3.5 S and exhibited two bands corresponding to apparent molecular weights of 12 000 and 21 000 on SDS-PAGE [1]. Lectin I was antigenically distinct from lectins II and III [1].

All three lectins exhibited heterogeneous profiles on isoelectric focusing, but the profiles of lectins II and III were similar to each other and distinct from lectin I.

BIOLOGICAL ACTIVITIES

Aaptos lectins II and III aggutinated human erythrocytes ABO non-specifically; *Aaptos* lectin I bound to, but did not agglutinate, human erythrocytes [1]. All the lectins were precipitated with various blood group substances. Lectin I reacted best with those possessing terminal non-reducing GlcNAc; tetraacetylchitotetraose ($[GlcNAc]_4$) was the best inhibitor. Lectin II was precipitated by blood group substances and other glycoconjugates containing terminal GalNAc,

GlcNAc or sialic acid residues; triacetylchitotriose and tetraacetylchitotetraose were the best inhibitors [1].

TISSUE AND/OR SUBCELLULAR DISTRIBUTION

These lectins were prepared from a homogenate of dried whole sponges. Nothing is known of the subcellular location.

POSSIBLE FUNCTION(S)

No relevant information is available concerning function.

COMMENTS

Less is known about the *Aaptos* lectins than those from *Axinella polypoides* (qv) or *Geodia cydonium* (qv). There is no reason to suppose the *Aaptos* lectins resemble those of either species, especially as the *Aaptos* lectins have a much larger number of cysteine residues/subunit [2].

REFERENCES

1. Bretting, H., Kabat, E.A., Liao, J. & Pereira, M.E.A. (1976). Purification and characterization of the agglutinins from the sponge *Aaptos papillata* and a study of their combining sites. *Biochemistry* **15**, 5029–5038.
2. Buck, F., Schulze, C., Breloer, M., Strupat, K. & Bretting, H. (1998). Amino acid sequence of the D-galactoside binding lectin II from the sponge *Axinella polypoides* (Schmidt) and identification of the carbohydrate binding site in lectin II and related lectin I. *Comparative Biochemistry and Physiology Part B* **121**, 153–160.

Achatinin$_H$

Other names: *Achatina fulica* sialic acid-binding lectin
Mol. wt: 240 000 (native); 15 000 (subunit)
Specificity: Sialic acid
Source: Haemolymph of the snail *Achatina fulica*

ISOLATION

Achatinin$_H$ was purified from clotted and clarified haemolymph by a single chromatography step on ovine submaxillary mucin–Sepharose, eluting with a citrate-containing buffer [1]. Similar results were obtained more readily by eluting the affinity column with the Ca^{2+}-free washing buffer at 30 °C [2]. The purified protein appeared homogeneous by electrophoresis, isoelectric focusing and other criteria.

STRUCTURE

Achatinin$_H$ is a glycoprotein (21% carbohydrate) with an unusually high xylose content [2]. Its amino acid composition is rich in acidic residues [2]. Its isoelectric point is 6.2 [1]. The lectin appears to be composed of 16 non-covalently bonded identical subunits of 15 000 determined by SDS-PAGE with or without reducing agent [1,2].

BIOLOGICAL ACTIVITIES

Achatinin$_H$ has a strong preference for rabbit erythrocytes but will also agglutinate guinea pig, hamster and rat red blood cells at an approximately 30-fold higher minimum concentration. Erythrocytes from many other species, including humans, horses, sheep, cats, mice and chickens, were not agglutinated at all [3].

The lectin activity is Ca^{2+}-dependent [1,2,4] and inhibited only by sialic acid and its derivatives, having an unusually narrow specificity for 9-*O*-acetylsialic acids [2,4,5].

Achatinin$_H$ is mitogenic for rat and human lymphocytes [2].

TISSUE AND/OR SUBCELLULAR DISTRIBUTION

Achatinin$_H$ has only been reported from haemolymph.

POSSIBLE FUNCTION(S)

A defensive function seems plausible but direct evidence for a particular function is lacking.

COMMENTS

Achatinin$_H$ agglutinates the erythrocytes of human subjects suffering from visceral leishmaniasis (kala-azar). This forms the basis of a simple but specific diagnostic test; patients with malaria or tuberculosis, like healthy subjects, give negative results [6]. This reactivity correlates with an increased expression of 9-*O*-acetylated derivatives of sialic acid in erythrocytes from kala-azar patients and is believed to reflect the presence of *Leishmania donovani* [6].

Achatinin$_H$ preferentially agglutinates lymphoblasts, and has potential for detection and monitoring of lymphoblasts in acute lymphoblastic leukaemia [7,8].

Another unrelated lectin has been found in the albumin glands of the same species. It agglutinates rabbit erythrocytes in a Ca^{2+}-independent manner and is inhibited by galactose and galactosides. Its most interesting feature is its temperature dependence, which resembles the human IgM autoantibodies known as 'cold agglutinins' [9].

REFERENCES

1. Basu, S., Sarkar, M. & Mandal, C. (1986). A single step purification of a sialic acid binding lectin (Achatinin$_H$) from *Achatina fulica* snail. *Molecular and Cellular Biochemistry* **71**, 149–157.
2. Basu, S., Mandal, C. & Allen, A.K. (1988). Chemical modification studies of a unique sialic acid binding lectin from the snail *Achatina fulica*. *Biochemical Journal* **254**, 195–202.
3. Mandal, C. & Basu, S. (1987). A unique specificity of a sialic acid binding lectin Achatinin$_H$, from the hemolymph of *Achatina fulica* snail. *Biochemical and Biophysical Research Communications* **148**, 795–801.
4. Mandal, C., Basu, S. & Mandal, C. (1989). Physiochemical studies on Achatinin$_H$, a novel sialic acid-binding lectin. *Biochemical Journal* **257**, 65–71.
5. Sen, G., Mandal, C. & Chowdhury, M. (1993). Characterization of combining site and cell surface sialoglycoconjugates using a lectin, achatinin$_H$. In: *Lectins – Biology, Biochemistry, Clinical Biochemistry*, Volume 9 (eds J. Basu, M. Kundu & P. Chakrabarti) pp. 110–115. Wiley Eastern, New Delhi.
6. Sharma, V., Chatterjee, M., Mandal, C., Sen, S. & Basu, D. (1998). Rapid diagnosis of Indian visceral leishmaniasis using achatinin$_H$, a 9-*O*-acetylated sialic acid binding lectin. *American Journal of Tropical Medicine and Hygiene* **58**, 551–554.
7. Sinha, D., Mandal, C. & Bhattacharya, D.K. (1999). Identification of 9-*O* acetyl sialoglycoconjugates (9-0AcSGs) as biomarkers in childhood acute lymphoblastic leukaemia using a lectin, achatinin$_H$, as a probe. *Leukemia* **13**, 119–125.
8. Sinha, D., Mandal, C. & Bhattacharya, D.K. (1999). Development of a simple, blood based lymphoproliferation assay to assess the clinical status of patients with acute lymphoblastic leukaemia. *Leukaemic Research* **23**, 433–439.
9. Sarkar, M., Bachhawat, B.K. & Mandal, C. (1984). A new cold agglutinin from *Achatina fulica* snails. *Archives of Biochemistry and Biophysics* **233**, 286–289.

Acorn barnacle isolectins

Other names: *Megabalanus rosa* isolectins; BRA-1, BRA-2 and BRA-3
Mol. wt: 64 000–330 000 (native); 22 000 (BRA-1 and BRA-2) & 18 000 (subunit)
Specificity: Sialic acid, galactose, etc.
Source: Coelomic fluid of the acorn barnacle *Megabalanus rosa*

ISOLATION

The agglutinins collectively were isolated from the coelomic fluid of *M. rosa* by affinity chromatography on acid-treated Sepharose 4B and elution with D-galactose. The agglutinins were then separated into three electrophoretically homogeneous peaks of protein by gel filtration on Sephadex G-200 [1]. These isolectins (in decreasing order of apparent Mr) were designated BRA-1, BRA-2 and BRA-3. The most abundant isolectin was BRA-2 (54% of the total lectin protein).

STRUCTURE

BRA-1, BRA-2 and BRA-3 eluted from the Sephadex G-200 column in volumes corresponding to apparent molecular weights of 350 000, 140 000 and 64 000, respectively. All were glycoproteins and all had amino acid compositions that were rich in acidic amino acids [1]. The isoelectric points of the denatured basic units (see below) were 5.0, 5.0 and 4.4, respectively, for BRA 1, BRA-2 and BRA-3 [1].

The subunit structure was investigated by SDS-PAGE both with and without 2-mercaptoethanol present. Both BRA-1 and BRA-2 were composed of a single 22 KDa basic unit (i.e. under reducing conditions) which forms disulphide bond-linked dimers. Although native BRA-2 consists simply of aggregates of these 44 KDa dimers, BRA-1 is a non-covalently bonded aggregate of both dimers and tetramers of the basic unit, the latter arising from additional disulphide bonding. The 22 KDa basic unit in BRA-1 and BRA-2 is very similar judged by amino acid composition, peptide mapping and circular dichroism spectra [1]. By the same criteria, the 18 KDa basic subunit of BRA-3 showed much less similarity. BRA-3 appears to consist of two identical subunits under non-reducing conditions, each consisting of two basic 18 KDa units.

The full amino acid sequences of BRA-3 [2] and BRA-2 [3] have been published. The former was confirmed as having four identical subunits (2 × dimers) of 138 amino acids (calculated Mr=15 713); it possesses two interchain and three intrachain disulphide bridges [2]. The latter is composed of six identical subunits of 173 amino acids (calculated Mr=19 550) which form dimers cross-linked by disulphide bonds [3]. Both BRA-3 and BRA-2 exhibited some limited sequence homology to the *Sarcophaga* lectin (qv) and other C-type lectins, but the strongest resemblance (25% identity) was to each other [3].

BIOLOGICAL ACTIVITIES

All three isolectins agglutinated human and lapine, but not ovine, erythrocytes. Although ABO non-specific in practical terms, BRA-1 and BRA-2 required higher minimum concentrations to agglutinate blood group A cells compared to B or O, while BRA-3 agglutinated type B cells more readily and type O cells least readily. All three lectins were Ca^{2+}-dependent, but BRA-3 alone exhibited the unusual property of being inhibited by high (>20 mM) concentrations of Ca^{2+} [1].

Lectin binding to rabbit erythrocyte ghosts demonstrated a very broad sugar specificity. Among monosaccharides, BRA-1 and BRA-2 were inhibited most effectively by *N*-acetylneuraminic acid, L-arabinose, D-galactose, D-xylose, D-glucose, *N*-acetylgalactosamine and D-mannose (in that order); L-fucose and GlcNAc were virtually without effect. BRA-3 had a narrower specificity, with *N*-acetylneuraminic acid and D-galactose being the most potent inhibitors, and all other monosaccharides having an insignificant effect [1].

TISSUE AND/OR SUBCELLULAR DISTRIBUTION

These lectins are major components of coelomic fluid but occur in many other tissues. Relatively high concentrations were found in homogenates of mantle, ovaries and haemocytes [4].

POSSIBLE FUNCTION(S)

The existence of multiple agglutinins potentially capable of recognising foreign substances and promoting phagocytosis is consistent with a defensive role, but no direct evidence has been reported. Other possible functions have also been mooted, including promotion of ovarian maturation and regulation of $CaCO_3$ crystallisation in the barnacle shell [4].

COMMENTS

It appears that BRA-1 and BRA-2 are true isolectins, composed of homologous subunits, and which differ mainly in the molecular arrangement of the basic 22 KDa subunits. BRA-3 is a separate minority lectin, with a scarcely related 18 KDa subunit which has a markedly different primary structure.

The unseparated mixture of isolectins can agglutinate certain cancer cells and induce binding of tumour cells to macrophages [1].

REFERENCES

1. Muramato, K., Ogata, K. & Kamiya, H. (1985). Comparison of the multiple agglutinins of the acorn barnacle, *Megabalanus rosa*. *Agricultural and Biological Chemistry* **49**, 85–93.
2. Muramato, K. & Kamiya, H. (1986). The amino-acid sequence of a lectin of the acorn barnacle, *Megabalanus rosa*. *Biochimica et Biophysica Acta* **874**, 285–295.
3. Muramato, K. & Kamiya, H. (1990). The amino-acid sequence of multiple lectins of the acorn barnacle, *Megabalanus rosa* and its homology with animal lectins. *Biochimica et Biophysica Acta* **1039**, 42–51.
4. Muller, K. (1996). 'Primitive' C-type lectins, an overview. In: *Lectins – Biology, Biochemistry, Clinical Chemistry*, Volume 11 (eds E. Van Driessche, P. Rougé, S. Beeckmans & T.-C. Bøg-Hansen) pp. 129–160. Textop, Hellerup.

Aggrecan

Other name: Cartilage proteoglycan
Mol. wt: >220 000 (for core protein)
Specificity: Fucose, galactose
Source: Mammalian cartilage

ISOLATION

Overlapping cDNA clones were first obtained for rat aggrecan [1], providing an entire coding sequence, while the corresponding human data were obtained from a combination of cDNA and genomic exon sequencing [2].

STRUCTURE

The complete nucleotide sequences of the aggrecan genes and their deduced amino acid sequences have been reported from rat [2] and human [2] sources; there is about 75% identity at the protein level between the species [2]. The polypeptide can be divided into three major regions. At the N-terminus two globular domains are joined by a rod-like interglobular domain with the N-terminal globule mediating the binding to hyaluronic acid; the long central region consists of a keratin sulphate domain and three chondroitin sulphate attachment domains made up of repeating sequences, which may contain up to 100 chondroitin sulphate chains; and the C-terminus has a globular C-type lectin domain [1,2]. The core protein has over 2000 amino acid residues and a molecular weight in excess of 200 000 [1,2]. Humans, but apparently not members of other species, possess extensive allelic polymorphism in the central region [3].

BIOLOGICAL ACTIVITIES

Aggrecan binds weakly to immobilised sugars in a Ca^{2+}-dependent manner; the highest affinities were for fucose and galactose [4].

Like other hyalectans, aggrecan can bind tenascin-R [5] via its C-type lectin domain in a Ca^{2+}-independent manner, as well as hyaluronic acid via its N-terminal domain [6].

Like versican, but unlike the brain-specific hyalectans, aggrecan binds with high affinity to the widely distributed extracellular matrix protein, fibulin-1, in a Ca^{2+}-dependent manner; this appears to be a protein–protein interaction, and fibulin-1 may be a physiological ligand [7].

TISSUE AND/OR SUBCELLULAR DISTRIBUTION

Aggrecan is abundantly expressed only in cartilage, but low levels may be expressed in other tissues, including brain [6].

POSSIBLE FUNCTION(S)

Aggrecan makes a major contribution to the physical nature of tissue architecture, including the resistance to compressive forces in joint surfaces.

COMMENTS

Aggrecan is the major proteoglycan of cartilage.

REFERENCES

1. Doege, K., Sasaki, M., Horigan, E., Hassell, J.R. & Yamada, Y. (1987). Complete primary structure of the rat cartilage proteoglycan core protein deduced from cDNA clones. *Journal of Biological Chemistry* **262**, 17757–17767.
2. Doege, K.J., Sasaki, M., Kimura, T. & Yamada, Y. (1991). Complete coding sequence and deduced primary structure of the human cartilage large aggregating proteoglycan, aggrecan. *Journal of Biological Chemistry* **266**, 894–902.
3. Doege, K.J., Coulter, S.N., Meek, L.M., Maslen, K. & Wood, J.G. (1997). A human-specific polymorphism in the coding region of the aggrecan gene. *Journal of Biological Chemistry* **272**, 13974–13979.
4. Halberg, D.F., Proulx, G., Doege, K.J., Yamada, Y. & Drickamer, K. (1988). A segment of the cartilage proteoglycan core protein has lectin-like activity. *Journal of Biological Chemistry* **263**, 9486–9490.
5. Aspberg, A., Muira, R., Bourdoulous, S., Shimonaka, M., Heinegård, D., Schachner, M., Ruoslahti, E. & Yamaguchi, Y. (1997). The C-type lectin domains of lecticans, a family of aggregating chondroitin sulphate proteoglycans, bind tenascin-R by protein–protein interactions independent of carbohydrate moiety. *Proceedings of the National Academy of Sciences USA* **94**, 10116–10121.
6. Ruoslahti, E. (1996). Brain extracellular matrix. *Glycobiology* **6**, 489–492.
7. Aspberg, A., Adam, S., Kostka, G., Timpl, R. & Heinegård, D. (1999). Fibulin-1 is a ligand for the C-type lectin domains of aggrecan and versican. *Journal of Biological Chemistry* **274**, 20444–20449.

Agkistrodon piscivorus venom lectin

Other name: APL
Mol. wt: 32 000 (native); 16 000 (subunit)
Specificity: D-Galactose
Source: Venom of the Eastern cottonmouth snake *Agkistrodon p. piscivorus*

ISOLATION

The lectin was isolated by affinity chromatography on Sephadex G-50; elution with lactose yielded a protein judged homogeneous by disc gel electrophoresis and reversed phase high pressure liquid chromatography [1].

STRUCTURE

The 135-amino acid sequence has been determined, and provides a calculated Mr of 16 200. Since SDS-PAGE analysis gave apparent Mr determinations of 15 500 and 29 000 in the presence and absence, respectively, of 2-mercaptoethanol, it is clear that APL is a disulphide-linked homodimer [1].

The primary sequence shows extensive homology with other snake venom lectins and C-type lectin-like proteins, and also with echinoidin (qv).

BIOLOGICAL ACTIVITIES

APL agglutinates lapine erythrocytes in a calcium-dependent manner. Haemagglutination is readily inhibited by D-galactose and L-rhamnose, but not by other common monosaccharides. Lactose was the only disaccharide tested, and was found to be inhibitory.

APL also agglutinates human red blood cells and shows some preference for group A cells over group B cells [1].

TISSUE AND/OR SUBCELLULAR DISTRIBUTION

APL has only been reported from venom.

POSSIBLE FUNCTION(S)

Although a contribution to the toxicity of venom has been discussed [1], no direct evidence concerning function has been obtained.

COMMENTS

APL is clearly a member of what might be called the thrombolectin (qv) group of snake venom agglutinins, a group of calcium-dependent, dimeric, D-galactoside-specific lectins with very similar primary structures. Significant, but considerably less, homology was noted with the platelet-binding, lectin-like proteins from various snake venoms [1].

REFERENCE

1. Komori, Y., Nikai, T., Tohkai, T. & Sugihara, H. (1999). Primary structure and biological activity of snake venom lectin (APL) from *Agkistrodon p. piscivorus* (Eastern cottonmouth). *Toxicon* **37**, 1053–1064.

Allomyrina dichotoma isolectins

Other names: Allo A-I and Allo A-II
Mol. wt: 66 000 (native); 20 000 and 19 000 or 17 000 (subunit)
Specificity: β-D-Galactosides
Source: Larvae of the Japanese beetle *Allomyrina dichotoma*

ISOLATION

Clarified haemolymph was applied to an affinity column of acid-treated Sepharose and the isolectins were eluted with lactose. They were then separated by ion-exchange chromatography on DEAE-cellulofine into Allo A-I (minor) and Allo A-II (major) forms. Each isolectin migrated as a single band on PAGE [1].

STRUCTURE

The isolectins are antigenically identical in relation to double diffusion against antisera to unseparated Allo A [1]. The molecular sizes and subunit structures are also very similar. The apparent molecular weights estimated by gel filtration on Sephadex G-100 were 65 000 (Allo A-I) and 66 500 (Allo A-II). The corresponding subunit values determined by SDS-PAGE under reducing conditions were 17 500/20 000 and 19 000/20 000. Based on SDS-PAGE analysis after cross-linking of subunits with glutaraldehyde, both isolectins were considered to be heterodimers of <40 kDa, possibly with a common 20 kDa subunit. The isolectric point of Allo A-I was 6.4; that of Allo A-II, 5.9 [1].

BIOLOGICAL ACTIVITIES

Both isolectins agglutinated human erythrocytes in an ABO non-specific manner. Neither was inhibited by EDTA or by monosaccharides (including galactose), but both were readily inhibited by lactose, lactulose and simple β-galactosides.

The saccharide specificity of Allo A I+II [2] and of Allo A-II [3] were investigated by lectin–agarose affinity chromatography. Both studies [2,3] indicated a primary interaction with Gal β1→4 GlcNAc, and that substitution of the terminal galactose by NeuAcα2,3(6) substantially augments the affinity of the binding. Although allo A bound sialylated complex-type and hybrid-type oligosaccharides, the isolectins had no affinity for mucin-type glycopeptides carrying the sialylated Gal β1→3 GalNAc sequence [2].

Allo A is strongly mitogenic for murine splenic T lymphocytes in a strictly adherent cell-dependent manner [4]. The isolated isolectins are both mitogenic

[4]. Allo A is also mitogenic for rat splenic lymphocytes, human peripheral blood lymphocytes and mature (but not immature) murine thymocytes [4].

TISSUE AND/OR SUBCELLULAR DISTRIBUTION

Lectin activity was highest in mature (third instar) larvae. Eggs were also a source of lectin, but not the haemocytes [1].

POSSIBLE FUNCTION(S)

Allo A exhibits a notable developmental regulation and has been suggested to have a role in surveillance of decomposing larval tissues [1].

COMMENTS

Allo A can be used to identify and/or separate complex-type oligosaccharides or glycopeptides. In this regard it has advantages over the plant lectin *Ricinus communis* agglutinin, which can also bind to mucin-type structures [2]. Allo A can also be used to separate sialylated complex-type sugar chains from sulphated sugar chains [2].

Immobilised Allo A-II has been used to separate complex-type oligosaccharides with the NeuAc α 2→6 Gal β1→4 GlcNAc group from the NeuAc α2→3 Gal β1→4 GlcNAc group [3].

Immobilised Allo A has been used to fractionate human serum into two major groups of proteins [5].

REFERENCES

1. Umetsu, K., Kosaka, S. & Suzuki, T. (1984). Purification and characterisation of a lectin from the beetle *Allomyrina dichotoma. Journal of Biochemistry* **95**, 239–245.
2. Sueyoshi, S., Yamamoto, K. & Osawa, T. (1988). Carbohydrate binding specificity of a beetle (*Allomyrina dichotoma*) lectin. *Journal of Biochemistry* **103**, 894–899.
3. Yamashita, K., Umetsu, K., Suzuki, T., Iwaki, Y., Endo, T. & Kobata, A. (1988). Carbohydrate binding specificity of immobilised beetle *Allomyrina dichotoma* lectin II. *Journal of Biological Chemistry* **263**, 17482–17489.
4. Kimura, S., Umetsu, K., Yamashita, T., Suzuki, T., Arai, S. & Sendo, F. (1987). T cell mitogenicity of a novel D-galatoside-specific lectin from the beetle *Allomyrina dichotoma* (allo A). *Immunopharmacology* **13**, 181–188.
5. Umetsu, K., Ikeda, N., Kashimura, S. & Suzuki, T. (1985). Affinity chromatography of human serum proteins using immobilised lectin from *Allomyrina dichotoma. Biochemistry International* **10**, 549–552.

Amphibian galectins

Mol. wt: Approx. 30 000 (native); approx. 15 000 (subunit)
Specificity: Lactose and other β-galactosides
Sources: Various tissues of frogs and toads

ISOLATION

Galectins have been isolated from several amphibian species by procedures based on affinity chromatography (usually involving immobilised lactose or asialofetuin), including the clawed toad *Xenopus laevis* [1,2], another toad, *Bufo arenarum* [3,4], and the frogs *Rana tigerina* [5] and *Rana catesbeiana* [6].

A cDNA has also been isolated from *X. laevis* [2].

STRUCTURE

The galectins from *R. catesbeiana* and *B. arenarum* have subunits of apparent Mr 14 000 estimated by SDS-PAGE, while those from *R. tigerina* and *X. laevis* have 16 000 subunits determined by the same method. Nevertheless, the two lectins that have been fully sequenced, from *X. laevis* [2] and *B. arenarum* [4], have true subunit molecular weights that are closer together at 15 524 (135 residues) and 14 797 (134 residues), respectively. All have apparent molecular weights of 30 000–32 000 based on gel filtration, indicating dimer formation in the native state.

The *Xenopus* lectin sequence has 39% identity to human galectin-1, 37% identity to chicken 14 000 galectin, 31% identity to chicken 16 000 galectin and 28% identity to electrolectin. The *Bufo* lectin most resembles bovine galectin-1 (48% identity) and is equally similar to the chicken 14 000 and 16 000 lectins (45%). However, the *Bufo* lectin shared only 38% identity with the *Xenopus* lectin.

The partial amino acid sequence obtained from *R. catesbeiana* galectin exhibits 48% identity to the *Bufo* galectin [6,4].

BIOLOGICAL ACTIVITIES

All of these galectins agglutinate trypsin-treated lapine erythrocytes, without a requirement for Ca^{2+}, as would be expected. The *R. tigerina* galectin can also agglutinate untreated rabbit erythrocytes [5]. The *R. catesbeiana* galectin is apparently stable in the absence of reducing agents [6], while the *B. arenarum* lectin requires either a reducing agent or a ligand for stability [4].

There are interesting differences in the fine carbohydrate-binding specificities of these galectins. Most particularly, the *Xenopus* lectin is more readily inhibited by lactose and the blood group A tetrasaccharide (see entry for Galectin-3) than by thiodigalactoside or *N*-acetylgalactosamine, while the *Bufo* lectin was most sensitive to thiodigalactoside and *N*-acetylgalactosamine, followed by lactose and then by the blood group A tetrasaccharide [4]. The *Bufo* lectin therefore resembles mammalian galectins-1 in fine specificity.

TISSUE AND/OR SUBCELLULAR DISTRIBUTION

The *Xenopus* galectin is abundant in the skin (100× more than in muscle) of adults, where it is concentrated in the cytoplasm of granular gland and mucous gland cells [1]; injection of adrenalin induces a massive secretion of the galectin from the granular gland cells [1]. The *R. tigerina* lectin was purified from skeletal muscle, which had more haemagglutinating activity than other tissues, including liver and brain [5]. The *R. catesbeiana* lectin was obtained from unfertilised eggs [6] and the *Bufo* galectin from ovaries [4].

POSSIBLE FUNCTION(S)

No significant evidence can be cited in support of any function.

COMMENTS

The *Bufo* galectin resembles mammalian galectins-1 more than the *Xenopus* lectin in primary sequence homology, fine carbohydrate specificity and in possessing three out of six cysteine residues conserved in all galectins-1 [4]. This may be an indication that structural and functional divergence may have occurred early in vertebrate evolution, and therefore the correlation between primary structure and phylogeny may be complicated by this consideration.

Lectins other than galectins have been detected from amphibian sources; some have blood group specificities (see Chapter 4).

REFERENCES

1. Bols, N.C., Robertson, M.M., Haywood-Reid, P.L., Cerra, R.F. & Barondes, S.H. (1986). Secretion of a cytoplasmic lectin from *Xenopus laevis* skin. *Journal of Cell Biology* **102**, 492–499.
2. Marschal, P., Herrmann, J., Leffler, H., Barondes, S.H. & Cooper, D.N.W. (1992). Sequence and specificity of a soluble lactose-binding lectin from *Xenopus laevis* skin. *Journal of Biological Chemistry* **267**, 12942–12949.
3. Fink de Cabutti, N.E., Caron, M., Joubert, R., Elola, M.T., Bladier, D. & Herkovitz, J. (1987). Purification and some characteristics of a β-galactoside binding soluble lectin from amphibian ovary. *FEBS Letters* **223**, 330–334.
4. Ahmed, H., Pohl, J., Fink, N.E., Strobel, F. & Vasta, G.R. (1996). The primary structure and carbohydrate specificity of a β-galactosyl-binding lectin from toad (*Bufo arenarum* Hensel) ovary reveal closer similarities to the mammalian galectin-1 than to the galectin from the clawed frog *Xenopus laevis*. *Journal of Biological Chemistry* **271**, 33083–33094.
5. Shet, M.S. & Madaiah, M. (1989). A lectin from thigh muscle of *Rana tigerina*. *Biochimica et Biophysica Acta* **991**, 465–469.
6. Ozeki, Y., Matsui, T., Nitta, K., Kawauchi, H., Takayanagi, Y. & Titani, K. (1991). Purification and characterization of β-galactoside binding lectin from frog (*Rana catesbeiana*) eggs. *Biochemical and Biophysical Research Communications* **178**, 407–413.

Amphoterin

Other name: Heparin binding p30
Mol. wt: 58 000 (native); 29 000 (monomer)
Specificity: Heparin; glucuronic acid-3-sulphate
Sources: Rat brain etc.

ISOLATION

Amphoterin was isolated from octyl glucoside extracts of young rat brain by sequential binding to columns of heparin–Sepharose and Affi-Gel blue and elutions with NaCl [1].

Amphoterin cDNA was later cloned from a cDNA library constructed from mRNA of rat brain, and translated *in vitro* [2].

STRUCTURE

Amphoterin is a protein or glycoprotein which derives its name from having both cationic and anionic regions. It appears to exist as a non-covalently linked dimer of very similar subunits which appear as a closely spaced doublet on electrophoresis and have the same amino acid sequence [1,2].

SDS-PAGE analysis yields a doublet corresponding to an Mr of just under 30 000 under reducing conditions, while a slightly smaller Mr is estimated under non-reducing conditions. Elution from a gel filtration column in the presence of 4 M urea indicates an apparent Mr of 58 000 [1].

The deduced primary sequence has 214 amino acid residues and a calculated molecular weight of 24 760. The sequence forms a dipolar structure with a large (184 residue) cationic, lysine-rich region and a shorter (30 residue) anionic region at the C-terminal end. A short sequence near the N-terminus (PKKPRG) conforms to the consensus sequence of heparin-binding proteins [2].

BIOLOGICAL ACTIVITIES

Amphoterin mediates the adhesion of cultured neurons to surfaces and promotes neurite outgrowth [1].

In addition to binding heparin, amphoterin can bind to sulphated glycolipids in which the glucuronic acid-3-sulphate structure appears to be recognised [3].

TISSUE AND/OR SUBCELLULAR DISTRIBUTION

Amphoterin is said to be found in many cell types, but has been studied mainly in rat brain, where it is found in the cytoplasm of cell soma, in cell processes and extracellularly. Most notably, a striking association with the plasma membrane filopodia of spreading and extending neurons was noted. Outside of brain, amphoterin is abundant in kidney, but lower amounts are found in liver and skeletal muscle (at least in 2-day postnatal rats) [2].

POSSIBLE FUNCTION(S)

Amphoterin is presumed to have a regulatory role in neuronal development [1,2].

COMMENTS

Amphoterin, both at the protein and mRNA levels, shows a definite developmental regulation, with higher levels in perinatal brain than in the corresponding adult tissue [1,2].

Another heparin-binding lectin of developing rat brain with neurite outgrowth-enhancing activity has been cloned; this 'heparin binding growth associated molecule' (or HB-GAM) has the heparin sulphate side chains of N-syndecan as endogenous ligands in which 2-*O*-sulphated iduronic acid units appear to be of prime importance for the interaction [4].

REFERENCES

1. Rauvala, H. & Pihlaskari, R. (1987). Isolation and some characteristics of an adhesive factor of brain that enhances neurite outgrowth in central neurons. *Journal of Biological Chemistry* **262**, 16625–16635.
2. Merenmies, J., Pihlaskari, R., Laitinen, J., Wartiovaari, J. & Rauvala, H. (1991). 30-kDa heparin-binding protein of brain (amphoterin) involved in neurite outgrowth. *Journal of Biological Chemistry* **266**, 16722–16729.
3. Mohan, P.S., Laitinen, J., Merenmies, J., Rauvala, H. & Jungalwala, F.B. (1992). Sulfoglycolipids bind to adhesive protein amphoterin (p30) in the nervous system. *Biochemical and Biophysical Research Communications* **182**, 689–696.
4. Kinnunen, T., Raulo, E., Nolo, R., Moccarana, M., Lindahl, U. & Rauvala, H. (1996). Neurite outgrowth in brain neurons induced by heparin-binding growth associated molecule (HB-GAM) depends on the specific interaction of HB-GAM with heparin sulphate at the cell surface. *Journal of Biological Chemistry* **271**, 2243–2248.

Anadara granosa heparin-binding lectin

Other name: Anadarin MS
Mol. wt: 300 000 (native); 60 000 (subunit)
Specificity: Heparin
Source: Haemolymph of the marine clam *Anadara granosa*

ISOLATION

The lectin was isolated from *A. granosa* plasma by affinity chromatography followed by gel filtration [1]. The plasma was applied to a column of heparin–Sepharose at 10 °C in a calcium-containing buffer and eluted at room temperature with 2 M NaCl. Fractions containing lectin were concentrated and applied to a column of Sepharose 6B. The lectin eluting in the second (minor) peak appeared homogeneous. Recovery of the lectin was quantitative with a 1500-fold increase in specific activity [1].

STRUCTURE

Anadarin MS is a protein or glycoprotein rich in aspartic acid, glutamic acid, glycine and histidine [1]. The apparent molecular weight was estimated at 300 000 by gel filtration, with a single subunit of apparent Mr 60 000 determined by SDS-PAGE [1]. It therefore appears to be a pentameric protein composed of identical subunits. Its isoelectric point is 6.2 [1].

BIOLOGICAL ACTIVITIES

Anadarin MS agglutinates lapine, but not human, erythrocytes. Lectin activity is calcium-dependent, but totally inhibited by magnesium or manganese ions [1].

Lectin activity cannot be inhibited by monosaccharides, but was readily inhibited by heparin and by dermatan sulphate [1]. It appears to have a strict specificity for glycosaminoglycans containing iduronic acid [1].

Anadarin MS also agglutinates infectious promastigotes of *Leishmania donovani* [1].

TISSUE AND/OR SUBCELLULAR DISTRIBUTION

The lectin is present in haemolymph plasma, but no other information was given about its distribution in tissues.

POSSIBLE FUNCTION(S)

It has been suggested that the lectin might function in tissue development [1], but evidence for a function is lacking.

COMMENTS

Anadarin MS failed to agglutinate *L. donovani* during the log (non-infectious) phase of growth, although it readily agglutinated infectious promastigotes [1]. Anadarin MS may therefore be used as a marker for the latter, and to study cell surface changes in *L. donovani* as the parasite acquires infectivity.

The plasma of *A. granosa* also contains a separate, galactose-specific lectin, Anadarin P (qv) [2].

REFERENCES

1. Dam, T.K., Bandyopadhyay, P., Sarkar, M., Ghosal, J., Bhattacharya, A. & Choudhury, A. (1994). Purification and partial characterization of a heparin-binding lectin from the marine clam *Anadara granosa. Biochemical and Biophysical Research Communications* **203**, 36–45.
2. Dam, T.K., Sarkar, M., Glosal, J. & Choudhury, A. (1992). A novel galactosyl-binding lectin from the plasma of the blood clam, *Anadara granosa* (L) and a study of its combining site. *Molecular and Cellular Biochemistry* **117**, 1–9.

Anadarin P

Other name: *Anadara granosa* galactose-binding lectin
Mol. wt: 130 000 (native); 17 000 and 16 000 (subunit)
Specificity: Arabinose; galactosides
Source: Plasma of the marine clam *Anadara granosa*

ISOLATION

Anadarin P was obtained by a single affinity chromatographic step on asialofetuin–Sepharose, eluting with galactose [1]. The purified lectin appeared homogeneous by electrophoresis, gel filtration and isoelectric focusing.

STRUCTURE

The apparent molecular weight estimated by gel filtration was 130 000. SDS-PAGE, with or without urea, and with or without 2-mercaptoethanol, yielded two bands corresponding to Mr values of 16 000 and 17 000 [1]. The native lectin may consist of four copies of each subunit non-covalently bound. The isoelectric point is 6.8.

BIOLOGICAL ACTIVITIES

Anadarin P agglutinates both lapine and human erythrocytes; activity was independent of calcium and other divalent cations tested [1].

Lectin activity was readily inhibited by L-arabinose, 2-deoxy-D-galactose, D-xylose, D-galactose and fucose (in that order of potency) among monosaccharides. Glucose and mannose were not inhibitory. Various α- and β-galactosides were also potent inhibitors, particularly *p*-aminophenyl-β-D-lactopyranoside.

TISSUE AND/OR SUBCELLULAR DISTRIBUTION

Anadarin P was discovered in haemolymph plasma [1].

POSSIBLE FUNCTION(S)

Nothing relevant to function was reported [1].

COMMENTS

A second lectin has been isolated from the same species, the *Anadara granosa* heparin-binding lectin (qv).

REFERENCE

1. Dam, T.K., Sarkar, M., Ghosal, J. & Choudhury, A. (1992). A novel galactosyl-binding lectin from the plasma of the blood clam, *Anadara granosa* (L) and a study of its combining site. *Molecular and Cellular Biochemistry* **117**, 1–9.

Anthocidaris crassispina egg lectin

Other names: Sea urchin egg lectin; SUEL
Mol. wt: 23 000 (native); 11 500 (subunit)
Specificity: D-galactosides
Source: Eggs of the sea urchin *Anthocidaris crassispina*

ISOLATION

A soluble extract of acetone powder prepared from dehydrated eggs was applied to a column of thiodigalactoside in the presence of EDTA. After washing, the lectin was eluted with lactose [1].

STRUCTURE

SUEL is a protein rich in aspartic and glutamic acids but lacking histidine, methionine and tryptophan [2]. Its native Mr was estimated to be 23 000 by both gel filtration and SDS-PAGE in the absence of reducing agent. SDS-PAGE under reducing conditions gives a single band of 11 500 [1]. The lectin has been completely sequenced and found to consist of 105 amino acid residues with a calculated Mr of 11 375 [2]. The isoelectric point is 5.3 [2].

BIOLOGICAL ACTIVITIES

SUEL agglutinates glutaraldehyde-fixed, trypsin-treated rabbit erythrocytes [1,2] independently of calcium or sulphydryl reagent [2]. However, haemagglutinating activity is slowly lost in the presence of 2-mercaptoethanol [2].

Although readily inhibited by D-galactose and D-fucose (but not GalNAc), SUEL is more sensitive to inhibition by galactosides (of both α and β configuration) [1].

Thiodigalactoside and *p*-nitrophenyl-α-D-galactoside were the most potent inhibitors of haemagglutination [1].

TISSUE AND/OR SUBCELLULAR DISTRIBUTION

SUEL is abundant in the cytoplasm of unfertilised eggs, appearing as randomly distributed particles by indirect immunofluorescence, but is found almost entirely at the periphery of fertilised eggs [1,3]. The particles correspond to small granules that migrate to the cortex within 10 minutes of fertilisation [3]. No SUEL is found in coelomic fluid [2].

POSSIBLE FUNCTION(S)

The redistribution of the lectin after fertilisation is consistent with a role in activation of eggs, or in subsequent differentiation. It has been suggested that SUEL may cross-link hyaline or hyaline components to embryo cells [3].

COMMENTS

SUEL is clearly a different lectin from echinoidin (qv) obtained from coelomic fluid of the same species. The saccharide specificity and the molecular size of SUEL are typical of galectins, but the amino acid sequence does not permit classification as a galectin [2]. Instead, it exhibits considerable homology with the trout egg lectins (qv).

REFERENCES

1. Sasaki, H. & Aketa, K. (1981). Purification and distribution of a lectin in sea urchin (*Anthocidaris crassispina*) eggs before and after fertilization. *Experimental Cell Research* **135**, 15–19.
2. Ozeki, Y., Matsui, T., Suzuki, M. & Titani, K. (1991). Amino acid sequence and molecular characterization of a D-galactoside-specific lectin purified from the sea urchin (*Anthocidaris crassispina*) eggs. *Biochemistry* **30**, 2391–2394.
3. Ozeki, Y., Yokota, Y., Kato, K.H., Titani, K. & Matsui, T. (1995). Developmental expression of D-galactoside-binding lectin in sea urchin (*Anthocidaris crassispina*) eggs. *Experimental Cell Research* **216**, 318–324.

Aplysia kurodai skin lectin

Other name: AKSL
Mol. wt: 200 000 (native); 28 000 and 26 000 (subunit)
Specificity: D-Galactose
Source: Skin of the sea hare *Aplysia kurodai*

ISOLATION

Particulate material obtained after homogenisation of skins of Japanese *A. kurodai* was extracted with lactose, and (after dialysis) subjected to affinity chromatography on lactose–Sepharose. The pure lectin was obtained by elution with galactose [1].

STRUCTURE

SDS-PAGE yielded a single band of apparent Mr 200 000 under non-reducing conditions, and two bands (28 000 and 26 000) under reducing conditions. AKSL therefore appears to be a disulphide-bonded heteromultimer.

BIOLOGICAL ACTIVITIES

AKSL strongly agglutinates lapine erythrocytes. This activity does not require divalent cations or disulphide bonds [1]. It is readily inhibited by galactose and galactosides but not by other monosaccharides.

AKSL promoted the specific attachment of human rhabdomyosarcoma cells to plastic plates; this activity was inhibited by galactose but not by glucose [1].

TISSUE AND/OR SUBCELLULAR DISTRIBUTION

AKSL was obtained from skin homogenate after removing the internal organs; the presence of this lectin in other tissues was apparently not investigated [1].

POSSIBLE FUNCTION(S)

No information was reported regarding function [1].

COMMENTS

Lectins are known to occur in the gonads of several species of *Aplysia*, and one from *Aplysia depilans* has been purified and partially characterised [2].

REFERENCES

1. Ozeki, Y. (1998). Purification and cell attachment activity of a D-galactoside-binding lectin from the skin of sea hare, *Aplysia kurodai*. *Biochemistry and Molecular Biology International* **45**, 989–995.
2. Gilboa-Garber, N., Susswein, A.J., Mizrahi, L. & Avichezer, D. (1985). Purification and characterisation of the gonad lectin of *Aplysia depilans*. *FEBS Letters* **181**, 267–270.

Asialoglycoprotein receptors

Other names: Mammalian hepatic lectin; chicken hepatic lectin
Mol. wt: 170 000 (native); 42 000 and 44 000 (subunit)
Specificity: Galactose, GalNAc (or GlcNAc:avian)
Sources: Mammalian (and avian) liver

ISOLATION

Asialoglycoprotein receptors have been purified by affinity chromatography from rabbit [1], rat [2,3] and human [4] liver homogenates. The cDNA encoding the human receptor was cloned from a hepatoma cell line cDNA library [5], and subsequently a second, homologous gene was cloned [6]. The mammalian asialoglycoprotein receptors I and II share about 60% primary sequence identity [6,7]. The chicken hepatic receptor appears to be the avian homologue with around 25% sequence identity [7], while another closely related gene product is expressed on mammalian macrophages [8].

STRUCTURE

The asialoglycoprotein receptors are glycoproteins (approximately 10% carbohydrate) composed of at least two homologous subunits that form heterogeneous non-covalently-linked complexes [9].

Typically, the major subunit has an apparent molecular weight of about 40 000 determined by SDS-PAGE, with the less abundant subunit(s) slightly heavier [1,3,4,6]. Each subunit has a similar structure: an N-terminal cytoplasmic domain, a transmembrane domain and a large extracellular region with a C-type lectin domain at the C-terminal end [5–7]. The latter is a typical recognition domain with the residues that form both calcium-binding sites well conserved. The human subunits have predicted polypeptide masses of 36 000 (H1) and 38 000 (H2), consistent with the SDS-PAGE estimates for glycosylated polypeptides [6]; the stoichiometry of the subunits in the functional receptor is uncertain, but some evidence exists for a 3 : 1 (H1 : H2) arrangement [9].

The mouse macrophage receptor possesses a 25-amino acid region which is not found in the receptors of hepatocyte origin [8]. The chicken homologue is much smaller, with a subunit molecular mass of 26 000 Da [10].

BIOLOGICAL ACTIVITIES

The asialoglycoprotein receptors recognise and bind to modified glycoproteins (e.g. asialo-orosomucoid) *in vivo* and *in vitro* in a calcium-dependent and pH-dependent manner [9,11]. Most act by binding to the terminal galactose residues exposed after neuraminidase treatment of the native glycoprotein, but the chicken hepatic lectin binds to GlcNAc-terminated glycoconjugates [10].

The major rat and human receptors show a strong preference for GalNAc over galactose [4], a property not shared with the homologous macrophage asialoglycoprotein receptor [12]. Each chain of the receptor binds galactose rather weakly but high affinity ligands have at least three terminal galactose residues that bind to distinct sites [12].

The rabbit receptor has been shown to agglutinate human (weakly A-specific) and lapine erythrocytes [11] and was mitogenic towards neuraminidase-treated human T lymphocytes, an activity greatly enhanced by the removal of accessory cells [13].

TISSUE AND/OR SUBCELLULAR DISTRIBUTION

The liver receptors are found in hepatocytes; cell surface receptors are constitutively recycled between the plasma membrane and endosomes via clustering in clathrin-coated pits. Most is present intracellularly, and ligand engagement at the cell surface causes downregulation of surface binding sites and doubles the rate of endocytosis [14].

The macrophage receptor is found on both tumoricidal and inflammatory macrophages [15].

POSSIBLE FUNCTION(S)

The function of these receptors is uncertain. It has long been assumed that they must function to remove modified (desialylated) plasma glycoproteins and cells (aged, damaged or apoptotic), but gene disruption data have not been outstandingly informative. Although knock-out mice are unable to clear circulating asialo-orosomucoid and other modified glycoproteins, these animals do not accumulate desialylated glycoproteins or lipoproteins in their circulations and appear normal [16,17]. There is some indirect evidence, however, that galactose-terminating minor plasma glycoproteins may accumulate in these animals [17].

COMMENTS

The hepatic asialoglycoprotein receptor provides a potential therapeutic application in the form of lectin-mediated drug targeting to hepatocytes. For example, asialo-IFNβ was found to have a greater anti-viral effect than the unmodified cytokine and reduced viraemia *in vivo* in hepatitis B virus-transfected nude mice [18].

Other members of this family include the Kupffer cell receptor (qv), the placental Gp120 receptor (qv) and CD23 (qv), the low affinity IgE receptor.

REFERENCES

1. Hudgin, R.L., Pricer, W.E. & Ashwell, G. (1974). The isolation and properties of a rabbit liver binding protein specific for asialoglycoproteins. *Journal of Biological Chemistry* **249**, 5536–5543.
2. Tanabe, T., Pricer, W.E. & Ashwell, G. (1979). Subcellular membrane topology and turnover of a rat hepatic binding protein specific for asialoglycoproteins. *Journal of Biological Chemistry* **254**, 1038–1043.
3. Schwarz, A.L., Marshak-Rothstein, A., Rup, D. & Lodish, H.F. (1981). Identification and quantification of the rat hepatocyte asialoglycoprotein receptor. *Proceedings of the National Academy of Sciences USA* **78**, 3348–3352.
4. Baenziger, J.U. & Maynard, Y. (1980). Human hepatic lectin. *Journal of Biological Chemistry* **255**, 4607–4613.
5. Spiess, M., Schwartz, A.L. & Lodish, H.F. (1979). Sequence of human asialoglycoprotein receptor cDNA. *Journal of Biological Chemistry* **260**, 1979–1982.
6. Spiess, M. & Lodish, H.F. (1985). Sequence of a second human asialoglycoprotein receptor: conservation of two receptor genes during evolution. *Proceedings of the National Academy of Sciences USA* **82**, 6465–6469.
7. Drickamer, K., Mamon, J.F., Binns, G. & Leung, J.O. (1984). Primary structure of the rat liver asialoglycoprotein receptor. *Journal of Biological Chemistry* **259**, 770–778.
8. Sato, M., Kawakomi, K., Osawa, T. & Toyoshima, S. (1992). Molecular cloning and expression of cDNA encoding a galactose/*N*-acetylgalactosamine-specific lectin on mouse tumoricidal macrophages. *Journal of Biochemistry* **111**, 331–336.
9. Lodish, H.F. (1991). Recognition of complex oligosaccharides by the multi-subunit asialoglycoprotein receptor. *Trends in Biochemical Science* **16**, 374–377.
10. Drickamer, K. (1981). Complete amino acid sequence of a membrane receptor for glycoproteins. *Journal of Biological Chemistry* **256**, 5827–5839.
11. Stockert, R.J., Morell, A.G. & Scheinberg, I.H. (1974). Mammalian hepatic lectin. *Science* **186**, 365–366.
12. Kolatkar, A.R., Leung, A.K., Isecke, R., Brossmer, R., Drickamer, K. & Weis, W.I. (1998). Mechanism of *N*-acetylgalactosamine binding to a C-type animal lectin carbohydrate-recognition domain. *Journal of Biological Chemistry* **273**, 19502–19508.
13. Novogrodsky, A. & Ashwell, G. (1977). Lymphocyte mitogenesis induced by a mammalian liver protein that specifically binds desialylated glycoproteins. *Proceedings of the National Academy of Sciences USA* **74**, 676–678.
14. Bider, M.D. & Spiess, M. (1998). Ligand-induced endocytosis of the asialoglycoprotein receptor: evidence for heterogeneity in subunit organization. *FEBS Letters* **434**, 37–41.
15. Kawakami, K., Yamamoto, K., Toyoshima, S., Osawa, T. & Tatsuro, I. (1994). Dual function of a macrophage galactose/*N*-acetylgalactosamine-specific lectins: glycoprotein uptake and tumoricidal cellular recognition. *Japanese Journal of Cancer Research* **85**, 744–749.
16. Ishibashi, S., Hammer, R.E. & Herz, J. (1994). Asialoglycoprotein receptor deficiency in mice lacking the minor receptor subunit. *Journal of Biological Chemistry* **269**, 27803–27806.
17. Braun, J.R., Willnow, T.E., Ishibashi, S., Ashwell, G. & Herz, J. (1996). The major subunit of the asialoglycoprotein receptor is expressed on the hepatocellular surface in mice lacking the minor receptor subunit. *Journal of Biological Chemistry* **271**, 21160–21166.
18. Eto, T. & Takahashi, H. (1999). Enhanced inhibition of hepatitis B virus production by asialoglycoprotein receptor-directed interferon. *Nature Medicine* **5**, 577–581.

Axinella polypoides agglutinins I and II

Other names: AP-I and AP-II
Mol. wt: 32 000 (native); 16 000 (subunit)
Specificity: D-Galactose
Source: The marine sponge *Axinella polypoides*

ISOLATION

The soluble fraction of a saline extract of *A. polypoides* homogenate was applied to Sepharose 4B and subsequently eluted with D-galactose to yield a mixture of agglutinins. AP-I was subsequently separated by application to DEAE-cellulose and elution at pH 3.5, while AP-II was obtained by preparative polyacrylamide gel electrophoresis [1].

STRUCTURE

Both agglutinins are proteins with no attached carbohydrate, have broadly similar amino acid compositions, and have isoelectric points around pH 3.9. However, the agglutinins are antigenically distinct; moreover AP-II is rich in tyrosine and contains some histidine, while these amino acids are absent from AP-I.

Both agglutinins migrate as single bands of apparent Mr 15 000 in SDS-PAGE under reducing conditions, although a slower band (Mr 21 000) is apparent with AP-I in the absence of reducing agent. In the analytical ultracentrifuge, AP-I and AP-II sedimented as single peaks with $S_{20,W}$ of 2.6 and 2.8, respectively [1].

AP-I is a homodimer of Mr 34 000 that has been fully sequenced [2]. Each monomer consists of 144 amino acids with a calculated Mr of 15 847 and one intrachain disulphide bridge.

AP-II has also been fully sequenced; it has 147 amino acids (Mr 16 228) and 65% homology with AP-I [3]. Like AP-I, AP-II possesses a disulphide loop between positions 4 and 46. This loop is an immunologically dominant epitope in both lectins, and its integrity is critical for haemagglutinating activity [3]. Neither boiling in SDS nor incubation with 2-mercaptoethanol alone significantly affects carbohydrate-binding ability, but the combination destroyed lectin activity completely. It seems that provided the ring or loop between residues 4 and 46 remains intact (even if the ring is opened), lectin activity is maintained [3].

BIOLOGICAL ACTIVITIES

Both AP-I and AP-II agglutinated human erythrocytes (ABO non-specifically), although AP-I was approximately 10 times more potent gravimetrically [1]. Cell agglutination and precipitation of glycoconjugates were most effectively inhibited by terminal non-reducing D-galactose glycosidically linked β1-6, or by *p*-nitrophenyl-β-D-galactose. Free D-galactose and D-fucose were the best monosaccharide inhibitors. The saccharide inhibition profiles for AP-I and AP-II were very similar but some differences were observed especially with disaccharides. The lectin activity of AP-I was calcium-independent [1].

AP-I readily agglutinates human peripheral blood mononuclear cells and is strongly mitogenic towards them, especially T cells; AP-II is a weak leukagglutinin and very weakly mitogenic [4]. AP-I was also anti-mitogenic when co-cultured with con A or PHA [4].

TISSUE AND/OR SUBCELLULAR DISTRIBUTION

Both AP-I and AP-II are stored in vesicles of the spherulous cells; they do not occur at the surface of any cell type [5].

POSSIBLE FUNCTION(S)

It has been suggested that these lectins may be involved in the formation of sponge fibres [2], but no direct evidence is available.

COMMENTS

Axinella polypoides contains at least five different lectins, four of them D-galactose-specific, the other specific for hexuronic acids [2,3]. None of them seems to be closely related to other known sponge lectins, nor do they seem to belong to any other known lectin families [2,3].

REFERENCES

1. Bretting, H. & Kabat, E.A. (1976). Purification and characterization of the agglutinins from the sponge *Axinella polypoides* and a study of their combining sites. *Biochemistry* **15**, 3228–3236.
2. Buck, F., Luth, C., Strupat, K. & Bretting, H. (1992). Comparative investigations on the amino acid sequences of different isolectins from the sponge *Axinella polypoides* (Schmidt). *Biochimica et Biophysica Acta* **1159**, 1–8.
3. Buck, F., Schulze, C., Breloer, M., Strupat, K. & Bretting, H. (1998). Amino acid sequence of the D-galactose binding lectin II from the sponge *Axinella polypoides* (Schmidt) and identification of the carbohydrate binding site in lectin II and related lectin I. *Comparative Biochemistry and Physiology B* **121**, 153–160.
4. Phillips, S.G., Bretting, H. & Kabat, E.A. (1976). A galactose-inhibitable mitogen for human lymphocytes from the sponge, *Axinella polypoides. Journal of Immunology* **117**, 1226–1232.
5. Bretting, H. & Königsmann, K. (1979). Investigations on the lectin-producing cells in the sponge *Axinella polypoides* (Schmidt). *Cell and Tissue Research* **201**, 487–497.

Biomphalaria glabrata lectin-I

Mol. wt: 67 000 (native); 17 000 and 15 000 (subunit)
Specificity: Sialic acid, mannose, GlcNAc
Source: The spawn of the water snail *Biomphalaria glabrata*

ISOLATION

B. glabrata lectin was isolated from saline extracts of spawn by affinity chromatography on hog gastric mucin–Sepharose (elution with GlcNAc), followed by gel filtration. The purified lectin appeared homogeneous by electrophoresis, immunoelectrophoresis and isoelectric focusing [1].

STRUCTURE

B. glabrata lectin-I is a glycoprotein (8.6% carbohydrate) with an amino acid composition not dissimilar to that of the other snail lectins, including 14 half-cysteine residues per molecule. SDS-PAGE, with or without mercaptoethanol, yielded two bands of apparent molecular weights 17 000 and 15 000 with the approximate relative intensity of 3 : 1. It is suggested, therefore, that the native lectin consists of the three 17 000 subunits non-covalently attached to one 15 000 subunit. The isoelectric point is 7.6 [1].

BIOLOGICAL ACTIVITIES

B. glabrata lectin-I agglutinates human erythrocytes, but not those of various other species tested, including rabbit, mouse and pigeon. Among human cells, a preference for blood group A_1 is apparent, although blood group A_2 and blood group B cells can be agglutinated at minimum concentrations approximately 10-fold higher; human blood group O cells are not agglutinated. The lectin can also precipitate hog and human blood group A substances from solution. The best inhibitors of the precipitation reactions were sialic acids, although some measure of inhibition was noted with a wide variety of monosaccharides [1].

TISSUE AND/OR SUBCELLULAR DISTRIBUTION

The lectin was isolated from the eggs of the snail; it is synthesised in cells of the albumin gland [1].

POSSIBLE FUNCTION(S)

Its function is not known.

COMMENTS

A comparison with other invertebrate sialic acid-binding lectins is presented in Table 2.5.

REFERENCE

1. Bretting, H., Stanislawski, E., Jacobs, G. & Becker, W. (1983). Isolation and characterisation of a lectin from the snail *Biomphalaria glabrata* and a study of its combining site. *Biochimica et Biophysica Acta* **749**, 143–152

Blaberus discoidalis β-1,3-glucan-specific lectin

Other name: GSL
Mol. wt: 520 000 (native); 82 000 (subunit)
Specificity: 1,3-Glucans
Source: Haemolymph of the cockroach *Blaberus discoidalis*

ISOLATION

The lectin was purified to homogeneity by low pH elution from a column of laminarin–Sepharose after a partial separation from other serum proteins by gel filtration and passage over Blue Sepharose CL-6B [1].

The lectin can also be isolated by immuno-affinity chromatography [2].

STRUCTURE

GSL is a glycoprotein with an aspartic acid-rich composition broadly similar to that of other β-1,3-glucan-specific insect lectins. The first 28 N-terminal residues were sequenced and found to possess some homology with the *Limulus polyphemus* C-reactive protein [1].

SDS-PAGE analysis indicated subunit apparent Mr values of 80 000 and 82 000 under non-reducing and reducing conditions, respectively. The native Mr estimated by gel filtration was 520 000, perhaps most compatible with six subunits. The isoelectric point was found to be 4.9 [1].

By electron microscopy and negative staining, the molecules of GSL appeared as large aggregates of Y-shaped monomers. These aggregates can be dissociated into monomers and dimers by incubation with laminarin [1].

BIOLOGICAL ACTIVITIES

GSL agglutinates yeast (*Saccharomyces cerevisiae*) cells and murine erythrocytes in the presence of calcium, but not red blood cells of human (group AB), ovine, equine, bovine, chicken, guinea pig, caprine or lapine origin [1]. Agglutination of yeast cells was potently inhibited by laminarin and bovine mucin, but not by fetuin or any of the monosaccharides commonly found in glycoconjugates [1].

GSL enhanced prophenoloxidase activity of haemocyte lysate and this activity was increased in the presence of laminarin [1].

GSL can opsonise the phagocytosis of yeast by *B. discoidalis* haemocyte monolayers; this activity could be inhibited by laminarin [3]. Phenylthiourea, an inhibitor of the prophenoloxidase system, also inhibited GSL-mediated phagocytosis [3].

TISSUE AND/OR SUBCELLULAR DISTRIBUTION

The lectin has been detected in haemocytes as well as free in haemolymph plasma [1]. GSL can be detected on the haemocyte surface [2].

POSSIBLE FUNCTION(S)

GSL is thought to function in defence as part of a team of opsonic, prophenoloxidase-enhancing lectins [3].

COMMENTS

Very similar lectins can be detected in four other species of cockroach (*Blaberus crancifer*, *Leucophaea maderae*, *Gromphadorhina portentosa* and *Periplaneta americana*) by cross-reactivity with rabbit antibodies to GSL, and three of them have been isolated by immunoaffinity chromatography with the same antibody. Each has a native molecular weight of >500 000, a subunit Mr of 80 000, is localised in haemocytes (cytoplasm and cell surface), can enhance the laminarin-mediated activation of prophenoloxidase, and has a characteristic Y-shape under the electron microscope [2]. However, GSL is antigenically unrelated to either the β-1,3-glucan-specific lectin from *Galleria mellonella* (qv) or the galactoside-binding *Blaberus discoidalis* lectins-1 and -2 (qv) from its own species of origin [1,2].

GSL is very similar to arylphorin, a member of the arthropod hexamerin family [1].

REFERENCES

1. Chen, C., Rowley, A.F., Newton, R.P. & Ratcliffe, N.A. (1999). Identification, purification and properties of a β1,3-glucan specific lectin from the serum of the cockroach *Blaberus discoidalis* which is implicated in immune defence reactions. *Comparative Biochemistry and Physiology B* **122**, 309–319.
2. Chen, C., Rowley, A.F. & Ratcliffe, N.A. (1998). Detection, purification by immunoaffinity chromatography and properties of β-1,3-glucan specific lectins from the sera of several insect species. *Insect Biochemistry and Molecular Biology* **28**, 721–731.
3. Wilson, R., Chen, C. & Radcliffe, N.A. (1999). Innate immunity in insects: the role of multiple, endogenous serum lectins in the recognition of foreign invaders in the cockroach, *Blaberis discoidalis*. *Journal of Immunology* **162**, 1590–1596.

Blaberus discoidalis lectin-1

Other name: BDL-1
Mol. wt: 390 000 (native); 36 000 (subunit)
Specificity: Mannose
Source: Haemolymph of the cockroach *Blaberus discoidalis*

ISOLATION

The starting material was a 30–50% ammonium sulphate cut of haemolymph serum. The key step was passage over a DEAE-cellulose column, to which BDL-1 did not bind but other lectins did. Finally, the BDL-1 was isolated in homogeneous form by affinity chromatography on AT–Sepharose 4B, eluting with glucose [1].

STRUCTURE

BDL-1 is a glycoprotein with an apparent molecular weight of 390 000 estimated by gel filtration [1]. SDS-PAGE gave Mr values of 158 000 and 36 000 in the absence and presence, respectively, of 2-mercaptoethanol [1].

The isoelectric point is 6.0. BDL-1 is antigenically distinct from *Blaberus discoidalis* lectin-2 (qv) [1].

BIOLOGICAL ACTIVITIES

BDL-1 agglutinates lapine erythrocytes at high titre and is also active towards human, bovine and chicken, but not ovine, red blood cells. Haemagglutination is calcium-dependent, although magnesium can substitute for calcium to some extent [1].

Haemagglutination is most readily inhibited by mannose. D-Fucose and GalNAc are also inhibitory, as are galactose and glucose to a lesser extent [1]. GlcNAc had no inhibitory effect [1].

BDL-1 can activate the prophenoloxidase system and can enhance activation mediated by laminarin; the effects of BDL-1 and laminarin were roughly additive. This activity could not be inhibited by sugars that inhibit haemagglutination [2].

BDL-1 can opsonise baker's yeast and *Escherichia coli* (but not *Bacillus cereus*) phagocytosis by haemocytes; this activity could be inhibited by mannose [3].

TISSUE AND/OR SUBCELLULAR DISTRIBUTION

BDL-1 is found in haemocytes by immunofluorescence, where the strongest staining is found in the granules [1]. It seems likely therefore that BDL-1 is synthesised in haemocytes and released into haemolymph.

POSSIBLE FUNCTION(S)

BDL-1 probably functions in defence (see **Comments** section)

COMMENTS

The haemolymph of *B. discoidalis* contains multiple, structurally unrelated lectins, including BDL-2 (qv), BDL-3 [1] and *Blaberus discoidalis* β-1,3-glucan-specific lectin (qv) [4]. BDL-3 may not have been isolated in homogeneous form, but it is a calcium-dependent galactose/GalNAc-specific opsonic lectin [1,3]. As a group, the different saccharide specificities of these lectins constitute an effective innate immune system with the collective ability to enhance dramatically the phagocytosis of a wide variety of different micro-organisms [3].

REFERENCES

1. Chen, C., Ratcliffe, N.A. & Rowley, A.F. (1993). Detection, isolation and characterisation of multiple lectins from the haemolymph of the cockroach *Blaberus discoidalis. Biochemical Journal* **294**, 181–190.
2. Chen, C., Durrant, H.J., Newton, R.P. & Ratcliffe, N.A. (1995). A study of novel lectins and their involvement in the activation of the prophenoloxidase system in *Blaberus discoidalis. Biochemical Journal* **310**, 23–31.
3. Wilson, R., Chen, C. & Ratcliffe, N.A. (1999). Innate immunity in insects: the role of multiple, endogenous serum lectins in the recognition of foreign invaders in the cockroach, *Blaberus discoidalis. Journal of Immunology* **162**, 1590–1596.
4. Chen, C., Rowley, A.F. & Ratcliffe, N.A. (1998). Detection, purification by immunoaffinity chromatography and properties of β-1,3-glucan specific lectins from the sera of several insect species. *Insect Biochemistry and Molecular Biology* **28**, 721–731.

Blaberus discoidalis lectin-2

Other names: BDL-2
Mol. wt: 140 000 (native); 23 000 (subunit)
Specificity: GalNAc, D-fucose, D-galactose
Source: Haemolymph of the cockroach *Blaberus discoidalis*

ISOLATION

The starting material was a 30–50% ammonium sulphate cut of haemolymph serum. Separation from other lectins was then achieved by application to a DEAE-cellulose column and elution with an NaCl gradient. Finally, a homogeneous lectin was obtained by affinity chromatography on GalNAc–Sepharose 6B, eluting with lactose [1].

STRUCTURE

BDL-2 is a glycoprotein with an apparent molecular weight of 140 000 estimated by gel filtration. SDS-PAGE gave Mr values of 67 000 and 23 000 in the absence and presence, respectively, of 2-mercaptoethanol [1].

The isoelectric point is 5.2. BDL-2 is antigenically distinct from *Blaberus discoidalis* lectin-1 (qv) [1].

BIOLOGICAL ACTIVITIES

BDL-2 agglutinated murine erythrocytes at high titre and, much less readily, human, equine and lapine cells. Ovine, bovine and chicken cells were not agglutinated. Haemagglutination is Ca^{2+}-dependent; Mg^{2+} cannot substitute to any extent [1].

Haemagglutination was most readily inhibited by GalNAc, D-fucose and D-galactose (in that order); GlcNAc was much less effective, and mannose was not inhibitory at all [1]. BDL-2 can activate the prophenoloxidase system and can enhance activation mediated by laminarin; the effects of BDL-2 and laminarin were roughly additive. This activity could not be inhibited by sugars effective at inhibiting haemagglutination [2].

BDL-2 opsonises *Bacillus cereus* for phagocytosis by haemocytes. It also possesses weak opsonic activity towards *Escherichia coli*, but none towards yeast. Opsonic activity was inhibited by GalNAc [3].

TISSUE AND/OR SUBCELLULAR DISTRIBUTION

BDL-2 is found in haemolymph and in haemocytes, where immunofluorescent staining is strongest in the granules [1]. Presumably BDL-2 is synthesised by haemocytes and released into the haemolymph plasma.

POSSIBLE FUNCTION(S)

BDL-2 probably functions in defence (see **Comments** section).

COMMENTS

BDL-2 is one of several (at least four) separate lectins found in *B. discoidalis* haemolymph. Together they appear to form an effective defence mechanism, collectively recognising and opsonising a wide variety of different micro-organisms [3].

REFERENCES

1. Chen, C., Ratcliffe, N.A. & Rowley, A.F. (1993). Detection, isolation and characterisation of multiple lectins from the haemolymph of the cockroach *Blaberus discoidalis. Biochemical Journal* **294**, 181–190.
2. Chen, C., Durrant, H.J., Newton, R.P. & Ratcliffe, N.A. (1995). A study of novel lectins and their involvement in the activation of the prophenoloxidase system in *Blaberus discoidalis. Biochemical Journal* **310**, 23–31.
3. Wilson, R., Chen, C. & Ratcliffe, N.A. (1999). Innate immunity in insects: the role of multiple, endogenous serum lectins in the recognition of foreign invaders in the cockroach, *Blaberus discoidalis. Journal of Immunology* **162**, 1590–1596.

Bombyx mori β-1,3-glucan-binding protein

Other names: Silkworm β-1,3-glucan receptor protein
Mol. wt: 62 000
Specificity: β-1,3-Glucans
Source: Haemolymph of the silkworm *Bombyx mori*

ISOLATION

Fifth-instar larvae of *B. mori* were bled and the haemolymph obtained added to saturated ammonium sulphate solution to provide the starting material. The lectin was then purified by further ammonium sulphate fractionation, DEAE–Toyopearl chromatography, heparin–Affigel affinity chromatography, FPLC using a Mono Q column, and finally FPLC with a Superose 12 column [1].

The lectin cDNA has been cloned from a haemocyte cDNA library [2].

STRUCTURE

This lectin is a protein (probably without attached carbohydrate) with an amino acid composition rich in glycine and proline as well as glutamic and aspartic acids. Essentially the same molecular weight estimated by sedimentation equilibrium (61 000) and SDS-PAGE under reducing conditions (62 000) indicates that it exists as a single polypeptide monomer. The isoelectric point was 4.3 [1].

The deduced primary sequence has 479 amino acid residues and a calculated Mr of 54 000. The carbohydrate-binding domain appears to be located in the N-terminal region [2].

BIOLOGICAL ACTIVITIES

The lectin binds β-1,3-glucan in the form of curdlan-type polysaccharide or zymosan in a calcium-independent manner and thereby triggers the prophenoloxidase activating system. It does not recognise dextrans, cellulose or peptidoglycan [1,3].

TISSUE AND/OR SUBCELLULAR DISTRIBUTION

This lectin is found in haemolymph; however, the corresponding mRNA was found to be expressed in haemocytes, epithelial cells and the fat body [2].

POSSIBLE FUNCTION(S)

A role in defence is assumed, since the definitive activity of this lectin is to activate the prophenoloxidase system [1,3]. Synthesis is upregulated following challenge with bacteria or yeasts [2].

COMMENTS

An LPS-binding haemolymph protein has been reported in the same species with an amino acid sequence indicative of the C-type lectin family [4]. However, as yet there is no evidence that this molecule is a true lectin. In contrast, the β-1,3-glucan-binding protein described here clearly recognises specific carbohydrate structures and is functionally related to the β-1,3-binding lectins from *Galleria mellonella* (qv), *Pacifasticus leniusculus* (qv) and *Blaberus* species (qv).

A developmentally regulated, heparin/glucuronic acid-specific agglutinin has also been found in the haemolymph of *B. mori* [5].

REFERENCES

1. Ochiai, M. & Ashida, M. (1988). Purification of a β-1,3-glucan recognition protein in the prophenoloxidase activating system from haemolymph of the silkworm, *Bombyx mori*. *Journal of Biological Chemistry* **263**, 12056–12062.
2. Ochiai, M. & Ashida, M. (2000). A pattern-recognition protein for β-1,3-glucan. *Journal of Biological Chemistry* **275**, 4995–5002.
3. Yoshida, H., Ochiai, M. & Ashida, M. (1986). β-1,3-glucan receptor and peptidoglycan receptor are present as separate entities within insect prophenoloxidase activating system. *Biochemical and Biophysical Research Communications* **141**, 1177–1184.
4. Koizumi, N., Imamura, M., Kadotani, T., Yaoi, K., Iwahana, H. & Sato, R. (1999). The lipopolysaccharide-binding protein participating in haemocyte nodule formation in the silkworm *Bombyx mori* is a novel member of the C-type lectin superfamily with two different tandem carbohydrate-recognition domains. *FEBS Letters* **443**, 139–143.
5. Suzuki, T. & Natori, S. (1983). Identification of a protein having hemagglutinating activity in the hemolymph of the silkworm, *Bombyx mori*. *Journal of Biochemistry* **93**, 583–590.

Botryllus schlosseri galectin-like lectin

Mol. wt: 67 000 (native); 17 000 and 19 000 (subunit)
Specificity: Lactose
Source: Haemolymph of the colonial ascidian *Botryllus schlosseri*

ISOLATION

Soluble proteins of haemocyte lysate of *B. schlosseri* were subjected to affinity chromatography on acid-treated Sepharose; a single protein peak was obtained after elution with D-galactose [1].

STRUCTURE

This lectin is a polypeptide (no attached carbohydrate detectable) with a relative abundance of hydrophilic amino acids. Two subunits where detected by SDS-PAGE; the apparent molecular weights were 17 000 and 15 000 under reducing conditions, but 15 000 and 16 000 under non-reducing conditions. This behaviour is typical of disulphide bridge formation between the thiols of cysteine, which is an abundant component of the amino acid composition [1]. The molecular size was estimated at 67 000 by gel filtration [2], perhaps indicating a tetramer with two of each type of subunit present.

BIOLOGICAL ACTIVITIES

The lectin agglutinates lapine erythrocytes in a divalent cation-independent manner, and the titre is increased eight-fold if the erythrocytes have been treated with trypsin. Weak haemagglutinating activity is also apparent against ovine, bovine and human cells. Haemagglutination can be inhibited by D-galactose, but not by other monosaccharides. Lactose and other β-galactosides are also inhibitory [1].

The lectin also agglutinates baker's yeast cells, and promotes the phagocytosis of the yeast by haemocytes. This opsonic activity is also inhibited by galactose [1].

TISSUE AND/OR SUBCELLULAR DISTRIBUTION

This lectin is found both in blood plasma and in haemocyte lysate, but no other location has been reported [1].

POSSIBLE FUNCTION(S)

It has been suggested that this lectin may function as an opsonin, but no direct supporting evidence is available [1].

COMMENTS

This lectin is one of several known to occur in *B. schlosseri* [1–4]. It has been suggested that it may belong to the galectin family on the basis of its specificity, lack of requirement for calcium and blocked N-terminus [1], but confirmation will require primary sequence determination.

REFERENCES

1. Ballarin, L., Tonello, C., Guidolin, L. & Sabbadin, A. (1999). Purification and characterization of a humoral opsonin, with a specificity for D-galactose, in the colonial ascidian *Botryllus schlosseri*. *Comparative Biochemistry and Physiology B* **123**, 115–123.
2. Coombe, D.R., Ey, P.L. & Jenkin, C.R. (1984). Ascidian haemagglutinins: incidence in various species, binding specificities and preliminary characterization of selected agglutinins. *Comparative Biochemistry and Physiology B* **77**, 811–819.
3. Vasta, G.R., Warr, G.W. & Marchalonis, J.J. (1982). Tunicate lectins: distribution and specificity. *Comparative Biochemistry and Physiology B* **73**, 887–900.
4. Pancer, Z., Diehl-Seifert, B., Rinkevich, B. & Muller, W.E.G. (1997). A novel tunicate (*Botryllus schlosseri*) putative C-type lectin features an immunoglobulin domain. *DNA and Cell Biology* **16**, 801–806.

Brevican

Mol. wt: 100 000 (core protein)
Specificity: Sulphated glycolipids
Source: Mammalian brain

ISOLATION

Brevican was cloned from a bovine brain cDNA library as a result of a strategy based on polyclonal antisera to multiple proteoglycan core proteins [1].

STRUCTURE

The full-length cDNA sequence predicts a protein of 912 amino acid residues with a calculated molecular mass of 99 510 for a 145 000 core protein [1]. Full-length brevican has the typical hyalectan structural arrangement (Chapter 3), with a homologous N-terminal arrangement with an immunoglobulin domain and two link protein modules. The C-terminal segment is also homologous to the other hyalectans, but differs in having only one epidermal growth factor-like repeat sequence. The middle (glycosaminoglycan-attachment) region of the molecule shows little sequence homology to the other hyalectans and differs from the others in being much shorter and in having an unusually high glutamic acid content [1,2]. Moreover, a significant quantity of brevican is present in the brain without any glycosaminoglycan chains [1]. Brevican exists either as a full-length proteoglycan or as a proteolytically truncated form consisting only of the selectin-like C-terminal region [1]. Conversely, a membrane-bound homologue exists without the C-terminal region, but this form can hardly be regarded as a lectin [2].

BIOLOGICAL ACTIVITIES

Brevican, like other hyalectans, binds tenascin-R, but does so with a very much higher affinity than the others [3]. Although calcium-dependent, brevican–tenascin-R interaction is protein–protein mediated via the fibronectin type III domain of tenascin-R [3].

However, brevican can bind to sulphated glycolipids in a calcium-dependent manner and can promote cell adhesion by interacting with cell-surface glycolipids [4].

Like other hyalectans, brevican can bind hyaluronic acid via its N-terminal region [1,2,5].

TISSUE AND/OR SUBCELLULAR DISTRIBUTION

Brevican mRNA was detected in the brain by Northern blotting, but not in any of several other tissues. Within the brain, brevican is expressed in cerebellar astrocytes but not in neurons [1].

POSSIBLE FUNCTION(S)

It has been suggested that brevican in extracellular matrices might regulate adhesion and migration of cells [4,5].

COMMENTS

Brevican appears to be the most abundant hyalectan in brain [2].

REFERENCES

1. Yamada, H., Watanabe, K., Shimonika, M. & Yamaguchi, Y. (1994). Molecular cloning of brevican, a novel brain proteoglycan of the aggrecan/versican family. *Journal of Biological Chemistry* **269**, 10119–10126.
2. Iozzo, R.V. (1998). Matrix proteoglycans: from molecular design to cellular function. *Annual Review of Biochemistry* **67**, 609–652.
3. Aspberg, A., Miura, R., Bourdoulous, S., Shimonaka, M., Heinegard, D., Schachner, M., Ruoslahti, E. & Yamaguchi, Y. (1997). The C-type lectin domains of lecticans, a family of aggregating chrondroitin sulphate proteoglycans, bind tenascin-R by protein–protein interactions independent of carbohydrate moiety. *Proceedings of the National Academy of Sciences USA* **94**, 10116–10121.
4. Miura, R., Aspberg, A., Ethell, I.M., Hagihara, K., Schnaar, R.L., Ruoslahti, E. & Yamaguchi, Y. (1999). The proteoglycan lectin domain binds sulphated cell surface glycolipids and promotes cell adhesion. *Journal of Biological Chemistry* **274**, 11431–11438.
5. Ruoslahti, E. (1996). Brain extracellular matrix. *Glycobiology* **6**, 489–492.

Caenorhabditis elegans galectin

Other name: Nematode 32 K galectin
Mol. wt: 32 000
Specificity: Lactose
Source: The nematode worm *Caenorhabditis elegans*

ISOLATION

Sonicated worms were extracted with lactose in a buffer containing 2-mercaptoethanol. After affinity chromatography on asialofetuin–Sepharose, the major lectin was separated from a second lectin by reversed-phase chromatography on a Tosoh phenyl-5PW-RP column [1].

A full-length cDNA was subsequently cloned from a *C. elegans* cDNA library using specific antiserum to the purified galectin [2]. A recombinant fusion protein was expressed in *Escherichia coli* [2].

STRUCTURE

C. elegans 32 K galectin is a non-glycosylated polypeptide monomer with an apparent molecular weight on gel filtration of about 30 000 [1,2]. The subunit size estimated by SDS-PAGE (32 000) is in excellent agreement with a calculated Mr of 31 850 [2]. The N-terminus is blocked, and the sequence of 279 amino acids, which lacks cysteine [1,2], exhibited ⩽30% identity to vertebrate galectins and consisted of two homologous (32% identity) domains arranged in tandem [2]. Recombinant forms of each domain have been expressed in *E. coli*: they appear to have somewhat different carbohydrate-binding properties [3].

BIOLOGICAL ACTIVITIES

The nematode galectin agglutinates trypsin-treated rabbit erythrocytes (rather weakly) and binds to asialofetuin–Sepharose; both actions are inhibited by lactose [1].

The affinity of the intact molecule for asialofetuin was much greater than that of either domain alone; moreover, the C-terminal domain had a stronger affinity than its N-terminal homologue [3].

TISSUE AND/OR SUBCELLULAR DISTRIBUTION

This galectin is most abundantly expressed in the cuticle [4].

POSSIBLE FUNCTION(S)

No function is known, but it has been suggested that the lectin may play a role in cuticle formation [2,4].

COMMENTS

The *C. elegans* 32 K galectin was the first non-vertebrate galectin to be recognised as such, and the first example of a tandem-repeat-type structure to be discovered [2]. It is the best characterised of several galectins now known to occur in *C. elegans* [4].

REFERENCES

1. Hirabayashi, J., Sahot, M., Ohyama, Y. & Kasai, K. (1992). Purification and characterisation of β-galactoside-binding proteins from *Caenorhabditis elegans*. *Journal of Biochemistry* **111**, 553–555.
2. Hirabayashi, J., Sohot, M. & Kasai, K. (1992). Evidence that *Caenorhabditis elegans* 32 KDa β-galactoside-binding protein is homologous to vertebrate β-galactoside binding lectins. *Journal of Biological Chemistry* **267**, 15485–15490.
3. Atara, Y., Hirabayashi, J. & Kasai, K. (1997). The two lectin domains of the tandem-repeat 32 KDa galectin of the nematode *Caenorhabditis elegans* have different binding properties. Studies with recombinant protein. *Journal of Biochemistry* **121**, 1002–1009.
4. Hirabayashi, J., Atara, Y. & Kasai, K. (1997). Galectins from the nematode *Caenorhabditis elegans* and the genome project. *Trends in Glycoscience and Glycotechnology* **9**, 113–122.

Calreticulin (and calnexin)

Other names: Calregulin; CRP55; CaBP3 etc.
Mol. wt: 46 000
Specificity: Monoglucosylated oligomannose
Sources: Endoplasmic reticulum of mammalian cells etc.

ISOLATION

During the 1980s a number of Ca^{2+}-binding proteins were purified by various means from different tissues and given a variety of names. Only after molecular cloning and N-terminal sequence analyses was it apparent that they were all the same highly conserved protein now universally known as calreticulin (reviewed by Michalak *et al.* [1]).

A membrane-bound homologue, calnexin (or pp90), was purified and cloned from canine pancreatic endoplasmic reticulum [2].

STRUCTURE

Calreticulin is a protein or glycoprotein with separate Ca^{2+}- and Zn^{2+}-binding sites. Its molecular size determined by SDS-PAGE varies from 55 000 to 63 000 depending on conditions, and the lower value is similar to that determined by sedimentation equilibrium. However, the Mr deduced from the primary sequence is only about 46 000. Similarly, the calculated pI (4.1) is slightly different from the measured value (4.7) [1].

The N-terminal globular domain is followed by a proline-rich region and an acidic C-terminal region rich in aspartic and glutamic acids. The full-length primary sequence, which contains a nuclear localisation sequence and an endoplasmic reticulum retention sequence, has been deduced from cDNAs from numerous species, and a high degree of sequence conservation throughout the animal kingdom is evident [1].

Calnexin is a non-glycosylated type I membrane protein of 573 amino acids and a calculated Mr of 65 000. Its primary sequence differs most obviously from calreticulin in having a hydrophobic transmembrane domain, which partly accounts for its larger molecular size. Otherwise, the two proteins are highly homologous and together constitute a separate family of animal lectins [2–4].

BIOLOGICAL ACTIVITIES

Both calreticulin and calnexin bind calcium, and calreticulin binds zinc with an associated dramatic change in conformation [1]. Both high affinity/low capacity (proline-rich domain) and low affinity/high capacity (C-domain) binding sites for calcium appear to be present in calreticulin [1].

Both calreticulin and calnexin bind transiently to a wide range of immature and folding forms of transmembrane and secretory glycoproteins [4]. This binding is largely of a lectin–carbohydrate nature, and depends on the recognition of *N*-glycans containing monoglycosylated oligomannosides [5–7]. Notably, the α1,6-mannose branch point of the oligosaccharide core appears to be essential [7]. Although calreticulin and calnexin have identical lectin specificities, endogenous substrate binding is not identical [8]. It is uncertain whether lectin–carbohydrate binding is followed by more stable protein–protein interactions.

Calreticulin binds to integrins via protein–protein interactions [9]. Cell-surface calreticulin promotes the spreading of mouse melanoma cells *in vitro*; this process requires cooperation between calreticulin and integrins [5].

Recombinant human calreticulin inhibits the proliferation of bovine endothelial cells *in vitro* and suppresses angiogenesis and tumour growth *in vivo* [10].

TISSUE AND/OR SUBCELLULAR DISTRIBUTION

Calreticulin and calnexin are very widely distributed in the animal kingdom, from protozoa like *Trypanosoma cruzi* [11] to their better-known mammalian sources. They reside mainly in the endoplasmic reticulum of all cell types, but calreticulin has also been found in other cellular compartments, including nuclear and cell surface locations [1,4,5].

POSSIBLE FUNCTIONS

Calreticulin and calnexin function as molecular chaperones in the endoplasmic reticulum, transiently binding to newly synthesised glycoproteins to promote the correct folding and oligomeric assembly, and thus contribute to the 'quality control' of glycoprotein synthesis [4]. Calreticulin serves as a major storage protein for calcium and participates in calcium signalling [12].

Cell-surface calreticulin may also mediate signal transduction events in association with integrins [5]. The leukocyte cell-surface form appears to function as a collectin receptor [13].

COMMENTS

Calreticulin was discovered and rediscovered as a calcium-binding protein before it was recognised as a lectin. Conversely, however, a lectin from bovine placenta was identified as such and then found to share identity with calreticulin over the first 13 residues from its N-terminus [14]. This actin-binding, Ca^{2+}-independent, sialic acid-specific lectin may be the ovine equivalent of the sialic acid-specific lectin found in human placenta [15]; however, more extensive sequence data are required to establish their relationships to each other and to calreticulin.

The anti-mitogenic activity of calreticulin towards endothelial cells and its related effects on angiogenesis and tumour growth hold promise that calreticulin could be used therapeutically to treat pathological angiogenesis in cancer and other diseases [10].

Calreticulin has been reported to be identical to the 52 kDa human Ro (SS-A) autoantigen, against which autoantibodies are formed in Sjögren's syndrome and other inflammatory disorders [16]. Although calreticulin appears to be a human autoantigen in both rheumatic disorders [17] and celiac disease [18], its putative identity to Ro has been disputed [17]. It may be relevant to autoimmunity that calreticulin shares peptide sequences with α-gliadin [18] and HLA-DR4 [19].

REFERENCES

1. Michalak, M., Milner, R.E., Burns, K. & Opas, M. (1992). Calreticulin. *Biochemical Journal* **285**, 681–692.
2. Wada, I., Rindress, D., Cameron, P.H., Ou, W.-J., Doherty, J.J., Louvard, D., Bell, A.W., Dignard, D., Thomas, D.Y. & Bergeron, J.J.M. (1991). SSRα and associated calnexin are major calcium binding proteins of the endoplasmic reticulum membrane. *Journal of Biological Chemistry* **266**, 19599–19610.
3. Helenius, A. (1994). How N-linked oligosaccharides affect glycoprotein folding in the endoplasmic reticulum. *Molecular Biology of the Cell* **5**, 253–265.
4. Helenius, A., Trombetta, S., Hebert, D.N. & Simons, J.F. (1997). Calnexin, calreticulin and the folding of glycoproteins. *Trends in Cell Biology* **7**, 193–200.
5. White, T.K., Zhu, Q. & Tanzer, M.L. (1995). Cell surface calreticulin is a putative mannoside lectin which triggers mouse melanoma cell spreading. *Journal of Biological Chemistry* **270**, 15926–15929.
6. Ware, F.E., Vassilakos, A., Peterson, P.A., Jackson, M.R., Lehrman, M.A. & Williams, D.B. (1995). The molecular chaperone calnexin binds $Glc_1Man_9GlcNAc_2$ oligosaccharide as an initial step in recognizing unfolded glycoproteins. *Journal of Biological Chemistry* **270**, 4697–4704.
7. Spiro, R.G., Zhu, Q., Bhoyroo, V. & Söling, H.-D. (1996). Definition of the lectin-like properties of the molecular chaperone, calreticulin, and demonstration of its copurification with endomannosidase from rat liver Golgi. *Journal of Biological Chemistry* **271**, 11588–11594.
8. Vassilakos, A., Michalak, M., Lehrmann, M.A. & Williams, D.B. (1998). Oligosaccharide binding characteristics of the molecular chaperones calnexin and calreticulin. *Biochemistry* **37**, 3480–3490.
9. Leung-Hagesteijn, C.Y., Milanko, K., Michalak, M., Wilkins, J. & Dedhar, S. (1994). Cell attachment to extracellular matrix substrates is inhibited upon downregulation of expression of calreticulin, an intracellular integrin α-subunit-binding protein. *Journal of Cell Science* **107**, 589–600.
10. Pike, S.E., Yao, L., Setsuda, J., Jones, K.D., Cherney, B., Appella, E., Sakaguchi, K., Nakhasi, H., Atreya, C.D., Teruya-Feldstein, J., Wirth, P., Gupta, G. & Tosato, G. (1999). Calreticulin and calreticulin fragments are endothelial cell inhibitors that suppress tumor growth. *Blood* **94**, 2461–2468.
11. Peterson, J.R. & Helenius, A. (1999). In vitro reconstitution of calreticulin–substrate interactions. *Journal of Cell Science* **112**, 2775–2784.
12. Krause, K.-H. & Michalak, M. (1997). Calreticulin. *Cell* **88**, 439–443.
13. Kishore, U., Sontheimer, R.D., Sastry, K.N., Zaner, K.S., Zappi, E.G., Hughes, G.R.V., Khamashta, M.A., Strong, P., Reid, K.B.M. & Eggleton, P. (1997). Release of calreticulin from neutrophils may alter C1q-mediated immune functions. *Biochemical Journal* **322**, 543–550.
14. Iglesias, M.M., Cymes, G.D. & Wolfenstein-Todel, C. (1996). A sialic acid-binding lectin from ovine placenta: purification, specificity and interaction with actin. *Glycoconjugate Journal* **13**, 967–976.
15. Ahmed, H. & Gabius, H.-J. (1989). Purification and properties of a Ca^{2+}-independent sialic acid-binding lectin from human placenta with preferential affinity to *O*-acetylsialic acids. *Journal of Biological Chemistry* **264**, 18673–18678.
16. McCauliffe, D.P., Lux, F.A., Lieu, T.-S., Sanz, I., Hanke, J., Newkirk, M.M., Bachinski, L.L., Itoh, Y., Siciliano, M.J., Reichlin, M., Sontheimer, R.D. & Copra, J.D. (1990). Molecular cloning, expression and chromosome 19 localization of a human Ro/SS-A autoantigen. *Journal of Clinical Investigation* **85**, 1379–1391.
17. Routsias, J.G., Tzioufas, A.G., Sakarellos-Daitsiotis, M., Sakarellos, C. & Moutsopoulos, H.M. (1993). Calreticulin synthetic peptide analogues: anti-peptide antibodies in autoimmune rheumatic diseases. *Clinical and Experimental Immunology* **91**, 437–441.
18. Karska, K., Tučková, L., Steiner, L., Tlaskalová–Hogenová, H. & Michalak, M. (1995). Calreticulin – the potential autoantigen in celiac disease. *Biochemical and Biophysical Research Communications* **209**, 597–605.
19. Verreck, F.A.W., Elferink, D., Vermeulen, C.J., Amons, R., Breedveld, F., DeVries, R.R.P. & Koning, F. (1995). DR4Dw4/DR53 molecules contain a peptide from the autoantigen calreticulin. *Tissue Antigens* 45, 270–275.

Cancer antennarius lectin

Other name: Crab lectin
Mol. wt: 70 000 (native); 36 000 (subunit)
Specificity: *O*-acetylsialic acids
Source: Haemolymph of Californian marine crab *Cancer antennarius*

ISOLATION

The lectin was isolated from the soluble fraction of clotted haemolymph by affinity chromatography on bovine submaxilliary mucin–agarose [1]. The haemolymph was applied at 4 °C and washed at 4 °C and 32 °C. Finally, the lectin was eluted at 32 °C with EDTA [1].

STRUCTURE

Based on SDS-PAGE in the presence and absence of reducing agents, the crab lectin appears to be a dimer of 36 000 subunits [1].

BIOLOGICAL ACTIVITIES

The crab lectin agglutinates lapine, equine and murine erythrocytes in increasing order of titre, but does not agglutinate human, ovine or chicken erythrocytes [1]. Agglutinating activity towards erythrocytes of different species correlates well with the *O*-acetyl sialic acid content of the cells, and is calcium-dependent [1].

Agglutination was inhibited by bovine and equine submaxilliary mucins, but not by ovine submaxilliary mucin (in contrast to limulin or the *Limax* lectin), fetuin, thyroglobulin or transferrin. It was more readily inhibited by free 9-*O*-acetyl-*N*-acetylneuraminic acid and 4-*O*-acetylneuraminic acid than by *N*-acetylneuraminic acid or *N*-glycolylneuraminic acid [1]. Other monosaccharides were not inhibitory. It appears to have a high specificity for 9-*O*-acetylsialic acid and 4-*O*-acetyl sialic acid.

TISSUE AND/OR SUBCELLULAR DISTRIBUTION

This lectin has been found in haemolymph but no information is available concerning other locations.

POSSIBLE FUNCTION(S)

It has been suggested that the crab lectin binds to sialic acid on bacteria and serves as an opsonin [1].

COMMENTS

Human melanoma cells are unusual in synthesising large quantities of gangliosides. G_{M3} is the major ganglioside of normal melanocytes, but G_{D3} increases after malignant transformation. The crab lectin is able to bind *O*-acetyl-G_{D3} on melanoma cell lines, a structure not expressed on normal tissues [2]. The crab lectin can therefore be used to detect a tumour marker for malignant melanoma.

REFERENCES

1. Ravindranath, M.H., Higa, H.H., Cooper, E.L. & Paulson, J.C. (1985). Purification and characterisation of an O-acetyl sialic acid-specific lectin from a marine crab, *Cancer antennarius. Journal of Biological Chemistry* **260**, 8850–8856.
2. Ravindranath, M.H., Paulson, J.C. & Irie, R.F. (1988). Human melanoma antigen *O*-acetylated ganglioside G_{D3} is recognised by *Cancer antennarius* lectin. *Journal of Biological Chemistry* **263**, 2079–2066.

Carcinoscorpin

Other names: CNS; *C. rotunda* lectin
Mol. wt: 420 000 (native); 27 000 and 28 000 (subunit)
Specificity: Sialic acid
Source: Haemolymph of the Indian horseshoe crab *Carcinoscorpius rotunda cauda*

ISOLATION

CNS was isolated from clotted haemolymph by affinity chromatography on ovine submaxilliary mucin–Sepharose, passage through DEAE–Sephadex to remove haemocyanin, and lectin affinity chromatography on immobilised concanavalin A. This procedure resulted in a 1860-fold increase in specific lectin activity with a 14% recovery of an apparently homogeneous lectin [1].

STRUCTURE

CNS is a glycoprotein (5.8% carbohydrate) with a composition rich in acidic amino acids, broadly similar to that of limulin, from which CNS is antigenically distinct [2]. The apparent molecular weight estimated by gel filtration is 420 000 [1], and it has a sedimentation coefficient of 12.7 [2]. Two glycosylated subunits of Mr 27 000 and 28 000 were detected by SDS-PAGE [1,2], but only one N-terminal amino acid (leucine) could be detected [2].

BIOLOGICAL ACTIVITIES

CNS agglutinates both lapine and equine erythrocytes, but the latter can be agglutinated at a 250-times lower concentration (0.12 ng/ml) of CNS [2]. Lectin activity is calcium-dependent [1,3].

CNS precipitates fetuin but not asialofetuin [1]. Fetuin binding and erythrocyte agglutination are inhibited more effectively by *N*-glycolylneuraminic acid than by *N*-acetylneuraminic acid, while *N*-acetylneuraminyl galactitol was the most potent inhibitor tested [3]. D-Glucuronic acid, GalNAc and GlcNAc all had a modest inhibitory effect [3].

TISSUE AND/OR SUBCELLULAR DISTRIBUTION

CNS has been reported only in the soluble fraction of haemolymph. CNS is not present in haemocytes [4].

POSSIBLE FUNCTIONS

CNS is presumed to function in host defence; it has been shown to act as an opsonin for *Escherichia coli* during experimental infection [5].

COMMENTS

CNS, immobilised on Sepharose, can be used to separate isoenzymes differing in sialic acid content, and to fractionate incompletely desialylated fetuin derivatives [6].

Another haemagglutinin has been isolated from haemocytes of the same species [4]. This protein, which is specific for human blood group A, yields a single band on SDS-PAGE (apparent Mr 24 000) under both reducing and non-reducing conditions, is not inhibited by sialic acid, and is antigenically distinct from CNS [4].

REFERENCES

1. Bishayee, S. & Dorai, D.T. (1980). Isolation and characterisation of a sialic acid-binding lectin (carcinoscorpin) from Indian horseshoe crab *Carcinoscorpius rotunda cauda. Biochimica et Biophysica Acta* **623**, 89–97.
2. Dorai, D.T., Bachhawat, B.K., Bishayee, S., Kannan, K. & Rao, D.R. (1981). Further characterisation of the sialic acid-binding lectin from the horseshoe crab *Carcinoscorpius rotundra cauda. Archives of Biochemistry and Biophysics* **209**, 325–333.
3. Mohan, S., Dorai, D.T., Srimal, S. & Bachhawat, B.K. (1982). Binding studies of a sialic acid-specific lectin from the horseshoe crab *Carcinoscorpius rotunda cauda* with various sialoglycoproteins. *Biochemical Journal* **203**, 253–261.
4. Srimal, D., Dorai, D.T., Somasundaran, M., Bachhawat, B.K. & Miyata, T. (1985). A new haemagglutinin from the amoebocytes of the horseshoe crab *Carcinoscorpius rotundicauda. Biochemical Journal* **230**, 321–327.
5. Sarbadhikari, S.B., Datta, D. & Bhadra, R. (1999). Role of carcinoscorpin, a haemolymph lectin of horseshoe crab *Carcinoscorpius rotunda cauda* as humoral factor. *Journal of Biosciences* **24**, 471–476.
6. Dorai, D.T., Bachhawat, B.K. & Bishayee, S. (1981). Fractionation of sialoglycoproteins on an immobilised sialic acid-binding lectin. *Analytical Biochemistry* **115**, 130–137.

CD11b/CD18

Other names: CR3; Mac-1; $\alpha_M\beta_2$ integrin
Mol. wt: 265 000 (native); 170 000 and 95 000 (subunit)
Specificity: β-Glucans
Source: Mammalian leukocytes

ISOLATION

First identified (as Mac-1) by an antibody specific for murine phagocytes [1], CD11b/CD18 was subsequently recognised as a complement (iC3b) receptor and as an adhesion molecule in humans and in rodents [2]. Murine Mac-1 was purified by immunoaffinity chromatography and the subunits separated by preparative SDS-PAGE [3]. Human CD11b/CD18 was purified by immunoaffinity chromatography from the neutrophils of a patient with chronic myeloid leukaemia, and the subunits separated using high performance liquid chromatography on a gel filtration column [4].

STRUCTURE

CR3 is a phosphorylated, glycosylated integrin, a heterodimer (non-covalent) of the α_M integrin, CD11b (Mr 170 000), and the β_2 integrin, CD18 (Mr 95 000). The amino acid sequences of both subunits of the complex have been determined [2,5]. CD11b has a constitutively phosphorylated cytoplasmic domain, a transmembrane domain and a large extracellular region containing an 'I domain' *inserted* among seven homologous repeats. CD18 has four cysteine-rich repeats in the extracellular region and eight potential phosphorylation sides in the cytoplasmic domain [2].

BIOLOGICAL ACTIVITIES

CR3 mediates the adhesion of monocytes and granulocytes to endothelial cells via ICAM-1 binding, and promotes the extravasation of neutrophils to sites of inflammation [2,5]. CR3 is also involved in phagocytosis mediated by immune complexes opsonised with iC3b [2,5]. These activities require the presence of divalent cations, and a calcium-dependent conformational change may permit a distinction between the two [6]. The binding to ICAM-1, iC3b and other protein ligands appears to be carbohydrate-independent with binding sites located within the I domain.

CR3 also has a binding site for yeasts and lapine erythrocytes [7]. This binding can be inhibited by various β-glucans, and by GlcNAc, mannosides and glucosides [8]. The lectin (β-glucan receptor) site is calcium-independent and located C-terminal to the I domain of CD11b, although lectin site-dependent signalling through CD18 probably requires the heterodimer [8–10].

TISSUE AND/OR SUBCELLULAR DISTRIBUTION

This integrin is found exclusively on certain leukocytes: macrophages, monocytes, granulocytes and some other myeloid cells [2,5]; it is stored in intracellular vesicles and rapidly transported to the cell surface in response to chemoattractants [2].

POSSIBLE FUNCTION(S)

CD11b/CD18 functions as an adhesion and signalling molecule, mediating monocyte and neutrophil adherence to endothelium and to the extracellular matrix [2,5].

CD11b/CD18 is also a factor involved in phagocytosis and degranulation-cytotoxicity, and both I and lectin domains have been implicated in its interaction with *Candida albicans*. It is also the primary leukocyte receptor for several other pathogens [9].

COMMENTS

Like other integrins, CD11b/CD18 is a dimeric complex of (in this instance) an α subunit encoded on human chromosome 16 (CD11b) and a β subunit encoded on chromosome 21. It is an adhesion molecule with a crucial role in host defence. It appears to have evolved to distinguish self from non-self through recognition of iC3b-opsonised micro-organisms while ignoring iC3b-opsonised host cells [11].

It has been suggested that CD11b/CD18 might be a suitable target for polysaccharide drugs, either to prime it for cytotoxic activation in response to iC3b-opsonised tumour cells or, alternatively, to block neutrophil diapedesis and hence tissue damage during inflammatory processes [10,11].

REFERENCES

1. Springer, T., Galfre, G., Secher, D.S. & Milstein, C. (1979). Mac-1: a macrophage differentiation antigen identified by monoclonal antibody. *European Journal of Immunology* **9**, 301–306.
2. Pigott, R. & Power, C. (1994). *The Adhesion Molecule Facts Book*, pp. 93–95, 105–108. Academic Press, London.

3. Springer, T.A., Teplow, D.B. & Dreyer, W.J. (1985). Sequence homology of the LFA-1 and Mac-1 leukocyte adhesion glycoproteins and unexpected relation to leukocyte interferon. *Nature* **314**, 540–542.
4. Hickstein, D.D., Ozols, J., Williams, S.A., Baenziger, J.U., Locksley, R.M. & Roth, G.J. (1987). Isolation and characterization of the receptor on human neutrophils that mediates cellular adherence. *Journal of Biological Chemistry* **262**, 5576–5580.
5. Barclay, A.N., Birkeland, M.L., Brown, M.H., Beyers, A.D., Davis, S.J., Somoza, C. & Williams, A.F. (1993). *The Leukocyte Antigens Facts Book*, pp. 124–127, 140–141. Academic Press, London
6. Graham, I.L. & Brown, E.J. (1991). Extracellular calcium results in a conformational change in Mac-1 (CD11b/CD18) on neutrophils. *Journal of Immunology* **146**, 685–691.
7. Ross, G.D., Cain, J.A. & Lachmann, P.J. (1985). Membrane complement receptor type three (CR_3) has lectin-like properties analogous to bovine conglutinin and functions as a receptor for zymosan and rabbit erythrocytes as well as a receptor for iC3b. *Journal of Immunology* **134**, 3307–3315.
8. Thornton, B.P., Větvička, V., Pitman, M., Goldman, R.C. & Ross, G.D. (1996). Analysis of the sugar specificity and molecular location of the β-glucan-binding lectin site of complement receptor type 3 (CD11b/CD18). *Journal of Immunology* **156**, 1235–1246.
9. Forsyth, C.B., Plow, E.F. & Zhang, L. (1998). Interaction of the fungal pathogen *Candida albicans* with integrin CD11b/CD18: recognition by the I domain is modulated by the lectin-like domain and the CD18 subunit. *Journal of Immunology* **161**, 6198–6205.
10. Xia, Y. & Ross, G.D. (1999). Generation of recombinant fragments of CD11b expressing the functional β-glucan-binding lectin site of CR3 (CD11b/CD18). *Journal of Immunology* **162**, 7285–7293.
11. Xia, Y., Větvička, V., Yan, J., Hanikýrová, M., Mayadas, T. & Ross, G.D. (1999). The β-glucan-binding site of mouse CR3 (CD11b/CD18) and its function in generating a primed state of the receptor that mediates cytotoxic activation in response to iC3b-opsonised target cells. *Journal of Immunology* **162**, 2281–2290.

CD22

Other names: Siglec-2; B lymphocyte cell adhesion molecule
Mol. wt: 140 000 (subunit)
Specificity: NeuAc-α2,6-Gal
Sources: Human and murine B lymphocytes

ISOLATION

Human CD22 was first cloned from a Daudi cell-expression library [1]. A larger cDNA was later isolated from a B lymphocyte cDNA library that encodes a protein that appears to be the predominant isoform (β) on normal B cells [2].

STRUCTURE

CD22 is a glycoprotein with an apparent Mr of 130 000–140 000 determined by SDS-PAGE [3]. The structure of the β isoform consists of a cytoplasmic C-terminal domain, a transmembrane region, and a large extracellular region composed of seven immunoglobulin domains (six C2-set domains and an N-terminal V-set domain) [2]. The polypeptide has a total of 847 amino acid residues and a calculated molecular mass of 95 kDa [2]. (The α variant is similar but has two fewer immunoglobulin-type domains.) The CD22 sequence shows significant homology with other I-type lectins, especially myelin-associated glycoprotein [1,2].

The cytoplasmic tail sequence can be phosphorylated [4]. On the cell surface, CD22 appears to exist as a multimer [5].

BIOLOGICAL ACTIVITIES

CD22 mediates the adhesion of B lymphocytes to other B cells and some T cells, and (to a weaker degree) to neutrophils, monocytes and erythrocytes [1,3,6].

CD22 has been shown to bind to sialylated cellular glycoproteins including CD45 and CD75 [3], and to the plasma glycoproteins, IgM and haptoglobin [7]. The structure recognised by CD22 on glycoconjugates is usually the trisaccharide, (NeuAc-α2,6-Gal β1,4-GlcNAc) but the minimum structure required appears to be NeuAc α2,6-Hex(Ac), where 'Hex(Ac)' can be galactose, GalNAc or GlcNAc [5].

TISSUE AND/OR SUBCELLULAR DISTRIBUTION

CD22 is found exclusively on B cells [1–3]; it is expressed on the surface of a major subset of normal B lymphocytes and in the cytoplasm of pre-B and mature B cells [1]. Expression is lost on activation [1].

POSSIBLE FUNCTION(S)

CD22 is thought to be an adhesion molecule with a signalling function which in some way modulates B lymphocyte activation.

Studies on CD22 knock-out mice are consistent with a role for CD22 in regulating B cell receptor (sIgM) responses. One study described mice with hyperresponsive B cells, with augmented cellular proliferation and autoantibody production [8]. Another group of gene-targeted mice had some diminished B cell responses (but an increased response to lipopolysaccharide) including abnormal responses to thymus-independent antigens [9].

CD22 also appears to mediate the homing of recirculating B cells to the bone marrow [10].

COMMENTS

CD22 lectin activity seems to be regulated via masking and unmasking by endogenous ligands on the B lymphocyte surface [11]. It has been suggested that this is a general mechanism by which each of the siglec group of lectins regulates its exposure and function in the midst of abundant glycoprotein inhibitors, permitting lectin activity only in selected settings [11].

REFERENCES

1. Stamenkovic, I. & Seed, B. (1990). The B cell antigen CD22 mediates monocyte and erythrocyte adhesion. *Nature* **345**, 74–77.
2. Wilson, G.L., Fox, C.H., Fauci, A.S. & Kehrl, J.H. (1991) cDNA cloning of the B cell membrane protein CD22: a mediator of B–B cell interactions. *Journal of Experimental Medicine* **173**, 137–146.
3. Stamenkovic, I., Sgroi, D., Aruffo, A., Sy, M.S. & Anderson, T. (1991). The B lymphocyte adhesion molecule CD22 interacts with leukocyte common antigen CD45RO on T cells and α2-6 sialytransferase, CD75, on B cells. *Cell* **66**, 1133–1144.
4. Schulte, R.J., Campbell, M.-A., Fischer, W.H. & Sefton, B.M. (1992). Tyrosine phosphorylation of CD22 during B cell activation. *Science* **258**, 1001–1004.
5. Powell, L.D., Jain, R.K., Matta, K.L., Sabesan, S. & Varki, A. (1995). Characterization of sialyloligosaccharide binding by recombinant soluble and native cell-associated CD22. *Journal of Biological Chemistry* **270**, 7523–7532.
6. Kelm, S., Schauer, R. & Crocker, P.R. (1996). The sialoadhesions – a family of sialic acid-dependent cellular recognition molecules within the immunoglobulin superfamily. *Glycoconjugate Journal* **13**, 913–926.
7. Hanasaki, K., Powell, L.D. & Varki, A. (1995). Binding of human plasma sialoglycoproteins by the B cell-specific lectin CD22. *Journal of Biological Chemistry* **270**, 7543–7550.
8. O'Keefe, T.L., Williams, G.T., Davies, S.L. & Neuberger, M.S. (1996). Hyperresponsive B cells in CD22-deficient mice. *Science* **274**, 798–801.
9. Otipoby, K.L., Andersson, K.B., Draves, K.E., Klaus, S.J., Farr, A.G., Kerner, J.D., Perlmutter, R.M., Law, C.-L. & Clark, E.A. (1996). CD22 regulates thymus-independent responses and the lifespan of B cells. *Nature* **384**, 634–637.
10. Nitschke, L., Floyd, H., Ferguson, D.J.P. & Crocker, P.R. (1999). Identification of CD22 ligands on bone marrow sinusoidal endothelium implicated in CD22-dependent homing of recirculating B cells. *Journal of Experimental Medicine* **189**, 1513–1518.
11. Razi, N. & Varki, A. (1998). Masking and unmasking of the sialic acid-binding lectin activity of CD22 (Siglec-2) on B lymphocytes. *Proceedings of the National Academy of Sciences USA* **95**, 7469–7474.

CD23

Other name: Low-affinity IgE receptor (Fc_{ε} RII)
Mol. wt: 45 000
Specificity: Galactose
Sources: B lymphocytes and B cell lines

ISOLATION

Human CD23 was isolated by immunoaffinity chromatography using a monoclonal antibody [1], and sequence information obtained led to the cloning and expression of cDNA in COS-7 cells [1]. Two other groups reported cloning the human cDNA at much the same time [2,3].

The murine homologue has also been purified by affinity chromatography on IgE–Sepharose [4], and cloned from a B cell cDNA library [5].

STRUCTURE

CD23 is a glycoprotein of apparent Mr 45 000 determined by SDS-PAGE. The human molecule consists of 321 amino acids with a calculated protein molecular mass of 36 100, and has the general arrangement of a type II transmembrane receptor. The short (just over 20 residues) cytoplasmic domain is followed by a hydrophobic transmembrane domain of similar size, then a larger extracellular regional including a C-terminal C-type lectin domain. The latter shows most homology with the asialoglycoprotein receptors [1,2].

The slightly larger (331 amino acid residues) murine CD23 is a very similar molecule [5], with 57% primary sequence identity (or 74% homology if conservative substitutions are considered). The isolectric points of both lie within the pH range 4.5–5 [4].

BIOLOGICAL ACTIVITIES

CD23 binds to IgE via its C-type lectin domain, but the binding appears to be protein–protein in nature [6].

CD23 can also bind to CD21, the complement and Epstein–Barr virus receptor, which may be a physiological ligand [7]; this interaction is Ca^{2+}-dependent and glycosylation-dependent, and can be partially inhibited by fucose-1-phosphate [8].

CD23 can bind to asialofetuin–Sepharose and be eluted with EDTA, galactose or lactose; other sugars were much less effective [9]. Moreover, cell-surface CD23 on B cells lines may mediate cell adhesion via lectin–carbohydrate interactions [10].

TISSUE AND/OR SUBCELLULAR DISTRIBUTION

In mouse and human, CD23 is a B lymphocyte associated antigen, but some expression on platelets, eosinophils and IL-4 stimulated monocytes has been reported [11].

POSSIBLE FUNCTION(S)

CD23 is a B-cell signalling molecule thought to have a role in immune regulation and IgE-mediated antigen presentation [12].

COMMENTS

CD23 has received considerable attention as a B lymphocyte marker and as an IgE-binding protein. This literature has been reviewed by Conrad [11] and by Gordon [12].

REFERENCES

1. Ikuta, K., Takami, M., Kim, C.W., Honjo, T., Miyoshi, T., Tagaya, Y., Kawabe, T. & Yodoi, J. (1987). Human lymphocyte Fc receptor for IgE: sequence homology of its cloned cDNA with animal lectins. *Proceedings of the National Academy of Sciences USA* **84**, 819–823.
2. Kikutani, H., Inui, S., Sato, R., Barsumian, E.L., Owaki, H., Yamasaki, K., Kaisho, T., Uchibayashi, N., Hardy, R.R., Hirano, T., Tsunasawa, S., Sakiyama, F., Suemura, M. & Kishimoto, T. (1986). Molecular structure of human lymphocyte receptor for immunoglobulin E. *Cell* **47**, 657–665.
3. Lüdin, C., Hofstetter, H., Sarfati, M., Levy, C.A., Suter, U., Alaimo, D., Kilchher, E., Frost, H. & Delespesse, G. (1987). Cloning and expression of the cDNA coding for a human lymphocyte IgE receptor. *EMBO Journal* **6**, 109–114.
4. Conrad, D.H. & Peterson, L.H. (1984). The murine lymphocyte receptor for IgE. I. Isolation and characterization of the murine B cell Fc_ε receptor and comparison with Fc_ε receptors from rat and human. *Journal of Immunology* **132**, 796–803.
5. Bettler, B., Hofstetter, H., Rao, M., Yokoyama, W.M., Kilchherr, F. & Conrad, D.H. (1989) Molecular structure and expression of the murine lymphocyte low affinity receptor for IgE (Fc_ε RII). *Proceedings of the National Academy of Sciences USA* **86**, 7566–7570.
6. Bettler, B., Maier, R., Rüegg, D. & Hofstetter, H. (1989). Binding site for IgE of the human lymphocyte low-affinity Fcε receptor (Fcε RII/CD23) is confined to the domain homologous with animal lectins. *Proceedings of the National Academy of Sciences USA* **86**, 7118–7122.
7. Aubry, J.-P., Pochon, S., Graber, P., Jansen, K.U. & Bonnefoy, J.-Y. (1992). CD21 is a ligand for CD23 and regulates IgE production. *Nature* **358**, 505–507.
8. Pochon, S., Graber, P., Yeager, M., Jansen, K., Bernard, A.R., Aubry, J.-P. & Bonnefoy, J.-Y. (1992). Demonstration of a second ligand for the low affinity receptor for immunoglobulin E (CD23) using recombinant CD23 reconstituted into fluorescent liposomes. *Journal of Experimental Medicine* **176**, 389–397.
9. Kijimoto-Ochiai, S., Horimoto, E. & Uede, T. (1994). Demonstration of the interaction between the CD23 molecule and the galactose residue of glycoproteins. *Immunology Letters* **40**, 49–53.
10. Kijimoto-Ochiai, S. & Uede, T. (1995). CD23 molecule acts as a galactose-binding lectin in the cell aggregation of EBV-transformed human B-cell lines. *Glycobiology* **5**, 443–448.
11. Conrad, D.H. (1990). Fc_ε RII/CD23: the low affinity receptor for IgE. *Annual Review of Immunology* **8**, 623–645.
12. Gordon, J. (1994). B cell signalling via the C-type lectins CD23 and CD72. *Immunology Today* **15**, 411–417.

CD31

Other name: Platelet–endothelial cell adhesion molecule-1 (PECAM-1)
Mol. wt: 130 000
Specificity: Heparin, chondroitin-6-sulphate
Sources: Human leukocytes and endothelial cells

ISOLATION

The pan-leukocyte antigen designated CD31 was only properly identified when antibodies used to define it were independently used to screen an endothelial λgt11 expression library [1], an umbilical vein endothelial cell cDNA library [2] and a COS cell expression library of stimulated HL60 cells [3].

STRUCTURE

CD31 is a glycoprotein (about 40% carbohydrate) with an apparent Mr of 130 000 determined by SDS-PAGE under reducing conditions. The nucleotide sequences of the cDNA clones predict a mature protein of 711 amino acid residues and a calculated polypeptide molecular mass of 80 000 Da. The medium-sized cytoplasmic region has the C-terminus, and is separated by a short transmembrane domain from the large extracellular region consisting of six (C2-set) immunoglobulin-like domains [1–3].

BIOLOGICAL ACTIVITIES

CD31 can mediate cell–cell adhesion [4]. CD31 qualifies as a lectin since the calcium-dependent aggregation of mouse L-cells stably transfected with CD31 cDNA is inhibited by heparin and chondroitin sulphate but not by other glycosaminoglycans [5].

TISSUE AND/OR SUBCELLULAR DISTRIBUTION

CD31 has been found in human endothelial cells, platelets and leukocytes (monocytes, granulocytes and some lymphocytes) [1–3]; it has also been found on a subpopulation of human trophoblasts [6]. It is expressed on the cell surface and concentrated in regions of cell contact [2].

POSSIBLE FUNCTION(S)

CD31 is thought of as an adhesion molecule, and appears to have a role in diapedesis [7]. CD31 knock-out mice are healthy, although a minor abnormality in neutrophil migration in response to an inflammatory stimulus was noted [8].

There is also some evidence that CD31 can function as an inhibitory signal transduction molecule [9].

COMMENTS

CD31 might be considered a minor histocompatibility antigen, since donor–recipient incompatibility associated with an allelic variant of CD31 was a risk factor for graft-versus-host disease after bone marrow transplantation [10].

Soluble CD31 is significantly elevated in the circulation of patients with pre-eclampsia compared to normotensive pregnant controls [11].

REFERENCES

1. Newman, P.J., Berndt, M.C., Gorski, J., White, G.C., Lyman, S., Paddock, C. & Muller, W.A. (1990). PECAM-1 (CD31) cloning and relation to adhesion molecules of the immunoglobulin gene superfamily. *Science* **247**, 1219–1222.
2. Simmons, D.L., Walker, C., Power, C. & Pigott, R. (1990). Molecular cloning of CD31, a putative intercellular adhesion molecule closely related to carcinoembryonic antigen. *Journal of Experimental Medicine* **171**, 2147–2152.
3. Stockinger, H., Godal, S.J., Eher, R., Majdic, O., Schreiber, W., Kasinrerk, W., Strass, B., Schnabl, E. & Knapp, W. (1990). Molecular characterization and functional analysis of the leukocyte surface protein CD31. *Journal of Immunology* **145**, 3889–3897.
4. Albelda, S.M., Muller, W.A., Buck, C.A. & Newmann, P.J. (1991). Molecular and cellular properties of PECAM-1 (endo CAM/CD31): a novel vascular cell–cell adhesion molecule. *Journal of Cell Biology* **114**, 1059–1068.
5. De Lisser, H.M., Yan, H.C., Newman, P.J., Muller, W.A., Buck, C.A. & Albelda, S.M. (1993). Platelet/endothelial cell adhesion molecule-1 (CD31)-mediated cellular aggregation involves cell surface glycosaminoglycans. *Journal of Biological Chemistry* **268**, 16037–16046.
6. Coukos, G., Makrigiannakis, A., Amin, K., Albelda, S.M. & Coutifaris, C. (1998). Platelet–endothelial cell adhesion molecule-1 is expressed by a subpopulation of human trophoblasts: a possible mechanism for trophoblast–endothelial interaction during haemochorial placentation. *Molecular Human Reproduction* **4**, 357–367.
7. Liao, F., Schenkel, A.R. & Muller, W.A. (1999). Transgenic mice expressing different levels of soluble platelet/endothelial cell adhesion molecule-IgG display distinct inflammatory phenotypes. *Journal of Immunology* **163**, 5640–5648.
8. Duncan, G.S., Andrew, D.P., Takimoto, H., Kaufman, S.A., Yoshida, H., Spellberg, J., De La Pompa, J.L., Elia, A., Wakeham, A., Karan-Tamir, B., Muller, W.A., Senaldi, G., Zukowski, M.M. & Mak, T.W. (1999). Genetic evidence for functional redundancy of platelet/endothelial cell adhesion molecule-1 (PECAM-1): CD31-deficient mice reveal PECAM-1-dependent and PECAM-1-independent functions. *Journal of Immunology* **162**, 3022–3030.
9. Newton-Nash, D.K. & Newman, P.J. (1999). A new role of platelet–endothelial cell adhesion molecule-1 (CD31): inhibition of TCR-mediated signal transduction. *Journal of Immunology* **163**, 682–688.
10. Balduini, C.L., Noris, P., Giorgiani, G., Martinetti, M., Klersy, C., Spedini, P. & Belletti, S. (1999). Incompatibility for CD31 and human platelet antigens and acute graft-versus-host disease after bone marrow transplantation. *British Journal of Haematology* **106**, 723–729.
11. Krauss, T., Kuhn, W., Lakoma, C. & Augustin, H.G. (1997). Circulating endothelial cell adhesion molecules as diagnostic markers for the early identification of pregnant women at risk for development of pre-eclampsia. *American Journal of Obstetrics and Gynecology* **177**, 443–449.

CD33

Other name: Siglec-3
Mol. wt: 67 000
Specificity: NeuAc-α2,3-Gal
Sources: Human and murine myeloid progenitor cells

ISOLATION

Human CD33 was first identified with a monoclonal antibody generated against myeloid-associated differentiation antigens on haematopoietic stem and progenitor cells [1]. A cDNA encoding CD33 was subsequently isolated from a lymphoma cell line cDNA library [2]. Molecular cloning of two isoforms of the murine homologue has also been reported [3].

STRUCTURE

Human CD33 is a glycoprotein with an apparent Mr of 67 000 determined by SDS-PAGE. The cDNA predicts a sequence of 364 residues with a molecular mass of 40 000 kDa. The C-terminal cytoplasmic sequence of 82 residues is followed by a short transmembrane domain and a relatively large extracellular region. The latter contains two (C2+V) immunoglobulin-like domains, with extensive homology to myelin-associated glycoprotein (qv) and CD22 (qv). The cytoplasmic region contains a potential phosphorylation site [2].

BIOLOGICAL ACTIVITIES

CD33 (in a soluble recombinant form) can bind to sialic acid-derivatised human erythrocytes, with a preference for NeuAc-α2,3-Gal over NeuAc-α2,6-Gal [4]. CD33 could also mediate cell adhesion, but only if the CD33-transfected cells were treated with neuraminidase to remove endogenous ligands [4].

TISSUE AND/OR SUBCELLULAR DISTRIBUTION

Human CD33 is found on the surface of $CD34^+$ and $CD34^-$ myelomonocytic progenitor cells in bone marrow, on circulating monocytes and on tissue macrophages [1,5].

POSSIBLE FUNCTION(S)

No actual or potential function has been clearly identified. Obvious possibilities include a role in the differentiation of myeloid lineage cells and a signalling function.

COMMENTS

Like CD22 and the other 'siglecs', CD33 lectin activity may be masked by endogenous ligands and regulated by unmasking [4].

An important application of CD33 is in the diagnosis of acute myeloid leukaemia (AML), by distinguishing myeloid cells from the usually CD33-negative lymphoblastic leukaemia cells [4]. The detection of CD33 is particularly valuable in the more primitive forms of AML, and more generally CD33 may have some value as a positive prognostic factor [6].

CD33 has been the target of antibodies used for bone marrow purging [7], and a humanised radiolabelled anti-CD33 has been used therapeutically in AML with promising results [8].

Unmodified murine anti-CD33 antibodies and other CD33-based immunotoxins have been used in clinical trials; short-term, but not sustained, benefit has been obtained [9].

REFERENCES

1. Andrews, R.G., Tork-Storb, B. & Bernstein, I.D. (1983). Myeloid-associated differentiation antigens on stem cells and their progeny identified by monoclonal antibodies. *Blood* **62**, 124–132.
2. Simmons, D. & Seed, B. (1988). Isolation of a cDNA encoding CD33, a differentiation antigen of myeloid progenitor cells. *Journal of Immunology* **141**, 2797–2800.
3. Tchilian, E.Z., Beverley, P.C.L., Young, B.D. & Watt, S.M. (1994). Molecular cloning of two isoforms of the murine homolog of the myeloid CD33 antigen. *Blood* **83**, 3188–3198.
4. Freeman, S.D., Kelm, S., Barber, E.K. & Crocker, P.R. (1995). Characterization of CD33 as a new member of the sialadhesion family of cellular interaction molecules. *Blood* **85**, 2005–2012.
5. Pierelli, L., Teofili, L., Menichella, G., Rumi, C., Paoloni, A., Iovino, S., Puggioni, P.L., Leone, G. & Bizzi, B. (1993). Further investigations on the expression of HLA-DR, CD33 and CD31 surface antigens in purified bone marrow and peripheral blood CD34+ haematopoietic progenitor cells. *British Journal of Haematology* **84**, 24–30.
6. Kristensen, J.S. & Hokland, P. (1991). Monoclonal antibodies in myeloid diseases: prognostic use in acute myeloid leukaemia. *Leukaemia Research* **15**, 693–700.
7. Robertson, M.J., Soiffer, R.J., Freedman, A.S., Rabinowe, S.L., Anderson, K.C., Ervin, T.J., Murray, C., Dear, K., Griffin, J.D., Nadler, L.M. & Ritz, J. (1992). Human bone marrow depleted of CD33-positive cells mediates delayed but durable reconstitution of hematopoiesis: clinical trial of MY9 monoclonal antibody-purged autografts for the treatment of acute myeloid leukemia. *Blood* **79**, 2229–2236.
8. Caron, P.C., Schwartz, M.A., Co, M.S., Queen, C., Finn, R.D., Graham, M.C., Divgi, C.R., Larson, S.M. & Scheinberg, D.A. (1994). Murine and humanized constructs of monoclonal antibody M195 (anti-CD33) for the therapy of acute myelogenous leukemia. *Cancer* **73**, 1049–1056.
9. Appelbaum, F.R. (1999). Antibody-targeted therapy for myeloid leukemia. *Seminars in Hematology* **36** (Suppl. 6), 2–8.

CD44

Other names: gp90^{HERMES}; Pgp-1
Mol. wt: 85 000
Specificity: Hyaluronic acid
Source: Mammalian tissues

ISOLATION

CD44 was first identified as a human 90 kDa homing receptor, the Hermes antigen [1], although the murine homologue (Pgp-1) had previously been detected on a variety of cell types [2]. Human CD44 has been isolated by immunoaffinity chromatography [3], and its cDNA cloned from a lymphoblastoid cell line [4]. Other cDNA clones have been expressed in COS cells [5].

STRUCTURE

CD44 is an acidic, sulphated glycoprotein [3]. Alternative splicing of numerous exons produces at least six variants [6], while post-translational modification also contributes to extensive heterogeneity. The epithelial cell form, which is much larger and contains an additional 132 amino acids inserted into the extracellular region, is not a lectin and shall not be considered further [7]. The haematopoietic (lectin) forms have either a short [5] or extremely short [4] cytoplasmic C-terminal tail, a transmembrane region and a large extracellular region including tandem-repeat sequences in the N-terminal portion bearing approximately 30% identity with hyalectans and cartilage link proteins [4,5]. Depending on the size of the cytoplasmic tail, the deduced primary sequences would have 293 or 341 amino acid residues and calculated polypeptide molecular masses of 30 000 or 37 000 [4,5]. The much higher molecular weight determined by SDS-PAGE is due to extensive glycosylation [3].

BIOLOGICAL ACTIVITIES

CD44 is an adhesion molecule that can mediate the attachment of leukocytes to high endothelial venules, and does so by recognising hyaluronic acid [7,8]. CD44–hyaluronate interaction is also responsible for adhesion between B lineage hybridoma and bone marrow stromal cells [9].

TISSUE AND/OR SUBCELLULAR DISTRIBUTION

CD44 is expressed widely on the surface of haematopoietic cells (leukocytes, monocytes, granulocytes etc.) and also on a wide variety of other cell types (fibroblasts, myocytes etc.) [1,2,4,7].

POSSIBLE FUNCTION(S)

CD44 is believed to function generally in cell–cell and cell–matrix adhesion. It undoubtedly contributes to lymphocyte adhesion to high endothelial venules, but may act as a supporting, rather than a principal, homing factor. It may also function as a signalling molecule involved in cellular activation [10].

CD44-deficient gene-targeted mice are viable and show no overt abnormalities, but lymphocyte recirculation is altered, with impaired homing to peripheral lymph nodes and a particularly striking deficit in thymic homing [11].

COMMENTS

Anti-CD44 antibodies or hyaluronan can induce the differentiation of acute myeloid leukaemia (AML) blasts irrespective of differentiation stage subtype [12]. This CD44-targeted differentiation therapy seems to be as effective as retinoic acid and more widely applicable since the latter is effective only for the uncommon AML3 (promyelocytic) subtype [12].

REFERENCES

1. Jalkanen, S.T., Bargatze, R.F., Herron, L.R. & Butcher, E.C. (1986). A lymphoid cell surface glycoprotein involved in endothelial cell recognition and lymphocyte homing in man. *European Journal of Immunology* **16**, 1195–1202.
2. Hughes, E.N., Mengod, G. & August, J.T. (1981). Murine cell surface glycoproteins. Characterization of a major component of 80 000 Daltons as a polymorphic differentiation antigen of mesenchymal cells. *Journal of Biological Chemistry* **256**, 7023–7027.
3. Jalkanen, S., Jalkanen, M., Bargatze, R., Tammi, M. & Butcher, E.C. (1988). Biochemical properties of glycoproteins involved in lymphocyte recognition of high endothelial venules in man. *Journal of Immunology* **141**, 1615–1623.
4. Goldstein, L.A., Zhou, D.F.H., Picker, L.J., Minty, C.N., Bargatze, R.F., Ding, J.F. & Butcher, E.C. (1989). A human lymphocyte homing receptor, the Hermes antigen, is related to cartilage proteoglycan core and link proteins. *Cell* **56**, 1063–1072.
5. Stamenkovic, I., Amiot, M., Pesando, J.M. & Seed, B. (1989). A lymphocyte molecule implicated in lymph node homing is a member of the cartilage link protein family. *Cell* **56**, 1057–1062.
6. Jackson, D.G., Buckley, J. & Bell, J.I. (1992). Multiple variants of the human lymphocyte homing receptor CD44 generated by insertions at a single site in the extracellular domain. *Journal of Biological Chemistry* **267**, 4732–4339.
7. Stamenkovic, I., Aruffo, A., Amiot, M. & Seed, B. (1991). The hematopoietic and epithelial forms of CD44 are distinct polypeptides with different adhesion potentials for hyaluronate-bearing cells. *EMBO Journal* **10**, 343–348.
8. Aruffo, A., Stamenkovic, I., Melnick, M., Underhill, C.B. & Seed, B. (1990). CD44 is the principal cell surface receptor for hyaluronate. *Cell* **61**, 1303–1313.
9. Miyake, K., Underhill, C.B., Lesley, J. & Kincade, P.W. (1990). Hyaluronate can function as a cell adhesion molecule and CD44 participates in hyaluronate recognition. *Journal of Experimental Medicine* **172**, 69–75.
10. Pigott, R. & Power, C. (1994). *The Adhesion Molecule Facts Book*, pp. 46–48. Academic Press, London.
11. Protin, U., Schweighaffer, T., Jochum, W. & Hilberg, F. (1999). CD44-deficient mice develop normally with changes in subpopulations and recirculation of lymphoid subsets. *Journal of Immunology* **163**, 4917–4923.
12. Charrad, R.-S., Li, Y., Delpech, B., Balitrand, N., Clay, D., Jasmin, C., Chomienne, C. & Smadja-Joffe, F. (1999). Ligation of the CD44 adhesion molecule reverses blockage of differentiation in human acute myeloid leukemia. *Nature Medicine* **5**, 669–676.

CD54

Other name: Intercellular adhesion molecule-1 (ICAM-1)
Mol. wt: 90 000–115 000
Specificity: Hyaluronic acid
Source: Mammalian tissues

ISOLATION

ICAM-1 was originally defined by a monoclonal antibody inhibiting leukocyte aggregation, and isolated by immunoaffinity chromatography [1]. ICAM-1 cDNAs were cloned from HL-60 cells and from endothelial cells [2].

STRUCTURE

ICAM-1 is a heavily glycosylated protein, the apparent Mr of which, determined by SDS-PAGE, varies with differential glycosylation in different cells [1,2]. The mature protein has a primary sequence of 505 amino acid residues and a predicted Mr of just over 55 000 [2]. It is an integral membrane protein with a short C-terminal cytoplasmic tail and a large extracellular region with five (all C2-set) immunoglobulin-like domains exhibiting a high level of homology with N-CAM (qv) and myelin-associated glycoprotein (qv) [2].

BIOLOGICAL ACTIVITIES

ICAM-1 mediates calcium-dependent cell adhesion as the ligand for the integrin, LFA-1 (CD11a/CD18) [1–3]. Other ligands include Mac-1 (CD11b/CD18) [4] and CD43 [5].

Rat-liver ICAM-1 can bind hyaluronic acid, and indeed is the predominant receptor for hyaluronan on the surface of liver endothelial cells [6]. This is the only lectin activity reported for an ICAM-1.

TISSUE AND/OR SUBCELLULAR DISTRIBUTION

ICAM-1 is expressed on the surface of a wide variety of haematopoietic and non-haematopoietic cells (particularly after cytokine-induced upregulation) including monocytes, lymphocytes, dendritic cells, endothelial cells, fibroblasts and epithelial cells [1].

POSSIBLE FUNCTION(S)

ICAM-1 is undoubtedly an adhesion molecule which mediates cell-cell contract in a variety of circumstances [1–5]. The generation of gene targeted mice has demonstrated the role of ICAM-1 in neutrophil rolling/entry into inflamed tissues, and that it functions synergistically with L-selectin [7].

COMMENTS

ICAM-1 may have some value as a disease marker. For example, soluble ICAM-1 in the circulation is approximately doubled in concentration in pre-eclampsia patients, and may permit early identification of pregnant women at risk of subsequently developing the disease [8]. Similarly, circulating ICAM-1 concentration in apparently healthy men correlates with risk of future myocardial infarction [9].

REFERENCES

1. Dustin, M.L., Rothlein, R., Bhan, A.K., Dinarello, C.A. & Springer, T.A. (1986). Induction by IL-1 and interferon-γ: tissue distribution, biochemistry and function of a natural adherence molecule (ICAM-1). *Journal of Immunology* **137**, 245–254.
2. Staunton, D.E., Marlin, S.D., Stratowa, C., Dustin, M.L. & Springer, T.A. (1988). Primary structure of ICAM-1 demonstrates interaction between members of the immunoglobulin and integrin supergene families. *Cell* **52**, 925–933.
3. Dustin, M.L. & Springer, T.A. (1991). Role of lymphocyte adhesion receptors in transient interactions and cell locomotion. *Annual Review of Immunology* **9**, 27–66.
4. Diamond, M.S., Staunton, D.E., Marlin, S.D. & Springer, T.A. (1991). Binding of the integrin Mac-1 (CD11b/CD18) to the third immunoglobulin-like domain of ICAM-1 (CD54) and its regulation by glycosylation. *Cell* **65**, 961–971.
5. Rosenstein, Y., Park, J.K., Hahn, W.C., Rosen, F.S., Bierer, B.E. & Burakoff, S.J. (1991). CD43, a molecule defective in Wiskott–Aldrich syndrome, binds ICAM-1. *Nature* **354**, 233–235.
6. McCourt, P.A.G., Ek, B., Forsberg, N. & Gustafson, S. (1994). Intercellular adhesion molecule-1 is a cell surface receptor for hyaluronan. *Journal of Biological Chemistry* **269**, 30081–30084.
7. Steeber, D.A., Tang, M.L.K., Green, N.E., Zhang, X.-Q., Sloane, J.E. & Tedder, T.F. (1999). Leukocyte entry into sites of inflammation requires overlapping interactions between the L-selectin and ICAM-1 pathways. *Journal of Immunology* **163**, 2176–2186.
8. Krauss, T., Kuhn, W., Lakoma, C. & Augustin, H.G. (1997). Circulating endothelial cell adhesion molecules as diagnostic markers for the early identification of pregnant women at risk for development of pre-eclampsia. *American Journal of Obstetrics and Gynecology* **177**, 443–449.
9. Ridker, P.M., Hennekens, C.H., Roitmann-Johnstone, B., Stampfer, M.J. & Allen, J. (1998). Plasma concentration of soluble intercellular adhesion molecule 1 and risks of future myocardial infarction in apparently healthy men. *Lancet* **351**, 88–92.

CD69

Other names: Early activation antigen (EA1); Leu-23
Mol. wt: 60 000 (native); 28 000 and 32 000 (subunit)
Specificity: GalNAc, GlcNAc
Sources: Human leukocytes and platelets

ISOLATION

CD69 was first identified on activated human leukocytes with a murine monoclonal antibody as a 60 KDa protein appearing within an hour of an activation stimulus and before the expression of IL-2 receptors [1]. Later a human cDNA encoding CD69 was cloned on the basis of its expression in COS cells [2].

STRUCTURE

CD69 is a phosphorylated glycoprotein dimer, with an apparent molecular weight of 60 000 determined by SDS-PAGE under non-reducing conditions, and subunits of 28 000 and 32 000 under reducing conditions [1,2]. The cDNA encodes a single 199-amino acid polypeptide with a predicted Mr of 22 559 [2]. Its structure is that of a type II membrane receptor with a 40-residue N-terminal region in the cytoplasm, a transmembrane domain of 21 amino acids and an extracellular C-type lectin domain of 138 amino acids at the C-terminal side [2]. This structure differs slightly from other type-II receptors in having a much reduced neck sequence (no 'stalk') between the transmembrane and lectin domains [2]; a cysteine (68) located in that sequence is essential for dimerisation [3].

BIOLOGICAL ACTIVITIES

Soluble CD69 dimers expressed in bacteria bind immobilised GlcNAc-BSA, but not other neoglycoproteins (BSA derivatised with fucose, mannose, galactose or glucose). This binding was strongly inhibited by GlcNAc and GalNAc [3].

CD69 has a very tight association with calcium ions, but lectin activity was shown to be calcium-dependent [3].

Cross-linking of CD69 with anti-CD69 promotes Ca^{2+} mobilisation, indicating a signal transducing capacity [4].

TISSUE AND/OR SUBCELLULAR DISTRIBUTION

CD69 is constitutively expressed on human platelets and a subset of thymocytes [4], as well as being induced in lymphocytes and natural killer cells upon activation [1,5]. It is expressed on the cell surface [1].

POSSIBLE FUNCTION(S)

The function of CD69 is unknown, although CD69 is assumed to have some role in the activation processes associated with both innate and acquired immunity [5].

COMMENTS

Another construct of the extracellular part of CD69 has been expressed in *Escherichia coli* [6]; this recombinant protein did not bind any of the monosaccharides tested, although some binding to fucoidin was detected.

The CD69 gene is broadly similar in structure to that of other type-II lectin-like receptors, and is located in the NK gene complex in both human and murine genomes [5].

REFERENCES

1. Hara, T., Jung, L.K.L., Bjorndahl, J.M. & Fu, S.M. (1986). Human T cell activation. III. Rapid induction of a phosphorylated 28 kD/32 kD disulphide linked early activation antigen (EA1) by 12-*O*-tetradecanoyl phorbol-13-acetate, mitogens and antigens. *Journal of Experimental Medicine* **164**, 1988–2005.
2. Hamann, J., Fiebig, H. & Strauss, M. (1993). Expression cloning of the early activation antigen CD69, a type II integral membrane protein with a C-type lectin domain. *Journal of Immunology* **150**, 4920–4927.
3. Bezouska, K., Nepovim, A., Horváth, O., Posípšil, M., Hamann, J. & Feizi, T. (1995). CD69 antigen of human lymphocytes is a calcium-dependent carbohydrate-binding protein. *Biochemical and Biophysical Research Communications* **208**, 68–74.
4. Testi, R., Pulcinelli, F., Frati, L., Gazzaniga, P.P. & Santoni, A. (1990). CD69 is expressed on platelets and mediates platelet activation and aggregation. *Journal of Experimental Medicine* **172**, 701–707.
5. Brown, M.G., Scalzo, A.A., Matsumoto, K. & Yokoyama, W.M. (1997). The natural killer gene complex: a genetic basis for understanding natural killer cell function and innate immunity. *Immunological Reviews* **155**, 53–65.
6. Childs, R.A., Galustian, C., Lawson, A.M., Dougan, G., Benwell, K., Frankel, G. & Feizi, T. (1999). Recombinant soluble human CD69 dimer produced in *Escherichia coli*: reevaluation of saccharide binding. *Biochemical and Biophysical Research Communications* **266**, 19–23.

Cepaea hortensis sialic acid-binding lectin

Other name: CHA-I
Mol. wt: 80 000 (native); 23 000 and 16 000 (subunit)
Specificity: Sialic acid
Source: Albumin glands of the snail *Cepaea hortensis*

ISOLATION

A clarified saline extract of albumin glands was first absorbed with group C streptococci to remove another lectin [1,2]. The sialic acid-binding lectin was then partially purified by gel filtration on Sephadex G100 followed by adsorption to a column of asialofetuin–Sepharose. The lectin eluted from the column in 6 M urea in glycine–HCl, pH 2.5 [2].

STRUCTURE

The lectin was eluted as a single symmetrical peak from a Sephacryl S-200 column in a volume corresponding to an apparent molecular weight of 80 000 [2]. SDS-PAGE in the presence of 2-mercaptoethanol yielded bands corresponding to 16 000 and 23 000 [2]. No evidence was available to establish whether lectin activity was associated with one or both bands.

BIOLOGICAL ACTIVITIES

The lectin agglutinated human (group A with a four-fold higher titre than groups B or O), murine, guinea-pig, rat, canine and caprine erythrocytes, but the highest titre was with ovine cells [2]. Bovine, equine, lapine and various simian erythrocytes were not agglutinated. This activity was calcium-dependent; manganese could substitute adequately for calcium but magnesium was much less effective. Haemagglutinating activity was retained in the presence of 6 M urea [2].

The lectin also agglutinated streptococcal group B strains containing a type-specific polysaccharide [1,2] and promoted the phagocytosis of group B streptococci by human granulocytes in the presence of complement in a dose-dependent manner [3].

The above biological activities are inhibited by sialic acid [1–4]. GlcNAc is the only other inhibitory monosaccharide, but is fifty-fold less potent [4]. 2-α-Methyl-9-*O*-acetylneuraminic acid was the very best inhibitor, followed closely by 2-α-methylneuraminic acid and 2-α-benzylneuraminic acid. The best glycoprotein inhibitors were those carrying O-chains only or preferentially, with the exception of Tamm–Horsfall glycoprotein (which carries only N-chains). The lectin therefore has a narrow specificity for α-linked *N*-acetylneuraminic acid but tolerates a number of modifications [2,4].

TISSUE AND/OR SUBCELLULAR DISTRIBUTION

The lectin is found in albumin glands, but may not have been investigated in other tissues.

POSSIBLE FUNCTION(S)

The function of this lectin is unknown, but an anti-bacterial defence role is an obvious possibility.

COMMENTS

The *C. hortensis* sialic acid-binding lectin has been successfully used to seek novel group B streptococcal serotypes among clinical isolates untypable by antisera to conventional types [5].

The albumin glands of *C. hortensis* contain a separate lectin with specificity for GalNAc which reacts with group C streptococci [1,2].

REFERENCES

1. Wagner, M. (1982). Agglutination of bacteria by a sialic acid-specific lectin of the snail *Cepaea hortensis. Acta Histochemica* **71**, 35–39.
2. Wagner, M. (1990). A sialic acid specific lectin from the snail *Cepaea hortensis*: purification and preliminary characterization. In: *Lectins – Biology, Biochemistry, Clinical Biochemistry*, Volume 7 (eds J. Kocourek & D.L.J. Freed) pp. 83–89. Sigma Chemical Company, St Louis, MO.
3. Holm, S.E., Bergholm, A.-M., Wagner, B. & Wagner, M. (1985). A sialic acid-specific lectin from *Cepaea hortensis* that promotes phagocytosis of a group-B, type Ia, streptococcal strain. *Journal of Medical Microbiology* **19**, 317–323.
4. Brossmer, R., Wagner, M. & Fischer, E. (1992). Specificity of the sialic acid-binding lectin from the snail *Cepaea hortensis. Journal of Biological Chemistry* **267**, 8752–8756.
5. Motlova, J., Wagner, M. & Jelinkova, J. (1986). A search for new group B streptococcal serotypes. *Journal of Medical Microbiology* **22**, 101–105.

Cerebellar soluble lectin(s)

Other name: CSL
Mol. wt: 1.2×10^6 (native); 33 000 and 31 500 (subunit)
Specificity: $(\text{Mannose})_6(\text{GlcNAc})_2$-
Source: Mammalian nervous tissues

ISOLATION

CSLs were first obtained in reasonably pure form by sequential extraction of a homogenate of cerebellum from 2–3-week-old rats, in which the CSL-containing fraction was solubilised with 0.5 M mannose in the absence of detergent. The two major subunits were separated by preparative gel electrophoresis and used to prepare rabbit antibodies, which in turn permitted the isolation of the CSL family by immunoaffinity chromatography [1].

Apparently equivalent molecules have been purified from porcine and human white matter by affinity chromatography on immobilised mannose [2].

STRUCTURE

The term 'cerebellar soluble lectin(s)' is used to describe a family of basic, lysine- and alanine-rich proteins with an apparent molecular weight of 1.2×10^6 estimated by gel filtration. SDS-PAGE analysis (in the presence or absence of reducing agent) yields two major subunits of 33 000 and 31 500, but higher molecular weight forms also exist. The smallest subunit has a blocked N-terminus, while the other major subunit has lysine as its first residue. The subunits are not serologically distinct, but apparent differences in immunohistochemical localisation suggest that each subunit must aggregate homotypically to form a polyvalent macromolecule [1].

BIOLOGICAL ACTIVITIES

CSLs have been found to bind to a variety of ligands in a Ca^{2+}-independent manner, including heparin, myelin-associated glycoprotein (qv), endogenous glycoproteins B1 and B2 transiently expressed in the cerebellum, and the major glycoprotein (P0) of the peripheral nervous system [2,3].

Mannose is a relatively weak inhibitor of CSL, but other common monosaccharides are not inhibitory at all. Glycopeptides isolated from the endogenous cerebellar ligands are potent inhibitors. The structure $(\text{mannose})_6$ $(\text{GlcNAc})_2$-asparagine is specifically recognised. CSL can agglutinate human red blood cells coated with such glycoasparagines at very high titre [2–4].

Rat or human CSL can agglutinate fresh or fixed resting human lymphocytes at high titre; this activity was divalent cation-independent and reversible with oligomannosidic inhibitors [5].

TISSUE AND/OR SUBCELLULAR DISTRIBUTION

Although predominantly associated with nervous tissue, CSL appears to occur much more widely, having been found in liver, kidney and intestine. CSL is an intracellular lectin which can be externalised and is found in many tight junctions. Within nervous tissues, CSL is found in granule cells of the cerebellum, astrocytes, oligodendrocytes and myelinating Schwann cells [2,3].

POSSIBLE FUNCTION(S)

Based on experiments with Fab fragments of anti-CSL antibodies, it appears that CSL is involved both in the contact guidance of neuronal migration and in the compaction of myelin [2,4]. Indeed, CSL appears to have several roles in the process of myelination, including that of growth factor for proliferating myelinating cells. This may be an example of a more general function as an endogenous amplifier of activation signals [5].

COMMENTS

Autoantibodies against CSL are present in the cerebrospinal fluid of around 90% of multiple sclerosis patients (compared with <15% of patients with other neurological diseases) [6,7]. The immunoblotting test used is of potential value in the diagnosis of the disorder.

Another Ca^{2+}-independent, mannose-binding lectin was recovered from rat cerebellum and named R1; it is a membrane-bound lectin insoluble in the absence of detergent and has a saccharide specificity similar but not identical to CSL [8].

REFERENCES

1. Zanetta, J.-P., Meyer, A., Kuchler, S. & Vincendon, G. (1987). Isolation and immunochemical study of a soluble cerebellar lectin delineating its structure and function. *Journal of Neurochemistry* **49**, 1250–1257.
2. Zanetta, J.-P., Kuchler, S., Lehmann, S., Badache, A., Maschke, S., Marschal, P., Dufourcq, P. & Vincendon, G. (1992). Cerebellar lectins. *International Review of Cytology* **135**, 123–154.
3. Zanetta, J.-P., Kuchler, S., Lehmann, S., Badache, A., Maschke, S., Thomas, D., Dufourco, P. & Vincendon, G. (1992). Glycoproteins and lectins in cell adhesion and cell recognition processes. *Histochemical Journal* **24**, 791–804.
4. Zanetta, J.-P. (1998). Structure and functions of lectins in the central and peripheral nervous system. *Acta Anatomica* **161**, 180–195.
5. Zanetta, J.-P., Wantyghem, J., Kuchler-Bopp, S., Badache, A. & Aubery, M. (1995). Human lymphocyte activation is associated with the early and high-level expression of the endogenous lectin CSL at the cell surface. *Biochemical Journal* **311**, 629–636.
6. Zanetta, J.-P., Warter, J.-M., Kuchler, S., Marschal, P., Rumbach, L., Lehmann, S., Tranchant, C., Reeber, A. & Vincendon, G. (1990). Antibodies to cerebellar soluble lectin CSL in multiple sclerosis. *Lancet* **335**, 1482–1484.
7. Zanetta, J.-P., Tranchant, C., Kuchler-Bopp, S., Lehmann, S. & Warter, J.-M. (1994). Presence of anti-CSL antibodies in the cerebrospinal fluid of patients: a sensitive and specific test in the diagnosis of multiple sclerosis. *Journal of Neuroimmunology* **52**, 175–182.
8. Zanetta, J.-P., Dontenwill, M., Meyer, A. & Roussel, G. (1985). Isolation and immunohistochemical localization of a lectin-like molecule from the rat cerebellum. *Developmental Brain Research* **17**, 233–243.

Charybdis japonica blood type B lectin

Other name: CJA-B
Mol. wt: 300 000 (native); 19 000 (subunit)
Specificity: α-Glucosides, α-galactosides
Source: Haemolymph of the Japanese crab *Charybdis japonica*

ISOLATION

CJA-B was isolated from filtered haemolymph by affinity chromatography on Sephadex G-200 using palatinose (glucose-1,6-fructose) to elute the adsorbed lectin [1]. This single step resulted in an 1300-fold increase in specific activity, and the product appeared virtually pure by gradient gel electrophoresis [1].

STRUCTURE

CJA-B is a glycoprotein relatively rich in hydroxy and acidic amino acids; the first 20 residues were determined, including a single N-terminal phenylalanine [1]. The native molecule (apparent Mr 300 000 by gradient PAGE) consists of non-covalently linked aggregates of (presumably eight) dimers of 38 kDa. The dimers are composed of identical 19 kDa subunits linked by disulphide bonds (established by SDS-PAGE in the presence and absence, respectively, of 2-mercaptoethanol). The isoelectric point was found to be 5.0 [1].

BIOLOGICAL ACTIVITIES

CJA-B agglutinated human group B erythrocytes specifically, showing no activity towards group A or O cells [1]. This activity was Ca^{2+}-dependent and could not be inhibited by monosaccharides, apart from a weak inhibitory effect by glucose and arabinose. The disaccharide palatinose and the tetrasaccharide stachyose were the best simple oligosaccharide inhibitors. Dextran was a potent inhibitor, as were blood group B saccharides coupled to silica. The primary sugar specificity seemed to be for non-reducing terminals of oligosaccharides bearing α-glucose and α-galactose residues [1].

TISSUE AND/OR SUBCELLULAR DISTRIBUTION

The lectin was found in haemolymph plasma, but other tissues do not seem to have been investigated [1].

POSSIBLE FUNCTION(S)

No evidence concerning function was reported [1].

COMMENTS

The blood group B specificity of this lectin is shared by the lectin from another crab species, *Scylla serrata* (qv) [2]. However, the latter has some weak activity towards blood group O cells, and paradoxically is highly specific for *N*-glycolylneuraminic acid, not galactose. CJA-B is an exclusively group B lectin, and this characteristic has a rational foundation since it clearly recognises Gal-α1,3-Gal.

The haemolymph of *C. japonica* also contains another lectin, with specificity for *N*-acetyl-aminosugars [1].

REFERENCES

1. Umetsu, K., Yamashita, K. & Suzuki, T. (1991). Purification and carbohydrate-binding specificities of a blood group B binding lectin from hemolymph of a crab (*Charybdis japonica*). *Journal of Biochemistry* **109**, 718–721.
2. Mercy, P.D. & Ravindranath, M.H. (1993). Purification and characterization of *N*-glycolylneuraminic acid specific lectin from *Scylla serrata*. *European Journal of Biochemistry* **215**, 697–704.

Chicken galectins

Other names: CLL-I and CLL-II; C-16 and C-14
Mol. wt: 30 000 (native); 15 000 (subunit)
Specificity: *N*-acetyllactosamine
Source: Chicken (especially embryonic) tissues

ISOLATION

Haemagglutinating activity was first identified in extracts obtained by homogenising embryonic tissues in a buffer containing lactose and 2-mercaptoethanol and then dialysing the soluble fraction to remove the lactose [1,2]. A mixture of lectins could be obtained by affinity chromatography on asialofetuin–Sepharose or *p*-aminophenyl-β-D-lactoside–Sepharose, eluting with lactose [3–8]. The C-16 and C-14 lectins can be separated and isolated in purified form by isoelectric focusing or by ion-exchange chromatography on DEAE–cellulose [8].

The cDNA for the 14 kDa lectin has been cloned and the nucleotide sequence determined [9]. An incomplete cDNA of the 16 kDa lectin was used in conjunction with peptide sequences to derive a complete amino acid sequence [10].

STRUCTURE

Both C-16 and C-14 consist of non-glycosylated polypeptides of 134 amino acid residues. The sequences have been published in full, and characteristics include an acetylated N-terminus and no obvious signal sequence [8–10]. C-14 has a few small regions homologous to discoidin I (qv) near the N-terminus [9]. Although C-14 and C-16 have apparent molecular weight values of 14 000 and 16 000, respectively, determined by SDS-PAGE, their true molecular weights are virtually identical (14 973 and 14 975, respectively). The two lectins are immunologically distinct and have very different peptide maps, yet share about 47% sequence identity with each other. Both have approximately 50% and 40% sequence identity, respectively, to mammalian galectins-1 and -2. The isoelectric points are approximately pH 4.1 and pH 6.3, respectively, for C-16 and C-14 [6].

C-16 exists as a homodimer; C-14 is usually found as a monomer [6].

BIOLOGICAL ACTIVITIES

Both C-14 and C-16 agglutinate glutaraldehyde-fixed, trypsin-treated lapine erythrocytes in a divalent cation-independent manner, but the specific activity of C-16 is very much greater [6], perhaps on account of a greater tendency towards dimer formation. Haemagglutination by both lectins is inhibited by galactose, but β-galactosides (thiogalactoside, lactose etc.) are much more potent [1,4,6]. Haemagglutinating activity is stable only in the presence of reducing agents [3].

Both C-14 and C-16 were weakly mitogenic for murine non-adherant, Thy-1-negative spleen cells [11]; this activity also appeared to be mediated by the carbohydrate-binding site.

TISSUE AND/OR SUBCELLULAR DISTRIBUTION

Both C-14 and C-16 occur in many chicken organs and tissues, especially during pre-natal development. C-14 is abundantly expressed in embryonic skin [10] and kidney [7], and above all in adult intestine [7]. C-16 is most abundant in embryonic skeletal muscle but adult liver is also a rich source [7].

Within the cell, C-16 has been found in the cytoplasm, on the cell surface and outside the cell [4,12,13]. A similar range of widespread locations has been reported for C-14 [14]. Additionally, both C-14 and C-16 have been detected in nuclei of dermal fibroblasts [15].

POSSIBLE FUNCTION(S)

No function can be ascribed to these lectins with confidence, but their developmental regulation (see **Comments** section) and apparent externalisation suggest a role in the organisation of the extracellular environment and thereby an influence on differentiation and morphogenesis. It is a regularly expressed opinion of reviewers that these and other galectins have different functions at different times and locations [16,17].

COMMENTS

C-16 first attracted attention on account of its dramatic developmental regulation in embryonic pectoral muscle [1]. Although the most marked changes are apparent in skeletal muscle, alterations in activity with developmental age are a feature of both C-16 and C-14 in other tissues. In general, the levels of both C-16 and C-14 are higher in embryonic compared to adult tissues [7]; however, C-16 is more abundant in adult liver (and, to a lesser extent, brain), while C-14 is vastly more abundant in adult intestine. Both lectins are present in embryonic skin, where the expression of each lectin is regulated independently and differently between dermis and epidermis [15,18].

Embryonic chick muscle also contains an entirely separate glycosaminoglycan-binding lectin, chicken heparin lectin [19,20].

REFERENCES

1. Nowak, T.P., Haywood, P.L. & Barondes, S.H. (1976). Developmentally regulated lectin in embryonic chick muscle and a mitogenic cell line. *Biochemical and Biophysical Research Communications* **68**, 650–657.

2. Kobiler, D. & Barondes, S.H. (1977). Lectin activity from embryonic chick brain, heart and liver: changes with development. *Developmental Biology* **60**, 326–330.
3. Den, H. & Malinyak, D.A. (1977). Isolation and properties of β-D-galactoside-specific lectin from chick embryo thigh muscle. *Journal of Biological Chemistry* **252**, 5444–5448.
4. Nowak, T.P., Kobiler, D., Roel, L.E. & Barondes, S.H. (1977). Developmentally regulated lectin from embryonic chick pectoral muscle. *Journal of Biological Chemistry* **252**, 6026–6030.
5. Kobiler, D., Beyer, E.C. & Barondes, S.H. (1978). Developmentally regulated lectins from chick muscle, brain and liver have similar chemical and immmunological properties. *Developmental Biology* **64**, 265–272.
6. Beyer, E.C., Zweig, S.C. & Barondes, S.H. (1980). Two lactose binding lectins from chicken tissues. *Journal of Biological Chemistry* **255**, 4236–4239.
7. Beyer, E.C. & Barondes, S.H. (1982). Quantitation of two endogenous lactose-inhibitable lectins in embryonic and adult chicken tissues. *Journal of Cell Biology* **92**, 23–27.
8. Hirabayashi, J., Kawasaki, H., Suzuki, K. & Kasai, K. (1987). Complete amino acid sequence of the 14 kDa β-galactoside-binding lectin of chick embryo. *Journal of Biochemistry* **101**, 775–787.
9. Ohyama, Y., Hirabayashi, J., Oda, Y., Ohno, S., Kawasaki, H., Suzuki, K. & Kasai, K.-I. (1986). Nucleotide sequence of chick 14 K β-galactoside-binding lectin mRNA. *Biochemical and Biophysical Research Communications* **134**, 51–56.
10. Sakakura, Y., Hirabayashi, J., Oda, Y., Ohyama, Y. & Kasai, K. (1990). Structure of chicken 16 kDa β-galactoside-binding lectin. *Journal of Biological Chemistry* **265**, 21573–21579.
11. Lipsick, J.S., Beyer, E.C., Barondes, S.H. & Kaplan, N.O. (1980). Lectins from chicken tissues are mitogenic for Thy-1 negative murine spleen cells. *Biochemical and Biophysical Research Communications* **97**, 56–61.
12. Barondes, S.H. & Haywood-Reid, P.L. (1981). Externalization of an endogenous chicken muscle lectin with in vivo development. *Journal of Cell Biology* **91**, 568–572.
13. Beyer, E.C., Tokuyasu, K.T. & Barondes, S.H. (1979) Localization of an endogenous lectin in chicken liver, intestine and pancreas. *Journal of Cell Biology* **82**, 565–571.
14. Beyer, E.C. & Barondes, S.H. (1982). Secretion of endogenous lectin by chicken intestinal goblet cells. *Journal of Cell Biology* **92**, 28–33.
15. Akimoto, Y., Obinata, A., Hirabayashi, J., Sakakura, Y., Endo, H., Kasai, K. & Hirano, H. (1995). Changes in expression of two endogenous β-galactoside-binding isolectins in the dermis of chick embryonic skin during development in ovo and in vitro. *Cell and Tissue Research* **279**, 3–12.
16. Barondes, S.H. (1984). Soluble lectins: a new class of extracellular proteins. *Science* **223**, 1259–1264.
17. Kasai, K. & Hirabayashi, J. (1996). Galectins: a family of animal lectins that decipher glycocodes. *Journal of Biochemistry* **119**, 1–8.
18. Akimoto, Y., Obinata, A., Hirabayashi, J., Sakakura, Y., Endo, Y., Endo, H., Kasai, K. & Hirano, H. (1993). Secretion of endogenous 16 kDa β-galactoside-binding lectin from vitamin A-pretreated chick embryonic cultured skin. *Experimental Cell Research* **205**, 251–260.
19. Kobiler, D. & Barondes, S.H. (1979). Lectin from embryonic chick muscle that interacts with glycosaminoglycans. *FEBS Letters* **101**, 257–261.
20. Ceri, H., Kobiler, D. & Barondes, S.H. (1981). Heparin-inhibitable lectin. *Journal of Biological Chemistry* **256**, 390–394.

Collectin-43

Other names: CL-43
Mol. wt: 120 000 (native); 40 000 (subunit)
Specificity: Mannose
Source: Bovine serum

ISOLATION

CL-43 was isolated after co-purification with bovine mannan-binding lectin (MBL) (qv) on immobilised mannan; it was separated from MBL by ion-exchange chromatography [1]. A cDNA clone coding for CL-43 has also been isolated from bovine liver [2].

STRUCTURE

CL-43 is a protein (no attached carbohydrate) rich in glycine and other collagen-suggestive amino acids [1]. The complete amino acid sequence, derived from partial polypeptide sequencing, cDNA cloning and RT-PCR techniques, has been published [2]. The 301 amino acid residues provide a calculated Mr of 31 500; this is inexplicably lower than the value of 43 000 obtained by SDS-PAGE [1,3]. The apparent Mr value of 120 000 in SDS-PAGE under non-reducing conditions [1,3] is consistent with a native Mr of 119 000–138 000 determined by density gradient ultracentrifugation; this may be taken as an indication that CL-43 normally exists as a monomeric triplet, and indeed only monomers were seen by electron microscopy [3].

Like other collectins, each polypeptide chain has a short N-terminal region followed by a collagen-like sequence then a 'neck' region linking the collagenous tail to the carbohydrate-recognition domain at the C-terminal end [2]. The collagen-like region is somewhat shorter than in conglutinin or SP-D; the carbohydrate recognition domain possessed 81% identity to conglutinin and 63% identity to bovine SP-D (overall identity >70% to both). The isoelectric point was at pH 4.9 [3].

BIOLOGICAL ACTIVITIES

CL-43 binds to immobilised mannan in a Ca^{2+}-dependent manner [1]. This binding was inhibited by various monosaccharides, of which mannose was the most effective. In order of potency, the following were also inhibitory: fucose>GlcNAc>glucose>galactose>>GalNAc [1].

CL-43 (like bovine conglutin and SP-D) does not activate complement under conditions permitting activation by MBL [3]. Unlike conglutinin, CL-43 did not bind to solid-phase or soluble iC3b [3]. However, CL-43 can bind to the same collectin receptor as the other collectins [3].

CL-43 has anti-viral activity, and was more active than conglutinin against rotaviruses [4].

TISSUE AND/OR SUBCELLULAR DISTRIBUTION

Cl-43 is synthesised in bovine liver [2] and secreted into the blood. Northern blot analyses failed to detect any synthesis in heart, spleen, lung or muscle tissue [2]. No cross-hybridising signals were detected in mRNA from various other species.

POSSIBLE FUNCTION(S)

A defensive function is presumed, by analogy with other collectins [1–4].

COMMENTS

CL-43 and conglutinin have only been detected with certainty in the Bovidae, and it has been suggested that the genes encoding them are derived from an ancestral SP-D-like gene which duplicated during evolution [2].

REFERENCES

1. Holmskov, U., Teisner, B., Willis, A.C., Reid, K.B.M. & Jensenius, J.C. (1993). Purification and characterization of a bovine serum lectin (CL-43) with structural homology to conglutinin and SP-D and carbohydrate specificity similar to mannan-binding protein. *Journal of Biological Chemistry* **268**, 10120–10125.
2. Lim, B.-L., Willis, A.C., Reid, K.B.M., Lu, J., Laursen, S.B., Jensenius, J.C. & Holmskov, U. (1994). Primary structure of bovine collectin-43 (CL-43). *Journal of Biological Chemistry* **269**, 11820–11824.
3. Holmskov, U., Laursen, S.B., Malhotra, R., Weidemann, H., Timpl, R., Stuart, G.R., Tornoe, I., Madsen, P.S., Reid, K.B.M. & Jensenius, J.C. (1995). Comparative study of the structural and functional properties of a bovine plasma C-type lectin, collectin-43, with other collectins. *Biochemical Journal* **305**, 889–896.
4. Reading, P.C., Holmskov, U. & Anders, E.M. (1998). Antiviral activity of bovine collectins against rotaviruses. *Journal of General Virology* **79**, 2255–2263.

Conglutinin

Mol. wt: 430 000 (native); 35 700 (subunit)
Specificity: GlcNAc
Source: Serum of ungulates

ISOLATION

Conglutinin can be isolated by a combination of affinity chromatography/adsorption and ion-exchange [1,2]. Conglutinin cDNA has been isolated from a bovine liver cDNA library [3]. Recombinant conglutinin has been expressed in CHO cells [4].

STRUCTURE

Conglutinin is a protein or glycoprotein with the typical collectin triplet building block (see Chapter 3). In this case, the native molecule is a homotetramer of those triplet subunits which assumes a cruciform shape [5]. Each polypeptide chain of the triplet has 351 amino acid residues and has been fully sequenced. There are a few discrepancies between the chemically-determined sequence [2] and the sequence deduced from the nucleotide sequence [3]. The first 25 residues at the N-terminus are followed by a collagen-like domain of 171 residues; the N-terminal half is connected to the C-terminal carbohydrate-binding domain by a short link or neck sequence [2]. The entire sequence shows 78% identity with bovine surfactant protein D (SP-D) (qv) [3].

The calculated Mr of the basic polypeptide is 35 700, much smaller than the values (43 000–48 000) obtained by SDS-PAGE [2], but consistent with the value obtained by sedimentation equilibrium in the presence of guanidine–HCl [5]. The apparent Mr estimated by gel filtration is also significantly higher than the true Mr [5].

BIOLOGICAL ACTIVITIES

Conglutinin binds to immune complexes, via carbohydrate (a high mannose structure containing terminal α 1,2 linkages) on the complement component derivative, iC3b [6]. This lectin activity enables it to agglutinate complement-coated erythrocytes or bind to zymosan and mannan from yeast extracts. Calcium is essential for activity, which can be inhibited by GlcNAc, mannose or fucose [7]. GlcNAc was the best inhibitor, followed by L-fucose [8,9].

Conglutinin has anti-bacterial activity against Gram-negative bacteria that can be demonstrated *in vitro* and *in vivo* [1]. Conglutinin can also neutralise viral infectivity, including that due to influenza A virus [10,11] and rotaviruses [11].

TISSUE AND/OR SUBCELLULAR DISTRIBUTION

Northern blotting analysis indicated that conglutinin is synthesised in bovine and ovine livers, but not in the corresponding tissue from humans, rats or mice. The liver was the only bovine source giving a positive signal [3].

POSSIBLE FUNCTION(S)

The concentration of conglutinin is largely inherited, with low levels associated with increased frequency of infections in calves [12]. Although there is no evidence that conglutinin activates the complement system, it seems to act as an opsonin by binding to complement-coated bacteria and viruses and thereby promoting their attachment to phagocytic cells via C1q/collectin receptors [10,13].

COMMENTS

Although a conglutinin-like activity has been detected and partially purified from human serum [14,15], the lack of a conglutinin RNA signal from human liver [3] casts doubt on the very existence of human conglutinin. Indeed, it seems more likely that conglutinin evolved from duplication of the SP-D gene within the Bovidae.

Bovine conglutinin is present over a wide concentration range, from below the limit of detection (1.25 μg/ml) to 47 μg/ml; interestingly, the levels are generally higher in male cattle compared to females [12]. There may also be some selective advantage in low conglutinin concentrations: in an experimental system, conglutinin, by providing an additional port of entry, actually enhanced the infection of murine liver with herpes simplex virus type-2 [16].

The ability of conglutinin to bind iC3b has been exploited in some assays for circulating immune complexes [17].

REFERENCES

1. Friis-Christiansen, P., Thiel, S., Sveghag, S.-E., Dessau, R., Svendsen, P., Anderson, O., Laursen, S.B. & Jensenius, J.-C. (1990). In vivo and in vitro antibacterial activity of conglutinin, a mammalian plasma lectin. *Scandinavian Journal of Immunology* **31**, 453–460.

2. Lee, Y.M., Leiby, K.R., Allar, J., Paris, K., Lerch, B. & Okarma, T.B. (1991). Primary structure of bovine conglutinin, a member of the C-type animal lectin family. *Journal of Biological Chemistry* **266**, 2715–2723.
3. Lu, J., Laursen, S.B., Thiel, S., Jensenius, J.C. & Reid, K.B.M. (1993). The cDNA cloning of conglutinin and the identification of liver as a primary site of synthesis of conglutinin in members of the Bovidae. *Biochemical Journal* **292**, 157–162.
4. Suzuki, Y., Eda, S., Kawai, T., Ohtani, K., Kasem T., Sakamotom T., Wakamiya, N. (1997). Characterization of recombinant bovine conglutinin expressed in a mammalian cell. *Biochemical and Biophysical Research Communications* **238**, 856–860.
5. Strang, C.J., Slayter, H.S., Lachmann, P.J. & Davis, D.L. (1986). Ultrastructure and composition of ovine conglutinin. *Biochemical Journal* **234**, 381–389.
6. Young, N.M. & Leon, M.A. (1987). The carbohydrate specificity of conglutinin and its homology to proteins in the hepatic lectin family. *Biochemical and Biophysical Research Communications* **143**, 645–651.
7. Loveless, R.W., Feizi, T., Childs, R.A., Mizuochi, T., Stoll, M.S., Oldroyd, R.G. & Lachmann, P.J. (1989). Bovine serum conglutinin is a lectin which binds non-reducing terminal *N*-acetylglucosamine, mannose and fucose residues. *Biochemical Journal* **258**, 109–113.
8. Lu, J., Willis, A.C. & Reid, K.B.M. (1992). Purification, characterization and cDNA cloning of human lung surfactant protein D. *Biochemical Journal* **284**, 795–802.
9. Haurum, J.S., Thiel, S., Haagsman, H.P., Laursen, S.B., Larsen, B. & Jensenius, J.C. (1993). Studies on the carbohydrate-binding characteristics of human pulmonary surfactant-associated protein A and comparison with two other collectins: mannan-binding protein and conglutinin. *Biochemical Journal* **293**, 873–878.
10. Hartshorn, K.L., Sastry, K., Brown, D., White, M.R., Okarma, T.B., Lee, Y.-M. & Tauber, A.I. (1993). Conglutinin acts as an opsonin for influenza A viruses. *Journal of Immunology* **151**, 6265–6273.
11. Reading, P.C., Holmskov, U. & Anders, E.M. (1998). Antiviral activity of bovine collectins against rotaviruses. *Journal of General Virology* **79**, 2255–2263.
12. Holskov, U., Jensenius, J.C., Tornoe, I. & Lovendahl, P. (1998). The plasma levels of conglutinin are heritable in cattle and low levels predispose to infection. *Immunology* **93**, 431–436.
13. Hirani, S., Lambris, J.D. & Muller-Eberhard, H.J. (1985). Localization of the conglutinin binding site on the third component of human complement. *Journal of Immunology* **134**, 1105–1109.
14. Baatrup, G., Thiel, S., Isagar, H., Svehag, S.E. & Jensenius, J.C. (1987). Demonstration in human plasma of a lectin activity analogous to that of bovine conglutinin. *Scandinavian Journal of Immunology* **26**, 355–361.
15. Thiel, S., Baatrup, G., Friis-Christiansen, P., Svehag, S.-E. & Jensenius, J.C. (1987). Characterization of a lectin in human plasma analogous to bovine conglutinin. *Scandinavian Journal of Immunology* **26**, 461–468.
16. Fischer, P.B., Ellermann-Eriksen, S., Thiel, S., Jensenius, J.C. & Mogensen, S.C. (1994). Mannan binding protein and bovine conglutinin mediate enhancement of *Herpes simplex* virus type-2 infection in mice. *Scandinavian Journal of Immunology* **39**, 439–445.
17. McDougal, J.S. & McDuffie, F.C. (1985). Immune complexes in man : detection and clinical significance. *Advances in Clinical Chemistry* **24**, 1–60.

C-reactive protein

Other name: CRP
Mol. wt: 120 000 (native); 24 000 (subunit)
Specificity: Galactose-6-phosphate
Source: Blood plasma of vertebrates

ISOLATION

CRP was first identified as an anti-pneumococcal C polysaccharide precipitin (and presumed to be an antibody) [1]. The CRP could be readily crystallised and this formed the basis of an early purification scheme which also involved precipitation with ammonium sulphate and pneumococcal C polysaccharide [2]. Superior methods based on affinity chromatography with solid-phase ligands have been described such as that involving C-polysaccharide–Sepharose, acrylamide–agarose and Sephacryl S-300 [3].

The gene for CRP was first isolated from two λ phage libraries containing inserted human fetal liver DNA fragments using synthetic oligonucleotide probes [4].

Homologues of human CRP have been isolated from many vertebrate species [5,6].

STRUCTURE

CRP is a non-glycosylated, ring-shaped pentamer of identical subunits with a broadly similar amino acid composition to that of the immunoglobulins [7]. Each subunit consists of 205 amino acid residues [4] with 51% sequence identity to serum amyloid P component (qv) [8].

The apparent subunit and native Mr values for CRP determined by SDS-PAGE and ultracentrifugation, respectively, are 24 000 and 120 000 [8,9].

The three-dimensional crystal structure of human CRP has been resolved at 0.3 nm [10]. The tertiary fold of CRP has a 'jelly roll' topology similar to SAP and the legume lectins, but there are significant structural differences from SAP at all levels.

BIOLOGICAL ACTIVITIES

CRP binds pneumococci and other micro-organisms in a calcium-dependent manner [1,11]; this interaction is mediated by high-affinity binding to phosphorylcholine [12]. However, CRP binding to the lipophosphoglycan on

Leishmania donovani is via repeated phosphorylated disaccharides [13], and this interaction can be inhibited most readily by galactose-6-phosphate [14]. This is consistent with reported binding to galactans [15].

CRP binds to cell nuclei and more specifically to small nuclear ribonucleoproteins *in vitro* [16], and can retard the development of autoimmune disease and anti-nuclear antibodies *in vivo* [17].

Complexed CRP can activate the classical pathway of complement *in vitro* [18] and perhaps *in vivo* [19] via binding to C1q [20]. However, CRP inhibits activation of the alternative pathway of complement and can antagonise mannan-binding lectin (qv)-mediated complement cytolysis by modulating alternative pathway recruitment through factor H (qv) [21].

CRP can stimulate the production of pro-inflammatory cytokines from tissue macrophages [22]. CRP can also interact with neutrophils and NK cells (reviewed by Gewurz *et al.* [23]).

CRP had anti-tumour activity in an experimental animal model [24].

TISSUE AND/OR SUBCELLULAR DISTRIBUTION

CRP is synthesised by hepatocytes and the distribution is mainly intravascular [25]. However, some CRP may be synthesised by leukocytes [26,27].

POSSIBLE FUNCTION(S)

In the absence of a deficiency state or polymorphism, the precise function of CRP has not been defined. However human CRP transgenic mice exhibit an increased resistance to infection with *Streptococcus pneumoniae* associated with a dramatic reduction of bacteremia [28]. Similarly, transgenic mice expressing high levels of rabbit CRP were partially protected from a lethal challenge of bacterial (*Escherichia coli*) lipopolysaccharide compared to litter mates in which CRP expression had been suppressed [29].

It seems likely that CRP functions to regulate the inflammatory response, promoting opsonisation of pathogens initially and later assisting the clearance of cell debris and thus minimising the risk of tissue damage or autoimmunity. Its role may be more complementary to, than overlapping with, that of serum amyloid P component.

COMMENTS

CRP is the prototypical acute-phase response protein, increasing up to 1000-fold from IL-6-induced synthesis but with a biological half-life of only 19 hours (in humans) [5,6,23,25,30]. It is the biochemical marker of choice for monitoring inflammation, and may provide the only evidence of an inflammatory event; however, measurement of CRP is particularly useful for assessing disease activity and response to treatment [30]. CRP measurement can also be used as a crude means of distinguishing viral (low levels) from bacterial (higher levels) infections, and is of major value in differentiating viral from bacterial meningitis [31].

Elevated CRP concentrations are associated with coronary heart disease [32], and CRP measurement has prognostic value in stable and especially unstable angina [33,34]. Mildly increased CRP concentrations within the normal range were higher in apparently healthy men who subsequently had a myocardial infarction or an ischaemic stroke than in men without subsequent vascular events [35]. A large eight-year follow-up study has confirmed the prognostic significance of CRP values for occurrence of a first major coronary event in initially healthy middle-aged men [36]. Mildly elevated CRP concentrations have also been found in association with high body mass index/obesity in normal American subjects [37].

CRP concentrations are generally much higher in rheumatoid arthritis than in other idiopathic rheumatic disorders [38], and serial measurements may be of prognostic value in that condition [39].

Inflammation is a major cause of resistance to erythropoietin treatment for anaemia in haemodialysis patients; high CRP concentration is a strong predictor of resistance to erythropoietin [40].

Hormone replacement therapy was associated with an increase in CRP in healthy women [41] and a decrease in CRP in diabetic women [42].

On the basis of very recent studies on complement regulation by CRP, it has been suggested that the function of CRP is to target functionally active complement factor H to damaged tissues and thereby mitigate complement-mediated tissue injury while permitting enhanced complement activity against invading microbes in blood during the acute phase [43].

REFERENCES

1. Tillett, W.S. & Francis, T. (1930). Serological reactions in pneumonia with a non-protein somatic fraction of pneumococcus. *Journal of Experimental Medicine* **52**, 561–571.
2. Wood, H.F., McCartney, M. & Slater, R.J. (1954). The occurrence during acute infections of a protein not normally present in the blood. V. Physical-chemical properties of the C-reactive protein crystallized by a modified technique. *Journal of Experimental Medicine* **100**, 71–79.
3. De Beer, F.C. & Pepys, M.B. (1982). Isolation of human C-reactive protein and serum amyloid P component. *Journal of Immunological Methods* **50**, 17–31.
4. Lei, K.-J., Lui, T., Zon, G., Soravia, E., Liu, T.-Y. & Goldman, N.D. (1985). Genomic DNA sequence for human C-reactive protein. *Journal of Biological Chemistry* **260**, 13377–13383.
5. Pepys, M.B. & Baltz, M.L. (1983). Acute phase proteins with special reference to C-reactive protein and related proteins (pentraxins) and serum amyloid A protein. *Advances in Immunology* **34**, 141–212.
6. Steele, D.M. & Whitehead, A.S. (1994). The major acute phase reactants : C-reactive protein, serum amyloid P component and serum amyloid A protein. *Immunology Today* **15**, 81–88.
7. Oliveira, E.B., Gotschlich. E.C. & Liu, T. (1979). Primary structure of human C-reactive protein. *Journal of Biological Chemistry* **254**, 489–502.

8. Woo, P., Korenberg, J.R. & Whitehead, A.S. (1985). Characterization of genomic and complementary DNA sequence of human C-reactive protein, and comparison with the complementary DNA sequence of serum amyloid P component. *Journal of Biological Chemistry* **260**, 13384–13388.
9. Kushner, I. & Summerville, J.A. (1970). Estimation of the molecular size of C-reactive protein and Cx-reactive protein in serum. *Biochimica et Biophysica Acta* **207**, 105–114.
10. Shrive, A.K., Cheetham, G.M.T., Holden, D., Myles, D.A.A., Turnell, W.G., Volanakis, J.E., Pepys, M.B., Bloomer, A.C. & Greenhough, T.J. (1996). Three dimensional structure of human C-reactive protein. *Nature Structural Biology* **3**, 346–354.
11. Gotschlich, E.C. (1982). Binding of C-reactive protein to C-carbohydrate and PC-substituted protein. *Annals of the New York Academy of Sciences* **389**, 163–171.
12. Schwalbe, R.A., Dahlback, B., Coe, J.E. & Nelsestuen, G.L. (1992). Pentraxin family of proteins interact specifically with phosphorylcholine and/or phosphorylethanolamine. *Biochemistry* **31**, 4907–4915.
13. Culley, F., Harris, R.A., Kaye, P.M., McAdam, K.P.W.J. & Raynes, J.G. (1996). C-reactive protein binds to a novel ligand on *Leishmania donovani* and increases uptake into human macrophages. *Journal of Immunology* **156**, 4691–4696.
14. Bodman-Smith, K.B., Culley, F., Nikolaev, A.V., Ferguson, M.A.J., Shantilal, N. & Raynes, J.G. (1998). C-reactive protein binds to phosphorylated carbohydrates. *Immunology* **95** (Suppl. 1), 37.
15. Uhlenbruck, G., Sölter, J., Janssen, E. & Haupt, H. (1981). Two new, additional combining sites of C-reactive protein–lectin specificity of the anti-galectin type and anti-hemocyanin reactivity. *Hoppe-Seyler's Zeitschrift für Physiologische Chemie* **362**, 1167–1169.
16. Pepys, M.B., Booth, S.E., Tennent, G.A., Butler, P.J.G. & Williams, D.G. (1994). Binding of pentraxins to different nuclear structures: C-reactive protein binds to small nuclear ribonucleoprotein particles, serum amyloid P component binds to chromatin and nucleoli. *Clinical and Experimental Immunology* **97**, 152–157.
17. Du Clos, T.W., Zlock, L.T., Hicks, P.S. & Mold, C. (1994). Decreased autoantibody levels and enhanced survival of (NZBxNZW) F_1 mice treated with C-reactive protein. *Clinical Immunology and Immunopathology* **70**, 22–27.
18. Volanakis, J.E. (1982). Complement activation by C-reactive protein complexes. *Annals of the New York Academy of Sciences* **389**, 235–250.
19. Wolbink, G.J., Brouwer, M.C., Buysmann, S., Ten Berge, I.J.M. & Hack, C.E. (1996). CRP-mediated activation of complement in vivo. *Journal of Immunology* **157**, 473–479.
20. Jiang, H., Robey, F.A. & Gewurz, H. (1992). Localization of sites through which C-reactive protein binds and activates complement to residues 14–26 and 76–92 of the human C1q A chain. *Journal of Experimental Medicine* **175**, 1373–1379.
21. Suankratay, C., Mold, C., Zhang, Y., Potempa, L.A., Lint, T.F. & Gewurz, H. (1998). Complement regulation in innate immunity and the acute-phase response: inhibition of mannan-binding lectin-initiated complement cytolysis by C-reactive protein. *Clinical and Experimental Immunology* **113**, 353–359.
22. Rochemonteix, B.G.-d., Wiktorowicz, K., Kushner, I. & Dayer, J.-M. (1993). C-reactive protein increases production of IL-1_α, IL-1_β, and TNF_α, and expression of mRNA by human alveolar macrophages. *Journal of Leukocyte Biology* **53**, 439–445.
23. Gewurz, H., Zhang, X.-H. & Lint, T.F. (1995). Structure and function of the pentraxins. *Current Opinion in Immunology* **7**, 54–64.
24. Barna, B.P., Eppstein, D.A., Thomassen, M.J., Nestor, J.J., Ho, T., Mendendorp, S.V. & Deodhar, S.D. (1993). Therapeutic effects of a synthetic peptide of C-reactive protein in pre-clinical tumor models. *Cancer Immunology and Immunotherapy* **36**, 171–176.
25. Viguishin, D.M., Pepys, M.B. & Hawkins, P.N. (1993). Metabolic and scintigraphic studies of radioiodinated human C-reactive protein in health and disease. *Journal of Clinical Investigation* **91**, 1351–1357.
26. Murphy, T.M., Baum, L.L. & Beaman, K.D. (1991). Extrahepatic transcription of human C-reactive protein. *Journal of Experimental Medicine* **173**, 495–498.

27. Khattri, R., Hansen, B., Nichols, T.C. & Palmer, K.L., Gilman-Sachs, A. & Baum, L.L. (1994). Anti-C-reactive protein inhibits cytoskeletal rearrangement without altering calcium influx in natural killer cell activation. *Cellular Immunology* **155**, 457–475.
28. Szalai, A.J., Briles, D.E. & Volanakis, J.E. (1995). Human C-reactive protein is protective against fatal *Streptococcus pneumoniae* infection in transgenic mice. *Journal of Immunology* **155**, 2557–2563.
29. Xia, D. & Samols, D. (1997). Transgenic mice expressing rabbit C-reactive protein are resistant to endotoxemia. *Proceedings of the National Academy of Sciences USA* **94**, 2575–2580.
30. Gabay, C. & Kushner, I. (1999). Acute-phase proteins and other systemic responses to inflammation. *New England Journal of Medicine* **340**, 448–454.
31. Sormunen, P., Kallio, M.J.T., Kilpi, T. & Peltola, H. (1999). C-reactive protein is useful in distinguishing Gram stain-negative bacterial meningitis from viral meningitis in children. *Journal of Pediatrics* **134**, 725–729.
32. Mendall, M.A., Patel, P., Ballam, L., Strachan, D. & Northfield, T.C. (1996). C-reactive protein and its relation to cardiovascular risk factors: a population based cross sectional study. *British Medical Journal* **312**, 1061–1064.
33. Liuzzo, G., Biasucci, L.M., Gallimore, J.R., Grillo, R.L., Rebuzzi, A.G., Pepys, M.B. & Maseri, A. (1994). The prognostic value of C-reactive protein and serum amyloid A protein in severe unstable angina. *New England Journal of Medicine* **331**, 417–424.
34. Haverkate, F., Thompson, S.G., Pyke, S.D.M., Gallimore, J.R. & Pepys, M.B. (1997). Production of C-reactive protein and risk of coronary events in stable and unstable angina. *Lancet* **349**, 462–466.
35. Ridker, P.M., Cushman, M., Stampfer, M.J., Tracy, R.P. & Hennekens, C.H. (1997). Inflammation, aspirin and the risk of cardiovascular disease in apparently healthy men. *New England Journal of Medicine* **336**, 973–979.
36. Koenig, W., Sund, M., Fröhlich, M., Fischer, H.-G., Löwel, H., Döring, A., Hutchinson, W.L. & Pepys, M.B. (1999). C-reactive protein, a sensitive marker of inflammation, predicts future risk of coronary heart disease in initially healthy middle-aged men. *Circulation* **99**, 237–242.
37. Visser, M., Bouter, L.M., McQuillan, G.M., Wener, M.H. & Harris, T.B. (1999). Elevated C-reactive protein levels in overweight and obese adults. *Journal of the American Medical Association* **282**, 2131–2135.
38. Hassan, A.B., Rönnelid, J., Gunnarsson, I., Karlsson, G., Berg, L. & Lundberg, I. (1998). Increased serum levels of immunoglobulins, C-reactive protein, type 1 and type 2 cytokines in patients with mixed connective tissue disease. *Journal of Autoimmunity* **11**, 503–508.
39. Van Leeuwen, M.A., Van Der Heijde, D.M.F.M., Van Rijswijh, M.H., Houtman, P.M., Van Riel, P.L.C.M., Van de Putte, L.B.A. & Limburg, P.C. (1994). Interrelationship of outcome measures and process variables in early rheumatoid arthritis: a comparison of radiological damage, physical disability, joint counts, and acute phase reactants. *Journal of Rheumatology* **21**, 425–429.
40. Bárány, P., Divino-Filho, J.C. & Bergström, J. (1997). High C-reactive protein is a strong predictor of resistance to erythropoietin in hemodialysis patients. *American Journal of Kidney Diseases* **29**, 565–568.
41. Van Baal, W.M., Kenemans, P., Van der Mooren, M., Kessel, H., Emeis, J.J. & Stehouwer, C.D.A. (1999). Increased C-reactive protein levels during short term hormone replacement therapy in healthy post-menopausal women. *Thrombosis and Haemostasis* **81**, 925–928.
42. Sattar, N., Perera, M., Small, M. & Lumsden, M.-A. (1999). Hormone replacement therapy and sensitive C-reactive protein concentrations in women with type-2 diabetes. *Lancet* **354**, 487–488.
43. Jarva, H., Jokiranta, T.S., Hellwage, J., Zipfel, P.F. & Meri, S. (1999). Regulation of complement activation by C-reactive protein: targeting the complement inhibitory activity of Factor H by an interaction with short consensus repeat domains 7 and 8–11. *Journal of Immunology* **163**, 3957–3962.

Crotalus atrox venom lectin

Other name: Rattlesnake venom lectin
Mol. wt: 120 000 (native); 16 000 (subunit)
Specificity: D-Galactose
Source: Venom of the western diamondback rattlesnake *Crotalus atrox*

ISOLATION

The lectin has been purified from a salt extract of commercial venom by affinity chromatography on agarose; the lectin could be eluted by galactose or EDTA [1,2].

STRUCTURE

The rattlesnake lectin is a non-glycosylated protein with a high content of hydrophilic amino acids (notably lysine) and half-cysteine residues [2]. Its basic unit is a homodimer, but there is evidence from gel permeation chromatography [2] and from isoelectric focusing [1] that it undergoes aggregation and the major native form may be 120 000.

Molecular weight determination by SDS-PAGE yields a single band of 28 000–32 000 under non-reducing conditions, and a band of 14 000–16 000 under reducing conditions [1,2]. The reduced subunit has 135 amino acid residues with four intra-subunit disulphide bridges and one inter-subunit disulphide bridge. The calculated Mr was 16 317 [2]. Isoelectric focusing yielded a broad 'smear' between pH 7 and pH 7.6 [1].

The amino acid sequence showed extensive homology with the carbohydrate-recognition domains of various C-type lectins [2].

BIOLOGICAL ACTIVITIES

The rattlesnake lectin agglutinated formalin-fixed erythrocytes from bovine, equine, human and lapine sources [1]. It was also found to agglutinate fresh lapine erythrocytes [2].

Haemagglutination could be inhibited by EDTA or dithiothreitol; the former could be reversed with additional Ca^{2+} [1]. Haemagglutination was also readily inhibited by galactose and galactosides; thiodigalactoside and lactose were the most potent inhibitors. D-fucose and L-rhamnose were of similar potency to galactose, while (surprisingly) L-glucose was also an effective inhibitor.

The rattlesnake lectin was mitogenic for bovine lymph node mononuclear cells [3], and could aggregate human platelets [4]; both activities could be inhibited by lactose.

TISSUE AND/OR SUBCELLULAR DISTRIBUTION

The lectin has been reported only in venom.

POSSIBLE FUNCTION(S)

The lectin may protect the epithelial cells of the venom glands from damage by proteolytic enzymes by binding to mucin [1]. Alternatively, another possibility is an offensive role in defence, by contributing to the tissue injury of prey animals [2].

COMMENTS

The *C. atrox* lectin is very similar to thrombolectin (qv) and to the lectins found in venoms of the southern copperhead, *Agkistrodon contortrix*, and the western cottonmouth moccasin, *Agkistrodon piscivorus* [1]. All have similar molecular weights, saccharide specificities, dependence on disulphide bonds and a requirement for Ca^{2+}. Those four lectins exhibit immunological cross-reactivity, but not identity, and also differ in their isoelectric points. Very similar lectins have been purified by affinity chromatography from the venoms of *Bothrops jararaca* [5] and *B. jararacussu* [6]. The lectin from the related crotalid species *Lachesis muta* (qv) has slightly different characteristics, while haemagglutinins present in crude venom extracts from species of other families had obviously distinct properties [7].

REFERENCES

1. Gartner, T.K. & Ogilvie, M.L. (1984). Isolation and characterization of three Ca^{2+}-dependent β-galactoside-specific lectins from snake venoms. *Biochemical Journal* **224**, 301–307.
2. Hirabayashi, J., Kusunoki, T. & Kasai, K. (1991). Complete primary structure of a galactose-specific lectin from the venom of the rattlesnake, *Crotalus atrox*. *Journal of Biological Chemistry* **266**, 2320–2326.
3. Mastro, A.M., Hurley, D.J., Winning, R.K., Filipowski, R., Ogilvie, M.L. & Gartner, T.K. (1986). Mitogenic activity of snake venom lectins. *Cell and Tissue Kinetics* **19**, 557–566.
4. Ogilvie, M.L., Wilson-Byl, J. & Gartner, T.K. (1989). Platelet aggregation is stimulated by lactose-inhibitable snake venom lectins. *Thrombosis and Haemostasis* **62**, 704–707.
5. Ozeki, Y., Matsui, T., Hamako, J., Suzuki, M., Fujimura, Y., Yoshida, E., Nishida, S. & Titani, K. (1994). C-type galactoside-binding lectin from *Bothrops jararaca* venom: comparison of its structure and function with those of botrocetin. *Archives of Biochemistry and Biophysics* **308**, 306–310.
6. Carvalho, D.D., Marangoni, S., Oliveira, B. & Novello, J.C. (1998). Isolation and characterization of a new lectin from the venom of the snake, *Bothrops jararacussa*. *Biochemistry and Molecular Biology International* **44**, 933–938.
7. Ogilvie, M.L. & Gartner, T.K. (1984). Identification of lectins in snake venoms. *Journal of Herpetology* **18**, 285–290.

Cucumaria echinata lectins I and IV

Other names: CEL-I and CEL-IV
Mol. wt: 27 000 (I) and 68 000 (IV) (native); 16 000 and 18 000 (subunit)
Specificity: GalNAc
Source: The sea cucumber *Cucumaria echinata*

ISOLATION

A mixture of lectins was separated from clarified *C. echinata* homogenate by application to a column of lactosyl–Sepharose 4B and elution with EDTA. CEL-IV was separated from the other lectins and recovered as a second (minor) peak. CEL-I was separated from the quantitatively minor lectin, CEL-II, by ion-exchange chromatography after separation from CEL-III by gel filtration on Sephacryl S-200 [1].

STRUCTURE

CEL-I is apparently a homodimer; the apparent molecular weights estimated by SDS-PAGE in the presence and absence of reducing agents were 16 000 and 27 000, respectively [1]. CEL-I has been partially sequenced and found to belong to the C-type lectin family (unpublished, but referred to in [2]).

CEL-IV is a homotetramer; the apparent molecular weights estimated by SDS-PAGE in the presence and absence of reducing agent were 18 000 and 68 000, respectively [1]. The complete amino acid sequence of CEL-IV has been published [2]. The subunit comprises 157 amino acid residues with a calculated Mr of 17 098, with glycine, serine and leucine the most abundant amino acids. Considerable homology was noted with various C-type lectins, including 40% identity with *Stichopus japonicus* lectin I (SJL-I) (qv), 33% identity with echinoidin (qv) and 22% identity with L-selectin (qv) [2].

BIOLOGICAL ACTIVITIES

Both CEL-I and CEL-IV agglutinate lapine and human erythrocytes strongly, but show little or no activity towards equine or chicken red blood cells [1]. The agglutination was Ca^{2+}-dependent and could be inhibited by GalNAc, galactose and galactose-related saccharides. Despite this general similarity, CEL-I and CEL-IV differed in detailed sugar specificity. CEL-I was exquisitely sensitive to GalNAc, equally sensitive to galactose and fucose, but insensitive to galactosamine. Although GalNAc was also the most effective monosaccharide inhibitor of CEL-IV, a 60-fold greater (compared to CEL-I) concentration was required; moreover galactose was a more potent inhibitor than fucose or galactosamine (which were equally effective) [1].

TISSUE AND/OR SUBCELLULAR DISTRIBUTION

Nothing appears to be known regarding the tissue or subcellular distribution of these lectins.

POSSIBLE FUNCTION(S)

Although it has been considered that these lectins might function in defence by neutralising and excluding foreign organisms, no definite information regarding function has been forthcoming.

COMMENTS

Of the four lectins originally described from *C. echinata* homogenate, *Cucumaria echinata* lectin III (CEL-III) (qv) has been the most extensively studied.

REFERENCES

1. Hatakeyama, T., Kohzaki, H., Nagatomo, H. & Yamasaki, N. (1994). Purification and characterization of four Ca^{2+}-dependent lectins from the marine invertebrate *Cucumaria echinata. Journal of Biochemistry* **116**, 209–214.
2. Hatakeyama, T., Ohuchi, K., Kuroki, M. & Yamasaki, N. (1995). Amino acid sequence of a C-type lectin CEL-IV from the marine invertebrate *Cucumaria echinata. Bioscience, Biotechnology and Biochemistry* **59**, 1314–1317.

Cucumaria echinata lectin III

Other name: CEL-III
Mol. wt: 48 000
Specificity: GalNAc
Source: The sea cucumber *Cucumaria echinata*

ISOLATION

A mixture of three lectins separated from clarified *C. echinata* homogenate by application to lactosyl–Sepharose and elution with EDTA was resolved into CEL-III and a separate mixture of the others by gel filtration on Sephacryl S-200 [1].

Alternatively, CEL-III with CEL-I can be obtained by binding to a lactosyl–Cellulofine column and elution with EDTA. The mixture is then applied to a column of GalNAc–Cellulofine; CEL-III is eluted with 0.1 M-lactose, leaving CEL-I still bound to the column [2].

STRUCTURE

CEL-III is a monomer with apparent molecular weights estimated at 45 000 and 51 000 by SDS-PAGE under non-reducing and reducing conditions, respectively [1]. Equilibrium dialysis data are consistent with the lectin's having two carbohydrate-binding sites per monomer, thus permitting agglutination of cells [3].

The primary sequence deduced from a cDNA consists of 432 amino acid residues, and has a calculated Mr of 47 457. It has six internal tandem repeats, each with 40–50 residues, comprising the N-terminal two-thirds of the molecule, which resemble the repeats found in the β-chains of ricin and abrin [4].

BIOLOGICAL ACTIVITIES

CEL-III agglutininates equine erythrocytes but lyses lapine and human erythrocytes in a calcium-dependent manner [1,2]. Haemolysis can be inhibited by several galactose-related sugars of which GalNAc was the most effective [1].

CEL is cytotoxic to various cell lines, including MDCK, HeLa and Vero. Cytotoxic activity towards different cell lines correlated with binding to and lysis of the cells [5]. For both haemolysis and cytotoxicity, activity correlated inversely with temperature (optimum at 4–10 °C) and positively with pH above 7 up to 10 [2,5]. Also, for both activities, there is evidence that a non-enzymic mechanism involves pore formation after the irreversible binding of CEL-III aggregates to the target cell plasma membrane [2,5,6]. Although lytic activity is optimal at alkaline pH, lectin activity is optimal around pH 5 [7]; it seems the lectin-receptor interaction is an essential first step, which is then followed by separate interactions with a membrane lipid bilayer.

CEL also interacts with liposomes made from erythrocyte membrane lipids, and this phenomenon has been studied in some detail. Pore formation arose after binding to lactosyl–ceramide and, to a lesser extent, after globoside binding [8].

TISSUE AND/OR SUBCELLULAR DISTRIBUTION

Nothing seems to be known about its distribution.

POSSIBLE FUNCTION(S)

CEL-III is presumably involved in defence since it has potentially the capacity both to recognise invading pathogens and to lyse them.

COMMENTS

CEL-III is the most abundant of the four Ca^{2+}-dependent GalNAc/galactose-specific lectins isolated from *C. echinata* homogenate, with 40 times more protein recovered than that of the least abundant, CEL-II [1].

CEL-III is both an agglutinin and a lysin, and both activities result from its lectin (carbohydrate-binding) activity. Haemagglutinating activity is surprising for a monomeric protein, although it appears to form oligomers on target cell surfaces in order to mediate pore formation, and it is possible that oligomerization is responsible for haemagglutination also [7]. However, there is some evidence that two carbohydrate-binding sites are found on the same polypeptide chain [2,6]; this arrangement is found on the macrophage mannose receptor (qv), but has not been described in invertebrates, although it would make sense for some agglutinins/lysins found in annelids which have not yet been well-studied at the molecular level.

REFERENCES

1. Hatakeyama, T., Kohzaki, H., Nagatomo, H. & Yamasaki, N. (1994). Purification and characterization of four Ca^{2+}-dependent lectins from the marine invertebrate *Cucumaria echinata. Journal of Biochemistry* **116**, 209–214.
2. Hatakeyama, T., Nagatomo, H. & Yamasaki, N. (1995). Interaction of the haemolytic lectin CEL-III from the marine invertebrate *Cucumaria echinata* with the erythrocyte membrane. *Journal of Biological Chemistry* **270**, 3560–3664.

3. Sallay, I., Hatakeyama, T. & Yamasaki, N. (1998). Studies on the carbohydrate binding sites of the hemolytic lectin CEL-III isolated from the marine invertebrate *Cucumaria echinata. Bioscience Biotechnology and Biochemistry* **62**, 1757–1761.
4. Nakano, M., Tabata, S., Sugihara, K., Kouzuma, Y., Kimura, M. & Yamasaki, N. (1999). Primary structure of hemolytic lectin CEL-III from murine invertebrate *Cucumaria echinata* and its cDNA: structural similarity to the β-chain from plant lectin, ricin. *Biochimica et Biophysica Acta* **1435**, 167–176.
5. Oda, T., Tsura, M., Hatakeyama, T., Nagatomo, H., Muramatsu, T. & Yamasaki, N. (1997). Temperature- and pH-dependent cytotoxic effect on the hemolytic lectin CEL-III from the marine invertebrate *Cucumaria echinata* on various cell lines. *Journal of Biochemistry* **121**, 560–567.
6. Hatakeyama, T., Matsuyama, Y., Funada, T., Fukuyama, S., Kuwahara, H., Aoyagi, H. & Yamasaki, N. (1998). Chemical modification of the hemolytic lectin CEL-III by succinic anhydrate: involvement of amino groups in the oligomerization process. *Bioscience, Biotechnology and Biochemistry* **62**, 1185–1189.
7. Hatakeyama, T., Furukawa, M., Nagatomo, H., Yamasaki, N. & Mori, T. (1996). Oligomerisation of the hemolytic lectin CEL-III from the marine invertebrate *Cucumaria echinata* induced by the binding of carbohydrate ligands. *Journal of Biological Chemistry* **271**, 16915–16920.
8. Hatakeyama, T., Sato, T., Taira, E., Kuwahara, H., Niidome, T. & Aoyagi, H. (1999). Characterization of the interaction of hemolytic lectin CEL-III from the marine invertebrate *Cucumaria echinata*, with artificial lipid membranes: involvement of neutral sphinoglycolipids in the pore-forming process. *Journal of Biochemistry* **125**, 277–284.

Didemnum candidum I and II isolectin families

Other names: DCL-I and DCL-II
Mol. wt: 57 000 (native); 14 000 (I) or 15 000 (II) (subunit)
Specificity: D-Galactose
Source: The tunicate *Didemnum candidum*

ISOLATION

When body-fluid supernatant (mainly plasma) of *D. candidum* was applied to a column of asialo-Sepharose and eluted with a gradient (0-50 mM) of galactose, a large peak (DCL-I) was eluted first followed by a smaller peak (DCL-II). DCL-I, but not DCL-II, can be isolated in a similar way by affinity chromatography on plain Sepharose 4B or acid-treated Sepharose 4B. DCL-I and DCL-II were apparently homogeneous by SDS-PAGE, immunoelectrophoresis and size exclusion HPLC, but not by isoelectric focusing [1].

STRUCTURE

Both DCL-I and DCL-II are polypeptide isolectin families which are antigenically distinct. DCL-I consists of a large group of acidic isolectins (pI range 3.8 to 5.2), while DCL-II consists of a smaller number of basic isolectins (pI range 9.2 to 10.2). The amino acid compositions of the two families are rather similar, but DCL-I has considerably more valine and phenylalanine while DCL-II is richer in glutamic acid and histidine. The native molecular weights estimated by sedimentation equilibrium were both approximately 57 000, with DCL-II being slightly larger. Both DCL-I and DCL-II are tetramers; the subunit sizes determined by SDS-PAGE and by size exclusion chromatography in guanidine–HCl were approximately 14 000 and 15 000, respectively [1].

The N-terminal amino acid sequence of DCL-I showed some homology with mammalian C-reactive protein, and DCL-I showed cross-reactivity in ELISA with human CRP using ovine anti-human CRP and lapine anti-DCL-I [1].

BIOLOGICAL ACTIVITIES

Both DCL-I and DCL-II agglutinated neuraminidase-treated red blood cells from all of the many species tested, but DCL-I exhibited a 12- to 400-fold greater titre on a gravimetric basis. DCL-I, but not DCL-II, agglutinated untreated duck, lapine and human (A and B) erythrocytes at low titre. Although both DCL-I and DCL-II showed a very slight preference for human O cells after neuraminidase treatment, DCL-II was highly group-B-specific after treatment with pronase, and DCL-I showed a modest preference for pronase-treated group B cells [2].

Haemagglutinating activity of both lectins was calcium-dependent and inhibited by asialoglycoproteins. The saccharide specificity of DCL-I was studied in some detail; galactose was the best monosaccharide inhibitor, followed by L-arabinose, while galacturonic acid, glucuronic acid and rhamnose were weakly inhibitory [2].

TISSUE AND/OR SUBCELLULAR DISTRIBUTION

The lectins were isolated from the released body fluids after slicing the tunicates with a scalpel; lectin activity was not investigated elsewhere [1].

POSSIBLE FUNCTION(S)

It has been suggested that these lectins may have a role in self/non-self discrimination [1], but, as for tunicate lectins in general [3], no definite information has been obtained regarding their biological function(s).

COMMENTS

Much less has been described for the minor (DCL-II) lectin compared to the major (DCL-I) lectin, but although they appear to be rather similar in size, structure and saccharide specificity, their amino acid compositions are sufficiently different for them to be considered distinct lectins. It should be noted that the two isolectins families have very different isoelectric points; the ratios of acidic to basic amino acids are similar, however, and therefore most of the acidic residues in DCL-II are probably in amidated form. The DCL-I family obviously has some structure in common with C-reactive proteins, but corresponding information concerning DCL-II has not been sought.

DCL-I appears to agglutinate red blood cells by binding to galactosyl residues exposed by neuraminidase treatment. The data with inhibitors indicate that the binding site is relatively small, and prefers β-linked to α-linked galactose residues [2].

An unexpected finding was the group B specificity of DCL-II for pronase-treated erythrocytes; it appears to be one of the most highly B-specific lectins known.

REFERENCES

1. Vasta, G.R., Hunt, J.C., Marchalonis, J.J. & Fish, W.W. (1986). Galactosyl-binding lectins from the tunicate *Didemnum candidum*. Purification and physiochemical characterization. *Journal of Biological Chemistry* **261**, 9174–9181.
2. Vasta, G.R. & Marchalonis, J.J. (1986). Galactosyl-binding lectins from the tunicate, *Didemnum candidum*. Carbohydrate specificity and characterization of the combining site. *Journal of Biological Chemistry* **261**, 9182–9186.
3. Saito, Y. (1996). Humoral factors in tunicates. In: *Invertebrate Immunology* (eds B. Rinkevich & W.E.G. Mueller). Springer-Verlag, Berlin, 1996.

Didemnum ternatanum lectin

Other name: DTL
Mol. wt: 28 000 (native); 10 000 (subunit)
Specificity: GlcNAc
Source: The ascidian *Didemnum ternatanum*

ISOLATION

DTL was isolated by affinity chromatography on cross-linked ovalbumin and gel filtration on Sephadex G100 [1].

STRUCTURE

The lectin is a glycoprotein (1.3% carbohydrate) containing relatively high proportions of lysine, alanine, aspartic acid, glutamic acid, leucine and valine. It appears to be a homotrimer since the apparent molecular weight was 28 000 from gel filtration and 10 000 by SDS-PAGE [1].

BIOLOGICAL ACTIVITIES

DTL agglutinated pronase-treated (but not fresh) human erythrocytes in a divalent cation-independent manner, showing a slight preference for blood group O over types A or B [1]. Haemagglutination was potently inhibited only by GlcNAc among monosaccharides, but $(GlcNAc)_2$ and $(GlcNAc)_3$ were slightly more effective. 4-Methylumbelliferyl-*N*-acetyl-β-D-glucosaminide and *p*-nitrophenyl-*N*-acetyl-β-D-gluosaminide were the most effective saccharide inhibitors, 40 times more potent than GlcNAc. Among glycoproteins, asialofetuin, asialo-IgG, and especially ovalbumin and ovomucoid were strongly inhibitory [1].

DTL promoted the spreading and attachment of embryonic sea urchin and mussel cells at the gastrula stage; cell adhesion was not inhibited by GlcNAc [2].

DTL can interact with collagen; this can be inhibited by (Arg-Gly-Asp) peptide and peptides derived from collagen [2].

TISSUE AND/OR SUBCELLULAR DISTRIBUTION

The distribution of this lectin has not been investigated.

POSSIBLE FUNCTION(S)

Although this lectin has some characteristics of an adhesion molecule, its function is unknown [2].

COMMENTS

A calcium-independent lectin is a minority finding among tunicate lectins described to date. Without knowledge of its primary structure it is impossible to classify, and its amino acid composition and physical measurements, indicating a preponderance of β-pleated sheets, are typical of lectins generally. Its precise saccharide specificity is somewhat unusual: although other lectins exclusively recognise GlcNAc and its oligomers, they generally show a strong preference for chito-oligosaccharides relative to the monosaccharide. DTL does not discriminate much between GlcNAc, $(GlcNAc)_2$ and $(GlcNAc)_3$ and it may prove to be a valuable tool for analysis of sugar chains rich in GlcNAc residues.

DTL increased the attachment and spreading of certain molluscan and echinoderm cells and also stimulated their growth [2]. It may prove useful as a reagent for maintaining cell lines from invertebrate sources.

REFERENCES

1. Belogortseva, N., Molchanova, V., Glazunov, V., Evtushenko, E., & Luk'yanov, P. (1998). *N*-Acetyl-D-glucosamine-specific lectin from the ascidian, *Didemnum ternatanum. Biochimica et Biophysica Acta* **1380**, 249–256.
2. Odintsova, N.A., Belogortseva, N.I., Ermak, A.V., Molchanova, V.I. & Luk'yanov, P.A. (1999). Adhesive and growth properties of lectin from the ascidian *Didemnum ternatanum* on cultivated marine invertebrate cells. *Biochimica et Biophysica Acta* **1448**, 381–389.

Diogenes affinis agglutinin

Other name: *Diogenes affinis* serum lectin
Mol. wt: 185 000 (native); 39 000, 42 000, 49 000 and 51 000 (subunit)
Specificity: Acetylated carbohydrates
Source: Haemolymph of the hermit crab *Diogenes affinis*

ISOLATION

The lectin was isolated from haemolymph serum by affinity chromatography on GlcNAc–Sepharose, eluting with GlcNAc [1].

STRUCTURE

The *D. affinis* agglutinin is a protein or glycoprotein of apparent molecular weight 185 000 estimated by gel filtration. It appears to be composed of four non-identical subunits (39, 42, 49 and 51 kDa) cross-linked by interchain disulphide bonds (on the basis of SDS-PAGE analysis under reducing conditions).

BIOLOGICAL ACTIVITIES

The lectin agglutinated rat erythrocytes at relatively high titre, and erythrocytes of various other species, including humans, sheep, rabbits and horses (in that order) at lower titres. Haemagglutination was calcium-dependent (neither Mg^{2+} nor Mn^{2+} could substitute), and ABO non-specific for human cells [1].

Haemagglutination was sensitive to inhibition by *N*-acetylated sugars including GlcNAc and its oligomers, GalNAc and *N*-acetylneuraminic acid. GlcNAc, $(GlcNAc)_2$ and $(GlcNAc)_3$, were equally inhibitory. *N*-Glycolylneuraminic acid was not inhibitory, nor were amino derivatives of hexoses. Haemagglutination could also be inhibited by acetylated amino acids and by bacterial lipopolysaccharides [1].

TISSUE AND/OR SUBCELLULAR DISTRIBUTION

The lectin was found in haemolymph serum, but was not investigated in other tissues [1].

POSSIBLE FUNCTIONS

It has been suggested that this lectin might function in host defence as an antibacterial agent [1].

COMMENTS

The *D. affinis* agglutinin is an unusual lectin that recognises the acetyl group rather than the attached carbohydrate; in this respect it resembles tachylectins-5A and -5B (qv).

REFERENCE

1. Murali, S., Mullainadhan, P. & Arumugam, M. (1999). Purification and characterization of a natural agglutinin from the serum of the hermit crab *Diogenes affinis*. *Biochimica et Biophysica Acta* **1472**, 13–24.

Discoidin I

Other names: Slime mould lectin/agglutinin
Mol. wt: 112 000 (native); 28 000 (subunit)
Specificity: GalNAc, fucose, galactose
Source: Differentiating amoebae of *Dictyostelium discoideum*

ISOLATION

Discoidin I was originally purified (with discoidin II as a minor contaminant) by affinity chromatography on Sepharose 4B and elution with galactose [1,2]. However, the affinity matrix, agarose-ε-aminocaproyl-fucosamine, has a 55-fold greater capacity for discoidin and yields similar results [3]. Discoidin I can be isolated free of discoidin II by binding to and elution from fixed ovine erythrocytes [2], or by ion-exchange chromatography [4].

STRUCTURE

Discoidin I is a protein with a high aspartic acid content [2]. It appeared to be a homotetramer on the basis of a single symmetrical peak of agglutinating activity (5.1 S) on sucrose density gradient centrifugation, and an Mr of 100 000 ± 2000 by sedimentation equilibrium, combined with subunit molecular sizes of 28 000 (sedimentation equilibrium) or 26 000 (SDS-PAGE).

Using recombinant DNA technology, Rowekamp and co-workers isolated chimeric plasmids carrying genomic DNA or cDNA sequences complementary to mRNA encoding discoidin I and discovered a homologous multi-gene family regulated at the level of transcription [5]. The same group identified four separate genes, three of which are expressed as isolectins. From the deduced amino acid sequence, the subunit Mr was calculated to be 28 100 [6]. Isolectins arise from the association of four of these gene products which are held together apparently without disulphide bonds [7].

Discoidin I shows sequence homology to the extracellular domain of a class of receptor tyrosine kinases which may be overexpressed in human cancer cells [8,9]. Discoidin I also shows some sequence homology with the B chain of ricin [10], and possesses a short sequence found within the cell-binding domain of fibronectin [6].

BIOLOGICAL ACTIVITIES

Discoidin I was first detected by the agglutination of formalin-treated ovine erythrocytes, but it readily agglutinates both fresh and fixed red blood cells from several species including humans. This lectin activity requires divalent cations [11]. Agglutinating activity can be inhibited by a variety of galactose-related monosaccharides, of which GalNAc is the most potent [4]. However, an apparently natural ligand, an (Xyl)(Man)$_3$GlcNAc oligosaccharide, inhibits with an approximately 100-fold lower association constant [12].

Discoidin I can also bind a 67 kDa glycoprotein from aggregating cells of its own species [13]. This reaction involves a fibronectin-like, non-lectin active site, since the interaction takes place in the presence of 0.3 M galactose and can be reversed with the synthetic peptide, gly-arg-gly-asp-his-asp.

TISSUE AND/OR SUBCELLULAR DISTRIBUTION

Discoidin I is found mainly in aggregates of and aggregating amoebae, not in differentiated stalk or spore cells [14]. Most of the lectin is located within the cells, although late in the life cycle much of it is externalised [14,15]. In the cytosol, discoidin I exists principally as a monomer, but in multilamellar bodies (food vacuoles which are subsequently externalised) discoidin I exists mainly as a tetramer [7].

Although a small proportion of discoidin I has been localised to the cell surface [14,16], this could not be confirmed by other workers, who suggested that the apparent cell surface location was an artefact of preparation [15].

POSSIBLE FUNCTION(S)

The function of discoidin I is obscure despite much work and the availability of much relevant information. The original suggestion that discoidin I mediated cell–cell cohesion between aggregating cells, despite some circumstantial evidence to support it, is no longer tenable. Mutant strains of *D. discoideum* that express <1% of the wild-type level of discoidin I have been obtained which undergo normal developmental events and terminal differentiation [17]. Several other mutants which misregulate discoidin I (synthesis induced at inappropriate developmental stages) were also able to complete development [18].

It has been suggested that discoidin I functions in cell to substatum attachment, spreading and migration, through its non-lectin, fibronectin-like binding site. Experimental support comes from the use of synthetic peptides to block organised cell streaming; the existence of low-discoidin mutants which fail to form streams; and the experimental inhibition of normal cell streaming with exogenously added discoidin or univalent anti-discoidin antibody fragments

[19]. The ability of discoidin I to bind cellulose, a component of the slime sheath, is also consistent with a cell–substratum adhesion role [20].

Alternatively, discoidin(s) may be necessary for the maintenance of the typical elongate cell morphology and cytoskeletal organisation normally observed during aggregation [21].

COMMENTS

Discoidin I exhibits a remarkable developmental regulation leading to an approximately 400-fold increase in specific activity during aggregation of amoebae until it comprises >1% of the total soluble protein [1,22]. The synthesis of discoidin I in bacterially-fed cells is induced by a 'prestarvation factor' which accumulates during growth in proportion to cell density [23]. The specific activity of discoidin I, however, is exceptionally high in axenically grown vegetative cells of the mutant strain, A3 [1].

REFERENCES

1. Rosen, S.D., Kafka, J.A., Simpson, D.L. & Barondes, S.H. (1973). Developmentally regulated, carbohydrate-binding protein in *Dictyostelium discoideum. Proceedings of the National Academy of Sciences USA* **70**, 2554–2557.
2. Simpson, D.L., Rosen, S.D. & Barondes, S.H. (1974). Discoidin, a developmentally regulated carbohydrate-binding protein from *Dictyostelium discoideum*. Purification and characterization. *Biochemistry* **13**, 3487–3493.
3. Kilpatrick, D.C., Schmidt, J.A. & Stirling, J.L. (1978). Purification of carbohydrate binding proteins from *Dictyostelium discoideum* by affinity chromatography on agarose-ε-aminocaproyl-fucosamine. *FEMS Microbiology Letters* **4**, 67–70.
4. Frazier, W.A., Rosen, S.D., Reitherman, R.W. & Barondes, S.H. (1975). Purification and comparison of two developmentally regulated lectins from *Dictyostelium discoideum.* Discoidin I and II. *Journal of Biological Chemistry* **250**, 7714–7721.
5. Rowekamp, W., Poole, S. & Firtel, R.A. (1980). Analysis of the multigene family coding the developmentally regulated carbohydrate-binding protein discoidin-I in *D. discoideum. Cell* **20**, 495–505.
6. Poole, S., Firtel, R.A., Lamar, E. & Rowekamp, W. (1981). Sequence and expression of the discoidin I gene family on *Dictyostelium discoideum. Journal of Molecular Biology* **153**, 273–289.
7. Fukuzawa, M. & Ochiai, H. (1993). Different subcellular localizations of discoidin I monomer and tetramer in *Dictyostelium discoideum* cells: using conformation-specific monoclonal antibodies. *Experimental Cell Research* **204**, 61–72.
8. Barker, K.T., Martindale, J.E., Mitchell, P.J., Kamalati, T., Page, M.J., Phippard, D.J., Dale, T.C., Gusterson, B.A. & Crompton, M.R. (1995). Expression patterns of the novel receptor-like tyrosine kinase DDR, in human breast tumours. *Oncogene* **10**, 609–618.
9. Playford, M.P., Butler, R.J., Wang, X.C., Katso, R.M., Cooke, I.E. & Ganesan, T.S. (1996). The genomic structure of discoidin receptor tyrosine kinase. *Genome Research* **6**, 620–627.
10. Robertus, J.D. & Ready, M.P. (1984). Ricin B chain and discoidin I share a common primitive protein fold. *Journal of Biological Chemistry* **259**, 13953–13956.
11. Alexander, S., Cibulsky, A.M. & Lerner, R.A. (1983). Ion dependence of the discoidin I lectin from *Dictyostelium discoideum. Differentiation* **24**, 209–212.

12. Amatayakul-Chantler, S., Ferguson, M.A.J., Dwek, R.A., Rademacher, T.W., Parekh, R.B., Crandall, I.E. & Newell, P.C. (1991). Cell surface oligosaccharides on *Dictyostelium* during development. *Journal of Cell Science* **99**, 485–495.
13. Gabius, H.-J., Springer, W.R. & Barondes, S.H. (1985). Receptor for the cell binding site of discoidin-I. *Cell* **42**, 449–456.
14. Barondes, S.H., Cooper, D.N. & Haywood-Reid, P.L. (1983). Discoidin I and discoidin II are localized differently in developing *Dictyostelium discoideum. Journal of Cell Biology* **96**, 291–296.
15. Erdos, G.W. & Whitaker, D. (1983). Failure to detect immunocytochemically reactive endogenous lectin on the cell surface of *Dictyostelium discoideum. Journal of Cell Biology* **97**, 993–1000.
16. Chang, C.-M., Reitherman, R.W., Rosen, S.D. & Barondes, S.H. (1975). Cell surface location of discoidin, a developmentally regulated carbohydrate-binding protein from *Dictyostelium discoideum. Experimental Cell Research* **95**, 136–142.
17. Alexander, S., Shinnick, T.M. & Lerner, R.A. (1983). Mutants of *Dictyostelium discoideum* blocked in expression of all members of the developmentally regulated discoidin multigene family. *Cell* **34**, 467–475.
18. Wetterauer, B., Jacobsen, G., Morandini, P. & MacWilliams, H. (1993). Mutants of *Dictyostelium discoideum* with defects in the regulation of discoidin-I expression. *Developmental Biology* **159**, 184–195.
19. Springer, W.R., Cooper, D.N.W. & Barondes, S.H. (1984). Discoidin-I is implicated in cell-substratum attachment and ordered cell migration of *Dictyostelium discoideum* and resembles fibronectin. *Cell* **39**, 557–564.
20. Eitle, E., Keller, T., Parish, C.R. & Parish, R.W. (1993). Polysaccharides influence the aggregation of *Dictyostelium discoideum* cells and bind to developmentally regulated cell surface proteins. *Experimental Cell Research* **205**, 374–382.
21. Alexander, S., Sydow, L.M., Wessels, D. & Soll, D.R. (1992). Discoidin proteins of *Dictyostelium* are necessary for normal cytoskeletal organisation and cellular morphology during aggregation. *Differentiation* **51**, 149–161.
22. Cooper, D.N.W. & Barondes, S.H. (1984). Colocalization of discoidin-binding ligands with discoidin in developing *Dictyostelium discoideum. Developmental Biology* **105**, 59–70.
23. Rathi, A., Kayman, S.C. & Clarke, M. (1991). Induction of gene expression in *Dictyostelium* by prestarvation factor secreted by growing cells. *Developmental Genetics* **12**, 82–87.

Discoidin II

Mol. wt: 114 000 (native); 28 600 (subunit)
Specificity: GalNAc, fucose, galactose
Source: Differentiating cells of *Dictyostelium discoideum*

ISOLATION

Discoidin II co-purifies with discoidin I when extracts of *D. discoideum* are subjected to affinity chromatography [1,2], but is not adsorbed out with formalin-treated ovine erythrocytes, unlike discoidin I [2]. Discoidin II can also be isolated from discoidin I by ion-exchange chromatography on DEAE-cellulose [3].

Discoidin II cDNA has been isolated from a cDNA library constructed from vegetative Ax2 cells by screening with anti-discoidin II antibodies [4].

STRUCTURE

Discoidin II is a tetrameric protein which migrates ahead of discoidin I during PAGE under non-dissociating conditions, and consists of subunits of apparent Mr 24 000 on SDS-PAGE [3]. Its isoelectric point is 6.9 (compared to 6.1 for discoidin I).

The complete amino acid sequence has been deduced from the cDNA nucleotide sequence [4]. This confirmed that discoidin II has much less threonine and much more serine than discoidin I [3,4]. The sequence is 49% identical to that of discoidin I and shares with the latter both a carbohydrate-binding site and a fibronectin-like Arg-Gly-Asp (RGD) sequence [4]. However, the sequence of discoidin-II did not resemble any other polypeptide in the data bank [4].

The calculated Mr was 28 574; it is noteworthy that this is slightly greater than the corresponding figure for discoidin I, despite the latter having a higher apparent Mr on SDS-PAGE and by sedimentation equilibrium.

BIOLOGICAL ACTIVITIES

Discoidin II readily agglutinates fresh and fixed human and (especially) lapine erythrocytes, but differs from discoidin I in having little activity towards ovine red blood cells [3]. Discoidin II recognises sugars with a galactose-related structure, but inhibition of agglutination by saccharides, while similar, is quantitatively different from discoidin I. The lectin activity of discoidin II does not require divalent cations [5].

Discoidin II is able to bind an endogenous polysaccharide (mainly of galactose) found in maturing aggregates, which may be a spore wall component [6]. This could be the natural ligand for discoidin II.

TISSUE AND/OR SUBCELLULAR DISTRIBUTION

Discoidin II is found appreciably only in differentiating (not vegetative) cells, especially pre-spore cells, possibly in vesicles [7,8].

POSSIBLE FUNCTION(S)

Mutant strains of *D. discoideum* expressing 1–2% of the normal (wild-type) level of discoidin II are capable of apparently normal development and differentiation into fruiting bodies [9].

It has been suggested by several authors that discoidin II might function in spore formation [4,7,8]. This suggestion is based on circumstantial evidence (developmental regulation and histochemical localisation) and has no direct experimental evidence to support it.

COMMENTS

Discoidin II is developmentally regulated in a manner broadly similar, but not identical, to discoidin I [3,7]. The highest specific activity is found during the culmination stage when terminal differentiation into either stalk or spore cells begins. Similarly, mRNA for discoidin II starts to increase at around 8 hours of the 24-hour developmental cycle; maximal accumulation is at 16 hours and this high expression is maintained throughout terminal differentiation [4].

REFERENCES

1. Rosen, S.D., Kafka, J.A., Simpson, D.L. & Barondes, S.H. (1973). Developmentally regulated, carbohydrate-binding protein in *Dictyostelium discoideum. Proceedings of the National Academy of Sciences USA* **70**, 2554–2557.
2. Simpson, D.L., Rosen, S.D. & Barondes, S.H. (1974). Discoidin, a developmentally regulated carbohydrate-binding protein from *Dictyostelium discoideum.* Purification and characterization. *Biochemistry* **13**, 3487–3493.
3. Frazier, W.A., Rosen, S.D., Reitherman, R.W. & Barondes, S.H. (1975). Purification and comparison of two developmentally regulated lectins from *Dictyostelium discoideum.* Discoidin I and II. *Journal of Biological Chemistry* **250**, 7714–7721.
4. Fukuzawa, M. & Ochiai, H. (1996). Molecular cloning and characterisation of the cDNA for discoidin-II of *Dictyostelium discoideum. Plant and Cell Physiology* **37**, 505–514.
5. Alexander, S., Cibulsky, A.-M. & Lerner, R.A. (1983). Ion dependence of the discoidin-I lectin from *Dictyostelium discoideum. Differentiation* **24**, 209–212.
6. Cooper, D.N.W., Lee, S.-C. & Barondes, S.H. (1983). Discoidin-binding polysaccharide from *Dictyostelium discoideum. Journal of Biological Chemistry* **258**, 8745–8750.
7. Cooper, D.N.W. & Barondes, S.H. (1984). Colocalization of discoidin-binding ligands with discoidin in developing *Dictyostelium discoideum. Developmental Biology* **105**, 59–70.
8. Fukuzawa, M. & Ochiai, H. (1993). Spatiotemporal patterning of discoidin I and II during development of *Dictyostelium discoideum. Development Growth and Differentiation* **35**, 11–23.
9. Alexander, S., Shinnick, T.M. & Lerner, R.A. (1983). Mutants of *Dictyostelium discoideum* blocked in expression of all members of the developmentally regulated discoidin multigene family. *Cell* **34**, 467–475.

Drosophila melanogaster lectin

Mol. wt: 41 000 (native); 14 000 (subunit)
Specificity: D-Galactose
Source: Pupae of the fruit fly *Drosophila melanogaster*

ISOLATION

A pupal extract was prepared by homogenisation and centifugation. The lectin was then purified from the soluble supernatant by affinity adsorption on fixed trypsin-treated bovine erythrocytes and elution either with galactose in a Ca^{2+}-containing buffer or EDTA in a Ca^{2+}-free buffer [1].

A cDNA for the lectin gene has also been isolated and cloned [1].

STRUCTURE

The lectin is a glycoprotein of Mr 41 000. It is a trimer of identical polypeptide subunits of apparent Mr 14 000 estimated by SDS-PAGE, although one of them is sometimes glycosylated, conferring an apparent Mr of 17 000 [1].

The *Drosophila* lectin gene is a single copy gene encoding a polypeptide of 183 amino acid residues. The first 20 residues make up a leader peptide with a signal sequence, while the remaining 163 residues constitute the 14 kDa subunit. The complete nucleotide sequence and the derived amino acid sequence have been published in full [1].

BIOLOGICAL ACTIVITIES

The *Drosophila* lectin has been detected by its ability to agglutinate trypsin-treated glutaraldehyde-fixed bovine erythrocytes. Activity is dependent on the presence of divalent cations, optimally calcium. Haemagglutination is readily inhibited by D-galactose but the only other inhibitory monosaccharide was glucose at a 100-fold greater concentration. Lactose was intermediate in effect between galactose and glucose.

TISSUE AND/OR SUBCELLULAR DISTRIBUTION

No information on distribution was provided [1].

POSSIBLE FUNCTION(S)

The response to injury (see **Comments** section) is consistent with a defence function [1]. The developmental regulation (see **Comments**), which is similar to that observed with the corresponding *Sarcophaga* lectin, led to the suggestion that the *Drosophila* lectin might also be involved in the regulation of imaginal disc differentiation [1].

COMMENTS

The *Drosophila* lectin is a C-type lectin. The sequence beyond the first 30 residues (from the amino terminal end) corresponds to the carbohydrate-recognition domain characteristic of C-type lectins and accounts for much of the overall 25% homology with the D-galactose-specific lectin from *Sarcophaga peregrina* (qv) [1].

Drosophila lectin gene activation occurs 4–6 hours after injury to the cell walls of larvae, similar to the situation with *Sarcophaga* lectin. Without injury, *Drosophila* lectin synthesis begins significantly in third-instar larvae, is increased in early pupae, and then subsides at the late pupal stage. The lectin gene was also expressed in adults; notably expression in males greatly exceeded that in females [1].

REFERENCE

1. Haq, S., Kubo, T., Kurata, S., Kobayashi, A. & Natori, S. (1996). Purification, characterisation and cDNA cloning of a galactose-specific C-type lectin from *Drosophila melanogaster*. *Journal of Biological Chemistry* **271**, 20213–20218.

Echinoidin

Other name: *Anthocidaris crassispina* agglutinin
Mol. wt.: 290 000 (native); 16 700 (subunit)
Specificity: Gal-β1,3-GalNAc
Source: Coelomic fluid of the sea urchin *Anthocidaris crassispina*

ISOLATION

The serum derived from coelomic fluid was applied to a DEAE–cellulose column and the lectin eluted with 0.5 M NaCl. The lectin was then purified to homogeneity by affinity adsorption to glutaraldehyde-fixed human erythrocyte ghosts and elution with EDTA [1].

STRUCTURE

Echinoidin is a glycoprotein (3.2% carbohydrate) rich in serine, glycine, glutamic acid and aspartic acid residues, but with a low content of methionine and basic amino acids [1]. The complete amino acid sequence has been published [2]. The native molecule is a multimer of Mr 290 000 made up of homodimers linked by disulphide bonds [1]. Although the apparent Mr of the subunits was estimated by SDS-PAGE to be 13 000, the calculated Mr from the 147-residue polypeptides was 16 671 [2].

The primary structure of echinoidin shows homology to those of the two mannan-binding lectins (qv) from rat liver, the *Sarcophaga peregrina* humoral lectin (qv), and other C-type lectins [2].

The N-terminal region contains the Arg-Gly-Asp sequence associated with the cell-binding domain of fibronectin [2].

BIOLOGICAL ACTIVITIES

Echinoidin agglutinated human (ABO non-specifically) and ovine, but not bovine or chicken, erythrocytes in a calcium-dependent manner [1]. Haemagglutination was not inhibited by any monosaccharide or disaccharide tested, nor by the glycoproteins, ovalbumin or thyroglobulin [1]. However, the lectin was inhibited by human glycophorin and by tryptic fragments of erythrocyte membranes. Experiments with such fragments indicated that O-glycosidically linked (Gal-β1,3-GalNAc) may be important for lectin recognition [1].

Echinoidin promoted the attachment of human rhabdomyosarcoma cells to a plastic surface; this was mediated by the RGD sequence within the lectin and not by the carbohydrate-binding site [3].

TISSUE AND/OR SUBCELLULAR DISTRIBUTION

Echinoidin was obtained from clotted coelomic fluid and therefore could have originated in the fluid itself or in cells activated or damaged by the clotting process [1].

POSSIBLE FUNCTION(S)

Echinoidin has the properties of a cell adhesion molecule and it has been conjectured that it could act as a bifunctional molecule mediating interaction between glycoconjugates and integrins [3].

COMMENTS

This sea urchin species possesses at least two separate lectins: the eggs of the same species contain a calcium-independent, galactoside-binding dimeric lectin with identical subunits of 11 500 [4].

Echinoidin has something rather more in common with discoidin I (qv), particularly in view of its cell adhesive property mediated by the Arg-Gly-Asp sequence independent of its lectin activity.

REFERENCES

1. Giga, Y., Sutoh, K. & Ikai, A. (1985). A new multimeric hemagglutinin from the coelomic fluid of the sea urchin *Anthocidaris crassispina*. *Biochemistry* **24**, 4461–4467.
2. Giga Y, Ikai, A. & Takahashi, K. (1987). The complete amino acid sequence of echinoidin, a lectin from the coelomic fluid of the sea urchin, *Anthocidaris crassispina*. *Journal of Biological Chemistry* **262**, 6197–6203.
3. Ozeki, Y., Matsui, T. & Titani, K. (1991). Cell adhesive activity of two animal lectins through different recognition mechanisms. *FEBS Letters* **289**, 145–147.
4. Ozeki, Y., Matsui, T., Suzuki, M. & Titani, K. (1991). Amino acid sequence and molecular characterization of a D-galactoside-specific lectin purified from the sea urchin (*Anthocidaris crassispina*) eggs. *Biochemistry* **30**, 2391–2394.

Echinonectin

Other name: *Lytechinus variegatus* adhesion molecule
Mol. wt: 220 000 (native); 110 000 (subunit)
Specificity: Sulphated glycans
Source: Eggs and embryos of the sea urchin *Lytechinus variegatus*

ISOLATION

A soluble extract was obtained from eggs or embryos homogenised in a detergent-containing buffer. The lectin was originally isolated by application to a column of gelatin–Sepharose and elution with 8 M urea [1]. Later it was simply applied to Sepharose and eluted with 1 M-galactose [2].

STRUCTURE

Echinonectin is a glycoprotein (6% carbohydrate) with a high proportion of polar residues and a low proportion of basic amino acids [1]. A native Mr of 205 000 derived from gradient PAGE is consistent with values of $220–230 \times 10^3$ Da estimated by SDS-PAGE under non-reducing conditions [1,2]. Under reducing conditions in SDS-PAGE, the lectin migrates as a single band corresponding to an Mr of 110 000–116 000, indicating a dimeric structure linked by disulphide bonds [1,2]. The structure forms an unusual bow-tie shape when visualised by electron microscopy. In the presence of urea, echinonectin was resolved into two charge isomers with isoelectric points of 5.0 and 5.2 [1].

BIOLOGICAL ACTIVITIES

Echinonectin binds to Sepharose (agarose). This activity can be inhibited by high concentrations of galactose and by much lower concentrations of α-1-*O*-methylgalactoside. The most potent inhibitors, however, were the high molecular weight polysaccharides, carrageenan (galactose sulphate) and fucoidan (fucose sulphate) [2].

Echinonectin also promotes the adhesion of dissociated embryonic cells to the surface of microtitre plates [1].

TISSUE AND/OR SUBCELLULAR DISTRIBUTION

Echinonectin is distributed evenly in the cytoplasm of unfertilised eggs, apparently in granules or vesicles, but appears on the cell surface after fertilisation [1]. Later, it becomes localised to the apical surface of ectoderm cells in the embryo [1].

POSSIBLE FUNCTION(S)

It has been suggested that echinonectin functions as a cell–substrate adhesion molecule during development [1].

COMMENTS

Echinonectin appears to be a substrate adhesion molecule functionally related to fibronectin [1,2].

REFERENCES

1. Alliegro, M.C., Ettensohn, C.A., Burdsal, C.A., Erickson, H.P. & McClay, D.R. (1988). Echinonectin: a new embryonic substrate adhesion protein. *Journal of Cell Biology* **107**, 2319–2327.
2. Veno, P.A., Strumski, M.A. & Kinsey, W.H. (1990). Purification and characterization of echinonectin, a carbohydrate-binding protein from sea urchin eggs. *Development, Growth and Differentiation* **32**, 315–319.

Eel agglutinins

Other names: AAA etc
Mol. wt: 120 000 (native); 40 000 (subunit)
Specificity: L-Fucose
Source: Serum of eels of the genus *Anguilla*

ISOLATION

The anti-blood group O 'antibody' found in eel (*Anguilla anguilla*) serum [1] can be purified by affinity chromatography on polyacrylamide-α-L-fucose [2]. A similar agglutinin was found in the serum of *Anguilla rostrata*: it was originally isolated by precipitation with 3-*O*-methyl-D-fucose [3].

STRUCTURE

The eel agglutinins are non-glycosylated proteins with similar amino acid compositions rich in aspartic and glutamic acids. Ultracentrifugal analyses revealed a single symmetrical boundary for each, with $S_{20,w}$ values of 6.5 (*A. anguilla* [2]) and 7.2 (*A. rostrata* [4]), suggesting a molecular weight in the vicinity of 120 000. The apparent Mr of the *A. anguilla* lectin estimated by gel filtration, however, was no more than 80 000 [2,5].

These molecules appear to consist of (two or three?) subunits of 40 000 based on SDS-PAGE analysis in the absence of reducing agent [2,4,5]. However, the reported SDS-PAGE analysis after reduction indicates two bands of approximate Mr 20 000 for the *A. anguilla* lectin [2,5] and (presumably four) bands of Mr 10 000 for the *A. rostrata* lectin [4]. N-terminal analysis and peptide mapping of the two 20 000 *A. anguilla* polypeptides indicates their probable identity [5]. In contrast, two separate N-terminal residues were found in the purified *A. rostrata* agglutinin, indicating two dissimilar polypeptides [3].

The isoelectric points of the lectins from both species are around pH 5.5 [2,4].

BIOLOGICAL ACTIVITIES

The eel serum lectins agglutinate human type O (and sometimes A_2) erythrocytes but not type A_1 or B cells. The more specific *A. anguilla* lectin agglutinates group O cells at 20 μg/ml, but a 250-fold higher concentration is required to agglutinate B, A_1 or even A_2 cells [2].

Glycoconjugates containing blood group H(O) substances are specifically precipitated, as are some polysaccharides without blood group activity [1,3,6].

These activities are most readily inhibited by L-fucose and its derivatives [1,7], but 3-*O*-methyl-D-galactose and 3-*O*-methyl-D-fucose also inhibit and precipitate the lectin from *A. rostrata* [3,6,7].

The *A. anguilla* lectin was non-mitogenic for rabbit lymph node lymphocytes [2].

TISSUE AND/OR SUBCELLULAR DISTRIBUTION

These lectins have been described only from serum.

POSSIBLE FUNCTION(S)

It is assumed that eel serum lectins are antibody-like, participating in antibacterial defence.

COMMENTS

These eel lectins show a high specificity for human blood group O, and historically contributed to the discovery of fucose as the immunodominant sugar in group O(H) [1].

The commercially available *A. anguilla* lectin has a long history of application as a blood typing and fucose-specific reagent. Recently, it was used successfully in a small lectin panel to differentiate strains of *Helicobacter pylori* [8].

REFERENCES

1. Watkins, W.M. & Morgan, W.T.J. (1952). Neutralization of the anti-H agglutinin in eel serum by simple sugars. *Nature* **169**, 825–826.
2. Hořejši, V. & Kocourek, J. (1978). Studies on lectins XXXVI. Properties of some lectins prepared by affinity chromatography on *O*-glycosyl polyacrylamide gels. *Biochimica et Biophysica Acta* **538**, 299–315.
3. Springer, G.F. & Desai, P.R. (1971). Monosaccharides as specific precipitinogens of eel anti-human blood group H(0) antibody. *Biochemistry* **10**, 3749–3761.
4. Bezkorovainy, A., Springer, G.F. & Desai, P.R. (1971). Physiochemical properties of the eel anti-human blood group H(0) antibody. *Biochemistry* **10**, 3761–3764.
5. Kelly, C. (1984). Physiochemical properties and N-terminal sequence of eel lectin. *Biochemical Journal* **220**, 221–226.
6. Springer, G.F., Takahashi, T., Desai, P.R. & Kolecki, B.J. (1965). Cross reactive human blood group H(0) specific polysaccharide from *Sassafras albidum* and characterization of its hapten. *Biochemistry* **4**, 2099–2113.
7. Springer, G.F. & Williamson, P. (1962). Immunochemical significance of L- and D-fucose derivatives. *Biochemical Journal* **85**, 282–291.
8. Hynes, S.O., Hirmo, S., Wadström, T. & Moran, A.P. (1999). Differentiation of *Helicobacter pylori* isolates based on lectin binding of cell extracts in an agglutination assay. *Journal of Clinical Microbiology* **37**, 1994–1998.

Electrolectin

Other name: *Electrophorus electricus* galectin
Mol. wt: 32 500 (native); 16 500 (subunit)
Specificity: *N*-acetyllactosamine
Source: The electric eel *Electrophorus electricus*

ISOLATION

Electrolectin was first partially purified by affinity chromatography on acid-treated Sepharose [1], and later purified to homogeneity on lactosyl–Sepharose [2]; the starting material was an extract of electric organ homogenate in saline containing EDTA and dithiothreitol.

STRUCTURE

Electrolectin is a homodimer of Mr 32 500 determined by equilibrium sedimentation. Each monomer (apparent Mr 16 500 by SDS-PAGE) consists of 134 amino acids including one tryptophan but no cysteine residue [1]. The amino acid sequence has approximately 40% identity to mammalian galectin-1 [3].

BIOLOGICAL ACTIVITIES

Electrolectin agglutinates fresh or glutaraldehyde-fixed trypsin-treated lapine erythrocytes in the absence of divalent cations [1,2]. Haemagglutinating activity is stable only in the presence of reducing agents, and presumably the oxidation of tryptophan leads to loss of activity [2]. Haemagglutination is inhibited by disaccharides containing non-reducing terminal β-D-galactose residues: D-galactopyranosyl-β-1,1-thiogalactopyranoside > lactose [1]. Lactose also protects the lectin from oxidative inactivation [2].

Electrolectin can agglutinate lapine lymphocytes and is weakly mitogenic towards them. Both properties can be inhibited by β-galactosides [4].

TISSUE AND/OR SUBCELLULAR DISTRIBUTION

Electrolectin has been found in both the electric organ and in swimming muscle tissue [1].

POSSIBLE FUNCTION(S)

The function of electrolectin is not known, but many suggestions have been made about this and other galectins (see Chapter 3).

COMMENTS

Electrolectin was the first vertebrate galectin to be discovered, and contemporaneously the existence of related lectins in a variety of tissues from other vertebrate species was realised [1].

Electrolectin was found to have a beneficial prophylactic and therapeutic action *in vivo* on experimental autoimmune myasthenia gravis in rabbits [4]. Prophylactic injections of electrolectin totally prevented the onset of disease symptoms, while administration to myasthenic animals led to a complete recovery within three weeks for most animals. Lymphocytes from animals immunised with electrolectin exhibited a diminished recall antigen response to stimulation *in vitro* with the inducing antigen (acetylcholine receptors) [4].

REFERENCES

1. Teichberg, V.I., Silman, I., Beitsch, D.D. & Resheff, G. (1975). A β-D-galactoside binding protein from electric organ tissue of *Electrophorus electricus. Proceedings of the National Academy of Sciences USA* **72**, 1383–1387.
2. Levi, G. & Teichberg, V.I. (1981). Isolation and physiochemical characterization of electrolectin, a β-D-galactoside binding protein from the electric organ tissue of *Electrophorus electricus. Journal of Biological Chemistry* **256**, 5735–5740.
3. Paroutaud, P., Levi, G., Teichberg, V.I. & Strosberg, A.D. (1987). Extensive amino acid homologies between animal lectins. *Proceedings of the National Academy of Sciences USA* **84**, 6345–6348.
4. Levi, G., Tarrab-Hazdai, R. & Teichberg, V.I. (1983). Prevention and therapy with electrolectin of experimental autoimmune myasthenia gravis in rabbits. *European Journal of Immunology* **13**, 500–507.

Entamoeba histolytica lectin

Other name: Amoebic adherence lectin
Mol. wt: 260 000 to 600 000 (native); 170 000 & 35 000 (subunit)
Specificity: GalNAc/Gal
Source: Trophozoites of *Entamoeba histolytica*

ISOLATION

Petrie *et al.* [1] isolated the adherence lectin by a selection of affinity methods: immobilised asialo-orosomucoid and elution with galactose; immobilised *p*-aminophenyl-thio-β-D-galactopyranoside and elution at pH 2.5; immobilised monoclonal anti-adherence lectin antibodies and elution at pH 2.5.

By using antibodies to the purified lectin, the corresponding cDNA was isolated from an expression library of a pathogenic *E. histolytica* isolate [2].

STRUCTURE

The amoebic adherence lectin is a glycoprotein with an isoelectric point of pH 6.2 and an apparent molecular weight of 600 000 determined by gel filtration [3]. Analysis of the purified lectin by SDS-PAGE reveals a single 260 kDa band under non-reducing conditions, and two subunits of 170 000 and 35 000 in the presence of mercaptoethanol [1,3]. The cDNA-deduced amino acid sequence corresponds to a molecular mass of 143 000 for the major subunit [2]. The gel filtration profile is most consistent with aggregates of two and mainly three molecules, each a heterodimer composed of a heavy lectin subunit covalently bound to a single light subunit [3].

The deduced primary structure of the heavy subunit [2] comprises four domains: a short C-terminal cytoplasmic region, a transmembrane region, an extracellular cysteine rich domain and an extracellular cysteine-poor domain. Computer searches indicate a low degree of sequence homology between the cysteine-rich region and wheat germ agglutinin. The carbohydrate-recognition domain, found in the cysteine-rich region of the heavy subunit, possesses some sequence homology to the receptor-binding domain of hepatocyte growth factor [4].

BIOLOGICAL ACTIVITIES

The amoebic adherence lectin can agglutinate erythrocytes from numerous species (humans, rabbits, guinea-pigs, rats, sheep) and adheres to human and rodent colonic mucosa [5–7]. These activities are inhibited only by GalNAc and galactose among monosaccharides, but are more susceptible to inhibition by glycoconjugates with terminal galactosyl residues like asialo-orosomucoid [1]. The synthetic neoglycoprotein $GalNAc_{39}BSA$ is an outstandingly potent inhibitor [8]. The lectin appears to recognise a unique pattern of spaced multiple GalNAc residues in preference to the less widely spaced GalNAc residues favoured by the mammalian hepatic lectin [8,9]. Ligand binding is calcium-dependent [4].

The lectin is mitogenic for human lymphocytes [1,10].

TISSUE AND/OR SUBCELLULAR DISTRIBUTION

E. histolytica exists as single cells. Some of the lectin is present at the cell surface [1].

POSSIBLE FUNCTION(S)

The characteristic pathology of invasive amoebiasis is mediated by the adherence lectin [5]. Binding to colonic epithelial cells is a necessary first step in a sequence of events including cytolysis and the amoebic phagocytosis of the dead or dying cells.

COMMENTS

Cellular and humoral arms of both systemic and mucosal immunity play a role in protecting humans from amoebic dysentery [11]. The human antibody response to the adherence lectin may either inhibit or enhance the adherence reaction [12]. An epitope provoking a protective antibody response has recently been described and has become a leading candidate for vaccine development [13]. Immunisation with the lectin completely protected 67% of immunised gerbils from experimental amoebic liver abscess formation, although liver abscess weights were significantly higher in the remaining animals compared to a control (non-immunised) group [14].

Active or passive immunisation of gerbils with antibodies to the carbohydrate-recognition domain inhibited adherence without any evidence of adherence-enhancing antibodies. Corresponding antibodies were found in humans with *E. histolytica*-mediated liver abscesses, but this response did not persist after recovery [4]. Passive immunisation with a monoclonal antibody to the lectin significantly reduced the incidence of liver abscesses in hamsters subsequently challenged intra-hepatically with trophozoites [15].

Although the 35 kDa subunit plays no part in adhesion, experiments with transfectants lacking the smaller subunit provide evidence that it contributes to the cytopathic and cytotoxic properties of the lectin [16].

REFERENCES

1. Petri, W.A., Smith, R.D., Schlesinger, P.H., Murphy, C.F. & Ravdin, J.I. (1987). Isolation of the galactose-binding lectin that mediates the in vitro adherence of *Entamoeba histolytica. Journal of Clinical Investigation* **80**, 1238–1244.
2. Tannich, E., Ebert, F. & Horstmann, R.D. (1991). Primary structure of the 170 kDa surface lectin of pathogenic *Entamoeba histolytica. Proceedings of the National Academy of Sciences USA* **88**, 1849–1853.
3. Petri, W.A., Chapman, M.D., Snodgrass, T., Mann, B., Broman, J. & Ravdin, J.I. (1989). Subunit structure of the galactose and *N*-acetyl-D-galactosamine-inhibitable adherence lectin of the *Entamoeba histolytica. Journal of Biological Chemistry* **264**, 3007–3012.
4. Dodson, J.M., Lenkowski, P.W., Eubanks, A.C., Jackson, T.F.G.H., Napodano, J., Lyerly, D.M., Lockhart, L.A., Mann, B.J. & Petri, W.A. (1999). Infection and immunity mediated by the carbohydrate recognition domain of the *Entamoeba histolytica* Gal/GalNAc lectin. *Journal of Infectious Diseases* **179**, 460–466.
5. Ravdin, J.I. (1989). *Entamoeba histolytica*: from adherence to enteropathy. *Journal of Infectious Diseases* **159**, 420–429.
6. Kobiler, D. & Mirelman, D. (1980). Lectin activity in *Entamoeba histolytica* trophozoites. *Infection and Immunity* **29**, 221–225.
7. Chadee, K., Petri, W.A., Innes, D.J. & Ravdin, J.I. (1987). Rat and human colonic mucins bind to and inhibit adherence lectin of *Entamoeba histolytica. Journal of Clinical Investigation* **80**, 1245–1254.
8. Adler, P., Wood, S.J., Lee, Y.C., Lee, R.T., Petri, W.A. & Schnaar, R.L. (1995). High affinity binding of the *Entamoeba histolytica* lectin to polyvalent *N*-acetylgalactosominides. *Journal of Biological Chemistry* **270**, 5164–5171.
9. Yi, D., Lee, R.T., Longo, P., Boger, E.T., Lee, Y.C., Petri, W.A., Schnaar, R.L. (1998). Substructural specificity and polyvalent carbohydrate recognition by the *Entamoeba histolytica* and rat hepatic *N*-acetylgalactosamine/galactose lectins. *Glycobiology* **8**, 1037–1043.
10. Salata, R.A. & Ravdin, J.I. (1985). *N*-acetyl-D-galactosamine-inhibitable adherence lectin of *Entamoeba histolytica II.* Mitogenic activity for human lymphocytes. *Journal of Infectious Diseases* **151**, 816–822.
11. Huston, C.D. & Petri, W.A. (1998). Host–pathogen interaction in amebiasis and progress in vaccine development. *European Journal of Clinical Microbiology and Infectious Diseases* **17**, 601–614.
12. Petri, W.A., Snodgrass, T.L., Jackson, T.F.H.G., Gathiram, V., Simjee, A.E., Chadee, K. & Chapman, M.D. (1990). Monoclonal antibodies directed against the galactose-binding lectin of *Entamoeba histolytica* enhance adherence. *Journal of Immunology* **144**, 4803–4809.
13. Lotter, H., Zhang, T., Seydel, K.B., Stanley, S.L. & Tannich, E. (1997). Identification of an epitope on the *Entamoeba histolytica* 70 kD lectin conferring antibody-mediated protection against invasive amebiasis. *Journal of Experimental Medicine* **185**, 1793–1801.
14. Petri, W.A. & Ravdin, J.I. (1991). Protection of gerbils from amebic liver abscess by immunisation with the galactose-specific adherence lectin of *Entamoeba histolytica. Infection and Immunity* **59**, 97–101.
15. Cheng, X.J., Tachibara, H. & Kaneda, Y. (1999). Protection of hamsters from amebic liver abscess formation by monoclonal antibody to a 150 kDa surface lectin of *Entamoeba histolytica. Parasitology Research* **85**, 78–80.
16. Ankri, S., Padilla-Vaca, F., Stolarsky, T., Koole, L., Katz, U. & Mirelman, D. (1999). Antisense inhibition of expression of the light subunit (35 kDa) of the Gal/GalNAc lectin complex inhibits *Entamoeba histolytica* virulence. *Molecular Microbiology* **33**, 327–337.

ERGIC-53

Other names: ER-Golgi intermediate compartment protein of 53 kDa; MR-60
Mol. wt: 110 000 (native); 53 000 (subunit)
Specificity: Mannose
Source: Mammalian cells

ISOLATION

ERGIC-53 was first identified in Caco-2 cells (by a monoclonal antibody to Golgi membranes) as an integral membrane protein and an intermediate compartment marker [1]. A cDNA was isolated [2] and later found to be identical to that of 'MR-60', a lectin isolated from human myelomonocytic HL60 cells [3]. Mutant cDNAs were engineered and transfected into COS cells, and the overexpressed protein was purified on immobilised mannose [4].

Homologous cDNAs have been isolated from rats, monkeys and the toad *Xenopus* [5].

STRUCTURE

ERGIC-53 is a non-glycosylated type-I transmembrane protein with a subunit molecular weight of 53 000–60 000 determined by SDS-PAGE under reducing conditions. It appears to form homo-oligomers, particularly dimers [1,2].

Each subunit is organised into three domains: a luminal domain of about 200 amino acids containing the carbohydrate-binding site, a transmembrane domain and a short cytoplasmic domain possessing sorting signals [2,4].

The most remarkable structural feature is the homology with leguminous plant lectins. The primary sequence has about 20% identity with several legume lectins, as well as even greater homology with VIP36 (qv), and the greatest similarity between ERGIC-53 and VIP36 was in the region showing homology with legume lectins [6]. Moreover, the predicted secondary structure contains a legume lectin-like fold, while the carbohydrate-binding site has the same critical amino acid residues that are conserved in all legume lectins [4].

BIOLOGICAL ACTIVITIES

Both natural MR60 and recombinant ERGIC-53 can bind to immobilised mannose and be eluted with the free sugar. This binding is calcium-dependent and neither Mg^{2+} nor Mn^{2+} can substitute for calcium ions. ERGIC-53 also exhibits some rather low affinity for glucose and GlcNAc, but none for galactose or D-fucose.

Experiments with fluorescein-tagged neoglycoproteins (BSA-mannose etc.) confirmed specificity for mannose [4].

TISSUE AND/OR SUBCELLULAR DISTRIBUTION

ERGIC-53 mRNA was expressed in all human tissues examined, including heart, lung, liver, kidney and skeletal muscle [5].

ERGIC-53 is found in the ER-Golgi intermediate compartment and continuously recycles between the cis-Golgi network and the endoplasmic reticulum [1,7].

POSSIBLE FUNCTION(S)

It has been suggested that ERGIC-53 functions as a sorting receptor in the secretory pathway. It may recognise fully folded high mannose glycoproteins in the ER and direct them to the intermediate compartment; the high calcium concentration in the lumen of the ER would favour binding, whereas the lower calcium level in the intermediate compartment might trigger release of ligands [4,8]. Alternatively, ERGIC-53 could redirect defective glycoproteins from the intermediate compartment or cis-Golgi back to the endoplasmic reticulum for repair [4].

COMMENTS

Mutations in the human gene for ERGIC-53 (leading to a deficiency of functional protein) are associated with combined factor V-factor VIII deficiency, a rare coagulation disorder inherited in an autosomal recessive manner [9].

REFERENCES

1. Schweizer, A., Fransen, J.A.M., Bächi, T., Ginsel, L. & Hauri, H.-P. (1988). Identification by a monoclonal antibody, of a 53 kD protein associated with a tubulo-vesicular compartment at the cis-side of the Golgi apparatus. *Journal of Cell Biology* **107**, 1643–1653.
2. Schindler, R., Itin, C., Zerial, M., Lottspeich, F. & Hauri, H.-P. (1993). ERGIC-53, a membrane protein of the ER-Golgi intermediate compartment, carries an ER retention motif. *European Journal of Cell Biology* **61**, 1–9.
3. Arar, C., Carpentier, V., Le Caer, J.-P., Monsigny, M., Legrand, A. & Roche, A.-C. (1995). ERGIC-53, a membrane protein of the endoplasmic reticulum-Golgi intermediate compartment, is identical to MR60, an intracellular mannose-specific lectin of myelomonocytic cells. *Journal of Biological Chemistry* **270**, 3551–3553.
4. Itin, C., Roche, A.-C., Monsigny, M. & Hauri, H.-P. (1996). ERGIC-53 is a functional mannose-selective and calcium-dependent human homologue of leguminous lectins. *Molecular Biology of the Cell* **7**, 483–493.

5. Sarnataro, S., Caporaso, M.G., Bonatti, S. & Remondelli, P. (1999). Sequence and expression of the monkey homologue of the ER-Golgi intermediate compartment lectin, ERGIC-53. *Biochimica et Biophysica Acta* **1447**, 334–340.
6. Fiedler, K. & Simons, K. (1994). A putative novel class of animal lectins in the secretory pathway homologous to leguminous lectins. *Cell* **77**, 625–626.
7. Lippincott-Schwartz, J., Donaldson, J.G., Schweizer, A., Berger, E.G., Hauri, H.-P., Yuan, L.C. & Klausner, R.G. (1990). Microtubule dependent retrograde transport of proteins into the ER in the presence of brefeldin A reveals an ER cycling pathway. *Cell* **60**, 821–836.
8. Appenzeller, C., Andersson, H., Kappeler, F. & Hauri, H.P. (1999). The lectin ERGIC-53 is a cargo transport receptor for glycoproteins. *Nature Cell Biology* **1**, 330–334.
9. Neerman-Arbez, M., Johnson, K.M., Morris, M.A., McVey, J.H., Peyvani, F., Nichols, W.C., Ginsburg, D., Rossier, C., Antonarakis, S.E. & Tuddenham, E.G.D. (1999). Molecular analysis of the ERGIC-53 gene in 35 families with combined factor V-factor VIII deficiency. *Blood* **93**, 2253–2260.

E-selectin

Other names: CD62E; ELAM-1; LECAM-2
Mol. wt: 115 000
Specificity: Sialyl-Lewis^x; sulphated-Lewis^a
Sources: Mammalian vascular endothelia

ISOLATION

Human E-selectin was first identified by immunoprecipitation using a monoclonal antibody recognising a cell surface antigen involved in leukocyte adhesion that was found on endothelial cells only after activation [1]. Shortly afterwards, the molecule was cloned from a cDNA library constructed with RNA obtained from endothelial cells that had been cultured with IL-1 [2]. The equivalent molecules have also been cloned from murine [3,4] and lapine [5] sources.

STRUCTURE

E-selectin is a glycoprotein with a high carbohydrate content [2,3]. Its human cDNA predicts a mature protein of 589 amino acids and a molecular mass of 64 000; the much higher apparent Mr of 115 000 estimated by SDS-PAGE is due to its carbohydrate content. E-selectin shares with other members of the family an N-terminal region with a C-type lectin domain followed by a shorter epidermal growth factor-like domain then a series of complement regulatory protein-like repeats, and a C-terminal cytoplasmic tail attached to a transmembrane region. The central segment is intermediate in size between L- and P-selectins, consisting of six repeating units [2]. The murine homologue has an overall sequence similarity of 81%, but has a pI of 6.9, compared to a pI of 5.0 for human E-selectin [3]. Both human and murine E-selectins possess approximately 40% sequence identity to the corresponding L- and P-selectins from the same species [6].

The three-dimensional structure of the ligand-binding region of human E-selectin has been determined at 0.2 nm resolution [7].

BIOLOGICAL ACTIVITIES

E-selectin is a calcium-dependent adhesion molecule that can mediate the binding of neutrophils [8], monocytes [9], eosinophils [10], T lymphocytes [11] and NK cells [12]. E-selectin collaborates with other (especially P-) selectins in mediating leukocyte attachment to endothelium at sites of inflammation *in vivo* [6,13].

Endogenous ligands for E-selectin include PSGL-1 (shared with P-selectin) [14], the cutaneous homing antigen in skin [15] and a 150 kDa glycoprotein of murine myeloid cells [16]. The physiological significance of these 'natural' ligands has not been clearly established [6].

The cellular interactions of E-selectin are readily inhibited by sialyl-Lewis^X [17,18], sulphated-Lewis^a tetra- and pentasaccharides [19] and related structures.

Ligand-dependent regulation of phosphorylation of the cytoplasmic tail region of E-selectin provides evidence for a transmembrane signalling function [20].

TISSUE AND/OR SUBCELLULAR DISTRIBUTION

The expression of E-selectin appears to be largely restricted to activated endothelial cells; synthesis is induced at transcriptional level and leads to cell surface expression [6]. However, E-selectin is also expressed in proliferating endothelial cells under non-inflammatory conditions [21].

POSSIBLE FUNCTION(S)

E-selectin contributes to the recruitment of leukocytes to vascular endothelium at sites of inflammation in concert with other members of the selectin family [6]. E-selectin may differ from the other selectins in having a role in angiogenesis, since antibodies against bovine E-selectin inhibited capilliary formation in an *in vitro* model system [22,23].

Knock-out mice lacking two or all three selectins in various models of inflammation demonstrate that all three selectins (but especially E- and P-selectins) have overlapping functions [6,24]. Although E-selectin alone can mediate some slow leukocyte rolling, it is inefficient at capturing leukocytes to the venule wall, and under more inflammatory conditions, it is much less effective at promoting leukocyte recruitment than P-selectin on its own [24].

COMMENTS

Activated T helper cell subsets can be distinguished on the basis of their selectin receptors; murine Th-1 cells, but not Th-2 cells, are able to bind E-selectin (and P-selectin) [25].

Increased E-selectin expression on vascular endothelium is found in association with many inflammatory conditions, including appendicitis, diverticulitis, ulcerative colitis, Crohn's disease, hepatitis, rheumatoid arthritis, sepsis and rejection of allogeneic transplants [26]. Soluble E-selectin is increased in the circulation in many diseases, including atopic dermatitis [27,28], psoriasis [27], septic shock [29], pre-eclampsia [30] and maturity onset (type 2) diabetes

[31]. In synovial fluid, the elevation of soluble E-selectin levels was greater with rheumatoid arthritis patients than with patients with osteoarthritis [32]. Soluble E-selectin concentration usually correlates with disease activity and could be used to monitor response to treatment.

A single patient has been described with recurrent infections and impaired pus formation in association with an absence of E-selectin on inflammed endothelia; the exact defect is unclear, however, since the patient possessed E-selectin mRNA and elevated levels of soluble E-selectin protein in her circulation [33].

A rat model of IgG-immune complex-induced inflammatory lung injury is E-selectin-dependent, and protective effects were observed with either sialyl-Lewis^{X}-containing oligosaccharides [34] or selectin-IgG chimeras [35].

The gene for E-selectin has been localised to chromosome 1 of humans and mice [36].

REFERENCES

1. Bevilacqua, M.P., Pober, J.S., Mendrick, D.L., Cotran, R.S. & Gimbrone, M.A. (1987). Identification of an inducible endothelial–leukocyte adhesion molecule. *Proceedings of the National Academy of Sciences USA* **84**, 9238–9242.
2. Bevilacqua, M.P., Stengelin, S., Gimbrone, M.A. & Seed, B. (1989). Endothelial leukocyte adhesion molecule 1: an inducible receptor for neutrophils related to complement regulatory proteins and lectins. *Science* **243**, 1160–1165.
3. Becker-André, M., Van Huijsduijnen, R.H., Losberger, C., Whelan, J. & Delamarter, J.F. (1992). Murine endothelial leukocyte-adhesion molecule 1 is a close structural and functional homologue of the human protein. *European Journal of Biochemistry* **206**, 401–411.
4. Weller, A., Isenmann, S. & Vestweber, D. (1992). Cloning of the mouse endothelial selectins. *Journal of Biological Chemistry* **267**, 15176–15183.
5. Larigan, J.D., Tsang, T.C., Rumberger, J.M. & Burns, D.K. (1992). Characterization of cDNA and genomic sequences encoding rabbit ELAM-1: conservation of structures and functional interactions with leukocytes. *DNA and Cell Biology* **11**, 149–162.
6. Kansas, G.S. (1996). Selectins and their ligands: current concepts and controversies. *Blood* **88**, 3259–3287.
7. Graves, B.J., Crowther, R.L., Chandran, C., Rumberger, J.M., Li, S., Huang, K.-S., Presky, D.H., Familletti, P.C., Wolitzky, B.A. & Burns, D.K. (1994). Insight into E-selectin/ligand interaction from the crystal structure and mutagenesis of the lec/EGF domains. *Nature* **367**, 532–538.
8. Lawrence, M.B. & Springer, T.A. (1993). Neutrophils roll on E-selectin. *Journal of Immunology* **151**, 6338–6346.
9. Carlos, T.M., Dobrina, A., Ross, R. & Harlan, J.M. (1990). Multiple receptors on human monocytes are involved in adhesion to cultured human endothelial cells. *Journal of Leukocyte Biology* **48**, 451–456.
10. Ulfman, L.H., Kuijper, P.H.M., Van Der Linden, J.A.M., Lammers, J.-W., Zwaginga, J.J. & Koenderman, L. (1999). Characterization of eosinophil adhesion to TNF-α-activated endothelium under flow conditions: α_4 integrins mediate initial attachment, and E-selectin mediates rolling. *Journal of Immunology* **163**, 343–350.
11. Picker, L.J., Kishimoto, T.K., Smith, C.W., Warnock, R.A. & Butcher, E.C. (1991). ELAM-1 is an adhesion molecule for skin-homing T cells. *Nature* **349**, 796–799.

12. Yago, T., Tsukuda, M., Fukushima, H., Yamaoka, H., Kurata-Miura, K., Nishi, T. & Minami, M. (1998). IL-12 promotes the adhesion of NK cells to endothelial selectins under flow conditions. *Journal of Immunology* **161**, 1140–1145.
13. Catalina, M.D., Estess, P. & Siegelman, M.H. (1999). Selective requirements for leukocyte adhesion molecules in models of acute and chronic cutaneous inflammation: participation of E- and P- but not L-selectin. *Blood* **93**, 580–589.
14. Asa, D., Raycroft, L., Ma, L., Aeed, P.A., Kaytes, P.S., Elhammer, A.P. & Geng, J.G. (1995). The P-selectin glycoprotein ligand functions as a common human leukocyte ligand for P- and E-selectins. *Journal of Biological Chemistry* **270**, 11662–11670.
15. Babi, L.F.S., Moser, R., Soler, M.T.P., Picker, L.J., Blaser, K. & Hauser, C. (1995). Migration of skin-homing T cell across cytokine activated human endothelial cell layers involves interaction of the cutaneous lymphocyte-associated antigen (CLA), the very late antigen-4 (VLA-4) and the lymphocyte function-associated antigen-1 (LFA-1). *Journal of Immunology* **154**, 1543–1550.
16. Levinivitz, A., Muhlhoff, J., Isenmann, S. & Vestweber, D. (1993). Identification of a glycoprotein ligand for E-selectin on mouse myeloid cells. *Journal of Cell Biology* **121**, 449–459.
17. Walz, G., Aruffo, A., Kolanus, W., Bevilacqua, M. & Seed, B. (1990). Recognition by ELAM-1 of the sialyl-Le^x determinant on myeloid and tumour cells. *Science* **250**, 1132–1135.
18. Thomas, V.H., Yang, Y. & Rice, K.G. (1999). In vivo ligand specificity of E-selectin binding to multivalent sialyl $Lewis^x$ N-linked oligosaccharides. *Journal of Biological Chemistry* **274**, 19035–19040.
19. Yuen, C.-T., Bezouska, K., O'Brien, J., Stoll, M., Lemoine, R., Lubineau, A., Kiso, M., Hasegawa, A., Bockovich, N.J., Nicolaou, K.C. & Feizi, T. (1994). Sulphated blood group $Lewis^a$. *Journal of Biological Chemistry* **269**, 1595–1598.
20. Yoshida, M., Szente, B.E., Kiely, J.-M., Rosenzweig, A. & Gimbrone, M.A. (1998). Phosphorylation of the cytoplasmic domain of E-selectin is regulated during leukocyte–endothelial adhesion. *Journal of Immunology* **161**, 933–941.
21. Luo, J., Paranya, G. & Bischoff, J. (1999). Noninflammatory expression of E-selectin is regulated by cell growth. *Blood* **93**, 3785–3791.
22. Nguyen, M., Strubel, N.A. & Bischoff, J. (1993). A role for sialyl Lewis-X/A glycoconjugates in capillary morphogenesis. *Nature* **365**, 267–269.
23. Koch, A.E., Halloran, M.M., Haskell, C.J., Shah, M.R. & Polverini, P.J. (1995). Angiogenesis mediated by soluble forms of E-selectin and vascular cell adhesion molecule-1. *Nature* **376**, 517–519.
24. Jung, U. & Ley, K. (1999). Mice lacking two or all three selectins demonstrate overlapping and distinct functions for each selectin. *Journal of Immunology* **162**, 6755–6762.
25. Austrup, F., Vestweber, D., Borges, E., Lohning, M., Brauer, R., Herz, U., Renz, H., Hallmann, R., Scheffold, A., Radbruch, A. & Hamann, A. (1997). P- and E-selectin mediate recruitment of T-helper-1 but not T-helper-2 cells into inflamed tissues. *Nature* **385**, 81–83.
26. Bevilacqua, M.P., Nelson, R.M., Mannori, G. & Cecconi, O. (1994). Endothelial-leukocyte adhesion molecules in human disease. *Annual Review of Medicine* **45**, 361–378.
27. Czech, W., Schopf, E. & Kapp, A. (1996). Soluble E-selectin levels in sera of patients with atopic dermatitis and psoriasis – correlation with disease activity. *British Journal of Dermatology* **134**, 17–21.
28. Furue, M., Koga, T. & Yamashita, N. (1999). Soluble E-selectin and eosinophil cationic protein are distinct serum markers that differentially represent clinical features of atopic dermatitis. *British Journal of Dermatology* **140**, 67–72.
29. Newman, W., Beall, L.D., Carson, C.W., Hunder, G.G., Braben, N., Randhawa, Z.I., Gopal, T.V., Wiener-Kronisch, J. & Matthay, M.A. (1993). Soluble E-selectin is found in

supernatants of activated endothelial cells and is elevated in the serum of patients with septic shock. *Journal of Immunology* **150**, 644–654.

30. Daniel, Y., Kupferminc, M.J., Baram, A., Jaffa, A.J., Wolman, I., Shenhav, M. & Lessing, J.B. (1998). Plasma soluble endothelial selectin is elevated in women with pre-eclampsia. *Human Reproduction* **13**, 3537–3541.
31. Cominacini, L., Pasini, A.F., Garbin, U., Campagnola, M., Davoli, A., Rigoni, A., Zenti, M.G., Pastorino, A.M. & Lo Cascio, V. (1997). E-selectin plasma concentration is influenced by glycaemic control in NIDDM patients: possible role of oxidative stress. *Diabetologia* **40**, 584–589.
32. Koch, A.E., Turkiewicz, W., Harlow, L.A. & Pope, R.M. (1993). Soluble E-selectin in arthritis. *Clinical Immunology and Immunopathology* **69**, 29–35.
33. De Lisser, H.M., Christofidou-Solomidou, M., Sun, J., Nakada, M.T. & Sullivan, K.E. (1999). Loss of endothelial surface expression of E-selectin in a patient with recurrent infections. *Blood* **94**, 884–894.
34. Mulligan, M.S., Lowe, J.B., Larsen, R.D., Paulsen, J., Zheng, Z., DeFrees, S., Maemura, K., Fukuda, M. & Ward, P.A. (1993). Protective effects of sialylated oligosaccharides in immune complex-induced acute lung injury. *Journal of Experimental Medicine* **178**, 623–631.
35. Mulligan, M.S., Watson, S.R., Fennie, C. & Ward, P.A. (1993). Protective effects of selectin chimeras in neutrophil-mediated lung injury. *Journal of Immunology* **151**, 6410–6417.
36. Watson, M.L., Kingsmore, S.F., Johnstone, G.I., Siegelman, M.H., Le Beau, M.M., Lemons, R.S., Bora, N.S., Howard, T.A., Weissman, I.L. & McEver, P.R. (1990). Genomic organization of the selectin family of leukocyte adhesion molecules on human and mouse chromosome 1. *Journal of Experimental Medicine* **172**, 263–272.

Factor H

Other names: Complement factor H; β_1H
Mol. wt: 155 000
Specificity: Sialic acid; heparin
Source: Mammalian blood plasma

ISOLATION

Factor H was initially recognised as a regulatory component of the alternative pathway of complement [1], and first purified from human plasma in high yield by a procedure involving five chromatographic steps including ion-exchange and gel filtration [2]. Isolation of overlapping cDNA clones permitted the derivation of the primary sequence of human factor H [3]; a factor H cDNA clone was also isolated from a murine liver cDNA library by screening with a human cDNA probe as the initial step in the cloning strategy [4].

STRUCTURE

Human factor H is a single-chain glycoprotein (9% carbohydrate) with an apparent Mr >160 000 determined by SDS-PAGE [2]. The Mr value of 155 000 from analytical ultracentrifugation, however, is in excellent agreement with the calculated polypeptide Mr of 137 000 and the likely extent of glycosylation [3].

The primary sequence of 1213 amino acid residues in human factor H shows 61% identity to its murine counterpart, and both human and murine forms are arranged in a series of 20 homologous domains of about 60 amino acid each, forming a 'string of beads' structure. Each domain has four invariant cysteine residues; similar short consensus repeats are found in other complement components (C2, factor B etc.) and some non-complement system proteins [3,4].

More detailed studies relating structure to activities have revealed the approximate locations of three separate C3b-binding sites, two heparin-binding sides and two CRP-binding sites [5,6].

Several allelic variants of human factor H exist, and the two major variants can be accounted for by alternative splicing [3].

BIOLOGICAL ACTIVITIES

Factor H binds complement component C3b; this interaction inhibits the binding of factor B to C3b, promotes factor I-mediated cleavage of C3b to iC3b (co-factor activity), and promotes the decay of the alternative pathway C3 convertase, C3bBb (decay accelerating activity). This interaction and associated activities leading to the inhibition of alternative pathway activation can be inhibited by sialic acids [7].

Factor H can also bind to polyanions, including heparin, and this binding can increase the affinity between factor H and C3b [8]. Factor H binds to the M-protein of *Streptococcus pyogenes* via a heparin-binding site.

Factor H binds to C-reactive protein, and this interaction may mediate CRP-associated inhibition of alternative pathway activation [6].

TISSUE AND/OR SUBCELLULAR DISTRIBUTION

Factor H is a major plasma protein that is synthesised by hepatocytes and monocytes [9]. It is also found in association with the plasma membrane of some tonsillar lymphocytes and other leukocytes [10]. Within monocytes, factor H was detected in cytoplasmic granules by immunoperoxidase localisation [9].

POSSIBLE FUNCTION(S)

Factor H acts as a negative regulator of the alternative pathway of complement [1]. Its fundamental function may be to distinguish host cells from activators of the alternative pathway.

COMMENTS

Factor H is of historical interest as the first member of what is now a sizeable family of proteins known to possess the complement regulatory-type short consensus repeat structural motif. Factor H itself is composed entirely of such homologous repeating domains linked together by short spacers of 3 to 8 amino acids to form a flexible, extended, 'string of beads' formation.

REFERENCES

1. Reid, K.B.M. & Porter, R.R. (1981). The proteolytic activation systems of complement. *Annual Review of Biochemistry* **50**, 433–464.
2. Sim, R.B. & Di Scipio, R.G. (1982). Purification and structural studies on the complement-system control protein β_1H (factor H). *Biochemical Journal* **205**, 285–293.
3. Ripoche, J., Day, A.J., Harris, T.J.R. & Sim, R.B. (1988). The complete amino acid sequence of human complement factor H. *Biochemical Journal*, **249**, 593–602.
4. Kristensen, T. & Tack, B.F. (1986). Murine protein H is comprised of 20 repeating units, 61 amino acids in length. *Proceedings of the National Academy of Sciences USA* **83**, 3963–3967.
5. Sharma, A.K. & Pangburn, M.K. (1996). Identification of three physically and functionally distinct binding sites for C3b in human complement factor H by deletion mutagenesis. *Proceedings of the National Academy of Sciences USA* **93**, 10996–11001.
6. Jarva, H., Jokiranta, T.S., Hellwage, J., Zipfel, P.F. & Meri, S. (1999). Regulation of complement activation by C-reactive protein: targeting the complement inhibitory activity of factor H by an interaction with short consensus repeat domains 7 and 8–11. *Journal of Immunology* **163**, 3957–3962.
7. Varki, A. (1997). Sialic acids as ligands in recognition phenomena. *FASEB Journal* **11**, 248–255.
8. Meri, S. & Pangburn, M.K. (1990). Discrimination between activators and nonactivators of the alternative pathway of complement: regulation via a sialic acid/polyanion binding site on factor H. *Proceedings of the National Academy of Sciences USA* **87**, 3982–3986.
9. Whaley, K. (1980). Biosynthesis of the complement components and the regulatory proteins of the alternative complement pathway by human peripheral blood monocytes. *Journal of Experimental Medicine* **151**, 501–516.
10. Schulz, T.F., Scheiner, O., Alsenz, J., Lambris, J.D. & Dierich, M.P. (1984). Use of monoclonal antibodies against factor H to investigate the role of a membrane-associated protein antigenically related to H in C3b-receptor function. *Journal of Immunology* **132**, 392–398.

Galectin-1

Other names: L14 etc.
Mol. wt: 28 000 (native); 14 000 (subunit)
Specificity: *N*-acetyllactosamine
Source: Mammalian tissues

ISOLATION

Members of the galectin-1 family have been isolated from many sources, including bovine heart [1], human brain [2], rat and human lungs [3], human and simian skeletal muscle [4], spleens of several species [5], and placentae of human [6] and ovine [7] origins. Typically, the lectin is extracted from homogenates with a buffer containing both lactose and a reductant, and subsequently purified by affinity chromatography. For the latter, adsorption to asialofetuin followed by elution with lactose is a common choice.

Both human [8] and rat [9] galectins-1 have been cloned. The cDNA has been isolated by several groups, and recombinant proteins expressed in bacteria and other cells have been isolated [10–13].

STRUCTURE

The galectins-1 comprise a homologous family sharing over 80% of their primary sequences [14,15]. The galectins-1 form homodimers, where each subunit is a non-glycosylated polypeptide of 134 amino acids containing no hydrophobic transmembrane domain or N-terminal signal sequence to direct secretion from the cell [14,15]. Sulphydryl groups are not essential for carbohydrate-binding activity [16,17], but oxidative inactivation is associated with a conformational change involving the formation of intramolecular disulphide bonds [18,19]. No structural change is induced by the addition of lactose [7]. Reported pI values vary somewhat but are always around pH 5 [14]. The tertiary structure consists of two β-sheets forming the 'jelly-roll' topology found also in pentraxins and legume lectins [20].

BIOLOGICAL ACTIVITIES

Galectins-1 agglutinate trypsin-treated lapine erythrocytes, fresh or fixed [3–6], in a divalent cation-independent manner. Although haemagglutination can be inhibited simply by D-galactose, disaccharides and higher oligosaccharides with terminal β-linked galactose are vastly more potent inhibitors [3,5–7,15]. There are minor differences between different family members, but all galectins-1 are highly sensitive to Gal-β1,4-GlcNAc (*N*-acetyllactosamine) as an inhibitor of carbohydrate-binding activity [1,3,5–7].

Galectins-1 have been at shown to bind to several macromolecules including laminin [20,21], fibronectin [22], ganglioside GM_1 [23], the leukocyte surface antigens CD43 and CD45 [24], the lysosome-associated membrane proteins and carcinoembryonic antigen [25]. The human brain galectin can also bind actin using a non-lectin activity [26].

Human galectin-1 can be either mitogenic or anti-mitogenic for human mononuclear cells *in vitro* depending on the conditions [27]. It has also been shown to stimulate human monocytes and a murine macrophage cell line to release cytokine(s) with cytotoxic activity [28]. Bovine lung galectin-1 was mitogenic for arterial smooth muscle and endothelial cells [29]; recombinant human galectin-1 (at low concentrations) was mitogenic for fibroblasts [12]. These mitogenic activities are mediated by the carbohydrate-binding site [12,28]. The galectins-1 from murine embryonic fibroblasts [11], activated T cells [30], ovine placenta [7] and humans (recombinant) at higher concentrations [12] are anti-mitogenic for murine fibroblasts, murine $CD8^+$ lymphocytes, rat spleen mononuclear cells and human cell lines (MPS fibroblasts and HEP-2 cells), respectively. Interestingly, the anti-mitogenic activity of the ovine placental galectin was neutralised by galectin-3, implying a reciprocal cross-regulation [7]. Anti-mitogenic/growth inhibitory galectin activities are independent of carbohydrate-binding [11,12], and are a manifestation of membership of the transforming growth factor-β (TGFβ) family of growth regulatory proteins [9,31]. While lectin/mitogenic activity is associated with the dimeric form, relatively low concentrations and a reducing environment, TGF (designated TGFγ2) activity is associated with a monomeric form [30,32], a broader and higher concentration range [12], an oxidising environment [19,27] and the presence of bound glycan [32]. Galectin-1 differs from many animal lectins, which combine a carbohydrate-recognition domain with an unrelated domain; rather, the same primary sequence can assume two mutually exclusive conformations and exist as a lectin conformer or a TGFγ2 conformer [19]. The relationship between structure and activity is further complicated by the consideration that, like other transforming growth factors, TGFγ2 may stimulate as well as inhibit growth under some circumstances: the rat galectin-1/TGFγ2 has mitogenic activity independent of lectin activity [9,19].

Human recombinant galectin-1 can induce apoptosis of activated human T lymphocytes [33] and proliferating thymocytes [34]. Rat macrophage-derived galectin-1 induced apoptosis in T lymphocytes and this was associated with carbohydrate-binding activity [35]; murine galectin-1 induced apoptosis of human B lymphocytes, and this was not associated with lectin activity [36].

TISSUE AND/OR SUBCELLULAR DISTRIBUTION

Galectin-1 is very widely distributed in mammalian organs and tissues, and is generally found where it is sought [14]. Organs that are good sources include heart, spleen, lung and placenta. Good tissue sources include smooth and skeletal muscle, neurological tissues and haematopoietic and lymphoid tissues.

Within cells, galectin-1 is invariably and predominantly expressed in the cytoplasm, much of it non-covalently attached to intracellular membranes [37–39]. However, galectin-1 has also been localised to nuclei [40,41], the cell surface [24,38] and the extracellular matrix [38–42].

The placenta is a complex organ, in which the distribution of galectin-1 has been investigated in several laboratories [39,43–45]. Discrepancies between the reports cited indicate the danger of artefacts and therefore the limitations of the technology used. Unquestionably, galectin-1 is abundantly expressed in the cytoplasm of the decidual stromal cells [9,44,45]. There are conflicting messages regarding presence [43,45] or absence [39,44) in the syncytiotrophoblast. A similar lack of agreement surrounds expression in the villous cytotrophoblast. However, there is a majority consensus view that there is weak transient expression of galectin-1 in columns of extravillous cytotrophoblast obtained during the first trimester of pregnancy [39,45]. The contradictory findings concerning the syncytiotrophoblast are puzzling and hard to reconcile, but the relative lack of galectin-1 activity in plasma membrane vesicles prepared from term placentae [27] is consistent only with the negative reports.

POSSIBLE FUNCTION(S)

A null mutation was introduced into the murine gene coding for galectin-1, but homozygous mutant mice developed normally and were fertile [46]. Moreover, no obvious abnormal phenotype was observed one month after the birth of mice in which both galectins-1 and -3 had been knocked out [47]. Therefore a function cannot be assigned with certainty, presumably because there is functional redundancy with other proteins compensating for the lack of galectins.

Nevertheless, there is circumstantial evidence for a role in (i) adhesion, (ii) cell growth regulation and (iii) regulation of cell survival [48]. Galectin-1 can both promote [24,49,50] and inhibit [51] the adhesion of cells to the extracellular matrix; likewise, it can both promote [9,12,19,27,29] and inhibit [7,11,12,27,30] cellular proliferation. Galectin-1 can induce the apoptosis of several lymphoid cell types and may be involved in thymic selection [33–36]. Additionally, it has been suggested that galectin-1 may have a role in pre-mRNA splicing [40], a function more readily associated with galectin-3.

COMMENTS

Like the chicken galectins, galectins-1 are sometimes developmentally regulated [52,53].

The lack of a recognised signal sequence implies the existence of a non-classical mechanism of secretion. Cooper and Barondes [54] have suggested a process of 'ectocytosis' for differentiating murine myoblasts, in which the lectin becomes concentrated in evaginations of plasma membrane that subsequently pinch off to form a lectin-rich extracellular vesicles. An alternative suggestion, based on immunohistochemical studies on human placentae, is that plasma membranes (of decidual stromal cells) simply degenerate when they make contact with a band of fibrinoid, releasing the lectin and other cytoplasmic constituents into the surrounding extracellular matrix [55].

The expression of galectin-1 is elevated in many malignant tissues, including several human carcinomas [14,48; see also Chapter 4]. For example, in a study of human thyroid tumours, galectin-1 expression was increased in most papillary carcinomas and anaplastic carcinomas, but not in follicular carcinomas or benign adenomas [56]. In human colorectal cancer, galectin-1 expression correlated with tumour progression [57]. Galectin-1 can also promote the attachment of human melanoma cells to laminin, so could participate in invasion and metastasis [58].

The immunomodulatory properties of recombinant human galectin-1 have been exploited in experimental autoimmune encephalomyelitis [59]. The lectin injected for 12 days (starting on the day of disease induction with myelin basic protein) completely prevented the appearance of clinical signs of the disease in 9 out of 10 rats and greatly reduced the extent of histological signs. Similarly, galectin-1 suppressed collagen-induced arthritis in mice, either by a single injection of fibroblasts engineered to secrete galectin-1, or by daily administration of murine galectin-1 [60].

Autoantibodies directed against galectin-1 were found in some patients with multiple sclerosis, and also in association with a miscellany of other diseases [61].

REFERENCES

1. De Waard, A., Hickman, S. & Kornfeld, S. (1976). Isolation and properties of β-galactoside binding lectins of calf heart and lung. *Journal of Biological Chemistry* **251**, 7581–7587.
2. Bladier, D., Joubert, A., Avellana-Adalid, V., Kéméný, J.-L., Doinel, C., Amouroux, J. & Caron, M. (1989). Purification and characterization of a galactoside-binding lectin from human brain. *Archives of Biochemistry and Biophysics* **269**, 433–439.
3. Powell, J.T. (1980). Purification and properties of lung lectin. *Biochemical Journal* **187**, 123–129.
4. Childs, R.A. & Feizi, T. (1979). β-Galactoside-binding muscle lectins of man and monkey show antigenic cross-reactions with those of bovine origins. *Biochemical Journal* **183**, 755–758.
5. Allen, H.J., Cywinski, M., Palmberg, R. & Di Cioccio, R.A. (1987). Comparative analysis of galactoside-binding lectins isolated from mammalian spleens. *Archives of Biochemistry and Biophysics* **256**, 523–533.
6. Hirabayashi, J. & Kasai, K. (1984). Human placental β-galactoside-binding lectin. Purification and some properties. *Biophysical and Biochemical Research Communications* **122**, 938–944.
7. Iglesias, M.M., Rabinovich, G.A., Ivanovic, V., Sotomayor, C. & Wolfenstein-Todel, C. (1998). Galectin-1 from ovine placenta. *European Journal of Biochemistry* **252**, 400–407.
8. Couraud, P.-O., Casentini-Borocz, D., Bringman, T.S., Griffithm J., McGrogan, M. & Nedivin, G.E. (1989). Molecular cloning, characterization and expression of a human 14-kDa lectin. *Journal of Biological Chemistry* **264**, 1310–1316.

9. Yamaoka, K., Ohno, S., Kawasaki, H. & Suzuki, K. (1991). Overexpression of a β-galactoside-binding protein causes transformation of BALB3T3 fibroblast cells. *Biochemical and Biophysical Research Communications* **179**, 272–279.
10. Hirabayashi, J., Ayaki, H., Soma, G. & Kasai, K. (1989). Production and purification of a recombinant human 14 kDa β-galactoside-binding lectin. *FEBS Letters* **250**, 161–165.
11. Wells, V. & Mallucci, L. (1991). Identification of an autocrine negative growth factor: mouse β-galactoside-binding protein is a cytostatic factor and cell growth regulator. *Cell* **64**, 91–97.
12. Adams, L., Scott, G.K. & Weinberg, C.S. (1996). Biphasic modulation of cell growth by recombinant human galectin-1. *Biochimica et Biophysica Acta* **1312**, 137–144.
13. Fouillit, M., Lévi-Strauss, M., Giudicelli, V., Lutomski, D., Bladier, D., Caron, M. & Joubert-Caron, R. (1998). Affinity purification and characterization of recombinant human galectin-1. *Journal of Chromatography B* **706**,167–171.
14. Harrison, L. (1991). Soluble β-galactoside-binding lectins in vertebrates. In: *Lectin Reviews* Volume 1 (eds D.C. Kilpatrick, E. Van Driessche & T.-C. Bøg-Hansen), pp. 17–39. Sigma Chemical Company, St Louis, MO.
15. Barondes, S.H., Cooper, D.N.W., Gitt, M.A. & Leffler, H. (1994). Galectins. *Journal of Biological Chemistry* **269**, 20807–20810.
16. Whitney, P.L., Powell, J.T. & Sanford, G.L. (1986). Oxidation and chemical modification of lung β-galactoside-binding-specific lectin. *Biochemical Journal* **238**, 683–689.
17. Hirabayashi, J. & Kasai, K. (1991). Effect of amino acid substitution by site-directed mutagenesis on the carbohydrate recognition and stability of human 14 Kda β-galactoside-binding lectin. *Journal of Biological Chemistry* **266**, 23648–23653.
18. Tracey, B.M., Feizi, T., Abbott, W.M., Carruthers, R.A., Green, B.N. & Lawson, A.M. (1992). Subunit molecular mass assignment of 14 654 kDa to the soluble β-galactoside-binding lectin from bovine heart muscle and demonstration of intramolecular disulphide bonding associated with oxidative inactivation. *Journal of Biological Chemistry* **267**, 10342–10347.
19. Yamaoka, K., Ingendoh, A., Tsubuki, S., Nagai, Y. & Sanai, Y. (1996). Structural and functional characterization of a novel tumor-derived rat galectin-1 having transforming growth factor (TGF) activity: the relationship between intramolecular disulphide bridges and TGF activity. *Journal of Biochemistry* **119**, 878–886.
20. Rini, J.M. (1995). Lectin structure. *Annual Review of Biophysical and Biomolecular Structure* **24**, 551–577.
21. Zhou, Q. & Cummings, R.D. (1990). The S-type lectin from calf heart tissue binds selectively to the carbohydrate chains of laminin. *Archives of Biochemistry and Biophysics* **281**, 27–35.
22. Ozeki, Y., Matsuim T., Yamamoto, Y., Funashashi, M., Hamako, J. & Titani, K. (1995). Tissue fibronectin is an endogenous ligand for galectin-1. *Glycobiology* **5**, 255–261.
23. Kopitz, J., von Reitzenstein, C., Burchert, M., Cantz, M. & Gabius, H.-J. (1998). Galectin-1 is a major receptor for ganglioside GM_1, a product of the growth-controlling activity of a cell surface ganglioside sialidase, on human neuroblastoma cells in culture. *Journal of Biological Chemistry* **273**, 11205–11211.
24. Baum, L.G., Pang, M., Perillo, N.L., Wu, T., Delegeane, A., Uittenbogaart, C.H., Fukuda, M. & Seilhamer, J.J. (1995). Human thymic epithelial cells express an endogenous lectin, galectin-1, which binds to core 2 *O*-glycans on thymocytes and T lymphoblastoid cells. *Journal of Experimental Medicine* **181**, 877–887.
25. Ohannesian, D.W., Lotan, D. & Lotan, R. (1994). Concomitant increases in galectin-1 and its glycoconjugate ligands (carcinoembryonic antigen, lamp-1 and lamp-2) in cultured human colon carcinoma cells by sodium butyrate. *Cancer Research* **54**, 5992–6000.
26. Joubert, R., Caron, M., Avellona-Adalid, V., Mornet, D. & Bladier, D. (1992). Human brain lectin: a soluble lectin that binds actin. *Journal of Neurochemistry* **58**, 200–203.

27. Kilpatrick, D.C. (1993). Is human placental lectin an immunoregulatory molecule? In: *Lectins – Biology, Biochemistry, Clinical Chemistry*, Volume 9 (eds J. Basu, M. Kudu & P. Chakrabarti), pp. 100–104. Wiley Eastern, New Delhi.
28. Kajikawa, T., Nakajama, Y., Hirabayashi, J., Kasai, K. & Yamazaki, M. (1986). Release of cytotoxin by macrophages on treatment with human placental lectin. *Life Sciences* **39**, 1177–1181.
29. Sanford, G.L. & Harris-Hooker, S. (1990). Stimulation of vascular cell proliferation by β-galactoside specific lectins. *FASEB Journal* **4**, 2912–2918.
30. Blaser, C., Kaufmann, M., Muller, C., Zimmermann, C., Wells, V., Mallucci, L. & Pircher, H. (1998). β-Galactoside-binding protein secreted by activated T cells inhibits antigen-induced proliferation of T cells. *European Journal of Immunology* **28**, 2311–2319.
31. Yamaoka, K., Ohno, S., Kawasaki, H. & Suzuki, K. (1993). Mammalian lectin as transforming growth factor. In: *Lectins and Glycobiology* (eds H.-J. Gabius & S. Gabius), pp. 492–499. Springer-Verlag, Berlin.
32. Wells, V. & Mallucci, L. (1992). Molecular expression of the negative growth factor murine β-galactoside binding protein (mGBP). *Biochimica et Biophysica Acta* **1121**, 239–244.
33. Perillo, N.L., Pace, K.E., Seilhamer, J.J. & Baum, L.G. (1995). Apoptosis of T cells mediated by galectin-1. *Nature* **378**, 736–739.
34. Perillo, N.L., Uittenbogaart, C.H., Nguyen, J.T. & Baum, L.G. (1997). Galectin-1, an endogenous lectin produced by thymic epithelial cells, induces apoptosis of human thymocytes. *Journal of Experimental Medicine* **185**, 1851–1858.
35. Rabinovich, G.A., Iglesias, M.M., Modesti, N.M., Castagna, L.F., Wofenstein-Todel, C., Rioera, C.M. & Sotomayor, C.E. (1998). Activated rat macrophages produce a galectin-1-like protein that induces apoptosis of T cells: biochemical and functional characterization. *Journal of Immunology* **160**, 4831–4840.
36. Murphy, J.J., Stevanin, T., Schoendorf, D.S., Wells, V. & Mallucci, L. (1997). β-Galactoside binding protein inhibits B cell growth and induces cell apoptosis. *Biochemical Society Transactions* **25**, 248S.
37. Sanford, G.L., Davis, L.D. & Powell, J.T. (1982). The subcellular localization of the β-galactoside-binding protein of rat lung. *Biochemical Journal* **204**, 97–102.
38. Harrison, F.L. & Catt, J.W. (1986). Intra- and extracellular distribution of an endogenous lectin during erythropoiesis. *Journal of Cell Science* **84**, 201–212.
39. Bevan, B.H., Kilpatrick, D.C., Liston, W.A., Hirabayashi, J. & Kasai, K. (1994). Immunohistochemical localization of a β-galactoside-binding lectin at the human maternofetal interface. *Histochemical Journal* **26**, 582–586.
40. Vyakarnam, A., Dagher, S.F., Wang, J.L. & Patterson, R.J. (1997). Evidence for a role for galectin-1 in pre-mRNA splicing. *Molecular and Cellular Biology* **17**, 4730–4737.
41. Vyakarnam, A., Lenneman, A.J., Lakkides, K.M., Patterson, R.J. & Wang, J.L. (1998). A comparative nuclear localization study of galectin-1 with other splicing components. *Experimental Cell Research* **242**, 419–428.
42. Cerra, R.F., Haywood-Reid, P.L. & Barondes, S.H. (1984). Endogenous mammalian lectin localized extracellularly in lung elastic fibres. *Journal of Cell Biology* **98**, 1580–1589.
43. Walzel, H., Neels, P., Bremer, H., Kohler, H., Raab, N., Barten, M. & Brock, J. (1995). Immunohistochemical and glycohistochemical localization of the β-galactoside-binding S-type lectin in human placenta. *Acta Histochemica* **97**, 33–42.
44. Maquoi, E., van den Brûle, F.A., Castronovo, V. & Foidart, J.-M. (1997). Changes in the distribution pattern of galectin-1 and galectin-3 in human placenta correlate with the differentiation pathways of trophoblasts. *Placenta* **18**, 433–439.
45. Vićovac, L., Janković, M. & Ćuperlović, M. (1998). Galectin-1 and galectin-3 in cells of the first trimester placental bed. *Human Reproduction* **13**, 730–735.
46. Poirier, F. & Robertson, E.J. (1993). Normal development of mice carrying a null mutation in the gene encoding the L14 S-type lectin. *Development* **119**, 1229–1236.
47. Colnot, C., Ripoche, M.-A., Scaerou, F., Fowlis, D. & Poirier, F. (1996). Galectins in mouse embryogenesis. *Biochemical Society Transactions* **24**, 1–599.

48. Perillo, N.L., Marcus, M.E. & Baum, L.G. (1998). Galectins: versatile modulators of cell adhesion, cell proliferation and death. *Journal of Molecular Medicine* **76**, 402–412.
49. Moiseeva, E.P., Spring, E.L., Baron, J.H. & de Bono, D.P. (1999). Galectin-1 modulates attachment, spreading and migration of cultured vascular smooth muscle cells via interactions with cellular receptors and components of extracellular matrix. *Journal of Vascular Research* **36**, 47–58.
50. Tenne-Brown, J., Puche, A.C. & Key, B. (1998). Expression of galectin-1 in the mouse olfactory system. *International Journal of Developmental Biology* **42**, 791–799.
51. Cooper, D.N.W., Massa, S.M. & Barondes, S.H. (1991). Endogenous muscle lectin inhibits myoblast adhesion to laminin. *Journal of Cell Biology* **115**, 1437–1448.
52. Powell, J.T. & Whitney, P.L. (1980). Postnatal development of rat lung. *Biochemical Journal* **188**, 1–8.
53. Choi, J.Y., van Wignen, A.J., Aslam, F., Leszyk, J.D., Stein, J.L., Stein, G.S., Lian, J.B. & Penman, S. (1998). Developmental association of the β-galactoside-binding protein galectin-1 with the nuclear matrix of rat calvarial osteoblasts. *Journal of Cell Science* **111**, 3035–3043.
54. Cooper, D.N.W. & Barondes, S.H. (1990). Evidence for export of a muscle lectin from cytosol to extracellular matrix and for a novel secretory mechanism. *Journal of Cell Biology* **110**, 1681–1691.
55. Kilpatrick, D.C., Bevan, B.H., Liston, W.A., Hirabayashi, J., Kasai, K. & Cooper, B.L. (1996). Lectins in the human feto-placental unit: a review. In: *Lectins – Biology, Biochemistry, Clinical Biochemistry* Volume 1 (eds E. Van Driessche, P. Rougé, S. Beeckmans & T.-C. Bøg-Hansen), pp. 161–167. Textop, Hellerup.
56. Chiariotti, L., Berlingieri, M.T., Battaglia, C., Benvenuto, G., Mortelli, M.L., Salvatore, P., Chiappetta, G., Bruni, C.B. & Fusco, A. (1995). Expression of galectin-1 in normal human thyroid gland and in differentiated and poorly differentiated thyroid tumours. *International Journal of Cancer* **64**, 171–175.
57. Sanjuán, X., Fernández, P.L., Castells, A., Castronovo, V., Van Den Brule, F., Liu, F.-T., Cardesa, A. & Campo, E. (1997). Differential expression of galectin-3 and galectin-1 in colorectal cancer progression. *Gastroenterology* **113**, 1906–1915.
58. Van Den Brule, F.A., Buicu, C., Baldet, M., Sobel, M.E., Cooper, D.N.W., Marschal, P. & Castronovo, V. (1995). Galectin-1 modulates human melanoma cell adhesion to laminin. *Biochemical and Biophysical Research Communications* **209**, 760–767.
59. Offner, H., Celnik, B., Bringman, T.S., Casentini-Borocz, D., Nedwin, G.E. & Vandenbark, A.A. (1990). Recombinant human β-galactoside-binding lectin suppresses clinical and histological signs of experimental autoimmune encephalomyelitis. *Journal of Neuroimmunology* **28**, 177–184.
60. Rabinovich, G.A., Daly, G., Dreja, H., Tailor, H., Riera, C.M., Hirabayashi, J. & Chernajovsky, Y. (1999). Recombinant galectin-1 and its genetic delivery suppress collagen-induced arthritis via T cell apoptosis. *Journal of Experimental Medicine* **190**, 385–397.
61. Lutomski, D., Joubert-Caron, R., Lefebure, C., Salama, J., Belin, C., Bladier, D. & Caron, M. (1997). Anti-galectin-1 autoantibodies in serum of patients with neurological diseases. *Clinica Chimica Acta* **262**, 131–138.

Galectin-2

Other name: L-14-II
Mol. wt: 29 000 (native); 14 650 (subunit)
Specificity: Lactose
Source: Mammalian alimentary tract

ISOLATION

The existence of human galectin-2 was first recognised from a partial cDNA clone during the cloning of galectin-1 [1]. The missing part of the cDNA was isolated by PCR from another hepatoma cDNA library, and the full-length cDNA was assembled and expressed in *Escherichia coli* [2]. Similarly, the rat equivalent was obtained as a recombinant protein synthesised in *E. coli* after cloning of the cDNA obtained from a rat stomach library [3].

STRUCTURE

Human galectin-2 has a calculated subunit molecular weight of 14 650; the sequence is 43% identical to that of human galectin-1 [2] and 67% identical to rat galectin-2 [3]. The apparent Mr estimated by gel filtration was 29 000, indicating homodimer formation [2].

The structure resolved by X-ray crystallography shows overall similarity to that of galectin-1 [4].

BIOLOGICAL ACTIVITIES

Galectin-2 binds to immobilised asialofetuin, from which it can be eluted with lactose [2,3]. Activity is independent of divalent cations and it is unstable in the absence of reducing agents.

TISSUE AND SUBCELLULAR DISTRIBUTION

Significant expression of human galectin-2 has been found only in epithelial cells of the gastrointestinal tract among adult tissues [5]; rat galectin-2 is predominantly expressed by gastric epithelial cells [3].

POSSIBLE FUNCTION(S)

Essentially nothing is known about the function of galectin-2.

COMMENTS

Galectin-2 is the product of a single gene which, in the human, is located on chromosome 22 [5].

REFERENCES

1. Gitt, M.A. & Barondes, S.H. (1986). Evidence that a human soluble β-galactoside-binding lectin is encoded by a family of genes. *Proceedings of the National Academy of Sciences USA* **83**, 7603–7607.
2. Gitt, M.A., Massa, S.M., Leffler, H. & Barondes, S.H. (1992). Isolation and expression of a gene encoding L-14-II, a new human soluble lactose-binding lectin. *Journal of Biological Chemistry* **267**, 10601–10606.
3. Oka, T., Murakomi, S., Arata, Y., Hirabayashi, J., Kasai, K.-I., Wada, Y. & Futai, M. (1999). Identification and cloning of rat galectin-2: expression is predominantly in epithelial cells of the stomach. *Archives of Biochemistry and Biophysics* **361**, 195–201.
4. Rini, J.M. (1995). Lectin structure. *Annual Review of Biophysics and Biomolecular Structure* **24**, 551–577.
5. Gitt, M.A., Jordan, E.T. & Leffler, H. (1997). Galectin-2, galectins-5 and 9, and galectins-4 and 6. *Trends in Glycoscience and Glycotechnology* **9**, 87–93.

Galectin-3

Other names: CBP-35; RL-29; HL-29; Mac-2; IgE-BP(εBP)
Mol. wt: 26 200 to 30 300
Specificity: *N*-acetyllactosamine
Source: Mammalian tissues

ISOLATION

Galectin-3 was first isolated from a detergent extract of murine 3T3 fibroblasts after purification of a mixture of *β*-galactoside-binding lectins by affinity chromatography on asialofetuin [1]. The CBP-35 was separated from the others by gel filtration on Sephadex G-150 [1]. A similar affinity step was employed by Cerra *et al.* [2] with rat lung homogenate before the RL-29 was isolated by ion-exchange chromatography [2].

Galectin-3 has also been cloned from several sources, including mouse fibroblasts [3], rat basophilic leukaemia cells [4], murine macrophages [5], human (HeLa) neoplastic cells [6] and human monocytes [7].

STRUCTURE

The galectin-3 subfamily is a highly homologous group of non-glycosylated polypeptides usually consisting of more than 260 amino acids; the sequences of several galectins-3 are given in full by Harrison [8].

Unlike any other known galectins, galectin-3 has a hybrid structure (chimera-type) composed of two distinctive parts. The larger part (140 residues) on the carboxyl-terminal end of the molecule is a typical S-type lectin domain, with around 30% sequence identity to galectin-1. The other part consists of a short (20 residues) N-terminal sequence followed by a series of repeated sequences (of around nine amino acids) rich in proline and glycine residues [8–10]. The length of the repeating sequence domain varies slightly between species, resulting in molecular sizes ranging from 26 000 to 30 000 [10]; the somewhat higher apparent Mr range (29 000–35 000) estimated by SDS-PAGE probably results from anomalous electrophoretic behaviour rather than post-translational modification [10]. There is no classical signal sequence, nor a transmembrane domain [8–10]. Galectin-3 is a basic protein with an isoelectric point of pH 8.7, which changes to pH 8.2 on phosphorylation [11]. The carbohydrate-binding ability does not depend on the presence of reducing agents [12], but reports conflict regarding stability under oxidising conditions [1,12].

Galectin-3 exists as a monomer in solution but can form multimers on binding to appropriate surfaces; the non-lectin part of the molecule is required for this [13,14].

Galectin-3 possesses some homology with proteins from heterogeneous nuclear ribonucleoprotein particles [3].

BIOLOGICAL ACTIVITIES

Galectin-3 can agglutinate fresh or fixed trypsin-treated lapine erythrocytes [1,12,13]; this activity is independent of divalent cations and can be inhibited by lactose.

Glycoprotein ligands include asialofetuin [1,2,7] and laminin [7,14,15]; IgE is also bound species non-specifically, although some human IgE myeloma proteins did not bind until treated with neuraminidase [5,6]. Other known glycoconjugate ligands are listed by Perillo *et al.* [16]. This lectin activity is inhibited most effectively by *N*-acetyllactosamine > thiodigalactoside > lactose and to a much lesser extent by D-galactose and L-arabinose [12]. Unlike galectin-1, however, galectin-3 has a high affinity for the more extended blood group A structure in which the galactose moiety of *N*-acetyllactosamine has been substituted with GalNAc α1,3 and fucose α1,2 [17].

Human recombinant galectin-3 induced DNA synthesis in and cell proliferation of a human lung fibroblast cell line; this mitogenic activity could be inhibited by lactose [18].

Galectin-3 activates granulocytes. Human recombinant galectin-3 has been shown to stimulate superoxide production by neutrophils [19] and by monocytes [20]. Galectin-3 can also activate mast cells [21] (on which it can bind to both IgE and IgE receptors) and eosinophils, downregulating IL-5 gene expression [22]. These activities are apparently mediated by the carbohydrate-binding site.

Human leukaemia T cells transfected with galectin-3 cDNA exhibit enhanced growth rates and a resistance to apoptosis induced by anti-Fas and staurosporine [23]. Galectin-3 has a motif that is highly conserved among members of the Bcl-2 family; moreover, galectin-3 can interact with Bcl-2 through its carbohydrate-binding site [23]. The anti-apoptosis activity of galectin-3 might therefore be mediated in some way by Bcl-2.

Galectin-3 can bind to RNA and ssDNA and is essential for pre-mRNA splicing in a cell-free system [24,25].

Galectin-3 is a substrate for tissue-type transglutaminase [26].

Immobilised recombinant galectin-3 can bind *Salmonella minnesota* R7 lipopolysaccharide via an interaction involving the N-terminal region; this second binding site recognises the lipid A/inner core region of lipopolysaccharide and is independent of the carbohydrate-binding site [27].

TISSUE AND SUBCELLULAR DISTRIBUTION

Galectin-3 is very widely distributed in organs and tissues, being found in lung, liver, heart, muscle etc. [2]; although first identified in fibroblasts [1,28], it has been particularly well studied in endothelium, epithelium and monocytes–macrophages [16].

Within cells, galectin-3 has been found within the nucleus, in the cytoplasm and on the cell surface, as well as within the extracellular milieu [5,10,28].

Galectin-3 is externalised by a mechanism independent of the endoplasmic reticulum–Golgi pathway. Instead, it aggregates beneath the plasma membrane and is released in vesicles by membrane blebbing (ectocytosis) [29].

POSSIBLE FUNCTION(S)

Knock-out mice lacking both galectins-1 and -3 have an apparently normal phenotype [30], yielding no clue to their functions, which are presumably assumed by other factors. However, the recruitment of granulocytes into the peritoneal cavity during acute (thioglycolate-induced) peritonitis was defective in galectin-3 null mutant mice [31]. The association of galectin-3 with activated macrophages [20], neutrophils [19] basophils [21] and eosinophils [32] is also consistent with a role in the inflammatory response, and the case for modulating allergic inflammation in particular has been argued [33].

Any role in inflammation is likely to be mediated at least in part by the adhesive properties of the lectin. Galectin-3 has been found, under varying circumstances, either to promote or inhibit cell adhesion [34–40]. Its complex associations with, and alleged involvement in, the spread of malignant cells (see Comments section) was also discussed in Chapter 4.

Galectin-3 is a necessary component for the splicing of pre-mRNA in the nucleus, and this function is inhibited by saccharides known to bind the lectin [24,20]. Galectin-3 must therefore function as a splicing factor or a regulator of splicing.

COMMENTS

With so much circumstantial evidence linking galectin-3 to cell adhesion and cell growth regulation, it is unsurprising that many studies have investigated its role in cancer. Overexpression of galectin-3 in fibrosarcoma cells [41], introduction of galectin-3 cDNA into null expressing breast cancer cells [42], and genetic upregulation of galectin-3 in colon cancer cells [43] have been shown to promote malignant disease. These experimental findings are consistent with observations on clinical material that galectin-3 expression is significantly increased in thyroid [44–46] and liver [47] cancers, and correlates with the malignant potential of CNS tumours [48]. On the other hand, the relationship of the lectin to breast cancer and colon cancer is not so straightforward. Indeed, there is evidence that galectin-3 expression is downregulated in breast [49] and colonic [50,51] cancers, and for an inverse correlation with neoplastic progression. Regarding colon cancer, however, another study reaches the opposite conclusion [52], one more in keeping with the experimental data referred to above [43]. Regarding breast carcinoma, it has been suggested that the pro-tumorigenic activity of galectin-3 requires the co-expression of the L1 retrotransposon gene product, L1 [53].

Nevertheless, galectin-3 may be considered a marker of human thyroid carcinomas, particularly papilliary carcinomas, of potential use in pre-surgical diagnosis from needle aspiration biopsies [45]. Galectin-3 may also be a useful immunohistochemical marker for anaplastic large-cell lymphoma, since it was detected in 8/8 cases of the disease but occurred infrequently in other kinds of lymphoma [54].

Finally, in uterine cancer, the levels of expression of galectin-3 and its binding sites determined by computer-assisted microscopy can aid the differentiation of benign (leiomyomas) from malignant (leiomyosarcomas) smooth muscle tumours [55].

As with other galectins, galectin-3 is sometimes developmentally regulated. For example, galectin-3 is readily detected in several tissues of new-born rats but is not found in adults [2], and similar findings have been made comparing tissues of embryonic and adult mice [56]. Galectin-3 is also upregulated by infection with human T lymphotropic virus-1 [57] and human immunodeficiency virus-1 [58]. The prominent presence of galectin-3 in the foam cells of atherosclerotic lesions [59] may simply reflect an upregulation by inflammation.

REFERENCES

1. Raff, C.F. & Wang, J.L. (1983). Endogenous lectins from cultured cells. *Journal of Biological Chemistry* **258**, 10657–10663.
2. Cerra, R.F., Gitt, M.A. & Barondes, S.H. (1985). Three soluble β-galactoside-binding lectins. *Journal of Biological Chemistry* **260**, 10474–10477.
3. Jia, S. & Wang, J.L. (1988). Carbohydrate binding protein 35. *Journal of Biological Chemistry* **263**, 6009–6011.
4. Albrandt, K., Orida, N.K. & Liu, F.-T. (1987). An IgE-binding protein with a distinctive repetitive sequence and homology with an IgG receptor. *Proceedings of the National Academy of Sciences USA* **84**, 6859–6863.
5. Cherayil, B.J., Weiner, S.J. & Pillai, S. (1989). The Mac-2 antigen is a galactose-specific lectin that binds IgE. *Journal of Experimental Medicine* **170**, 1959–1972.
6. Robertson, M.W., Albrandt, K., Keller, D. & Liu, F.-T. (1990). Human IgE-binding protein: a soluble lectin exhibiting a highly conserved recognition of IgE glycoforms. *Biochemistry* **29**, 8093–8100.
7. Cherayil, B.J., Chaitovitz, S., Wong, C. & Pillai, S. (1990). Molecular cloning of a human macrophage lectin specific for galactose. *Proceedings of the National Academy of Sciences USA* **87**, 7324–7328.
8. Harrison, L. (1991). Soluble β-galactoside-binding lectins in vertebrates. In: *Lectin Reviews*, Volume 1 (eds D.C. Kilpatrick, E. Van Driessche & T.-C. Bøg-Hansen), pp. 17–39. Sigma Chemical Company, St Louis, MO.
9. Barondes, S.H., Cooper, D.N.W., Gitt, M.A. & Leffler, H. (1994). Galectins. *Journal of Biological Chemistry* **269**, 20807–20810.
10. Herrmann, J., Turck, C.W., Atchison, R.E., Huflejt, M.E., Poulter, L., Gitt, M.A., Burlingame, A.L., Barondes, S.H. & Leffler, H. (1993). Primary structure of the soluble lactose binding lectin L-29 from rat and dog and interaction of its non-collagenous proline, -glycine-, tyrosine-rich sequence with bacterial and tissue collagenase. *Journal of Biological Chemistry* **268,** 26704–26711.

11. Cowles, E.A., Agrwal, N., Anderson, R.L. & Wang, J.L. (1990). Carbohydrate-binding protein 35. Isoelectic points of the polypeptide and a phosphorylated derivative. *Journal of Biological Chemistry* **265**, 17706–17712.
12. Frigeri, L.G., Robertson, M.W. & Liu, F.-T. (1990). Expression of biologically active recombinant rat IgE-binding protein in *Escherichia coli. Journal of Biological Chemistry* **265**, 20763–20769.
13. Hsu, D.K., Zuberi, R.I. & Lui, F.-T. (1992). Biochemical and biophysical characterization of human recombinant IgE-binding protein, an S-type animal lectin. *Journal of Biological Chemistry* **267**, 14167–14874.
14. Massa, S.M., Cooper, D.N.W., Leffler, H. & Barondes, S.H. (1993). L-29, an endogenous lectin, binds to glycoconjugate ligands with positive cooperativity. *Biochemistry* **32**, 260–267.
15. Ohannesian, D.W., Lotan, D., Thomas, P., Jessup, J.M., Fukuda, M., Gabius, H.-J. & Lotan, R. (1995). Carcinoembryonic antigen and other glycoconjugates act as ligands for galectin-3 in human colon carcinoma cells. *Cancer Research* **55**, 2191–2199.
16. Perillo, N.L., Marcus, M.E. & Baum, L.G. (1998). Galectins: versatile modulators of cell adhesion, cell proliferation, and cell death. *Journal of Molecular Medicine* **76**, 402–412.
17. Leffler, H. & Barondes, S.H. (1986). Specificity of binding of three soluble rat lung lectins to substituted and unsubstituted mammalian β-galactosides. *Journal of Biological Chemistry* **261**, 10119–10126.
18. Inohara, H., Akahani, S. & Raz, A. (1998). Galectin-3 stimulates cell proliferation. *Experimental Cell Research* **245**, 294–302.
19. Yamaoka, A., Kuwabara, I., Frigeri, G. & Liu, F.-T. (1995). A human lectin, galectin-3 (εbp/Mac-2), stimulates superoxide production by neutrophils. *Journal of Immunology* **154**, 3479–3487.
20. Liu, F.-T., Hsu, D.K., Zuberi, R.I., Kuwabara, I., Chi, E.Y. & Henderson, W.R. (1995). Expression and function of galectin-3, a beta-galactoside binding lectin, in human monocytes and macrophages. *American Journal of Pathology* **147**, 1016–1028.
21. Frigeri, L.G., Zuberi, R.I. & Liu, F.-T. (1993). εBP, a β-galactoside-binding animal lectin, recognises IgE receptor (FcεRI) and activates mast cells. *Biochemistry* **32**, 7644–7649.
22. Cortegano, I., del Pozo, V., Cárdaba, B., de Andrés, B., Gallardo, S., del Amo, A., Arrieta, I., Jurado, A., Palomino, P., Liu, F.-T. & Lahoz, C. (1998). Galectin-3 down-regulates IL-5 gene expression on different cell types. *Journal of Immunology* **161**, 385–389.
23. Yang, R.Y., Hsu, D.K. & Liu, F.-T. (1996). Expression of galectin-3 modulates T cell growth and apoptosis. *Proceedings of the National Academy of Sciences USA* **93**, 6737–6742.
24. Dagher, S.F., Wang, J.L. & Patterson, R.J. (1995). Identification of galectin-3 as a factor in pre-mRNA splicing. *Proceedings of the National Academy of Sciences USA* **92**, 1213–1217.
25. Patterson, R.J. & Wang, J.L. (1997). Galectin-3: structure and function. *European Journal of Cell Biology* **74** (Suppl. 46), 36.
26. Mehul, B., Bawumia, S. & Hughes, R.C. (1995) Cross-linking of galectin-3, a galactose-binding protein of mammalian cells, by tissue-type transglutaminase. *FEBS Letters* **360**, 160–164.
27. Mey, A., Leffler, H., Hmama, Z., Normier, G. & Revillard, J.-P. (1996). The animal lectin galectin-3 interacts with bacterial lipopolysaccharides via two independent sites. *Journal of Immunology* **156**, 1572–1577.
28. Moutsatsos, I.K., Davis, J.M. & Wang, J.L. (1986). Endogenous lectins from cultured cells: subcellular localization of carbohydrate-binding protein 35 in 3T3 fibroblasts. *Journal of Cell Biology* **102**, 477–483.
29. Mehul, B. & Hughes, R.C. (1997). Plasma membrane targeting, vesicular budding and release of galectin-3 from the cytoplasm of mammalian cells during secretion. *Journal of Cell Science* **110**, 1169–1178.
30. Colnot, C., Ripoche, M.-A., Scaeron, F., Fowlis, D. & Poirier, F. (1996). Galectins in mouse embryogenesis. *Biochemical Society Transactions* **24**, 141–146.

31. Colnot, C., Ripoche, M.-A., Milon, G., Montagutelli, X., Crocker, P.R. & Poirier, F. (1998). Maintainance of granulocyte numbers during acute peritonitis is defective in galectin-3-null mutant mice. *Immunology* **94**, 290–296.
32. Truong, M.-J., Gruant, V., Liu, F.-T., Prin, L., Capron, A. & Capron, M. (1993). Human eosinophils express functional IgE-binding molecules (Mac-2/εBP) of the S-type lectin family. *European Journal of Immunology* **23**, 3230–3235.
33. Liu, F.-T. (1993). S-type lectins in allergic inflammation. *Immunology Today* **14**, 486–490.
34. Sato, S. & Hughes, R.C. (1992). Binding specificity of a baby hamster kidney lectin for H type I and II chains; polylactosamine glycans, and appropriately glycosylated forms of laminin and fibronectin. *Journal of Biological Chemistry* **267**, 6983–6990.
35. Le Marer, N. & Hughes, R.C. (1996). Effects of the carbohydrate-binding protein galectin-3 on the invasiveness of human breast carcinoma cells. *Journal of Cellular Physiology* **168**, 51–58.
36. Kuwabara, I. & Liu, F.-T. (1996). Galectin-3 promotes adhesion of human neutrophils to laminin. *Journal of Immunology* **156**, 3939–3944.
37. Inohara, H. & Raz, A. (1995). Functional evidence that cell surface galectin-3 mediates homotypic cell adhesion. *Cancer Research* **55**, 3267–3271.
38. Ochieng, J., Leite-Browning, M.L. & Warfield, P. (1998). Regulation of cell adhesion to extracellular matrix proteins by galectin-3. *Biochemical and Biophysical Research Communications* **246**, 788–791.
39. Van den Brule, F.A., Buici, C., Sobel, M.E., Liu, F.-T. & Castronovo, V. (1995). Galectin-3, a laminin binding protein, fails to modulate adhesion of human melanoma cells to laminin. *Neoplasma* **42**, 215–219.
40. Pesheva, P., Kuklinski, S., Schmitz, B. & Probstmeier, R. (1998). Galectin-3 promotes neural cell adhesion and neurite growth. *Journal of Neuroscience Research* **54**, 639–654.
41. Raz, A., Zhu, D.G., Hogan, V., Shah, N., Raz, T., Karkash, R., Pazerini, G. & Carmi, P. (1990). Evidence for the role of 34 kDa galactoside-binding lectin in transformation and metastasis. *International Journal of Cancer* **46**, 871–877.
42. Nangia-Makker, P., Thompson, E., Hogan, C., Ochieng, J. & Raz, A. (1995). Induction of tumorigenicity by galectin-3 in a non-tumorigenic human breast carcinoma cell line. *International Journal of Oncology* **7**, 1079–1087.
43. Bresalier, R.S., Mazurek, N., Sternberg, L.R., Byrd, J.C., Yunker, C.K., Nangia-Makker, P. & Raz, A. (1998). Metastasis of human colon cancer is altered by modifying expression of the β-galactoside-binding protein galectin-3. *Gastroenterology* **115**, 287–296.
44. Fernandez, P.L., Merino, M.J., Gomez, M., Campo, E., Medina, T., Castronovo, V., Sanjuan, X., Cardesa, A., Lui, F.-T. & Sobel, M.E. (1997). Galectin-3 and laminin expression in neoplastic and non-neoplastic thyroid tissue. *Journal of Pathology* **181**, 80–86.
45. Orlando, F., Saggiorato, E., Pivano, G., Puligheddu, B., Termine, A., Cappia, S., De Giuli, P. & Angeli, A. (1998). Galectin-3 is a presurgical marker of human thyroid carcinoma. *Cancer Research* **58**, 3015–3020.
46. Cvejic, D., Savin, S., Paunovic, I., Tatic, S., Havelka, M. & Sinadinovic, J. (1998). Immunohistochemical localization of galectin-3 in malignant and benign human thyroid tissue. *Anticancer Research* **18**, 2637–2642.
47. Hsu, D.K., Dowling, C.A., Jeng, K.C.G., Chen, J.T., Yang, R.Y. & Liu, F.-T. (1999). Galectin-3 expression is induced in cirrhotic liver and hepatocellular carcinoma, *International Journal of Cancer* **81**, 519–526.
48. Bresalier, R.S., Yan, P.S., Byrd, J.C., Lotan, R. & Raz, A. (1997). Expression of the endogenous galactose-binding protein galectin-3 correlates with the malignant potential of tumors in the central nervous system. *Cancer* **80**, 776–787.
49. Castronovo, V., van den Brule, F.A., Jackers, P., Clausse, N., Liu, F.-T., Gillet, C. & Sobel, M.E. (1996). Decreased expression of galectin-3 is associated with progression of human breast cancer. *Journal of Pathology* **179**, 43–48.

50. Sanjuan, X., Fernandez, P.L., Castello, A., Castronova, V., van den Brule, F., Liu, F.-T., Cardessa, A. & Campo, E. (1997). Differential expression of galectin-3 and galectin-1 in colorectal cancer progression. *Gastroenterology* **113**, 1906–1915.
51. Lotz, M.M., Andrews, C.W., Korzelius, C.A., Lee, E.C., Steele, G.D., Clarke, A. & Mercurio, A.M. (1993). Decreased expression of Mac-2 (carbohydrate-binding protein 35) and loss of its nuclear localization are associated with the neuroplastic progression of colon carcinoma. *Proceedings of the National Academy of Sciences USA* **90**, 3466–3470.
52. Schoeppner, H.L., Raz, A., Ho, S.B. & Bresalier, R.S. (1995). Expression of an endogenous galactose-binding lectin correlates with neoplastic progression in the colon. *Cancer* **75**, 2818–2826.
53. Nangria-Makker, P., Sarvis, R., Visscher, D.W., Bailey-Penrod, J., Raz, A. & Sarkar, F.H. (1998). Galectin-3 and L1 retrotransposons in human breast carcinomas. *Breast Cancer and Treatment* **49**, 171–183.
54. Konstantinov, K.N., Robbins, B.A. & Liu, F.-T. (1996). Galectin-3, a β-galactoside-binding animal lectin, is a marker of anaplastic large-cell lymphoma. *American Journal of Pathology* **148**, 25–30.
55. Schwarz, G., Remmelink, M., Decaestecker, C., Gielen, I., Budel, V., Burchert, M., Darro, F., Danguy, A., Gabius, H.-J., Salmon, I. & Kiss, R. (1999). Galectin fingerprinting in tumour diagnosis. *American Journal of Clinical Pathology* **111**, 623–631.
56. Crittenden, S.L., Raff, C.F. & Wang, J.L. (1984). Carbohydrate binding protein 35: identification of the galactose-specific lectin in various tissues of mice. *Molecular and Cellular Biology* **4**, 1252–1259.
57. Hsu, D.K., Hammes, S.R., Kuwabara, I., Greene, W.C. & Liu, F.-T. (1996). Human T lymphotropic virus-1 infection of human T lymphocytes induces expression of the β-galactoside-binding lectin, galectin-3. *American Journal of Pathology* **148**, 1661–1670.
58. Fogel, S., Guittaut, M., Legrand, A., Monsigny, M. & Hebert, E. (1999). The Tat protein of HIV-1 induces galectin-3 expression. *Glycobiology* **9**, 383–387.
59. Nachtigal, M., Al-Assaad, Z., Mayer, E.P., Kim, K. & Monsigny, M. (1998). Galectin-3 expression in human atherosclerotic lesions. *American Journal of Pathology* **152**, 1199–1208.

Galectin-4 (and galectin-6)

Other name: L-36
Mol. wt: 36 000 (34 000 for galectin-6)
Specificity: β-Galactosides
Source: Mammalian intestine

ISOLATION

Although fragments containing one domain of galectin-4 had been recovered from rat intestinal tissue [1,2], the complete molecule was first obtained by cloning using PCR-based protocols and a rat intestine cDNA library [2]. Galectins-4 from murine [3], porcine [4] and human [5] sources have also been cloned.

The closely related galectin-6 was discovered and cloned during the cloning of murine galectin-4 [3].

STRUCTURE

Galectin-4 was the first tandem-repeat-type galectin found in mammals [2]; the two carbohydrate-recognition domains found on the single polypeptide are approximately 40% identical in amino acid sequence to each other and about 20% identical to galectin-1. The complete cDNA nucleotide and 320 amino acid protein sequences of rat intestinal galectin-4 most closely resemble (apart from galectin-6) those of galectin-3 and galectin-5 [2].

Galectin-6 is very similar in overall structure to galectin-4, with about 80% primary sequence identity [3]. The most obvious difference between galectins-4 and -6 is a 24 amino acid sequence missing from the link region of galectin-6.

BIOLOGICAL ACTIVITIES

Both domains of galectin-4 have carbohydrate-binding activity that can be inhibited by lactose and other β-galactosides [2]. The detailed saccharide inhibition profile of binding to lactosyl–Sepharose was typical of, but not identical to, other galectins [2]. Galectin-6 has also been shown to bind to lactosyl–Sepharose and be eluted with lactose [3].

TISSUE AND SUBCELLULAR DISTRIBUTION

Galectins-4 and -6 are so similar they cannot be reliably distinguished by the usual localisation methods [6]. Galectins-4 and/or -6 are expressed in epithelial cells of the alimentary canal in all species examined, but the extent of this expression may vary from species to species [6].

Like other galectins, galectin-4/-6 is found in the cytosol and also at the cell surface [6–8].

POSSIBLE FUNCTION(S)

It has been suggested that galectin-4/-6 plays a role in cell adhesion by cross-linking glycoconjugate ligands [6]. An alternative suggestion is that galectin-4/-6 functions intracellularly to help stabilise the luminal plasma membrane [8].

COMMENTS

The tandem-repeat structure of galectins-4 and -6 is now known to be common both in mammals and in nematodes.

Human galectin-4 is downregulated in colorectal cancer, probably during the early stages of tumour formation [5].

REFERENCES

1. Leffler, H., Masiarz, F.R. & Barondes, S.H. (1989). Soluble lactose-binding and vertebrate lectins: a growing family. *Biochemistry* **28**, 9222–9229.
2. Oda, Y., Herrmann, J., Gitt, M.A., Turck, C.W., Burlingame, A.L., Barondes, S.H. & Leffler, H. (1993). Soluble lactose-binding lectin from rat intestine with two different carbohydrate-binding domains in the same peptide chain. *Journal of Biological Chemistry* **268**, 5929–5939.
3. Gitt, M.A., Colnot, C., Poirier, F., Nani, K.J., Barondes, S.H. & Leffler, H. (1998). Galectin-4 and galectin-6 are two closely related lectins expressed in mouse gastrointestinal tract. *Journal of Biological Chemistry* **273**, 2954–2960.
4. Chiu, M.L., Parry, D.A., Feldman, S.R., Klapper, D.G. & O'Keefe, E.J. (1994). An adhesive junction protein is a member of the family of lactose-binding lectins. *Journal of Biological Chemistry* **269**, 31770–31776.
5. Rechreche, H., Mallo, G.V., Montalto, G., Dagorn, J.-C. & Iovanna, J.L. (1997). Cloning and expression of the mRNA of human galectin-4, an S-type lectin downregulated in colorectal cancer. *European Journal of Biochemistry* **248**, 225–230.
6. Gitt, M.A., Jordan, E.T. & Leffler, H. (1997). Galectins-2, galectins-5 and -9, and galectins-4 and -6. *Trends in Glycoscience and Glycotechnology* **9**, 87–93.
7. Huflejt, M.E., Jordan, E.T., Gitt, M.A., Barondes, S.H. & Leffler, H. (1997). A strikingly different localisation of galectin-3 and galectin-4 in human colon adenocarcinoma T84 cells: galectin-4 is localised at sites of cell adhesion. *Journal of Biological Chemistry* **272**, 14294–14303.
8. Wasano, K. & Hirakawa, Y. (1999). Two domains of rat galectin-4 bind to distinct structures of the intercellular borders of colorectal epithelia. *Journal of Histochemistry and Cytochemistry* **47**, 75–82.

Galectin-5

Other names: RL-18; erythroid developmental agglutinin (EDA)
Mol. wt: 16 000
Specificity: Lactose
Source: Mammalian erythroid cells

ISOLATION

Erythroid developmental agglutinin was purified from lactose extracts of rabbit bone marrow by gel filtration and affinity chromatography on mannose–Sepharose 4B [1]. RL-18 was separated by ion-exchange chromatography from other galectins obtained from lactose extracts of rat lung by affinity chromatography on asialofetuin–Sepharose [2]. The probable equivalence of EDA and RL-18 was realised after galectin-5 was cloned from a rat reticulocyte cDNA library [3]. Galectin-5 was subsequently isolated from rat erythrocyte extract by application to lactosyl–Sepharose and elution with lactose [3].

STRUCTURE

Galectin-5 is a typical proto-type galectin, the sequence of which has all the absolutely conserved residues found in galectins-1 to -4 [3]. The 144-residue deduced sequence has a calculated Mr of 16 200, and overall shows most homology with galectin-4 [3]. However, when galectin-9 (qv) was discovered, it was found to have a C-terminal domain highly homologous to galectin-5.

The apparent Mr estimated by gel filtration was 17 000 [3], indicating that galectin-5 normally behaves as a monomer, but its ability to agglutinate erythrocytes (see **Biological Activities** section) implies a capacity to dimerise under some circumstances.

The isoelectric point of galectin-5 is around pH 5.2 [3].

BIOLOGICAL ACTIVITIES

EDA agglutinates lapine erythroblasts, myeloid cells, reticulocytes and erythrocytes [1], while RL-18 agglutinates rat erythrocytes at relatively high concentrations [2]; all agglutinations are inhibited in the presence of lactose.

Galectin-5 binds to affinity matrices containing galactosides and can be eluted with lactose [1–3].

TISSUE AND/OR SUBCELLULAR DISTRIBUTION

Galectin-5 undoubtedly occurs in erythrocytes and other cells of the erythroid lineage, but apparent expression in other cells (including goblet cells) is uncertain on account of possible immunological cross-creativity with other galectins [4]. Within the cell, galectin-5 is abundant in the cytoplasm, but is also found on the cell surface and outside the cell [5].

POSSIBLE FUNCTION(S)

It has been suggested that EDA functions in erythroblast differentiation by promoting inter-erythroblast recognition and adhesion [1].

COMMENTS

Galectin-5 is unique among galectins in apparently being restricted to cells of the erythroid lineage. Like most galectins, it appears to be developmentally regulated in both rats [2] and rabbits [1].

The gene encoding the murine homologue has been mapped to chromosome 11, a region syntenic with human chromosome 17q11 [3].

REFERENCES

1. Harrison, F.L. & Chesterton, C.J. (1980). Erythroid developmental agglutinin is a protein lectin mediating specific cell–cell adhesion between differentiating rabbit erythroblasts. *Nature* **286**, 502–504.
2. Cerra, R.F., Gitt, M.A. & Barondes, S.H. (1985). Three soluble rat β-galactoside-binding lectins. *Journal of Biological Chemistry* **260**, 10474–10477.
3. Gitt, M.A., Wiser, M.F., Lefflon, H., Herrmann, J., Xia, Y.-R., Massa, S.M., Cooper, D.N.W., Lusis, A.J. & Barondes, S.H. (1995). Sequence and mapping of galectin-5, a β-galactoside-binding lectin, found in rat erythrocytes. *Journal of Biological Chemistry* **270**, 5032–5038.
4. Gitt, M.A., Jordon, E.T. & Leffler, H. (1997). Galectin 2, galectins-5 and 9, and galectins 4 and 6. *Trends in Glycoscience and Glycotechnology* **9**, 87–93.
5. Harrison, F.L. & Catt, J.W. (1986). Intra- and extracellular distribution of an endogenous lectin during erythropoiesis. *Journal of Cell Science* **84**, 201–212.

Galectin-7

Other name: IEF17 keratinocyte protein
Mol. wt: 15 000
Specificity: Lactose
Source: Epithelial cells of mammalian skin

ISOLATION

Human galectin-7 was first identified as an abundant keratinocyte protein (IEF17) and subsequently cloned by screening a λgt11 cDNA expression library using degenerate oligodeoxyribonucleotides that had been back-translated from a peptide sequence derived by microsequencing purified IEF17 [1]. Independent cloning from an epidermal cDNA library was also reported from another laboratory [2]. Recombinant galectin-7 was expressed in COS-1 cells [1] and a glutathione-*S*-transferase-galectin-7 fusion protein was expressed in bacteria [2]. The rat and mouse homologues have also been cloned [3].

STRUCTURE

Galectin-7 is a 136-amino acid, proto-type galectin with a molecular mass of 15 075 and an isoelectric point around pH 7.5 [1,2]. It has 38% primary sequence identity to galectin-1, and 27% identity to galectin-2 [2].

The crystal structure both in free form and in the presence of saccharide ligands has been determined at high (0.19–0.20 nm) resolution [4]. The topology consists of the 'jelly-roll' motif similar to that of galectins-1 and -2. The fold is also similar to that of galectin-3 but shows greatest similarity to the Charcot–Leyden crystal protein (galectin-10). The normally monomeric galectin-7 exists as a dimer in the crystals, and adopts a different packing arrangement from galectins-1 and 2 [4].

BIOLOGICAL ACTIVITIES

Galectin-7 can bind to immobilised asialofetuin [1] and lactosyl–Sepharose 4B [2]; elution from both can be achieved with lactose.

TISSUE AND SUBCELLULAR DISTRIBUTION

Galectin-7 is expressed in the epidermis of the skin, but not to any appreciable extent in a wide variety of other tissues examined [1,2]. Although found mainly in stratified squamous epithelium [1], it can be considered a marker of all subtypes of keratinocytes [3]. Galectin-7 is found in the cytoplasm [3], but can be externalised [1].

POSSIBLE FUNCTION(S)

It has been suggested that galectin-7 has a role in cell–cell and cell–matrix interactions during keratinocyte development [4].

COMMENTS

The expression of galectin-7 in all stratified mammalian epithelia examined, whether keratinised or not, and at all stages of differentiation, appears to be unique [3,5].

Galectin-7 is dramatically downregulated in SV40-transformed K14-keratinocytes (that are anchorage-independent and unable to differentiate) and is not expressed in certain cell lines derived from squamous cell carcinomas [1,3].

Retinoic acid reduces, but does not abolish, the synthesis of human galectin-7 mRNA in cultured keratinocytes [2]; expression of mRNA was found to be altered in psoriatic epidermis [2].

Ultraviolet B irradiation induces a high level of galectin-7 expression and this appears to contribute to the apoptotic process in sunburn keratinocytes [6].

The human gene for galectin-7 was mapped to chromosome 19 [1].

REFERENCES

1. Madsen, P., Rasmussen, H.H., Flint, T., Gromov, P., Kruse, T.A., Honoré, B., Vorum, H. & Celis, J.E. (1995). Cloning, expression and chromosome mapping of human galectin-7. *Journal of Biological Chemistry* **270**, 5823–5829.
2. Magnoldo, T., Bernerd, F. & Darmon, M. (1995). Galectin-7, a human 14 kDa S-lectin; specifically expressed in keratinocytes and sensitive to retinoic acid. *Developmental Biology* **168**, 259–271.
3. Magnoldo, T., Fowlis, D. & Darmon, M. (1998). Galectin-7, a marker of all types of stratified epithelia. *Differentiation* **63**, 159–168.
4. Leonidas, D.D., Vatzaki, E.H., Vorum, H., Celis, J.E., Madsen, P. & Acharya, K.R. (1998). Structured basis for the recognition of carbohydrates by human galectin-7. *Biochemistry* **37**, 13930–13940.
5. Timmons, P.M., Colnot, C., Cail, I., Poirier, F. & Magnaldo, T. (1999). Expression of galectin-7 during epithelial development coincides with onset of stratification. *International Journal of Developmental Biology* **43**, 229–235.
6. Bernerd, F., Sarasin, A. & Magnaldo, T. (1999). Galectin-7 overexpression is associated with the apoptotic process in UVB-induced sunburn keratinocytes. *Proceedings of the National Academy of Sciences USA* **96**, 11329–1134.

Galectin-8

Other names: Prostate carcinoma tumour antigen-1; PCTA-1
Mol. wt: 35 000
Specificity: Lactose
Source: Various mammalian tissues

ISOLATION

Galectin-8 was cloned from a rat liver cDNA expression library, a rather fortuitous consequence of screening with antibodies directed against a peptide from the insulin-receptor substrate-1 [1]. Galectin-8 was also obtained from rat liver homogenate by affinity chromatography, and recombinant galectin-8 was expressed in bacteria as a tag-free protein or as a glutathione-*S*-transferase-fusion protein [1].

Human galectin-8 has been cloned from a human brain hippocampus library, using the rat cDNA as a probe [2].

STRUCTURE

Galectin-8 has a tandem-repeat-type structure, in which two homologous (38% primary sequence identity) domains of very roughly 140 amino acids are joined by a 32 amino acid link region [1]. The deduced 316 amino acid polypeptide has several potential glycosylation sites, but no direct evidence was obtained of attached carbohydrate [1]. The overall sequence is 34% identical to galectin-4 (qv) and 31% identical to *Caenorhabditis elegans* galectin (qv), but the link region does not resemble that of either [1]. Unlike galectin-4, the sequence contains four cysteine residues. The N-terminal domain contains a unique motif implicated in sugar binding, predicting some difference in specificity between the domains [1,2].

Human galectin-8 has 97% sequence identity with a human prostate carcinoma tumour antigen, PCTA-1, which may represent a mutant form in prostate cancer cells [2].

The predicted isoelectric points of the rat and human galectin-8 are 9.5 and 8.6, respectively [2].

BIOLOGICAL ACTIVITIES

Galectin-8 can agglutinate fixed, trypsinised lapine erythrocytes, and can bind to lactosyl–Sepharose and be eluted with lactose [1].

Galectin-8 remains soluble and maintains its sugar-binding capacity when extracted and purified in the absence of reducing agents [2].

Galectin-8 can bind to an $\alpha_3\beta_1$ integrin via its sugar-binding site, and can inhibit binding of several cell types to plastic coated with integrins, fibronectin or laminin [3].

Galectin-8 can also induce apoptosis [3].

TISSUE AND/OR SUBCELLULAR DISTRIBUTION

In the rat, the highest expression of galectin-8 was found in lung tissue, but substantial expression was also found in the liver, kidney, heart, spleen and brain [1].

There is evidence that galectin-8 is secreted, but a high proportion of the secreted protein appears to remain bound to the cell surface [3].

POSSIBLE FUNCTION(S)

It has been suggested that galectin-8 functions to modulate integrin interactions with the extracellular matrix to regulate cell adhesion and survival [3].

COMMENTS

Galectin-8 seems to be developmentally regulated, since very low levels of expression were found in rat embryos while high levels were detected in various adult tissues [1,2].

Galectin-8/PCTA-1 is expressed in prostate carcinoma cells (but not in normal or benign hyperplastic prostates) and lung carcinoma cells [2].

REFERENCES

1. Hadari, Y.R., Paz, K., Dekel, R., Mestrovic, T., Accili, D. & Zick, Y. (1995). Galectin 8. A new rat lectin, related to galectin-4. *Journal of Biological Chemistry* **270**, 3447–3453.
2. Yaron, H., Eisenstein, M., Rina, Z. & Zick, Y. (1997). Galectin-8: on the road from structure to function. *Trends in Glycoscience and Glycotechnology* **9**, 103–112.
3. Hadari, Y.R., Arbel-Goren, R., Levy, Y., Amsterdam, A., Alon, R., Zakut, R. & Zick, Y. (2000). Galectin-8 binding to integrins inhibits cell adhesion and induces apoptosis. *Journal of Cell Science* **113**, 2385–2397.

Galectin-9

Other name: Hodgkin's disease galectin antigen
Mol. wt: 36 000
Specificity: Lactose
Source: Mammalian tissues

ISOLATION

Murine galectin-9 cDNA was isolated from embryonic kidney using a degenerate primer PCR cloning strategy, and was expressed as a fusion protein in bacteria [1]. Human galectin-9 was discovered as a Hodgkin's disease-specific antigen and the cDNA isolated by immmunoscreening an expression library derived from a Hodgkin's-involved spleen; however, an identical cDNA sequence was cloned from normal peripheral blood lymphocytes, and a recombinant protein expressed in CHO cells [2]. A closely related human galectin, 'ecalectin', perhaps an allelic variant, was partially purified from a T cell line by a complex series of fractionation steps; the gene was cloned from a cDNA expression library and the recombinant protein expressed in both COS-1 cells and insect cells [3].

STRUCTURE

Galectin-9 is a tandem-repeat galectin, with two mutually homologous (39% primary sequence identity) carbohydrate-recognition domains linked by a sequence of about 30 amino acids that shows no similarity to the link region of other galectins [1,2]. The N-terminal domain exhibits moderate homology with other galectins, particularly galectin-4 (48% primary sequence identity); the C-terminal domain, however, is distinctively similar to the proto-type galectin-5, with over 80% identity between the two domains from rodent sources [1,2].

Both human and murine galectins-9 have a predicted amino acid sequence of just over 320 residues and a calculated Mr of just under 36 000 [1,2]. In the mouse, an alternative splicing isoform (with a 31-residue insertion) is exclusively expressed in the small intestine [1]. The human variant isoform, ecalectin, differs from the standard form by only five amino acid residues [3]. What appears to be a human homologue of the mouse isoform (with a 32 amino acid insertion) has been identified more recently, but does not appear to be tissue-restricted [4].

BIOLOGICAL ACTIVITIES

All recombinant forms of galectin-9, as well as endogenous galectin-9 partially purified from tissues, can bind to immobilised lactose and be eluted with the free disaccharide [1–3].

The murine galectin-9 fusion protein induced apoptosis in 30–60% of murine thymocytes and, at higher concentrations, agglutinated thymocytes; these activities could be inhibited by lactose [5].

Ecalectin is a potent chemoattractant for eosinophils, but has no effect on neutrophils, lymphocytes or monocytes [3].

TISSUE AND/OR SUBCELLULAR DISTRIBUTION

Murine galectin-9 is very widely distributed in organs and tissues, but is most abundant in liver, small intestine and thymus [1,5]. Human galectin-9 is particularly associated with leukocytes and lymphoid tissues, but has also been detected in the stomach and other organs [2,3,4].

Expression is predominantly intracellular, but galectin-9 is also present at the cell surface [5].

POSSIBLE FUNCTION(S)

One suggested function is a role in thymocyte–epithelial cell interactions relevant to clonal selction and deletion. It appears that thymic epithelial cells synthesise and secrete galectin-9, which then binds to cellular receptors on thymocytes, promoting adhesion and perhaps conveying signals [5].

Human T-cell-derived ecalectin may function in the regulation of eosinophil recruitment to tissues during inflammatory episodes [3].

COMMENTS

There is evidence, at least in mice, for developmental regulation of galectin-9 [1,5]. The very high expression in thymic epithelial cells during embryonic life has not been described for any other galectin. Human ecalectin is upregulated by stimulation by antigens [3].

Although autoantibodies to galectin-9 were first described exclusively in the sera of Hodgkin's disease patients, natural antibodies to galectin-9 and its isoform have been detected in 34/34 normal, healthy human subjects and 10/10 diabetic patients [4].

REFERENCES

1. Wada, J. & Kanwar, Y.S. (1997). Identification and characterization of galectin-9, a novel β-galactoside-binding mammalian lectin. *Journal of Biological Chemistry* **272**, 6078–6086.
2. Tureci, O., Schmitt, H., Fadle, N., Pfreundschuh, M. & Sahin, U. (1997). Molecular definition of a novel human galectin which is immunogenic in patients with Hodgkin's disease. *Journal of Biological Chemistry* **272**, 6416–1422.
3. Matsumoto, R., Matsumoto, H., Seki, M., Hata, M., Asano, Y., Kanegasaki, S., Stevens, R.L. & Hirashima, M. (1998). Human ecalectin, a variant of human galectin-9, is a novel eosinophil chemoattractant produced by T lymphocytes. *Journal of Biological Chemistry* **273**, 16976–16984.
4. Suk, K., Hwang, D.-Y. & Lee, M.-S. (1999). Natural autoantibody to galectin-9 in normal human sera. *Journal of Clinical Immunology* **19**, 158–165.
5. Wada, J., Ota, K., Kumar, A., Wallner, E.I. & Kanwar, Y.S. (1997). Developmental regulation, expression and apoptotic potential of galectin-9, a beta-galactoside binding lectin. *Journal of Clinical Investigation* **99**, 2452–2461.

Galectin-10

Other name: Charcot–Leyden crystal protein (CLC, CLCP)
Mol. wt: 17 000
Specificity: Mannose; agarose
Source: Human eosinophils and basophils

ISOLATION

Charcot–Leyden crystal protein was first purified to apparent homogeneity from an eosinophil-rich leukocyte homogenate by spontaneous crystallisation [1]. It was later isolated biochemically from a similar source by a series of column chromatographic procedures [2]. The isolation and sequencing of a cDNA clone revealed the homology of CLCP with the galectin family, but no sequence similarity to any known mammalial lysophospholipases [3]. A recombinant protein has also been produced in bacteria and purified [4].

STRUCTURE

The deduced 142 amino acid sequence, lacking a signal peptide and having a high proportion (37%) of hydrophobic and aromatic amino acids, yields a predicted Mr of 16 500 and a pI of 7.3 [3]. The primary sequence shows significant similarity to the galectin family, particularly galectin-3 [3,4]. The crystal structure determined at 0.18 nm, however, indicated a 'jelly-roll' fold most similar to that of galectin-7 [5].

The apparent Mr estimated by gel filtration is 17 000–20 000 [1,2], indicating a monomer, but galectin-10 can also form homodimers [4].

BIOLOGICAL ACTIVITIES

Recombinant galectin-10 bound to lactose-conjugated agarose, and this binding was weakly inhibited by lactose and to a lesser extent by fucose, but not by arabinose [4]. However, this binding could be attributed to interaction with the agarose, and indeed galectin-10 fails to bind lactosamine-containing sugars and instead binds mannose when in the crystalline form [6]. Somewhat conflicting results on sugar binding may be an indication that carbohydrate-binding activity might vary between the soluble and crystalline forms.

Galectin-10/CLCP co-purifies with lysophospholipase [1,2] and was assumed to have this enzymatic activity, but recently CLCP has been affinity purified without lysophosphatase activity, which apparently belongs to a separate protein [6].

TISSUE AND/OR SUBCELLULAR DISTRIBUTION

Galectin-10 is expressed in eosinophils and basophils but not neutrophils, monocytes or lymphocytes [3]. Within the cell, it is found principally in granules, but also in additional locations in nucleus and cytoplasm, and can be secreted [3].

POSSIBLE FUNCTION(S)

Even less can be said with confidence about the function(s) of galectin-10 than about that of the other galectins. A role in the regulation of eosinophil or basophil development [3] is as good a conjecture as any, but for more comprehensive speculation the reader is referred to the paper by Swaminathan *et al.* [6].

COMMENTS

The hexagonal, dipyramidal Charcot–Leyden crystals have been recognised in histology since the middle of the 19th century, occurring in tissues and secretions in association with eosinophilia, and have served as useful indicators of eosinophilic inflammation or proliferation. Circulating galectin-10 is also dramatically elevated in sera of patients with eosinophilia-associated disorders compared to sera of healthy individuals [1].

Galectin-10 is the major (if not the sole) protein constituent of Charcot–Leyden crystals and accounts for $\geqslant$7% of the total eosinophil protein content [2].

CLCP was originally regarded as a galectin-like protein rather than a true galectin until its carbohydrate-binding activity was demonstrated [4]. In view of its specificity for mannose [6], there must again be some doubt whether it should be considered a true galectin.

REFERENCES

1. Ackerman, S.J., Loegering, D.A. & Gleich, G.J. (1980). The human eosinophil Charcot–Leyden crystal protein: biochemical characteristics and measurement by radio-immunoassay. *Journal of Biological Chemistry* **125**, 2118–2126.
2. Weller, P.F., Bach, D.S. & Austen, K.F. (1984). Biochemical characterization of human eosinophil Charcot–Leyden crystal protein (lysophosphatase). *Journal of Biological Chemistry* **259**, 15100–15105.
3. Ackerman, S.J., Corrette, S.E., Rosenberg, H.F., Bennett, J.C., Mastrianni, D.M., Nicholson-Weller, A., Weller, P.F., Chin, D.T. & Tenen, D.G. (1993). Molecular cloning and characterization of human eosinophil Charcot–Leyden crystal protein (lysophospholipase). *Journal of Immunology* **150**, 456–468.
4. Dyer, K.D. & Rosenberg, H.F. (1996). Eosinophil Charcot–Leyden crystal protein binds to beta-galactoside sugars. *Life Sciences* **58**, 2073–2082.
5. Leonidas, D.D., Vatzaki, E.H., Vorum, H., Celis, J.E., Madsen, P. & Acharya, K.R. (1998). Structural basis for the recognition of carbohydrates by human galectin-7. *Biochemistry* **37**, 13930–13940.
6. Swaminathan, G.J., Leonidas, D.D., Savage, M.P., Ackerman, S.J. & Acharya, K.R. (1999). Selective recognition of mannose by the human eosinophile Charcot–Leyden crystal protein (galectin-10): a crystallographic study at 1.8 Å resolution. *Biochemistry* **38**, 13837–13843.

Galleria mellonella β-1,3-glucan-binding lectin

Mol. wt: 38 000
Specificity: Glucose
Source: Haemocytes from wax moth (*Galleria mellonella*) larvae

---

ISOLATION

The lectin was isolated from a clarified homogenate of *G. mellonella* haemocytes by affinity chromatography on zymosan–Sepharose, eluting with zymosan [1].

STRUCTURE

The lectin is a protein or glycoprotein with a high content of glycine as well as the acidic amino acids. On SDS-PAGE under non-reducing conditions it appears as a single polypeptide with an apparent molecular weight of 38 000 [1].

BIOLOGICAL ACTIVITIES

The lectin agglutinates lapine erythrocytes. Haemagglutination could be inhibited by D-glucose and to a lesser extent by L-rhamnose, but not by other monosaccharides. Zymosan and laminarin were potent inhibitors [1].

TISSUE AND/OR SUBCELLULAR DISTRIBUTION

By both indirect immunoperoxidase staining and by immunoelectron microscopy, the lectin was found in plasmatocytes, on the cell surface, and in Golgi complexes and associated tubules [2].

POSSIBLE FUNCTION(S)

It has been suggested that the lectin mediates the recognition of non-self as a first stage of a defensive response [1,2].

COMMENTS

The localisation of this lectin both on the cell surface of plasmatocytes and in their Golgi bodies suggests it is synthesised by, rather than passively adsorbed on to, plasmatocytes.

Lectin-like β-1,3-glucan-binding proteins had been described in several insect species, but this was the first to be shown to be a haemagglutinin.

REFERENCES

1. Matha, V., Grubhoffer, L. & Söderhäll, K. (1990). New glucose and zymosan specific lectin from *Galleria mellonella* hemocyte lysate. In: *Lectins – Biology, Biochemistry, Clinical Chemistry*, Volume 7 (eds J. Kocourek & D.L.J. Freed), pp. 91–95. Sigma Chemical Company, St Louis, MO.
2. Matha, V., Grubhoffer, L., Weyda, F. & Hermanová, L. (1990). Detection of β-1,3-glucan-specific lectin on the surface of plasmatocytes, immunocompetent cells of great wax moth, *Galleria mellonella*. *Cytobios* **64**, 35–42.

Geodia cydonium galectin

Other name: Sponge galectin
Mol. wt: 30 000 (dimer); 15 000 (subunit)
Specificity: *N*-acetyllactosamine
Source: The marine sponge *Geodia cydonium*

ISOLATION

The lectin was first purified by various agarose-based affinity chromatographic methods using elution with D-galactose or lactose [1–3], resulting in the isolation of isolectins erroneously thought to possess covalently attached carbohydrate. More recently, the cDNA for the galectin was cloned after screening a cDNA library from *G. cydonium* with monoclonal antibodies to the lectin [4]. Recombinant galectin was obtained by expression as a fusion protein in *Escherichia coli,* affinity purification on glutathione–Sepharose 4B, cleavage with thrombin and affinity chromatography with immobilised anti-lectin antibodies [4].

STRUCTURE

The nucleotide and deduced amino acid sequences of the cloned lectin cDNA have been published [4]. The deduced primary structure contains a carbohydrate-recognition domain homologous to those of vertebrate galectins. The full-length protein has an Mr of 20 877 and an estimated isoelectric point of 4.3, but in its processed form the corresponding values are 15 313 and 5.6. The purified recombinant lectin has an Mr of 21 000. The galectin in sponge extracts consists of three isolectins with apparent Mr values of 12 000–15 000 [2–4]; these polymerise in the presence of Ca^{2+} [5].

BIOLOGICAL ACTIVITIES

The sponge galectin agglutinates neuraminidase-treated erythrocytes from sheep, cows and humans; it is ABO non-specific towards human cells [1,3]. Haemagglutination can be inhibited by GalNAc, lactose and galactans from the snails *Helix pomatia* and *Achatina fulica* [1]. The same galactans as well as pneumococcus type XIV polysaccharide, but not *Geodia* polysaccharide, are precipitated from solution [1].

The sponge galectin is mitogenic for human peripheral blood lymphocytes [6].

TISSUE AND/OR SUBCELLULAR DISTRIBUTION

Immunohistological studies have indicated that the cytoplasm of spherulous cells contain and presumably synthesise the galectin; the latter is also located on the outer surface extending into the extracellular matrix [4].

POSSIBLE FUNCTION(S)

The galectin enhances the aggregation of sponge cells mediated by the aggregation factor, a large multiglycoprotein complex. In the model of sponge cell aggregation currently proposed by Muller and colleagues, the galectin forms intercellular bridges between components of the extracellular aggregation factor and aggregation receptors on the cell surface. This is achieved by a Ca^{2+}-dependent dimerisation of the galectin; however, the binding of one galectin monomer to the aggregation receptor and the binding of the other galectin monomer to the aggregation factor are both Ca^{2+}-independent [4].

COMMENTS

The aggregation of sponge cells has been studied for most of the 20th century and advances in its molecular basis have progressed steadily since the 1960s [4,7]. Calcium-dependent aggregation factors have been studied in some detail in *Microciona prolifera* [8] as well as *G. cydonium*. In both species the aggregation factors are high molecular weight complexes which bind to cell surface structures (aggregation receptors/baseplates). The evidence supporting the model now proposed, that the aggregation factors and receptors/baseplates are linked by the galectin in *G. cydonium*, is a major step forward in understanding the complex process of aggregation in these 'simple' multicellular organisms.

REFERENCES

1. Vaith, P., Uhlenbruck, G., Muller, W.E.G. & Holz, G. (1979). Sponge aggregation factor and sponge hemagglutinin: possible relationships between two different molecules. *Developmental and Comparative Immunology* **3**, 399–416.
2. Muller, W.E.C., Conrad, J., Schroder, C., Zahn, R.K., Kerelec, B., Dreesbach, K. & Uhlenbruch, G. (1983). Characterisation of the trimeric, self-recognizing *Geodia cydonium* lectin I. *European Journal of Biochemistry* **133**, 263–267.
3. Hanisch, F.-G., Saur, A., Muller, W.E.G., Conrad, J. & Uhlenbruck, G. (1984). Further characterisation of a lectin and its in vivo receptor from *Geodia cydonium*. *Biochimica et Biophysica Acta* **801**, 388–395.
4. Wagner-Hülsmann, C., Bachinski, N., Diehl-Seifert, B., Blumbach, B., Steffen, R., Pancer, Z. & Muller, W.E.G. (1996). A galectin links the aggregation factor to cells in the sponge (*Geodia cydonium*) system. *Glycobiology* **6**, 785–793.
5. Diehl-Seifert, B., Uhlenbruck, G., Geisent, M., Zahn, R.K. & Muller, W.E.G. (1985). Physiochemical and functional characterization of the polymerization process of the *Geodia cydonium* lectin. *European Journal of Biochemistry* **147**, 517–523.
6. Bretting, H., Phillips, S.G., Klumpart, H.J. & Kabat, E.A. (1981). A mitogenic lactose-binding lectin from the sponge, *Geodia cydonium*. *Journal of Immunology* **127**, 1652–1658.
7. Winzler, R.J. (1970). Carbohydrates in cell surfaces. In: *International Review of Cytology*, Volume 29 (eds G.H. Bourne & J.F. Danielli), pp. 77–125. Academic Press, New York.
8. Weinbaum, G. & Burger, M.M. (1973). A two-component system for surface guided reassociation of animal cells. *Nature* **244**, 510–512.

Granulocytin

Other name: *Sarcophaga peregrina* granulocyte lectin
Mol. wt: 60 000 (native); 20 000 (subunit)
Specificity: Unknown
Source: Larvae of the flesh fly *Sarcophaga peregrina*

ISOLATION

Granulocytin was isolated from an embryonic cell line (NIH-Sape-4) of *Sarcophaga peregrina* by affinity chromatography using an immobilised monoclonal antibody to a ligand of the *Sarcophaga peregrina* humoral lectin (qv). The material recovered from the affinity column with 8 M urea was subjected to fast protein liquid chromatography on a Mono-Q column; the lectin protein in the flow-through peak was electrophoretically pure [1].

Polyclonal antibodies against the above-purified protein were used to screen a NIH-Sape-4 cDNA library to obtain granulocytin cDNA [1].

STRUCTURE

Granulocytin is a glycoprotein of apparent Mr 60 000 consisting of a trimer of identical subunits of apparent Mr 20 000 determined by SDS-PAGE (14 000 after deglycosylation) [1].

The nucleotide and deduced amino acid sequences from granulocytin cDNA have been published in full. The cDNA encodes a 151 amino acid protein plus a 22 amino acid leader peptide. The deduced sequence contains a carbohydrate-binding domain-like sequence with around 25% similarity to the corresponding domain from the humoral lectin of the same species [1].

BIOLOGICAL ACTIVITIES

Granulocytin agglutinates lapine erythrocytes, but has no activity towards ovine, porcine, equine, bovine, human, chicken or guinea-pig red blood cells. Calcium was essential for haemagglutination, which could be 50% inhibited by several monosaccharides tested (glucose, galactose, fucose, NeuAc etc.) but only at 100 mM. These results suggest that granulocytin is a genuine lectin, but that its sugar specificity has yet to be discovered. Bovine submaxilliary gland mucin, but not porcine gastric mucin, was inhibitory, as was LPS from *Escherichia coli* to a lesser extent [1].

TISSUE AND/OR SUBCELLULAR DISTRIBUTION

Granulocytin has been detected in haemocytes, haemolymph and fat bodies of third-instar larvae. The lectin is synthesised in the haemocytes then secreted into the haemolymph. Within the haemocyte, the lectin is localised within granules of large granulocytes [1].

POSSIBLE FUNCTION(S)

By analogy with the *S. peregrina* humoral lectin (qv) and the *Drosophila melanogaster* lectin (qv), granulocytin is presumed to have a defence function and perhaps also a separate role in development (see **Comments** section).

COMMENTS

Granulocytin is developmentally regulated and behaves as an acute phase response component. During development, it is first synthesised in third-instar larvae, but synthesis ceases after the early pupal stage; synthesis is resumed in newly emerged adults. Synthesis is also significantly enhanced within a few hours of injury to the body walls of the larvae.

These characteristics resemble those of the better known *S. peregrina* humoral lectin and especially the *Drosophila* lectin. However, primary sequence analysis showed amino acid identities of less than 20% with those primitive C-type lectins [1].

Granulocytin was discovered because of immunological cross-reactivity with a ligand for the *Sarcophaga* humoral lectin; the biochemical nature of the shared epitope was not established [1].

REFERENCE

1. Fujita, Y., Kurata, S., Homma, K. & Natori, S. (1998). A novel lectin from *Sarcophaga*. Its purification, characterization and cDNA cloning. *Journal of Biological Chemistry* **273**, 9667–9672.

Hakata antigen

Other names: HA; thermolabile β_2-macroglobulin
Mol. wt: 520 000 (native); 35 000 (subunit)
Specificity: D-Fucose
Sources: Human blood, lungs and liver

ISOLATION

The Hakata antigen was originally identified and defined by autoantibodies in the sera of systemic lupus erythematosus patients, and purified by a complex procedure which included immunoaffinity chromatography [1,2].

The Hakata antigen cDNA sequence was identified from cDNA clones obtained using a human lung cDNA library, and transiently expressed in a human liver cell line [2].

STRUCTURE

The Hakata antigen is a glycoprotein (8% carbohydrate) of 275 amino acids with a blocked N-terminus; the determined and deduced primary sequences have been published in full [1,2]. The molecule consists of a collagen-like domain near the N-terminus connected to a C-terminal fibrinogen-like domain by a short neck region. Overall, there was 48% identity to the sequences from human ficolin-M and P35 (qv) [2].

HA is a homopolymer of uncertain size. The basic subunit has an Mr of 35 000 determined by SDS-PAGE under reducing conditions, but a ladder of bands representing various sizes is apparent under non-reducing conditions [1,2]. Estimates of native molecular mass vary from 520 kDa (ultracentrifugation) to 650 kDa (gel filtration) [1]. Although it was originally suggested that the native molecule consists of six triplets [2], it is perhaps more likely to be a dodecamer of four triplets [3]. The isoelectric point is between pH 6.7 and 7.0 [1].

By electron microscopy, HA is visualised as aggregates of structures with globular heads attached to thin rods [2].

BIOLOGICAL ACTIVITIES

HA agglutinates ovine (but not human) erythrocytes and (more readily) trypsin-treated ovine erythrocytes, in a calcium-independent manner. HA strongly agglutinated human red blood cells coated with LPS from *Salmonella tymphimurium* and weakly agglutinated cells coated with LPS from two other Gram-negative bacteria. Haemagglutination was potently inhibited by D-fucose, moderately inhibited by GlcNAc, and relatively weakly inhibited by GalNAc, but other common monosaccharides had no effect [2].

HA bound to GalNAc and GlcNAc immobilised on agarose and could be eluted with the corresponding sugars; HA did not bind to immobilised mannose or lactose.

TISSUE AND/OR SUBCELLULAR DISTRIBUTION

By Northern and Western blotting, it was shown that only the lung and liver express Hakata antigen out of many organs investigated [4]. In the lung, the lectin is found in the cytoplasm of type II alveolar cells and ciliated (cf. SP-A/SP-D) epithelial cells and is secreted into the bronchi. In the liver, the lectin is abundant in the cytoplasm of bile duct epithelial cells and is secreted into the bile ducts. Some Hakata antigen is also found in hepatocyte cytoplasm, and this is thought to be the source of the serum lectin [4].

POSSIBLE FUNCTION(S)

The function of HA is unknown but a defensive role as an opsonin within the bloodstream and lining the surface of the lung is a possibility. A separate function is presumably relevant in the bile [4].

COMMENTS

The Hakata antigen is a normal plasma protein present in Japanese blood donors at a concentration of 7–23 μg/ml [1]. Autoantibodies to HA occur in a small minority (4%) of patients with systemic lupus erythematosus and are extremely rare in subjects without that syndrome [5].

HA is clearly a member of the ficolin family, but has markedly less homology (50%) with the original porcine ficolins α and β than the other known members of the family. One of the murine ficolins is also something of an outlier in this respect (60% homology), but those two are not more closely related to each other than to the other ficolins [3].

REFERENCES

1. Yae, Y., Inaba, S., Sato, H., Okochi, K., Tokunaga, F. & Iwanoga, S. (1991). Isolation and characterization of a thermolabile β-2 macroglobulin ('thermolabile substance' or 'Hakata antigen') detected by precipitating (auto)antibody in sera of patients with systemic lupus erythematosus. *Biochimica et Biophysica Acta* **1078**, 369–176.
2. Sigimoto, R., Yae, Y., Akaiwa, M., Kitajima, S., Shibata, Y., Sato, H., Hirato, J., Okochi, K., Izuhara, K. & Hamasaki, N. (1998). Cloning and characterization of the Hakata antigen, a member of the ficolin/opsonin P35 lectin family. *Journal of Biological Chemistry* **273**, 20721–20727.
3. Ohashi, T. & Erickson, H.P. (1998). Oligomeric structure and tissue distribution of ficolins from mouse, pig and human. *Archives of Biochemistry and Biophysics* **360**, 223–232.
4. Akaiwa, M., Yae, Y., Sugimoto, R., Suzuki, S.O., Iwaki, T., Izuhara, K. & Hamasaki, N. (1999). Hakata antigen, a new member of the ficolin/opsonin p35 family, is a novel human lectin secreted into bronchus/alveolus and bile. *Journal of Histochemistry & Cytochemistry* **47**, 777–785.
5. Inaba, S., Okochi, K. & Yae, Y. (1990). Serological studies of an SLE-associated antigen–antibody system discovered as a precipitation reaction in agarose gel: the HAKATA antigen–antibody system. *Fukuoka Acta Medica* **81**, 284–291.

Halichondria okadai lectin I

Other name: HOL-I
Mol. wt: 84 000 (native); 21 000 (subunit)
Specificity: GlcNAc, GalNAc
Source: The toxic marine sponge *Halichondria okadai*

ISOLATION

The soluble fraction of *H. okadai* homogenised in saline was used as starting material. HOL-I (bound) was separated from HOL-II (unbound) by application to a column of bovine submaxilliary mucin coupled to Toyopearl. HOL-I was eluted from the column with 0.5 M acetic acid and subjected to gel filtration on Toyopearl 55F [1].

STRUCTURE

HOL-I is a glycoprotein (2% carbohydrate) with an amino acid composition broadly similar to that of *H. okadai* lectin II (HOL-II) (qv), but with a higher content of aspartic acid, threonine and trytophan residues. The N-terminus is blocked [1].

HOL-I appears to be a tetramer, since the apparent molecular weight estimated by gel filtration (84 000) is four times the apparent Mr of the single band obtained after SDS-PAGE in the presence or absence of reducing agent. The isoelectric point is 4.5 [1].

BIOLOGICAL ACTIVITIES

HOL-I did not agglutinate native human erythrocytes, but agglutinated pronase-treated cells in a divalent cation-independent manner. A slight preference for pronase-treated group A cells was shown. This activity was stable below 50 °C, and was considerably more thermo-stable than HOL-II [1].

Haemagglutination was most potently inhibited by GlcNAc and GalNAc, but sialic acids were also effective, as were a variety of acetylated saccharides. Bovine submaxilliary mucin and more especially its asialo-derivative were the most effective glycoprotein inhibitors. It appears that HOL-I mainly recognises *N*-acetyl groups.

TISSUE AND/OR SUBCELLULAR DISTRIBUTION

Nothing was reported concerning the distribution of this lectin [1].

POSSIBLE FUNCTION(S)

No possible functions were suggested [1].

COMMENTS

Both haemagglutination inhibition and NMR spectral analyses indicate that HOL-I recognises *N*-acetyl groups. The most potent inhibitor tested was the oligosaccharide, GlcNAc β1,4(GlcNAc β1,2)Manα-1-O $(CH_2)_2CH_3$. This unusually precise specificity suggests that HOL-I could be a valuable tool for oligosaccharide research.

REFERENCE

1. Kawagishi, H., Yamawaki, M., Isobe, S., Usui, T., Kimura, A. & Chiba, S. (1994). Two lectins from the marine sponge *Halichondria akadai*. *Journal of Biological Chemistry* **269**, 1375–1379.

Halichondria okadai lectin II

Other name: HOL-II
Mol. wt: 42 000
Specificity: *N*-acetyllactosamine
Source: The toxic marine sponge *Halichondria okadia*

ISOLATION

The soluble fraction of *H. okadia* homogenate which did not bind to BSM-Toyopearl (see HOL-I) was applied to a column of acid-treated Sepharose 4B. HOL-II was obtained by elution with distilled water [1].

STRUCTURE

HOL-II is a glycoprotein (7.5% carbohydrate) with an amino acid composition broadly similar to that of *H. okadia* lectin I (qv), but with a higher content of methionine and leucine. The N-terminus is blocked [1].

HOL-II is a monomer of Mr 42 000, the same value being obtain by gel filtration and by SDS-PAGE (in the presence or absence of reducing agent). The isoelectric point is 4.5 [1].

BIOLOGICAL ACTIVITIES

HOL-II did not agglutinate native human erythrocytes but agglutinated pronase-treated cells in a divalent cation-independent manner and with a very slight preference for blood group A. HOL-II was significantly more labile than HOL-I, exhibiting a rapid loss of activity at temperatures above 40 °C [1].

Haemagglutination was inhibited by relatively high concentrations of galactose and GalNAc, but was vastly more sensitive to *N*-acetyllactosamine and its analogues. Indeed, HOL-II appears to have a very precise specificity for Gal-β1,4-GlcNAc [1].

TISSUE AND/OR SUBCELLULAR DISTRIBUTION

No information concerning distribution was reported [1].

POSSIBLE FUNCTION(S)

The function of HOL-II is unknown.

COMMENTS

Although *N*-acetyllactosamine-specific lectins are hardly uncommon, the binding reactivity of HOL-II seems to be very strictly limited to Gal-β1,4-GlcNAc, since Gal-β1,3-GlcNAc, Gal-β1,6-GlcNAc, Gal-β1,3-GalNAc, Gal-β1,4-GalNAc and Gal-β1,6-GalNAc etc. were not inhibitory. This strict specificity may make it a particularly useful tool for oligosaccharide research. A lectin with rather similar specificity has also been found in a moth, *Phalera flavescens* (qv) [2].

REFERENCES

1. Kawagishi, H., Yamawaki, M., Isobe, S., Usui, S., Kimura, A. & Chiba, S. (1994). Two lectins from the marine sponge *Halichondria okadai*. *Journal of Biological Chemistry* **269**, 1375–1379.
2. Umetsu, K., Yamashita, K., Suzuki, J., Yamashita, T. & Suzuki, T. (1993). Purification and characterization of an *N*-acetyllactosamine-specific lectin from larvae of the moth *Phalera flavescens*. *Archives of Biochemistry and Biophysics* **301**, 200–205.

Halocynthia roretzi lectin

Other name: *H. roretzi* plasma lectin-1
Mol. wt: Approx. 600 000 (native); 41 000 (subunit)
Specificity: D-Galactose
Source: Haemolymph of the solitary ascidian *Halocynthia roretzi*

ISOLATION

The lectin was isolated from haemolymph plasma by the single step of affinity chromatography on acid-treated Sepharose and elution with lactose [1,2].

STRUCTURE

The lectin is a protein relatively rich in lysine as well as aspartic and glutamic acid residues [1]. It exists as a large molecule (sedimentation coefficients of 20–26 S were obtained under various experimental conditions [1,2]) which exhibits heterogeneity in SDS-PAGE under non-reducing conditions [2]. Nevertheless, a single peak was obtained on gel filtration, a single boundary on ultracentrifugation, and a single band on SDS-PAGE in the presence of reducing agent [1,2]. The molecular weight was determined to be $660\,000 \pm 30\,000$ from sedimentation equilibrium analysis and 540 000 by gel filtration on Sephacryl S300 [2]. The apparent Mr of the subunits detected by SDS-PAGE was 41 000. It seems therefore that the subunits are linked by a variable number of inter-subunit disulphide bonds and these molecular species aggregate into a single larger molecule under physiological conditions. Electron microscopy confirmed the presence of large molecules of different sizes [2].

The complete amino acid sequence has been reported, and verified by cDNA cloning [3]. The primary structure comprising 327 amino acid residues (calculated Mr = 36 464) had little overall sequence similarity to other proteins. However, the C-terminal two-thirds of the molecule exhibited some homology with C-type lectins, while residues 50–150 were homologous to the fibrinogen β-chain of humans and lampreys. The N-terminal region consisted of a repeated sequence of 16 residues, which was separated from the C-type lectin domain by the fibrinogen-like sequence [3].

BIOLOGICAL ACTIVITIES

The lectin agglutinated ovine erythrocytes in a calcium-dependent manner [1,2]. Neither Mn^{2+} nor Ni^{2+} could substitute for Ca^{2+}, while Mg^{2+} antagonised the effect of Ca^{2+} at high concentrations [2]. Haemagglutination was inhibited by

D-galactose and fucose, but not by other monosaccharides. Lactose was as effective, and melibiose considerably more effective, than galactose [1].

The lectin stimulated superoxide production by guinea-pig polymorphonuclear leukocytes; this activity had a roughly similar but broader inhibition profile by saccharides [2]. Phagocytosis of sheep red blood cells by *H. roretzi* haemocytes was marginally enhanced by physiological concentrations of the lectin [3].

The lectin bound to blue-Sepharose and to phenyl-Sepharose, and could be eluted with ethylene glycol but not by lactose [2].

TISSUE AND/OR SUBCELLULAR DISTRIBUTION

Indirect evidence indicates that the lectin may be synthesised by the hepatopancreas and secreted into the haemolymph plasma [3].

POSSIBLE FUNCTION(S)

On the flimsy basis that this lectin can activate mammalian neutrophils, it has been conjectured that the lectin may have a defensive, bactericidal function in the haemolymph [2], and this suggestion is supported by the weak stimulation of phagocytosis of sheep red blood cells by haemocytes of *H. roretzi* [3].

COMMENTS

The reported binding to columns of blue-Sepharose and phenyl-Sepharose may be an indication of a hydrophobic binding site in addition to its carbohydrate-binding site.

The closely related species *Halocynthia pyriformis* also possesses a lectin with 40 000 Da subunits, but it recognises sialic acid; this lectin has a very similar amino acid composition to those of the *Didemnum candidum* isolectins (qv), and (like the latter) to mammalian C-reactive proteins [4].

A minor lectin with GalNAc specificity has also been discovered in *H. roretzi* plasma [5]; and a lectin-like haemagglutinin which binds LPS, heparin and chondroitin sulphate, and agglutinates various marine Gram-negative bacteria, has been isolated from haemocytes of *H. roretzi* [6].

REFERENCES

1. Yokosawa, H., Sawada, H., Abe, Y., Numakurai, T. & Ishii, S. (1982). Galactose-specific lectin in the hemolymph of solitary ascidian, *Halocynthia roretzi*: isolation and characterization. *Biochemical and Biophysical Research Communications* **107**, 451–457.

2. Yokosawa, H., Harada, K., Igarashi, K., Abe, Y., Takahashi, K. & Ishii, S. (1986). Galactose-specific lectin in the hemolymph of solitary ascidian, *Halocynthia roretzi*. Molecular, binding and functional properties. *Biochimica et Biophysica Acta* **870**, 242–247.
3. Abe, Y., Tokuda, M., Ishimoto, R., Azumi, K. & Yokosawa, H. (1999). A unique primary structure; cDNA cloning and function of a galactose-specific lectin from ascidian plasma. *European Journal of Biochemistry* **261**, 33–39.
4. Vasta, G.R. & Marchalonis, J.J. (1983). Lectins from tunicates and cyclostomes. A biochemical characterization. In: *Lectins – Biology, Biochemistry, Clinical Chemistry*, Volume 3 (eds T.-C. Bög-Hansen & G.A. Spengler). Walter de Gruyter, Berlin.
5. Harada-Azumi, K., Yokosawa, H. & Ishii, S. (1987). *N*-acetylgalactosamine-specific lectin, a novel lectin in the haemolymph of the ascidian *Halocynthia roretzi*: isolation, characterization and comparison with galactose-specific lectin. *Comparative Biochemistry and Physiology* **88B**, 375–381.
6. Azumi, K., Ozeki, S., Yokosawa, H. & Ishii, S. (1991). A novel lipopolysaccharide-binding hemagglutinin isolated from hemocytes of the solitary ascidium, *Halocynthia roretzi*: it can agglutinate bacteria. *Developmental and Comparative Immunology* **15**, 9–16.

Helix pomatia agglutinin

Other names: HPA; vineyard/edible/Roman snail lectin
Mol. wt: 79 000 (native); 13 000 (subunit)
Specificity: GalNAc
Sources: Albumin glands and eggs

ISOLATION

HPA has been purified by affinity chromatography on blood group A substance [1] or GalNAc–Sepharose [2], and elution with GalNAc. It has also been isolated on unsubstituted Sephadex by elution with GalNAc, galactose or glucose [3].

STRUCTURE

HPA is a glycoprotein (7% carbohydrate) with a high content of acidic amino acids, serine and proline [1]. It is composed of six virtually identical subunits after reduction, each with a single binding site [4,5]. The native molecule is arranged as three non-covalently-linked sets of dimers, each dimer formed by a single disulphide bond [5]. A large number of isolectins has been reported [2].

BIOLOGICAL ACTIVITIES

HPA agglutinates human blood group A erythrocytes and precipitates blood group A substance [1,6]. Either purified or in crude extracts, it is one of the most potent and specific of the blood grouping lectins [6]. While recognising terminal GalNAc residues [1], its lectin activity is more readily inhibited by more complex saccharides, particularly the disaccharide sequence, GalNAc-α1,3-GalNAc [7]. These activities do not seem to be divalent cation-dependent [7].

HPA binds a majority of desialylated, but not untreated, human lymphocytes and is non-mitogenic for both untreated and desialated cells [8]. However, it binds preferentially to T lymphocytes [9], thus providing a means to isolate (by cell affinity chromatography) B lymphocytes suitable for use in serological HLA-DR typing [10]. HPA also binds preferentially to desialylated T cells in several other species [11]. HPA binds to a single human lymphocyte receptor [12], which might be an isoform of the leukocyte common antigen, CD45.

HPA binds specifically to human IgA_1, but not to other classes of immunoglobulin [13,14].

TISSUE AND/OR SUBCELLULAR DISTRIBUTION

HPA occurs in albumin glands and eggs, but apparently not in haemolymph.

POSSIBLE FUNCTION(S)

The function of HPA is unknown, but it presumably has some role in reproduction.

COMMENTS

HPA is by far the best known, most studied and most utilised of a group of snail lectins specific for blood group A. Others include *Otala lactea* (land snail) and *Euphadra periomphala* (Japanese snail), as well as the closely related species *H. aspera* and *H. hortensis* [6]. The lectins from all of these species are antigenically distinct [15].

An important application of HPA is in lectin histochemistry, where it has been found that HPA binding has prognostic value in breast cancer [16,17]. HPA binding has also been associated with a poorer prognosis in colorectal, gastric and oesophageal carcinomas [18; see also Chapter 4].

Since HPA can agglutinate *Bacillus mycoides* but not *B. anthracis*, it may also be of some value as a reagent in diagnostic bacteriology [19].

Two distinct lectins have been discovered on the surface of *H. pomatia* haemocytes [20].

REFERENCES

1. Hammarström, S. & Kabat, E.A. (1969). Purification and characterization of a blood-group A reactive hemagglutinin from the snail *Helix pomatia* and a study of its combining site. *Biochemistry* **8**, 2696–2705.
2. Vretblad, P., Hjorth, R. & Lååas, T. (1979). The isolectins of *Helix pomatia*. Separation by isoelectric focusing and preliminary characterization. *Biochimica et Biophysica Acta* **579**, 52–61.
3. Goldstein, I.J. & Poretz, R.D. (1986). Isolation and chemical properties of lectins. In: *The Lectins* (eds I.E. Liener, N. Sharon & I.J. Goldstein), pp. 134–137. Academic Press, Orlando, FL.
4. Hammarström, S. & Kabat, E.A. (1971). Studies on specificity and binding properties of the blood group A reactive hemagglutinin from *Helix pomatia*. *Biochemistry* **10**, 1684–1692.
5. Hammarström, S., Westöö, A. & Björk, I. (1972). Subunit structure of *Helix pomatia* A hemagglutinin. *Scandinavian Journal of Immunology* **1**, 295–309.
6. Kilpatrick, D.C. & Green, C. (1992). Lectins as blood typing reagents. In: *Advances in Lectin Research*, Volume 5 (ed. H. Franz), pp. 51–94. Ullstein Mosby, Berlin.
7. Baker, D.A., Sugii, S., Kabat, E.A., Ratcliffe, R.M., Hermentin, P. & Lemieux, R.U. (1983). Immunochemical studies on the combining sites of Forssman hapten reactive hemagglutinins from *Dolichos biflorus*, *Helix pomatia* and *Wistaria floribunda*. *Biochemistry* **22**, 2741–2750.

8. Dillner, M.-L., Hammarström, S. & Perlmann, P. (1975). The lack of mitogenic response of neuraminidase-treated and untreated human blood lymphocytes to divalent, hexavalent, or insoluble *Helix pomatia* A hemagglutinin. *Experimental Cell Research* **96**, 374–382.
9. Hammarström, S., Hellström, U., Perlmann, P. & Dillner, M.-L. (1973). A new surface marker on T lymphocytes of human peripheral blood. *Journal of Experimental Medicine* **138**, 1270–1275.
10. Schrempf-Decker, G.E., Baron, D. & Wernet, P. (1980). *Helix pomatia* agglutinin (HpA) affinity chromatography: the isolation of pure B and T cell populations and their use for routine HLA-DR (Ia) serology. *Journal of Immunological Methods* **32**, 285–296.
11. Sharon, N. (1983). Lectin receptors as lymphocyte surface markers. *Advances in Immunology* **34**, 213–298.
12. Alesson, B., Kimura, A., Hammarström, S., Wigzell, H., Nilsson, K. & Mellstedt, H. (1978). *Helix pomatia* A hemagglutinin: selectivity of binding to lymphocyte surface glycoproteins on T cells and certain B cells. *European Journal of Imunology* **8**, 757–764.
13. Tomana, M., Matousovic, K., Julian, B.A., Radl, J., Konecny, K. & Mestecky, J. (1997). Galactose-deficient IgA1 in sera of IgA nephropathy patients is present in complexes with IgG. *Kidney International* **52**, 509–516.
14. Booth, J.R., Munks, R. & Sokol, R.J. (1995). Isolation of IgA1 from human serum by affinity chromatography using an immobilised extract of the albumin gland of *Helix pomatia*. *Transfusion Medicine* **5**, 117–121.
15. Schneider, H.A.W. (1985). Properties of lectins from snails of the genus *Helix* probed by monoclonal antibodies. *Zeitschrift für Naturforschung* **40c**, 254–261.
16. Leathem, A.J. & Brooks, S.A. (1987). Predictive value of lectin binding on breast cancer recurrence and survival. *Lancet* **1**, 1054–1056.
17. Schumacher, U. (1995). Lectins and cancer – an old friend revisited. In: *Lectins – Biomedical Perspectives* (eds A. Pusztai & S. Bardocz). pp. 211–224. Taylor & Francis, London.
18. Schumacher, U., Adam, E., Brooks, S.A. & Leathem, A.J. (1996). The use of *Helix pomatia* agglutinin in metastasis research. In: *Lectins – Biology, Biochemistry, Clinical Biochemistry* (eds E. Van Driessche, P. Rougé, S. Beeckmans & T.-C. Bög-Hansen). Textop, Hellerup.
19. Doyle, R. & Keller, K. (1984). Lectins in diagnostic microbiology. *European Journal of Clinical Microbiology* **3**, 4–9.
20. Richards, E.H. & Renwrantz, L.R. (1991). Two lectins on the surface of *Helix pomatia* haemocytes: a Ca^{2+}-dependent, GalNAc-specific lectin and a Ca^{2+}-independent mannose-6-phosphate-specific lectin which recognises activated homologous opsonins. *Journal of Comparative Physiology B* **161**, 43–54.

Heterometrus granulomanus lectin

Other name: Scorpin
Mol. wt: 500 000 (native); 15 000 (subunit)
Specificity: Sialic acid
Source: Haemolymph of the Indian scorpion *Heterometrus granulomanus*

ISOLATION

Scorpin was purified from clarified, clotted haemolymph in two steps. The first consisted of affinity chromatography on equine submandibular gland glycopeptides conjugated to agarose, and eluting with EDTA in a calcium-free buffer. The second step consisted of gel filtration on Sephadex G-200. The purified lectin exhibited a 400-fold increase in specific activity and appeared homogeneous by PAGE [1].

STRUCTURE

Scorpin is a glycoprotein containing 2.8% carbohydrate. It consists of a single subunit of apparent Mr 15 000 determined by SDS-PAGE in the presence of 2-mercaptoethanol. As scorpin is eluted ahead of ferritin on Sephadex G-200, the native molecule may consist of aggregates of 30 or more subunits [1].

BIOLOGICAL ACTIVITIES

Scorpin agglutinates human erythrocytes and this activity is calcium-dependent. Agglutination was inhibited by *N*-acetylneuraminic acid, *N*-glycolylneuraminic acid and D-glucuronic acid, but not by GlcNAc or GalNAc. Sialyllactose and other more complex saccharides containing terminal sialic acids were more potent inhibitors than free sialic acids. In particular, equine submaxilliary mucin-glycopeptide, containing mainly *N*-acetyl-4-*O*-acetylneuraminic acid, was the most effective inhibitor [1].

TISSUE AND/OR SUBCELLULAR DISTRIBUTION

No information regarding distribution (other than presence in haemolymph) was reported [1].

POSSIBLE FUNCTION(S)

The function is unknown, but a role in defence seems likely.

COMMENTS

This lectin was one of the first scorpion lectins to be isolated.

REFERENCE

1. Ahmed, H., Chatterjee, B.P., Kelm, S. & Schauer, R. (1986). Purification of a sialic acid-specific lectin from the Indian scorpion *Heterometrus granulomanus*. *Biological Chemistry Hoppe-Seyler* **367**, 501–506.

HIP/PAP

Other name: Pancreas associated protein
Mol. wt: 16 000
Specificity: Lactose
Sources: Mammalian pancreas and small intestine

ISOLATION

During an investigation of tumour-specific gene expression, the differential screening of a human hepatocellular carcinoma cDNA library using subtracted probes led to the discovery of a gene named HIP (hepatocellular carcinoma-intestine-pancreas) [1]. The HIP gene product is identical to the 'pancreatitis associated protein' (PAP) encoded by a gene cloned from a human pancreatic cDNA library [2]. The rat homologue of HIP/PAP had previously been purified from pancreatic juice by ion-exchange chromatography and gel filtration [3]. Recombinant human HIP/PAP has been produced in transgenic mice and in *Escherichia coli* [4].

STRUCTURE

The rat HIP/PAP lectin is rich in basic amino acids (giving it a pI of 8.2), and has an apparent Mr estimated by SDS-PAGE between 14 000 and 17 000 [3].

The human gene structure predicts a 175 amino acid polypeptide (71% identical to the rat homologue) consisting of a short signal peptide and a single domain homologous to that of many C-type lectins to which its sequence was compared [1,2].

BIOLOGICAL ACTIVITIES

Recombinant human HIP/PAP, in the form of glutathione-*S*-transferase fusion proteins, bound to immobilised lactose in a divalent cation (Ca^{2+} or Mg^{2+})-dependent manner. The lectin could only be recovered by elution with lactose (not other sugars) and no binding was observed with other immobilised saccharides [5].

Recombinant human HIP/PAP promotes the adhesion of rat hepatocytes *in vitro*, and binds strongly to laminin and fibronectin. The binding to laminin could not be inhibited with thiodigalactoside, suggesting that it involved protein–protein interaction [4].

TISSUE AND/OR SUBCELLULAR DISTRIBUTION

In normal human tissues and organs, HIP/PAP is found in the small intestine (mainly in Paneth cells) and pancreas (mainly in acinar cells, but also in some cells of the islets of Langerhans), but not in many other locations investigated [1,2,4]. It is also found at low concentration in serum [2]. It is not present in normal adult or fetal liver, but may be expressed at high levels in hepatocellular carcinoma cells [1].

POSSIBLE FUNCTION(S)

The function of HIP/PAP is obscure. There is no evidence to support the suggestion that it functions in defence against intestinal bacteria [2].

COMMENTS

HIP/PAP is present in the serum of healthy subjects at <0.6 pmol/ml, but increases to 15–82 pmol/ml in acute pancreatitis [2]. A similar 100-fold increase has been observed with experimentally-induced acute pancreatitis in rats [3]. Elevated serum levels were a feature of a minority of patients with chronic pancreatitis, but not of patients with other abdominal disorders. HIP/PAP might therefore serve as a useful biochemical marker of pancreatic inflammation [2].

Intracellular HIP/PAP expression was elevated in 7/29 cases of hepatocellular carcinoma and 2/2 cases of pancreatic cell hyperplasia (hyperinsulinism) [1].

REFERENCES

1. Lasserre, C., Christa, L., Simon, M.-T., Vernier, P. & Bréchot, C. (1992). A novel gene (HIP) activated in human primary liver cancer. *Cancer Research* **52**, 5089–5095.
2. Orelle, B., Keim, V., Masciotra, L., Dagorn, J.-C. & Iovanna, J.-L. (1992). Human pancreas-associated protein. *Journal of Clinical Investigation* **90**, 2284–2291.
3. Keim, V., Iovanmna, J.-L., Rohr, G., Usadel, K.H. & Dagorn, J.-C. (1991). Characterization of a rat pancreatic secreting protein associated with pancreatitis. *Gastroenterology* **100**, 775–782.
4. Christa, L., Carnot, F., Simon, M.-T., Levavasseur, F., Stinnakre, M.-G., Lasserre, C., Thepot, D., Clement, B., Devinoy, E. & Brechot, C. (1996). HIP/PAP is an adhesive protein expressed in hepatocarcinoma, normal Paneth and pancreatic cells. *American Journal of Physiology* **271**, G993–G1002.
5. Christa, L., Felin, M., Morali, O., Simon, M.-T., Lasserre, C., Brechot, C. & Sève, A.-P. (1994). The human HIP gene; overexpressed in primary liver cancer encodes for a C-type carbohydrate-binding protein with lactose binding activity. *FEBS Letters* **337**, 114–118.

Homarus americanus agglutinins I and II

Other names: Lobster lectins LAg-1 and LAg-2
Mol. wt: >1 × 10^6 (LAg-1) and 0.4 × 10^6 (LAg-2) (native); 70 000 (LAg-1) and 35 000 (LAg-2) (subunit)
Specificity: Sialic acid (LAg-1) and GalNAc (LAg-2)
Source: Haemolymph of the American lobster *Homarus americanus*

ISOLATION

The haemolymph of the American lobster contains a mixture of lectins, from which two major species can be vastly enriched by precipitation with ammonium sulphate to 35% saturation. These two lectins can then be largely separated from each other by gel filtration and/or preparative electrophoresis [1,2]. Better separation can be achieved by affinity chromatography on bovine submaxilliary mucin–Sepharose and sequential elution of the washed column with *N*-acetylmannosamine (LAg-1) and/or GalNAc (LAg-2) [3]. Alternatively, Lag-1 can be isolated in homogeneous form by application to a column of colominic acid–Sepharose and elution with N-acetylmannosamine [4].

STRUCTURE

Both lectins are glycoproteins [1] which may to some extent bind to each other in the native state [4]. Nevertheless, LAg-1 seems to exist mainly as a very large aggregate which can largely be separated from LAg-2 by gel filtration or ultracentrification (19 S v. 11 S) to yield lectins that are antigenically distinct [1]. The apparent subunit molecular weights determined by SDS-PAGE were 70 000 (LAg-1) and 35 000 (LAg-2) [4]. Both lectins were eluted in a similar volume from a column of Sephadex G-200 in the presence of urea; the elution volume was not altered by the presence or absence of mercaptoethanol [1].

BIOLOGICAL ACTIVITIES

LAg-1 agglutinated the erythrocytes of a wide variety of species (human, mouse, horse, hamster, sheep, chicken), while LAg-2 agglutinated only mouse, horse and hamster cells, and notably not human cells [1,2]. The activities of both lectins were Ca^{2+}-dependent and inactivated by 2-mercaptoethanol; both were labile at 56 °C [1]. LAg-1 was shown to bind sialic acids on human and murine erythrocytes, and could be inhibited by free *N*-acetylneuraminic acid and *N*-acetylglycolylneuraminic acid. Agglutination of murine erythrocytes by LAg-2 was inhibited most readily by GalNAc and to a much lesser extent by GlcNAc [2].

Both LAg-1 and Lag-2 agglutinated lymphocytes of human and murine origin. Bacteria of *Achromobacter* species were agglutinated by haemolymph, and erythrocyte adsorption experiments implicated LAg-1 [2].

TISSUE AND/OR SUBCELLULAR DISTRIBUTION

Lectin activity (undifferentiated) was present in the lymphatic gland, pericardium, testis, ovary, hepatopancreas (= liver) and green gland (= kidney) at low specific activity; it was more abundant in the haemolymph, but by far the highest specific activity was found in haemocyte extracts [1].

POSSIBLE FUNCTION(S)

The possibility of a defence function is supported by the observation that LAg-1 demonstrated some opsonic activity in relation to the phagocytosis of erythrocytes by haemocytes [2].

COMMENTS

LAg-1 agglutinated only a small proportion of normal murine thymocytes, but most cortisone-resistant thymocytes, as well as most splenic and peripheral blood T cells [5]. In so far as LAg-1 appears to bind to mature, but not immature, lymphocytes of the T lineage, it behaves in the opposite manner to peanut agglutinin, and, like the latter, may find application in lymphoid cell identification and fractionation.

REFERENCES

1. Hall, J.L. & Rowlands, D.T. (1974). Heterogeneity of lobster agglutinins I. Purification and physiochemical characterization. *Biochemistry* **13**, 821–827.
2. Hall, J.L. & Rowlands, D.T. (1974). Heterogeneity of lobster agglutinins II. Specificity of agglutinin–erythrocyte binding. *Biochemistry* **13**, 828–832.
3. Hartmann, A.L., Campbell, P.A. & Abel, C.A. (1978). An improved method for the isolation of lobster lectins. *Developmental and Comparative Immunology* **2**, 617–625.
4. Vanderwall, J., Campbell, P.A. & Abel, C.A. (1981). Isolation of a sialic acid-specific lobster lectin (LAg-1) by affinity chromatography on Sepharose-colominic acid beads. *Developmental and Comparative Immunology* **5**, 679–684.
5. Hartman, A.L., Despont, J.-P. & Able, C.A. (1976). Differential agglutination of murine lymphoid cells by an invertebrate lectin (LAg-1). In: *Leukocyte Membrane Determinants Regulating Immune Reactivity*, pp. 115–121. Academic Press, New York.

Interleukin-1β

Other names: IL-1β; lymphocyte activating factor (LAF); endogenous pyrogen (EP)
Mol. wt: 17 500
Specificity: (GalNAc β1,4[SO_4-3]Gal)
Sources: Mammalian cells

ISOLATION

The activity originally known as 'lymphocyte activating factor' was shown by cDNA cloning to arise from the products of two separate genes in mice and humans [1,2]. These IL-1α and IL-1β polypeptides have similar biological activities and bind to the same receptors, yet the primary sequence homology between the two is only about 20%.

Large amounts of recombinant IL-1 can be expressed in, and purified from, *Escherichia coli* [3].

STRUCTURE

IL-1β is a polypeptide of Mr 17 500 and has a pI of 7; it lacks a signal sequence and has an unremarkable primary structural composition. The amino acid sequences of human [1] and murine [2] IL-1β have been published in full; the former has 153 residues, the latter 159 residues, and there is about 70% sequence identity between the two.

BIOLOGICAL ACTIVITIES

IL-1β is a pleiotropic pro-inflammatory cytokine with multiple biological activities affecting the immune system [4–6]. These include induction of IL-2 synthesis, stimulation of B cell differentiation, the promotion of fever and the induction of the acute-phase response.

Lectin activity was first suggested by the observation that murine IL-1α could bind Tamm–Horsfall glycoprotein in a saccharide inhibitable manner [7], but evidence was later obtained that this interaction was artefactual [8]. Nevertheless, a decade later, recombinant human IL-1β was shown to bind to an immobilised oligosaccharide fraction derived from Tamm–Horsfall glycoprotein [9]. This interaction was highly specific for saccharides containing a sulphate residue and a β-GalNAc residue at the non-reducing end. Oligosaccharides containing high mannose-type structures, however, could not inhibit.

TISSUE AND/OR SUBCELLULAR DISTRIBUTION

Originally thought of as a monocyte/macrophage product, IL-1 is now known to be synthesised by a wide variety of immune and non-immune system cell types [6]. Within monocytes, IL-1β is found intracellularly, but (unlike IL-1α) little is detectable on the cell surface [10].

POSSIBLE FUNCTION(S)

IL-1 acts in concert with other cytokines to promote and amplify immune responses via receptors found on many cell types [4–6]. The specific role of its lectin activity is unknown.

COMMENTS

IL-1 is often used in combination with other cytokines as a growth or differentiation factor in culture; for example, for expanding haematopoietic progenitor cells [11].

Although IL-1 has potential therapeutic applications [12], specific antagonists of IL-1 are perhaps more promising as drugs. For example, an IL-1 receptor antagonist was found to be beneficial in an animal model of septic shock [13].

REFERENCES

1. March, C.J., Mosley, B., Larsen, A., Ceretti, D.P., Braedt, G., Price, V., Gillies, S., Henney, C.S., Kronheim, S.R., Grabstein, K., Conlon, P.J., Hopp, T.P. & Cosman, D. (1985). Cloning, sequence and expression of two distinct human interleukin-1 complementary DNAs. *Nature* **315**, 641–647.
2. Gray, P.W., Glaister, D., Chen, E., Goeddel, D.V. & Pennica, D. (1986). Two interleukin 1 genes in the mouse: cloning and expression of the cDNA for murine interleukin 1β. *Journal of Immunology* **137**, 3644–3648.
3. Kronheim, S.R., Cantrell, M.A., Deeley, M.C., March, C.J., Glackin, P.J., Anderson, D.M., Hemenway, T., Merriam, J.E., Cosman, D. & Hopp, T.P. (1986). Purification and characterization of human interleukin-1 expressed in *Escherichia coli*. *Biotechnology* **4**, 1078–1082.
4. Kampschmidt, R.F. (1984). The numerous postulated biological manifestations of interleukin-1. *Journal of Leukocyte Biology* **36**, 341–355.
5. Durum, S.K., Schmidt, J.A. & Oppenheim, J.J. (1985). Interleukin 1: an immunological perspective. *Annual Review of Immunology* **3**, 263–287.
6. Callard, R. & Gearing, A. (1994). *The Cytokine Facts Book*, pp. 31–38. Academic Press, London.
7. Muchmore, A.V. & Decker, J.M. (1987). Evidence that recombinant IL-1α exhibits lectin–lectin specificity and binds to homogenous uromodulin via N-linked oligosaccharides. *Journal of Immunology* **138**, 2541–2546.
8. Moonen, P., Gaffner, R. & Wingfield, P. (1988). Native cytokines do not bind to uromodulin (Tamm–Horsfall glycoprotein). *FEBS Letters* **226**, 314–318.
9. Tandai-Hiruma, M., Endo, T. & Kobata, A. (1999). Detection of novel carbohydrate-binding activity of interleukin-1. *Journal of Biological Chemistry* **274**, 4459–4466.
10. Conlon, P.J., Grabstein, K.H., Alpert, A., Prickett, K.S., Hopp, T.P. & Gillies, S. (1987). Localisation of human mononuclear cell interleukin 1. *Journal of Immunology* **139**, 98–102.
11. Koyayashi, M., Imamura, M., Gotohda, Y., Maeda, S., Iwasaki, H., Sakurada, K., Kasai, M., Hapel, A.J. & Miyazaki, T. (1991). Synergistic effects of interleukin-1β and interleukin-3 on the expression of human hematopoetic progenitor cells in liquid cultures. *Blood* **78**, 1947–1953.
12. Neta, R. & Oppenheim, J.J. (1988). Why should internists be interested in interleukin-1? *Annals of Internal Medicine* **109**, 1–3.
13. Ohlsson, K., Bjork, P., Bergenfeldt, M., Hageman, R. & Thompson, R.C. (1990). Interleukin-1 receptor antagonist reduces mortality from endotoxin shock. *Nature* **348**, 550–552.

Interleukin-2

Other names: IL-2; T cell growth factor
Mol. wt: 16 000
Specificity: Oligomannose; diacetylchitobiose
Source: Mammalian T lymphocytes

ISOLATION

Human IL-2 has been purified from culture media of various T cells and T cell lines via a variety of chromatographic combinations exemplified by the examples cited [1–3]. Murine IL-2 could be purified using a similar set of fractionation steps as those used for the human form [4].

Both the human [5] and murine [6,7] IL-2 genes have been cloned, characterised and expressed.

STRUCTURE

Human IL-2 is a variably glycosylated polypeptide of apparent Mr 15 000–20 000 determined by SDS-PAGE. The predicted Mr of its 133 amino acid residues is 15 400. The murine homologue possesses >60% primary sequence identity to the human form, but is slightly larger and significantly more acidic (pI 4.5 versus 8.2) [4–7]. Short regions of homology with mannan-binding lectin (qv) have been noted [8].

BIOLOGICAL ACTIVITIES

IL-2 has been intensively studied as a cytokine that not only stimulates the growth and differentiation of T lymphocytes as originally noticed, but also influences the development of other immune cells, including B lymphocytes, NK cells and monocytes. IL-2 is therefore central to the immune response but these activities are mediated by interactions with the IL-2 receptor complex and are unconnected with any lectin activity [4,9,10].

However, IL-2 has a second binding site enabling it to bind high-mannose structures in a calcium-independent manner [8,11]. The binding of immobilised recombinant human IL-2 to uromodulin, ovalbumin or yeast mannan could be prevented by high-mannose glycopeptides and also by some simpler sugars, especially diacetylchitobiose [8]. This lectin activity was confirmed by competitive binding studies with solid phase RNAse B glycans; *N*-glycans with five or six mannose residues were effective inhibitors [11].

TISSUE AND/OR SUBCELLULAR DISTRIBUTION

IL-2 is produced exclusively by T lymphocytes, particularly activated CD4-positive cells [12].

POSSIBLE FUNCTION(S)

IL-2 plays an essential role in orchestrating the acquired immune response, and its absence leads to a profound and fatal immune deficiency state [9,13].

It has been suggested that the lectin activity of IL-2 is responsible for the association of IL-2 receptor-γ with the T cell receptor complex leading to the phosphorylation of the IL-2 receptor [11].

COMMENTS

Recombinant IL-2 is routinely used as a growth factor for T cells and other leukocytes *in vitro*. It has also been extensively used therapeutically on an empirical basis to treat patients with various disorders, including AIDS, leprosy and (above all) cancers [10,14,15]. Such treatments have enjoyed rather limited success, and are limited by the toxicity of the cytokine *in vivo* [16].

It has been suggested that the lectin nature of IL-2 may be responsible for some of the immune deficiency observed in AIDS via interaction with the major envelope glycoprotein of the HIV-1 virus; several theoretical consequences of IL-2–gp120 interaction have been considered [17].

REFERENCES

1. Robb, R.J. (1982). Human T-cell growth factor: purification, biochemical characterization, and interaction with a cellular receptor. *Immunobiology* **161**, 21–50.
2. Mier, J.W. & Gallo, R.C. (1982). The purification and properties of human T cell growth factor. *Journal of Immunology* **128**, 1122–1127.
3. Korner, I.J., Dettmer, R., Kopp, J., Gaestel, M. & Malz, W. (1986). Simple preparative two-step purification of interleukin-2 from culture medium of lectin stimulated normal human lymphocytes. *Journal of Immunological Methods* **87**, 185–191.
4. Gillis, S. (1983). Interleukin 2: biology and biochemistry. *Journal of Clinical Immunology* **3**, 1–13.
5. Maeda, S., Nishino, N., Obaru, K., Mita, S., Nomiyama, H., Shimada, K., Fujimoto, K., Teranishi, T., Hirano, T. & Onoue, K. (1983). Cloning of interleukin 2 mRNAs from human tonsils. *Biochemical and Biophysical Research Communications* **115**, 1040–1047.
6. Kashima, N., Nishi-Takaoka, C., Fujita, T., Taki, S., Yamada, G., Hamuro, J. & Taniguchi, T. (1985). Unique structure of murine interleukin-2 as deduced from cloned cDNAs. *Nature* **313**, 402–406.
7. Yokota, T., Arai, N., Lee, F., Rennick, D., Mosmann, T. & Arai, K.-I. (1985). Use of a cDNA expression vector for isolation of mouse interleukin 2 cDNA clones: expression of T-cell growth factor activity after transfection of monkey cells. *Proceedings of the National Academy of Sciences USA* **82**, 68–72.

8. Sherblom, A.P., Sathyamoorthy, N., Decker, J.M. & Muchmore, A.V. (1989). IL-2, a lectin with specificity for high mannose glycopeptides. *Journal of Immunology* **143**, 939–944.
9. Smith, K.A. (1988). Interleukin-2: inception, impact, and implications. *Science* **240**, 1169–1176.
10. Kaplan, G., Cohn, Z.A. & Smith, K.A. (1992). Rational immunotherapy with interleukin 2. *Biotechnology* **10**, 157–162.
11. Zanetta, J.-P., Alonso, C. & Michalski, J.-C. (1996). Interleukin 2 is a lectin that associates its receptor with the T-cell receptor complex. *Biochemical Journal* **318**, 49–53.
12. Seder, R.A. & Paul, W.E. (1994). Acquisition of lymphokine-producing phenotype by $CD4^+$ T cells. *Annual Review of Immunology* **12**, 635–673.
13. Sugamura, K., Asao, H., Kondo, M., Nobuyuki, T., Ishii, N., Ohbo, K., Nakamura, M. & Takeshita, T. (1996). The interleukin-2 receptor γ chain: its role in the multiple cytokine receptor complexes and T cell development in XSCID. *Annual Review of Immunology* **14**, 179–205.
14. Riddell, S., Cheever, M.A. & Greenberg, P.D. (1988). Interleukin-2 in cancer immunotherapy. *ISI Atlas of Science: Immunology*, pp. 175–183.
15. Brenner, M.K. (1994). Haematological applications of interleukin-2 and other immunostimulatory cytokines. *Baillière's Clinical Haematology* **7**, 115–134.
16. Smith, K.A. (1993). Lowest dose interleukin-2 immunotherapy. *Blood* **81**, 1414–1423.
17. Zanetta, J.-P., Scior, T., Wantyghem, J., Wermuth, C., Aubery, M., Strecker, G. & Michalski, J.-C. (1996). Lectin activities of cytokines and growth factors: function and implications for pathology. *Histology and Histopathology* **11**, 1101–1108.

Kupffer cell receptor

Other name: Fucose-binding protein
Mol. wt: 88 000
Specificity: GalNAc, D-fucose
Source: Rat liver

ISOLATION

This lectin was first purified from a fraction of rat liver homogenate by procedures involving affinity chromatography with the neoglycoprotein, L-fucose-β-BSA, immobilised on agarose, employing strategies to remove asialoglycoprotein receptors and mannan-binding lectin [1]. A cDNA encoding the lectin was later isolated from a rat liver cDNA library [2].

STRUCTURE

The Kupffer cell receptor is a glycoprotein with apparent molecular weights determined by SDS-PAGE of 88 000 and 58 000, respectively, before and after deglycosylation with *N*-glycanase [1,2]. The predicted proline-scarce primary sequence contains 550 amino acid residues and has a calculated Mr of 61 104 [2]. The blocked N-terminal region of 42 amino acids is followed by a short (17 residues) hydrophobic transmembrane domain, a series of short repeating sequences, and finally a C-type lectin domain at the carboxyl-terminus. The lectin domain shows 40% identity with the corresponding domains of human and rat hepatic asialoglycoprotein receptors [2]. Analysis of the gene structure confirmed membership of the group II C-type lectins [3].

The native receptor may exist as a high molecular weight oligomer, since it migrated very slowly in SDS-PAGE under non-reducing conditions and did not migrate in PAGE without detergent under non-reducing conditions [1].

BIOLOGICAL ACTIVITIES

The Kupffer cell receptor binds to the neoglycoproteins BSA-fucose and BSA-galactose with high affinity, and to BSA-GlcNAc with much lower affinity, in a calcium- and pH-dependent manner. Of several divalent cations tested, only Sr^{2+} could partially replace Ca^{2+}; binding was optimal at alkaline pH, and negligible below pH 6. Among neoglycoprotein inhibitors, β-linked galactosides and fucosides were the best inhibitors. Among monosaccharide inhibitors, GalNAc and D-fucose where the most potent, but L-arabinose, D-galactose and D-mannose were more effective than L-fucose, and indeed a rather broad specificity was apparent. GlcNAc, however, was a very poor inhibitor [4].

TISSUE AND/OR SUBCELLULAR DISTRIBUTION

This lectin appears to be unique to Kupffer cells (liver macrophages), at least in rats. Within the Kupffer cells, some of the receptor is detectable on the cell surface [5].

POSSIBLE FUNCTION(S)

Although it is thought that this lectin should contribute to the uptake and endocytosis of abnormal glycoproteins containing terminal fucose residues, no direct evidence to support this assumption has been found [5,6]. Other hypothetical functions proposed include cell adhesion and signal transduction [6].

COMMENTS

The Kupffer cell receptor belongs to the asialoglycoprotein receptor family, which also includes the placental gp120 receptor (qv) and CD23 (qv). Unlike the other members of this family, the Kupffer cell receptor is confined to one cell type and possibly to one species. The possibility of murine and human homologues has been raised but not demonstrated [5].

REFERENCES

1. Lehrman, M.A. & Hill, R.L. (1986). The binding of fucose-containing glycoproteins by hepatic lectins. *Journal of Biological Chemistry* **261**, 7419–7425.
2. Hoyle, G.W. & Hill, R.L. (1988). Molecular cloning and sequencing of a cDNA for a carbohydrate-binding receptor unique to rat Kupffer cells. *Journal of Biological Chemistry* **263**, 7487–7492.
3. Hoyle, G.W. & Hill, R.L. (1991). Structure of the gene for a carbohydrate-binding receptor unique to rat Kupffer cells. *Journal of Biological Chemistry* **266**, 1850–1857.
4. Lehrman, M.A., Haltiwanger, R.S. & Hill, R.L. (1986). The binding of fucose-contaning glycoproteins by hepatic lectins. *Journal of Biological Chemistry* **261**, 7426–7432.
5. Haltiwanger, R.S., Lehrman, M.A., Eckhardt, A.E. & Hill, R.L. (1986). The distribution and localization of the fucose-binding lectin in rat tissues and the identification of a high affinity form of the mannose/*N*-acetylglucosamine-binding lectin in rat liver. *Journal of Biological Chemistry* **261**, 7433–7439.
6. Greene, J.D. & Hill, R.L. (1993). Isolation, detection, and ligand binding specificity of a lectin uniquely found in rat Kupffer cells. In: *Lectins and Glycobiology* (eds H.-J. Gabius & S. Gabius), pp. 55–64. Springer-Verlag, Berlin.

Lachesis muta venom lectin

Other name: Bushmaster lectin (BML)
Mol. wt: 32 000 (native); 16 000 (subunit)
Specificity: Lactose
Source: Venom of the bushmaster *Lachesis muta stenophyrs*

ISOLATION

BML was first isolated from an extract of venom by affinity binding to Sepharose 4B and subsequent elution with lactose [1]. Alternatively, it has been purified by a combination of gel filtration and ion-exchange chromatography [2].

STRUCTURE

BML is a homodimeric protein with a high content of glycine and serine residues. Each subunit has 135 amino acid residues and a calculated Mr of 16 223. This is slightly higher than the value of 14 000 obtained by SDS-PAGE under reducing conditions. The isoelectric point is around pH 4.5 [1,2].

The primary structure shows extensive homology with several C-type lectins including echinoidin (qv) and the rat asialoglycoprotein receptor (qv), with around 33% identity to both. The sequence is outstandingly similar to that of *Crotalus atrox* venom lectin (qv), with 92% identity [2].

BIOLOGICAL ACTIVITIES

BML readily agglutinates trypsin-treated, formalin-fixed erythrocytes of bovine and human origin, the latter ABO non-specifically. Haemagglutination was Ca^{2+}-dependent, and had a similar saccharide inhibition profile to that of thrombolectin (qv), being most readily inhibited by D-fucose, L-rhamnose and D-galactose among monosaccharides. Lactose and especially thiodigalactoside were even more potent inhibitors. In contrast to thrombolectin and other closely related lectins, haemagglutination by BML was not inhibited by 1 mM dithiothreitol [1].

BML induced the aggregation of washed human platelets; this activity could also be inhibited by lactose [3].

TISSUE AND/OR SUBCELLULAR DISTRIBUTION

BML has been described only in venom extracts.

POSSIBLE FUNCTION(S)

The function of this lectin is unknown.

COMMENTS

BML is clearly closely related to the thrombolectin (qv) group of lectins, but (unlike the others) is not affected by dithiothreitol, implying that each monomeric subunit has two or more binding sites. The venom lectin from Jameson's mamba, a member of the family Elapidae, is obviously even more distinct, since its haemagglutinating activity is calcium-independent and its predominant molecular form has an Mr of 26 000 [1].

REFERENCES

1. Ogilvie, M.L., Dockter, M.E., Wenz, L. & Gartner, T.K. (1986). Isolation and characterization of lactose-binding lectins from the venoms of the snakes *Lachesis muta* and *Dendroaspsis jamesonii*. *Journal of Biochemistry* **100**, 1425–1431.
2. Aragón-Ortiz, F., Mentele, R. & Auerswald, E.A. (1996). Amino acid sequence of a lectin-like protein from *Lachesis muta stenophyrs* venom. *Toxicon* **34**, 763–769.
3. Ogilvie, M.L., Wilson-Byl, J. & Gartner, T.K. (1989). Platelet aggregation is stimulated by lactose-inhibitable snake venom lectins. *Thrombosis and Haemostasis* **62**, 704–707.

Limax flavus lectin

Other names: Slug lectin; slug agglutinin; LFA
Mol. wt: 44 000 (native); 22 000 (subunit)
Specificity: Sialic acid
Source: The integument of the land slug *Limax flavus*

ISOLATION

A fraction obtained by ammonium sulphate precipitation of the soluble homogenate of eviscerated slugs was subjected to affinity chromatography on bovine submaxilliary mucin–Sepharose. An apparently homogeneous protein was eluted with *N*-acetylneuraminic acid [1].

Alternatively, mRNA was extracted and used to construct a cDNA library, which was then screened using a PCR product as a hybridisation probe. One of the cDNA clones obtained was used to construct a lectin expression vector which produced recombinant slug lectin in *Escherichia coli* [2].

STRUCTURE

LFA is a protein or glycoprotein with a high content of basic amino acids [1]. It is a homodimer (subunit Mr 22 000), with a tendency towards molecular aggregation at concentrations above 1 mg/ml. The isoelectric point is between 9 and 9.5 [1].

The primary amino acid sequence has been determined [2,3]. It has a region of homology with human tenascin-C, P35 (qv) and porcine ficolins; this homology resides in a fibrinogen-like domain [2].

BIOLOGICAL ACTIVITIES

LFA agglutinates human erythrocytes [1] and precipitates sialylated glycoconjugates [4]. The lectin activity can be inhibited by *N*-acetylneuraminic acid or *N*-glycolylneuraminic acid, but not by other monosaccharides [1]. The binding specificity has been studied in some detail and compared with other sialic acid-binding lectins [3,4]. LFA binds to *N*-acetylneuraminic acid in any linkage and in any position in a glycoconjugate; a preference for the α-ketosidic configuration at C2, an obligate *N*-acetamido group at C5, and tolerance for a substituted hydroxyl group at C4 were noted [3,4].

TISSUE AND/OR SUBCELLULAR DISTRIBUTION

LFA is not present to any appreciable extent in haemolymph or internal organs [1]. Within the body tissue, the lectin is found mainly on the surface of the epithelial cells of the skin and the mucous glands that penetrate into the connective tissue above the layer of muscle [2].

POSSIBLE FUNCTION(S)

The function of this lectin is not known.

COMMENTS

Although this 44 000 lectin is the only lectin to have been characterised from this species, agglutinating activity in crude extracts can be inhibited by GalNAc and GlcNAc, suggesting the presence of other, undiscovered lectins [1]. The large number of closely related but distinct cDNA clones obtained suggests that the LFA gene is a member of a multi-gene family [2].

REFERENCES

1. Miller, R.L., Collawn, J.F. & Fish, W.W. (1982). Purification and macromolecular properties of a sialic acid-specific lectin from the slug *Limax flavus*. *Journal of Biological Chemistry* **257**, 7574–7580.
2. Kurachi, S., Song, Z., Takagaki, M., Yang, Q., Winter, H.C., Kurachi, K. & Goldstein, I.J. (1998). Sialic acid binding lectin from the slug, *Limax flavus*. Cloning, expressing of the polypeptide, and tissue localization. *European Journal of Biochemistry* **254**, 217–222.
3. Knibbs, R.N., Osborne, S.E., Glick, G.D. & Goldstein, I.J. (1993). Binding determinants of the sialic acid-specific lectin from the slug *Limax flavus*. *Journal of Biological Chemistry* **268**, 18524–18531.
4. Knibbs, R.N., Goldstein, I.J., Ratcliffe, R.M. & Shibuya, N. (1991). Characterization of the carbohydrate-binding specificity of the leukoagglutinating lectin from *Maackia amurensis*. *Journal of Biological Chemistry* **266**, 83–88.

Limulin

Other names: *Limulus polyphemus* agglutinin; limulin-III
Mol. wt: Approx. 300 000 (native); 25 000 (subunit)
Specificity: Sialic acid
Source: Haemolymph of the horseshoe crab *Limulus polyphemus*

ISOLATION

All the early attempts at isolation, with hindsight, clearly failed to obtain a purified protein. The most successful of these early attempts involved affinity chromatography on bovine submaxilliary mucin–Sepharose and elution with a calcium-free buffer [1]. However, separation from the major pentraxin (*Limulus* C-reactive protein) requires two affinity stages: (i) on phosphorylethanolamine–agarose (to which both pentraxins bind) and (ii) on fetuin–Sepharose (to which limulin but not CRP binds) [2].

STRUCTURE

Limulin is apparently a glycoprotein containing approximately 3.5% carbohydrate [3,4]. Estimates of the apparent molecular weight range from 300 000 [2] to 500 000 [5]. There is better agreement that the sedimentation coefficient is around 13.6 S [1,3,4,6]. The subunit molecular weight estimated by SDS-PAGE has also been variously reported over a range from <18 000 to 33 000 [1–6]. The isoelectric point is approximately 4.9 [3,4].

Certainly, limulin consists of a single polypeptide subunit (unlike *Limulus* C-reactive protein, which is a heterodimer of two closely related peptides), and has an identical amino terminal sequence to *Limulus* C-reactive protein (which is highly homologous) [2]. Both *Limulus* pentraxins have approximately 25% sequence identity to mammalian C reactive proteins [7]. It is possible that both are arranged as double-stacked hexameric assemblies.

BIOLOGICAL ACTIVITIES

Limulin can agglutinate erythrocytes from many species [8,9]. Human erythrocytes are agglutinated (ABO non-specifically) at a ten-fold lower concentration than ovine erythrocytes, while equine red blood cells are agglutinated at an order of magnitude lower than human [8,9]. This reaction is dependent on terminal sialic acids [10], although sialylated glycoproteins are much more effective inhibitors than free sialic acids [3]. Ca^{2+} is necessary for, and may enhance, agglutination [3,4,6].

Limulin is mitogenic for human and peripheral blood mononuclear cells [9], and haemolytic for ovine erythrocytes [2].

Limulin can also bind to phosphorylethanolamine [2].

TISSUE AND/OR SUBCELLULAR DISTRIBUTION

Limulin has always been obtained from the haemolymph.

POSSIBLE FUNCTION(S)

Limulin probably functions in defence. Limulin possesses a cytotoxic activity which manifests itself *in vitro* by lysing ovine erythrocytes. This haemolytic activity is dependent on calcium and on sialic acid recognition, and is not shared by other known sialic acid-binding lectins or the other pentraxins in *L. polyphemus* haemolymph [2]. Limulin is the principal cytotoxic protein in *L. polyphemus* plasma and presumably acts by recognising foreign cells then lysing them.

COMMENTS

Limulin is of interest as the best-studied invertebrate analogue of the pentraxins. Until the recently achieved separation from the major *Limulus* pentraxin, limulin had not been isolated in genuinely pure form, and therefore most of its characteristics have been determined in partially purified preparations.

Although limulin will agglutinate leukocytes, a much higher concentration is required than for agglutinating erythrocytes [9]. This differential was exploited to separate red and white blood cells from multi-transfused cancer patients so that the leukocytes could be used as target cells for screening for leuko-agglutinins [8].

Limulin has been used to monitor alterations in cell surface sialic acid on lymphoma cells in culture [10]. Limulin, labelled with fluorescein, biotin, rhodamine or horseradish peroxidase, has been widely used in histochemistry, particularly as a probe to study changes in sialylated glycoconjugates in relation to normal differentiation and malignant transformation [11–14].

REFERENCES

1. Roche, A.C., Schauer, R. & Monsigny, M. (1975). Protein–sugar interactions. Purification by affinity chromatography of limulin, an *N*-acyl-neuramnidyl-binding protein. *FEBS Letters* **57**, 245–249.
2. Armstrong, P.B., Swarnaker, S., Srimal, S., Misquith, S., Hahn, E.A., Aimes, R.T. & Quigley, J.P. (1996). A cytolytic function for a sialic acid-binding lectin that is a member of the pentraxin family of proteins. *Journal of Biological Chemistry* **271**, 14717–14721.

3. Roche, A.C. & Monsigny, M. (1974). Purification and properties of limulin: a lectin (agglutinin) from hemolymph of *Limulus polyphemus*. *Biochimica et Biophysica Acta* **371**, 242–254.
4. Nowak, T.P. & Barondes, S.H. (1975). Agglutinin from *Limulus polyphemus*. Purification with formalinized horse erythrocytes as the affinity adsorbent. *Biochimica et Biophysica Acta* **393**, 115–123.
5. Robey, F.A. & Lui, T.-Y. (1981). Limulin: a C-reactive protein from *Limulus polyphemus*. *Journal of Biological Chemistry* **256**, 969–975.
6. Marchalonis, J.J. & Edelman, G.M. (1968). Isolation and characterization of a hemagglutinin from *Limulus polyphemus*. *Journal of Molecular Biology* **32**, 453–465.
7. Nguyen, N.Y., Suzuki, A., Boykins, R.A. & Liu, T.-Y. (1986). The amino acid sequence of *Limulus* C-reactive protein. *Journal of Biological Chemistry* **261**, 10456–10465.
8. Cohen, E., Rowe, A.N. & Wissler, F.C. (1965). Heteroagglutinins of the horseshoe crab, *Limulus polyphemus*. *Life Sciences* **4**, 2009–2016.
9. Roche, A.-C., Perrodon, Y., Halpern, B. & Monsigny, M. (1977). Limulin (*Limulus polyphemus* lectin): mitogenic effect of human peripheral lymphocytes. *European Journal of Immunology* **7**, 263–267.
10. Pardoe, G.I., Uhlenbruck, G. & Bird, G.W.G. (1970) Studies on some heterophile receptors of the Burkett EB2 lymphoma cell. *Immunology* **18**, 73–83.
11. Raedler, A. & Raedler, E. (1985). The use of lectins to study normal differentiation and malignant transformation. *Journal of Cancer Research and Clinical Oncology* **109**, 245–251.
12. Alroy, J., Ucci, A.A. & Pereira, M.E.A. (1984). Lectins: histochemical probes for specific carbohydrate residues. In: *Advances in Immunohistochemistry* (ed. R.A. De Lellis), pp. 67–88. Castle House Publications, UK.
13. Muresan, V., Iwanij, V., Smith, Z.D.J. & Jamiesan, J.D. (1982). Purification and use of limulin: a sialic acid specific lectin. *Journal of Histochemistry and Cytochemistry* **30**, 938–946.
14. Muresan, V., Sarras, M.P. & Jamieson, J.D. (1982). Distribution of sialoglycoconjugates on acinar cells of the mammalian pancreas. *Journal of Histochemistry and Cytochemistry* **30**, 947–955.

Liocarcinus depurator lectin

Other name: LD lectin
Mol. wt: 700 000 (native); 38 000 (subunit)
Specificity: *N*-acetylneuraminic acid
Source: Haemolymph of the marine crab *Liocarcinus depurator*

ISOLATION

The lectin was affinity purified from the soluble fraction of haemolymph serum in a one-step procedure by binding to bovine submaxilliary mucin–Sepharose and eluting with GlcNAc [1].

STRUCTURE

This lectin is a glycoprotein seemingly composed of aggregates of a single subunit of apparent molecular weight 38 000 determined by SDS-PAGE. Under non-reducing conditions, the lectin migrated slowly in SDS-PAGE, corresponding to a molecular size of approximately 700 000 [1]. The pI was about 6.5, but there was evidence of microheterogeneity [1].

BIOLOGICAL ACTIVITIES

LD lectin strongly agglutinated murine erythrocytes, and also had some weaker activity towards equine, rat, lapine, ovine and human (A,B or O) erythrocytes in decreasing order. Haemagglutination was Ca^{2+}-dependent and readily inhibited by bovine submaxilliary mucin, but not by its desialated derivative [1]. Among free sugars, GalNAc and GlcNAc were weakly inhibitory, but sialic acids were much more effective. *N*-acetylneuraminic acids, but not *N*-glycolylneuraminic acids, were potent inhibitors, and *O*-acetylation of hydroxyl groups was associated with an increase in potency.

LD lectin agglutinated human, but not murine, mononuclear cells, showing a preference for B cells over T cells. Similarly, LD lectin was weakly mitogenic for a human monocyte-depleted, B-cell-enriched lymphocyte preparation, but was non-mitogenic towards murine lymphocytes [1].

LD lectin agglutinated several Gram-negative bacteria (including *Escherichia coli*); *E. coli* LPS could inhibit LD lectin-mediated agglutination of murine red blood cells [1].

TISSUE AND/OR SUBCELLULAR DISTRIBUTION

LD lectin is synthesised in the hepatopancreas, but is also found in the large granular haemocytes. Within the haemocytes, the lectin occurs mainly within the granules [1].

POSSIBLE FUNCTION(S)

A defensive function against microbial pathogens has been suggested [1].

COMMENTS

This lectin is one of several sialic acid-specific lectins from invertebrate sources, and, while not unique in recognising *O*-acetylated neuraminic acid derivatives, has a distinct fine specificity evidenced by its own distinctive reaction pattern with erythrocytes of different species (see Table 2.5).

REFERENCE

1. Fragkiadakis, G.A. & Stratakis, E.K. (1997). The lectin from the crustacean *Liocarcinus depurator* recognises *O*-acetylsialic acids. *Comparative Biochemistry and Physiology* **117B**, 545–552.

Lophopsetta maculata agglutinin

Other name: Windowpane flounder agglutinin
Mol. wt: 68 000 (native); 16 000 (subunit)
Specificity: *N*-acetylneuraminic acid
Source: The body mucus of the flounder *Lophopsetta maculata*

ISOLATION

The major lectin was isolated from an extract of fish mucus by a series of steps including column chromatography on DEAE–Sephacel, hydroxyapatite and Sephadex G-200 [1].

STRUCTURE

The flounder agglutinin is a protein with an apparent Mr of 68 000 estimated by gel filtration. Its amino acid composition was determined, and glutamic acid, aspartic acid and glycine were found to be the most abundant amino acids. A single subunit of Mr 16 000 was detected by SDS-PAGE, but two sharp bands with a pI around 4.7 could be resolved by isoelectric focusing [1].

BIOLOGICAL ACTIVITIES

The purified major lectin agglutinated lapine erythrocytes at high titre and murine erythrocytes somewhat less readily. Low-titre agglutination of equine and human group B cells was also noted. (Sheep red blood cells were not agglutinated.) The lectin also agglutinated a marine yeast (*Metschnikowia reukaufii*) and a murine leukaemia cell line [1].

Haemagglutinating activity was inhibited by *N*-acetylneuraminic acid at a moderate concentration, but not by other monosaccharides.

TISSUE AND/OR SUBCELLULAR DISTRIBUTION

Agglutinins are found only in the body mucus of *L. maculata*, not in the ovaries or in serum [1].

POSSIBLE FUNCTION(S)

Although the lectin agglutinated the marine yeast, *M. reukaufii*, it had no growth inhibitory activity towards that species, and it had little or no agglutinating activity towards several marine bacteria [1]. There is therefore no circumstantial evidence to support the defence function which has been suggested as a possible role for skin mucus lectins generally [2].

COMMENTS

Several lectins have been isolated from roe and from serum of piscine species; this lectin represents the rarer examples of fish mucus lectins, which also includes the galectin congerin [2].

REFERENCES

1. Kamiya, H. & Shimizu, Y. (1980). Marine biopolymers with cell specificity. II. Purification and characterization of agglutinins from mucus of windowpane flounder, *Lophopsetta maculata*. *Biochimica et Biophysica Acta* **622**, 171–178.
2. Muramoto, K. & Kamiya, H. (1992). The amino-acid sequence of a lectin from conger eel, *Conger myriaster*, skin mucus. *Biochimica et Biophysica Acta* **1116**, 129–136.

L-selectin

Other names: CD62L; MEL-14 antigen; lymphocyte homing receptor; LECAM-I; Leu-8; LAM-1
Mol. wt: 74 000–100 000
Specificity: Sialyl-Lewis^X; mannose-6-phosphate
Sources: Mammalian leukocytes

ISOLATION

Murine L-selectin was first identified with a rat monoclonal antibody which prevented organ-specific homing of lymphocytes and precipitated a surface antigen (MEL-14 antigen) with an apparent Mr of around 80 000 on SDS-PAGE from mesenteric node lymphocytes [1]. The corresponding cDNA was subsequently cloned in two independent laboratories and reported in the same month [2,3]. The human cDNA for L-selectin has also been isolated [4,5].

STRUCTURE

L-selectin is a glycoprotein with a high (approximately 30%) carbohydrate content [2–5]. The murine cDNA encodes a mature protein of 334 amino acids with a calculated molecular weight of 37 000 [2,3]. The N-terminal region consists of a C-type lectin domain, while the C-terminal region has a transmembrane domain of about 20 residues and a slightly shorter cytoplasmic tail region. In between, there is an epidermal growth factor domain next to the lectin domain and two complement regulatory protein-like domains [2,3]. The human L-selectin has a similar overall structure, and 77% sequence identity to its murine homologue [4,5]. Both murine and human L-selectins possess approximately 40% sequence identity to E- and P-selectins from the same species [2–5].

BIOLOGICAL ACTIVITIES

L-selectin is a calcium-dependent adhesion molecule. L-selectin mediates the binding of circulating lymphocytes to the endothelium of peripheral lymph nodes *in vitro* and *in vivo* [1,6,7]. L-selectin may also have a less obvious role in the corresponding migration into Peyer's patches [8]. L-selectin mediates the adherence of granulocytes to vascular endothelium in inflammatory conditions *in vitro* and *in vivo* [7,9], and can mediate adhesion of monocytes to endothelial cells *in vitro* [10]. L-selectin has also been implicated in leukocyte–leukocyte adhesive interactions, including neutrophil–neutrophil [11] and lymphocyte–dendritic cell [12] interactions. Triggering lymphocytes via L-selectin can activate the RAS pathway via the tyrosine kinase p56 [13], leading to an altered arrangement of the cytoskeleton. L-selectin is therefore both an adhesion and a signalling molecule; surprisingly, the intracytoplasmic tail is essential for leukocyte rolling, which also depends on functional expression of p56 [14].

Endogenous ligands for L-selectin on murine high endothelial venules include GlyCAM-1 [15] and CD34 [16]; other glycoproteins have been found to bind to L-selectin, including the P-selectin ligand, PSGL-1 [17], but the physiological relevance of any individual ligand is unclear.

Interactions mediated by L-selectin can be inhibited by a variety of saccharide structures, including mannose-6-phosphate, fucoidan, and sialylated or sulphated fucose-containing tetrasaccharides [18,19]. Sialyl-Lewis^X (Figure 3.4) or a closely related structure is thought to be a minimal component for natural ligands of L- and other selectins, and synthetic analogues, aryl C-glycosides, were found to be extremely potent inhibitors [20]. Certain heparin-derived sulphated tetrasaccharides are unusual in inhibiting the function of L- (and P-) selectins, but not E-selectin [21].

L-selectin can also bind phospholipids in a Ca^{2+}-independent manner, and this can be inhibited by fucoidan or mannose-6-phosphate, but not by sialyl-Lewis-^{x} [22].

L-selectin can also bind the lipid A domain of LPS from Gram-negative bacteria [23], and L-selectin-deficient mice have an exceptional resistance to the deleterious effects of LPS [24].

TISSUE AND/OR SUBCELLULAR DISTRIBUTION

L-selectin is expressed on the surface of lymphoid cells, neutrophils, some NK cells, eosinophils and monocytes [1,25]. It is shed into the circulation in a soluble form on activation of lymphocytes [26]. Within lymph nodes, L-selectin is present in the mantle zone and paracortex, but is virtually absent from germinal centres [1,27]. On the cell surface, L-selectin is selectively localised to microvilli and ruffles [28].

POSSIBLE FUNCTION(S)

It is well established that L-selectin is an adhesion molecule contributing to the initial contact between leukocytes and vascular endothelium [29]. (Subsequent, firmer attachment appears to be mainly mediated by integrins.) L-selectin has also been implicated in post-adhesion signalling events [13,14], including events leading to a novel adhesion mechanism involving galectin-3 [12].

Studies on knock-out mice lacking L-selectin, either as the only selectin defect or in combination with other selectin defects, indicate an absence of leukocyte rolling after trauma, and reduced rolling after TNF-α-induced inflammation [30]. Nevertheless, the presence of any of the three selectins is sufficient to support some neutrophil recruitment; it seems that L-selectin has a distinct but overlapping function [30].

COMMENTS

The three members of the selectin family, L-selectin, E-selectin and P-selectin, exhibit cooperativity or synergy in fulfilling some functions; for example, the rolling of leukocytes [29].

Although L-selectin-rich naive T cells shed the lectin on activation [26], L-selectin levels are maintained on the murine splenic Th-1 subset, a phenomenon dependent on IL-12 [31]. In contrast, L-selectin appears to be a marker for Th-2 human memory cells [32]. Particularly high levels of L-selectin were found on the $CD56^{bright}$ subset of NK cells, while $CD56^{dim}$ cells expressed little or no L-selectin [3].

Soluble L-selectin is present in the circulation of normal subjects at relatively high concentrations (up to $1.7\,\mu g/ml$) and is elevated in certain diseases, including chronic lymphocytic leukaemia [34] and acute myeloid leukaemia [35]; lower than average concentrations were associated with systemic sclerosis and vasculitis [36]. An increased level in acute myeloid leukaemia has prognostic value and has been linked to extramedullary disease, lower chance of complete remission and shorter survival [35]. Elevated L-selectin levels in cerebrospinal fluid have been reported in children with meningoencephalitis [37] and in leukaemia patients with meningeal involvement [38]. Low surface expression of L-selectin is found in hairy cell leukaemia [39] and tumour infiltrating T cells [40].

In animal models of complement-mediated experimental lung injury [41] and ischaemia-induced reperfusion injury [42,43], blocking of L-selectin was found to reduce neutrophil rolling and/or consequent tissue damage.

The gene for L-selectin has been localised to chromosome 1 in both mice [44] and humans [4,44].

REFERENCES

1. Gallatin, W.M., Weissman, I.L. & Butcher, E.C. (1983). Cell-surface molecule involved in organ-specific homing of lymphocytes. *Nature* **304**, 30–34.
2. Siegelman, M.H., Van de De Rijn, M. & Weissman, I.L. (1989). Mouse lymph node homing receptor cDNA clone encodes a glycoprotein revealing tandem interaction domains. *Science* **243**, 1165–1172.
3. Lasky, L.A., Singer, M.S., Yednock, T.A., Dowbenko, D., Fennie, C., Rodriguez, H., Nguyen, T., Stachel, S. & Rosen, S.D. (1989). Cloning of a lymphocyte homing receptor reveals a lectin domain. *Cell* **56**, 1045–1055.
4. Tedder, T.F., Isaacs, C.M., Ernst, T.J., Demetri, G.D., Adler, D.A. & Disteche, C.M. (1989). Isolation and chromosomal localization of cDNAs encoding a novel human lymphocyte cell surface molecule, LAM-1. *Journal of Experimental Medicine* **170**, 123–133.
5. Camerini, D., James, S.P., Stomenkovic, I. & Seed, B. (1989). Leu-8/TQ1 is the human equivalent of the Mel-14 lymph node homing receptor. *Nature* **342**, 78–82.
6. Bradley, L.M., Watson, S.R. & Swain, S.L. (1994). Entry of naïve CD4 T cells into peripheral lymph nodes requires L-selectin. *Journal of Experimental Medicine* **180**, 2401–2406.

7. Arbones, M.L., Ord, D.C., Ley, K., Ratech, H., Maynard-Curry, C., Otten, G., Capon, D.J. & Tedder, T.F. (1994). Lymphocyte homing and leukocyte rolling and migration are impaired in L-selectin (CD62L) deficient mice. *Immunity* **1**, 247.
8. Steeber, D.A., Tang, M.L.K., Zhang, X.-Q., Müller, W., Wagner, N. & Tedder, T.F. (1998). Efficient lymphocyte migration across high endothelial venules of mouse Peyer's patches requires overlapping expression of L-selectin and β_7 integrin. *Journal of Immunology* **161**, 6638–6647.
9. Ley, K., Tedder, T.F. & Kansas, G.S. (1993). L-selectin can mediate rolling independent of E- and P-selectin in mesenteric venules in vivo. *Blood* **82**, 1632–1638.
10. Spertini, O., Luscinskas, F.W., Gimbrone, M.A. & Tedder, T.F. (1992). Monocyte attachment to activated vascular endothelium in vitro is mediated by leukocyte adhesion molecule-1 (L-selectin) under non-static conditions. *Journal of Experimental Medicine* **175**, 1789–1792.
11. Rochon, Y.P., Simon, S.I., Lynam, E.B. & Sklar, L.A. (1994). A role for lectin interactions during human neutrophil aggregation. *Journal of Immunology* **152**, 1385–1393.
12. Swarte, V.V.R., Mebius, R.E., Joziasse, D.H., Van Den Eijnden, D.H. & Kraal, G. (1998). Lymphocyte triggering via L-selectin leads to enhanced galectin-3-mediated binding to dendritic cells. *European Journal of Immunology* **28**, 2864–2871.
13. Brenner, B., Gulbins, E., Schlottman, K., Koppenhoeffer, U., Busch, G.L., Walzog, B., Steinhausen, M., Coggeshall, K.M., Linderkamp, O. & Lang, F. (1996). L-selectin activates the RAS pathway via the tyrosine kinase p56[eck]. *Proceedings of the National Academy of Sciences USA* **93**, 15376–15381.
14. Brenner, B., Gulbins, E., Koppenhoefer, U., Lang, F. & Linderkamp, O. (1997). L-selectin-mediated lymphocyte rolling of jurkat T lymphocytes depends on functional expression of the tyrosine kinase p561ck. *Cellular Physiology and Biochemistry* **7**, 107–118.
15. Lasky, L.A., Singer, M.S., Dowbenko, D., Imai, Y., Henzel, W.J., Grimley, C., Fernie, C., Gillett, N., Watson, S.R. & Rosen, S.D. (1992). An endothelial ligand for L-selectin is a novel mucin-like molecule. *Cell* **69**, 927–938.
16. Baumhueter, S., Singer, M.S., Henzel, W., Hemmerich, S., Renz, M., Rosen, S.D. & Lasky, L.A. (1993). Binding of L-selectin to the vascular sialomucin CD34. *Science* **262**, 436–438.
17. Tu, L.L., Chen, A.J., Delahunty, M.D., Moore, K.L., Watson, S.R., McEver, R.P. & Tedder, T.F. (1996). L-selectin binds to P-selectin glycoprotein ligand-1 on leukocytes – interactions between the lectin, epidermal growth factor, and consensus repeat domains of the selectins determine ligand binding specificity. *Journal of Immunology* **157**, 3995–4004.
18. Stoolman, L.M., Tenforde, T.S. & Rosen, S.D. (1984). Phosphomannosyl receptors may participate in the adhesive interaction between lymphocytes and high endothelial venules. *Journal of Cell Biology* **99**, 1535–1540.
19. Mebius, R.E. & Watson, S.R. (1993). L- and E-selectin can recognise the same naturally occurring ligands on high endothelial venules. *Journal of Immunology* **151**, 3252–3260.
20. Kuribayashi, T., Ohkawa, N. & Satoh, S. (1998). Aryl C-glycosides: physiologically stable glycomimetics of sialyl Lewis X. *Bioorganic and Medicinal Chemistry Letters* **8**, 3307–3310.
21. Nelson, R.M., Cecconi, O., Roberts, W.G., Aruffo, A., Linhardt, R.J. & Bevilacqua, M.P. (1993). Heparin oligosaccharides bind L- and P-selectin and inhibit acute inflammation. *Blood* **82**, 3253–3258.
22. Malhotra, R., Taylor, N.R. & Bird, M.I. (1996). Anionic phospholipids bind to L-selectin (but not E-selectin) at a site distinct from the carbohydrate-binding site. *Biochemical Journal* **314**, 219–303.
23. Malhotra, R., Priest, R. & Bird, M.I. (1996). Role for L-selectin in lipopolysaccharide-induced activation of neutrophils. *Biochemical Journal* **320**, 589–593.
24. Tedder, T.F., Steeber, D.A. & Pizeueta, P. (1995). L-selectin deficient mice have impaired leukocyte recruitment into inflammatory sites. *Journal of Experimental Medicine* **181**, 2259–2264.

25. Lewinsohn, D.M., Bargatze, R.F. & Butcher, E.C. (1987). Leukocyte–endothelial cell recognition: evidence of a common molecular mechanism shared by neutrophils, lymphocytes, and other leukocytes. *Journal of Immunology* **138**, 4313–4321.
26. Jung, T.M. & Dailey, M.O. (1990). Rapid modulation of homing receptors ($gp90^{mel-14}$) induced by activators of protein kinase C. Receptor shedding due to accelerated proteolytic cleavage at the cell surface. *Journal of Immunology* **144**, 3130–3136.
27. Reichert, R.A., Gallatin, M.W., Weissman, I.L. & Butcher, E.C. (1983). Germinal centre B cells lack homing receptors necessary for normal lymphocyte recirculation. *Journal of Experimental Medicine* **157**, 813–827.
28. Hasslen, S.R., Von Andrian, U.H., Butcher, E.C., Nelson, R.D. & Erlandsen, S.L. (1995). Spatial distribution of L-selectin (CD62L) on human lymphocytes and transfected murine L1-2 cells. *Histochemical Journal* **27**, 547–554.
29. Kansas, G.S. (1996). Selectins and their ligands: current concepts and controversies. *Blood* **88**, 3259–3287.
30. Jung, U. & Ley, K. (1999). Mice lacking two or all three selectins demonstrate overlapping and distinct functions for each selectin. *Journal of Immunology* **162**, 6755–6762.
31. Van Wely, C.A., Beverley, P.C.L., Brett, S.J., Britten, C.J. & Tite, J.P. (1999). Expression of L-selectin on Th1 cells is regulated by IL-12. *Journal of Immunology* **163**, 1214–1221.
32. Kanegare, H., Kasahara, Y., Niida, Y., Yachie, A., Sugii, S., Takatsu, K., Taniguichi, N. & Miyawaki, T. (1996). Expression of L-selectin (CD62L) discriminates Th1 and Th2-like cytokine-producing memory $CD4^+$ T cells. *Immunology* **87,** 186–190.
33. Frey, M., Packianathan, N.B., Fehniger, T.A., Ross, M.E., Wang, W.-C., Stewart, C.C., Caligiuri, M.A. & Evans, S.S. (1998). Differential expression and function of L-selectin on $CD56^{bright}$ and $CD56^{dim}$ natural killer cell subsets. *Journal of Immunology* **161**, 400–408.
34. Chen, J.R., Gu, B.J., Dao, L.-P., Bradley, C.J., Mulligan, S.P. & Wiley, J.S. (1999). Transendothelial migration of lymphocytes in chronic lymphocytic leukaemia is impaired and involves down-regulation of both L-selectin and CD23. *British Journal of Haematology* **105**, 181–189.
35. Extermann, M., Bacchi, M., Monai, N., Fopp, M., Fey, M., Tichelli, A., Schapira, M. & Spertini, O. (1998). Relationship between cleaved L-selectin levels and the outcome of acute myeloid leukaemia. *Blood* **92**, 3115–3122.
36. Blann, A.D., Sanders, P.A., Herrick, A. & Jayson, M.I.V. (1996). Soluble L-selectin in the connective tissue diseases. *British Journal of Haematology* **95**, 192–194.
37. Buhrer, C., Herold, R., Stibenz, D., Henze, G. & Obladen, M. (1996). Cerebrospinal fluid soluble L-selectin (sCD62L) in meningoencephalitis. *Archives of Disease in Childhood* **74**, 288–292.
38. Stucki, A., Cordey, A.-S., Monai, N., de Flaugergues, J.-C., Schapira, M. & Spertini, O. (1995). Cleaved L-selectin concentrations in meningeal leukaemia. *Lancet* **345**, 286–289.
39. Pettitt, A.R., Zezel, M. & Cawley, J.C. (1999). Hairy-cell leukaemia: biology and management. *British Journal of Haematology* **106**, 2–8.
40. Kjaergaard, J. & Shu, S. (1998). Tumor infiltration by adoptively transferred T cells is independent of immunologic specificity but requires down-regulation of L-selectin expression. *Journal of Immunology* **163**, 751–759.
41. Mulligan, M.S., Watson, S.R., Fennie, C. & Ward, P.A. (1993). Protective effects of selectin chimeras in neutrophil-mediated lung injury. *Journal of Immunology* **151**, 6410–6417.
42. Mihelcic, D., Schleiffenbaum, B., Tedder, T.F., Sharar, S.R., Harlan, J.M. & Winn, R.K. (1994). Inhibition of leukocyte L-selectin function with a mononuclear antibody attenuates reperfusion injury to the rabbit ear. *Blood* **84**, 2322–2328.
43. Kubes, P., Jutila, M. & Payne, D. (1995). Therapeutic potential of inhibiting leukocyte rolling in ischemia/reperfusion. *Journal of Clinical Investigation* **95**, 2510–2519.
44. Watson, M.L., Kingsmore, S.F., Johnstone, G.I., Siegelman, M.H., Le Beau, M.M., Lemons, R.S., Bora, N.S., Howard, T.A., Weissman, I.L. & McEver, P.R. (1990). Genomic organization of the selectin family of leukocyte adhesion molecules on human and mouse chromosome 1. *Journal of Experimental Medicine* **172,** 263–272.

Ly-49 family

Other name: A1 antigen
Mol. wt: 85 000 (native); 43 000 (subunit)
Specificity: Sulphated glycans
Source: Murine NK cells

ISOLATION

A member of the Ly-49 family (Ly-49A or -A1) was first identified by a monoclonal antibody to murine T cell lymphoma cells [1]. A cDNA clone encoding this antigen was isolated by immunoselection of a cDNA library in the eukaryotic expression vector CDM8, and COS cells transfected with this cDNA clone expressed the antigen [2]. The subsequent isolation of a series of homologous cDNAs from an IL-2-activated NK cell library indicated the existence of a substantial multi-gene family [3].

STRUCTURE

The first described member of the family (Ly-49A) is a glycosylated homodimer of 262 amino acid residues with a calculated Mr of 30 648 [2]. Its structure is that of a type-II membrane receptor, with an N-terminal cytoplasmic tail, a transmembrane domain, and a relatively large C-terminal extracellular region with homology to the C-type lectins (particularly the asialoglycoprotein receptors and CD23) [2]. The other eight members of the family are similar in structure, with 52–79% primary sequence identity to Ly-49A [3]. All appear to be homodimers [4]. The apparent Mr of Ly-49A determined by SDS-PAGE was 43 000 and 85 000 under reducing and non-reducing conditions, respectively [1].

BIOLOGICAL ACTIVITIES

Ly-49A recognises fucoidan, but not mannan [5]. Ly-49C recognises the sulphated glycans, fucoidan, λ-carrageenan and dextran sulphate, but not chondroitin sulphate, heparin sulphate or dextran [6]. There was no obvious requirement for exogenous Ca^{2+}, but the nature of any divalent cation requirement was not rigorously examined [6].

Ly-49A binds to MHC class 1 molecules, and recognises a carbohydrate-independent epitope on murine H-2D^d [5]. Ly-49A can also bind H-2D^k, while Ly-49C has a broader MHC-specificity [4]. LY-49A specifically interacts with the $\alpha 1/\alpha 2$ domains of H-2D^d [4,5].

TISSUE AND/OR SUBCELLULAR DISTRIBUTION

LY-49 family members are expressed on the cell surface of NK cell subsets and on some T lymphocytes [4].

POSSIBLE FUNCTION(S)

LY-49A and other family members are thought to function as inhibitory receptors on NK cells, distinguishing between normal (self) and aberrant ('missing self') cells on the basis of recognition of self MHC class 1. In the absence or dysregulation of MHC class 1, activation of NK cells leading to cytotoxicity is permitted [4].

COMMENTS

The Ly-49 family is encoded within the NK gene complex on mouse chromosome 6, and belongs to the group V C-type lectins, which includes NKR-P1 (qv) and CD69 (qv). There appears to be no functioning gene homologue of Ly-49 in humans (only a pseudogene), but the CD94/NKG2 family appears to be the corresponding group in humans. The monomorphic CD94 chain combines with one of several NKG2 gene products to form a dimer that recognises HLA class 1 and inhibits NK cell activity, but no lectin activity has been reported for any CD94/NKG2 molecule [4].

Ly-49C shares with the selectins an affinity for sulphated saccharides. However, Ly-49C binding to target cells is unaffected or enhanced by neuraminidase treatment of the cells, which is clearly distinct from the sialic acid-dependent recognition of selectins [6].

REFERENCES

1. Nagasawa, R., Gross, J., Kanagawa, D., Townsend, K., Lanier, L.L., Chiller, J. & Allison, J.P. (1987). Identification of a novel T cell surface disulphide-bonded dimer distinct from the α/β antigen receptor. *Journal of Immunology* **138**, 815–824.
2. Yokoyama, W.M., Jacobs, L.B., Kanagawa, O., Shevach, E.M. & Cohn, D.I. (1989). A murine T lymphocyte antigen belongs to a supergene family of type II integral membrane proteins. *Journal of Immunology* **143**, 1379–1386.
3. Smith, H.R.C., Karlhofer, F.M. & Yokoyama, W.M. (1994). Ly-49 multigene family expressed by IL-2 activated NK cells. *Journal of Immunology* **153**, 1068–1079.
4. Takei, F., Brennan, J. & Mager, D.L. (1997). The Ly-49 family: genes, proteins and recognition of class I MHC. *Immunological Reviews* **155**, 67–77.
5. Matsumoto, N., Ribaudo, R.K., Abastado, J.-P., Margulies, D.H. & Yokoyama, W.M. (1998). The lectin-like NK cell receptor Ly-49A recognises a carbohydrate-independent epitope on its MHC class 1 ligand. *Immunity* **8**, 245–254.
6. Brennan, J., Takei, F., Wong, S. & Mager, D.L. (1995). Carbohydrate recognition by a natural killer cell receptor, Ly-49C. *Journal of Biological Chemistry* **270**, 9691–9694.

Macrophage mannose receptor

Other name: MMR
Mol. wt: 175 000
Specificity: Fucose, mannose, GlcNAc
Sources: Tissue macrophages and related cells

ISOLATION

A membrane receptor known to bind and internalise mannosylated glycoproteins was first isolated in homogeneous form from detergent extracts of rabbit [1] and rat [2] alveolar macrophages by procedures based on affinity chromatography. The corresponding molecule was subsequently purified from human placenta by similar means [3].

Later the human MMR was cloned from placental cDNA libraries [4,5],and expressed in rat fibroblasts [4] and in COS-1 cells [5].

STRUCTURE

MMR is a glycoprotein type-I (C-terminus in the cytoplasm) transmembrane receptor with an apparent Mr of 175 000 estimated by SDS-PAGE [1–6]. The cDNA sequence predicts a mature protein of 1438 amino acids, an Mr of 164 135 and a structure consisting of 12 domains [4,5]. At the extracellular (N-terminal) end, there is a cysteine-rich domain followed by a fibronectin type-II repeat domain. The remainder of the extracellular portion consists of eight C-type lectin carbohydrate-recognition domains [4,5]. These domains show up to 34% identity to each other and resemble that of mannan-binding lectin [5]. The carbohydrate-binding activity, however, is confined to the region between and including the fourth and eighth domains, with only the fourth domain being able to bind mannose in the absence of other domains [7]. The tail region on the cytoplasmic side of the transmembrane domain is essential for internalisation of bound ligands [5]. MMR appears to exist as a monomer [8].

BIOLOGICAL ACTIVITIES

MMR is able to bind neoglycoproteins, like mannose-lactoperoxidase, mannan-BSA and fucose-BSA, in a divalent cation-dependent manner [1–5]. Binding is minimal below pH 5.5, and optimal at pH 6.5 and above [3]. Binding to neoglycoprotein ligands is readily inhibited by mannose, fucose and GlcNAc (but not by galactose), and by certain glycoproteins like mannan [1–5].

MMR mediates complement-independent phagocytosis of micro-organisms including *Candida albicans* [5] and *Pneumocystis carinii* [9].

MMR also mediates endocytosis of glycoproteins [1–7] and mycobacterially-derived glycolipids [10].

MMR may mediate induction of IFN-α after recognition of enveloped viruses [11].

TISSUE AND/OR SUBCELLULAR DISTRIBUTION

Expression of MMR is largely restricted to tissue macrophages and related cell types like immature dendritic cells [12] and epidermal Langerhans cells [13], but has also been reported on some endothelial cells [14] and on smooth muscle cells [15].

MMR is found in endosomes and on the cell surface; it is known to recycle constitutively from the plasma membrane to the endocytic apparatus [15,16].

POSSIBLE FUNCTION(S)

MMR is thought to be a pattern-recognition molecule that functions in first-line host defence by mediating phagocytosis, antigen processing and signal transduction, and thereby linking innate and adaptive immunity [17].

MMR may also regulate the bioavailability of sulphated glycoprotein hormones like thyroid stimulating hormone and lutropin, since such sulphated glycoproteins are rapidly removed from the circulation by MMR via a binding site independent of the sugar binding region [18]. There appears to be a major SO_4-4GalNAc-binding site localised to the cysteine-rich domain (which is unique to the MMR family), but additional interactions between mannose-type sugars and the carbohydrate-binding domains are also involved in hormone binding [19].

COMMENTS

MMR may be regarded as a membrane-bound functional homologue of mannan-binding lectin (MBL) (qv), with which it shares a rather similar saccharide specificity, enabling it to discriminate self from non-self and thereby to function in host defence. High-affinity binding to oligosaccharides is achieved by clustering of several active carbohydrate-recognition domains in a single polypeptide [7], analogous to the bringing together of multiple MBL carbohydrate recognition domains by oligomerisation.

Despite strong similarities in the C-type lectin domain, MMR and MBL exhibit structural differences overall. Whereas MBL is a collectin, MMR is the prototype of a small family of multi-lectin domain receptors; the other known members of this structurally defined group are the phospholipase A2 receptor, DEC-205 and mannose receptor X [17].

The selective uptake of macromolecules by MMR can be exploited for targeting of drugs [20].

REFERENCES

1. Wileman, T.E., Lennartz, M.R. & Stahl, P.D. (1986). Identification of the macrophage mannose receptor as a 175-kDa membrane protein. *Proceedings of the National Academy of Sciences USA* **83**, 2501–2505.
2. Haltiwanger, R.S. & Hill, R.L. (1986). The isolation of a rat alveolar macrophage lectin. *Journal of Biological Chemistry* **261**, 7440–7444.

3. Lennartz, M.R., Cole, F.S., Shepherd, V.L., Wileman, T.E. & Stahl, P.D. (1987). Isolation and characterization of a mannose-specific endocytosis receptor from human placenta. *Journal of Biological Chemistry* **262**, 9942–9944.
4. Taylor, M.E., Conary, J.T., Lennartz, M.R., Stahl, P.D. & Drickamer, K. (1990). Primary structure of the mannose receptor contains multiple motifs resembling carbohydrate-recognition domains. *Journal of Biological Chemistry* **265**, 12156–12162.
5. Ezekowitz, R.A.B., Sastry, K., Bailly, P. & Warner, A. (1990). Molecular characterization of the human macrophage mannose receptor: demonstration of multiple carbohydrate recognition-like domains and phagocytosis of yeasts in Cos-1 cells. *Journal of Experimental Medicine* **172**, 1785–1794.
6. Lennartz, M.R., Cole, F.S. & Stahl, P.D. (1989). Biosynthesis and processing of the mannose receptor in human macrophages. *Journal of Biological Chemistry* **264**, 2385–2390.
7. Taylor, M.E., Bezouska, K. & Drickamer, K. (1992). Contributions to ligand binding by multiple carbohydrate-recognition domains in the macrophage mannose receptor. *Journal of Biological Chemistry* **267**, 1719–1726.
8. Taylor, M.E. & Drickamer, K. (1993). Structural requirements for high affinity binding of complex ligands by the macrophage mannose binding receptor. *Journal of Biological Chemistry* **268**, 399–404.
9. Ezekowitz, R.A.B., Williams, D.J., Koziel, H., Armstrong, M.Y.K., Warner, A., Richards, F.F. & Rose, R.M. (1991). Uptake of *Pneumocystis carinii* mediated by the macrophage mannose receptor. *Nature* **351**, 155–158.
10. Prigozy, T.I., Sieling, P.A., Clemens, D., Stewart, P.L., Behar, S.M., Porcelli, S.A., Brenner, M.B., Modlin, R.L. & Kronenberg, M. (1997). The mannose receptor delivers lipoglycan antigens to endosomes for presentation to T cells by CD1b molecules. *Immunity* **6**, 187–197.
11. Milone, M.C. & Fitzgerald-Bocarsly, P. (1998). The mannose receptor mediates induction of IFN-α in peripheral blood dendritic cells by enveloped RNA and DNA viruses. *Journal of Immunology* **161**, 2391–2399.
12. Sallusto, F., Cella, M., Danieli, C. & Lanzavecchia, A. (1995). Dendritic cells use macropinocytosis and the mannose receptor to concentrate macromolecules in the major histocompatability complex class II compartment: downregulation by cytokines and bacterial products. *Journal of Experimental Medicine* **182**, 389–400.
13. Condaminet, B., Peguet-Navarro, J., Stahl, P.D., Dalbiez-Gauthier, C., Schmitt, D. & Berthier-Vergnes, O. (1998). Human epidermal Langerhans cells express the mannose-fucose binding receptor. *European Journal of Immunology* **28**, 3541–3551.
14. Magnusson, S. & Berg, T. (1993). Endocytosis of ricin by rat liver cells in vivo and in vitro is mainly mediated by mannose receptors on sinosoidal endothelial cells. *Biochemical Journal* **91**, 749–755.
15. Lew, D.B., Songu-Mize, E., Pontow, S.E., Stahl, P.D. & Rattezzi, M.C. (1994). A mannose receptor mediates mannosyl-rich glycoprotein-induced mitogenesis in bovine airway smooth muscle cells. *Journal of Clinical Investigation* **94**, 1855–1863.
16. Stahl, P.D., Schlesinger, P.H., Sigardson, E., Rodman, J.S. & Lee, Y.C. (1980). Receptor mediated pinocytosis of mannose glycoconjugates by macrophages: characterization and evidence for receptor recycling. *Cell* **19**, 207–215.
17. Stahl, P.D. & Ezekowitz, R.A.B. (1998). The mannose receptor is a pattern recognition molecule involved in host defense. *Current Opinion in Immunology* **10**, 50–55.
18. Fiete, D., Beronek, M.C. & Baenziger, J.U. (1997). The macrophage/endothelial cell mannose receptor cDNA encodes a protein that binds oligosaccharides terminating with SO_4-4-GalNAc β1,4 GlcNAcβ or Man at independent sites. *Proceedings of the National Academy of Sciences USA* **94**, 11256–11261.
19. Simpson, D.Z., Hitchen, P.G., Elmhirst, E.L. & Taylor, M.E. (1999). Multiple interactions between pituitary hormones and the mannose receptor. *Biochemical Journal* **343**, 403–411.
20. Liang, W.W., Shi, X., Deshpande, D., Malanga, C.J. & Rojanasakul, Y. (1996). Oligonucleotide targeting to alveolar macrophages by mannose receptor-mediated endocytosis. *Biochimica et Biophysica Acta* **1279**, 227–234.

Mannan-binding lectin

Other names:	MBL; mannan-binding protein; mannose-binding protein; mannose-binding lectin; Ra-reactive factor
Mol. wt:	90 000–700 000 (native); 28 000 (subunit)
Specificity:	GlcNAc, mannose, fucose
Sources:	Mammalian blood and liver; homologue found in chickens

ISOLATION

Early attempts at purification were based on affinity chromatography on mannose–Sepharose [1], mannan–Sepharose [2] and mannan–BioGel [3] using elution with EDTA, but contaminating proteins were always present. Later protocols typically included the use of two affinity steps on mannan–Sepharose (first elution with EDTA, second with mannose) plus additional steps [4,5]. Preparations of comparable purity have been achieved by means of two affinity steps on Emphaze–mannan [6], or acid elution from a column of immobilised anti-MBL antibodies followed by removal of contaminating immunoglobulins with protein G [7]. None of the above methods entirely removes the MBL-associated serine proteases, but this was apparently accomplished by Tan and co-workers [8] by gel filtration at pH 5 after a procedure including two steps of affinity chromatography.

The human MBL gene has been cloned [9]. Recombinant human MBLs have been produced in various eukaryotic expression systems [10–12]. The rat MBP-A and MBL-C forms have also been cloned [13].

STRUCTURE

Human MBL is a glycoprotein collectin (see Chapter 3) containing about 5% carbohydrate [14]. After denaturation and reduction, it consists of identical subunits; the apparent molecular size has often been reported as 31 000, but 28 000 [6] is probably more accurate, and the true molecular size of the polypeptide chain is only 24 000 [15]. The basic unit, however, is a trimer of apparent Mr 90 000, the polypeptides twisted into a collagenous tail with three C-type lectin domains joined by a short neck region forming a head cluster. The three carbohydrate-binding sites are separated by approx 5.4 nm, a distance permitting easy interaction with the repeating sugar residues typical of microbial surfaces while restricting interaction with mammalian high-mannose oligosaccharides [16], thus providing a simple but effective self from non-self discrimination. The basic trimer unit can self-associate into a series of oligomers, with increasing affinity for carbohydrate; there is evidence that the predominant forms in humans are dimers and trimers of the basic triplet [17] although other data indicate a predominance of trimers and tetramers [18]. The distribution of oligomers may vary between MBLs from different species [15], but irrespective of species the higher oligomers are better able to activate complement.

The structure and hence the lectin activity of MBL is genetically determined (see **Comments**). Rodents have two functional MBL genes (A and C) encoding separate proteins, which are principally located in blood and liver, respectively. Humans appear to have a single functional MBL gene, but the protein encoded by it exists in separate serum and liver forms, presumably as a result of post-translational modification [19]. The human isolectins have an identical amino acid sequence, and the most obvious difference between them is the size of the oligomers. Notably, as with the rodent homologues, human liver MBL is relatively small in native molecular size and has little ability to activate complement, while human serum MBL is larger and strongly activates complement [19].

The crystal structure of the carbohydrate-recognition domain of rat MBL-A at 0.17 nm resolution has been solved: Ca^{2+} was found to form coordination bonds with the carbohydrate ligand, and specificity was determined by a network of coordination and hydrogen bonds [20]. A change in specificity can be engineered by replacing Glu-185 and Asn-187 with Gln and Asp, resulting in a strong preference for galactose over mannose [21].

BIOLOGICAL ACTIVITIES

The ability of MBL to opsonise baker's yeast was known long before the identity of the protein responsible [22]. MBL can also promote the phagocytosis of other yeasts, viruses and sundry bacteria [23]. MBL can act as an opsonin directly, by binding to the microbial surface with its lectin site while attaching its collagenous tail region to a receptor on phagocytes [24]. Moreover, MBL can activate both the alternative [25] and the classical [26] pathways of complement. Indeed, MBL can activate the classical pathway by means of unique serine proteases, a route sometimes known as the 'lectin pathway of complement'. Two enzymes have been discovered, MASP-1 [27] and MASP-2 [28], which correspond to the C1r and C1s components of the classical complement pathway. The relative physiological relevance of these different pathways has not been established. Similarly the relative importance of different collectin receptors, such as the 126 kDa C1q receptor [29] and the cell surface calreticulin-like C1q receptor [30], is unclear. Although complement activation is presumed to be important for anti-microbial activity, MBL can neutralise influenza A virus without complement [31].

The binding of MBL to solid-phase mannan is most readily inhibited by GlcNAc, mannose and L-fucose, is moderately inhibited by glucose, but is very weakly inhibited by galactose and GalNAc [32–34].

Both rat [35] and human [36] MBLs have phospholipid-binding activities. The relative specificities for different phospholipids differed between rat serum (MBL-A) and rat liver (MBL-C) lectins, and both differed in this respect from the human MBL [36]. All three were similar in being susceptible to inhibition by

EDTA, and all three had a broadly similar (but by no means identical) pattern of inhibition by monosaccharides. Moreover, the saccharide inhibition profile with phospholipids showed some marked differences when compared to that obtained with mannan [35,36].

Human MBL may have anti-tumour activity [37,38]. MBL was found to bind to some colon adenocarcinoma cell lines via Lewis A and Lewis B moieties. A recombinant vaccinia virus carrying the MBL gene had potent growth inhibitory activity *in vivo* against human colorectal carcinoma cells transplanted into nude mice [38]. This cytotoxic activity was independent of complement activation, since mutant MBL with little ability to activate complement was nearly as effective therapeutically as the wild-type MBL [38].

TISSUE AND/OR SUBCELLULAR DISTRIBUTION

MBL is synthesised in the liver and secreted into the bloodstream [19]. It has also been found in amniotic fluid [39], synovial fluid [40] and fetally-derived first-trimester placental tissue [41].

POSSIBLE FUNCTION(S)

The widely accepted probable function is as an ante-antibody, providing first-line defence by distinguishing self from non-self and opsonising microbial surfaces for pathogen elimination by phagocytic cells [16,23]. This view is strongly supported by various data indicating that a very low concentration of MBL confers increased susceptibility to infection (see **Comments** section and Chapter 3). Evidence from a transgenic mouse model [42] and from immunohistochemical studies on virus-infected chickens [43] is consistent with the ante-antibody concept.

COMMENTS

Mannan-binding lectins/Ra-reactive factors constitute a family of proteins highly conserved among mammalian species [16], and which have a homologue in chickens [44]. The human MBL is the product of a single gene on chromosome 10, which has characteristics in its 5′ region suggesting that MBL might be an acute-phase protein [9]. Serial measurements of MBL protein, however, have only confirmed a modest (1.5- to 3-fold) increase in concentration in the acute phase [45]. Instead, individual variation in MBL concentration is largely under genetic control. The most important influence is genetic polymorphism within the first exon of the structural region of the gene. Point mutations at three separate loci (codons 52, 54 and 57) result in allelic variants which occur at different frequencies in different populations. The reader is referred to the tables of mutant gene frequencies given by Turner [16,23] and by the detailed explanations in Chapter 3. A more modest genetic influence was attributed to dimorphisms (H/L and X/Y) in the promoter region [46]. The complexity of the haplotypes listed in Table 3.4 is increased by consideration of a dimorphism (P/Q) located in the 5′ untranslated portion of the gene [47] and by the discovery of five additional promoter region mutations [48].

These complex genetic influences are responsible for a >1000-fold variation in circulating MBL in humans, from <0.01 μg/ml to 10 μg/ml. The level of MBL in amniotic fluid increases during gestation [39], but whether there is a corresponding rise in the fetal circulation is less clear. Cord-blood MBL concentrations are lower in babies born prematurely [49,50], but there was no obvious correlation between MBL concentrations and gestational age during the six weeks immediately preceding term in another study [51]. Incidentally, discordant values from twins indicated that cord-blood MBL is derived from the fetoplacental unit and not from the maternal circulation [51]. MBL concentrations at term in cord blood are in general significantly lower than adult values [49,51,52], but there is a rapid rise during the first week of life [52]. During childhood, MBL concentrations are generally a little higher than those found during adulthood [4,53,54]. Although estimates vary somewhat, the median concentration in healthy adults is approximately 1.2 μg/ml [55].

Despite a 1000-fold variation in apparently healthy adults, the distribution of MBL concentrations may be considered bimodal [51,55]. Approximately 5–10% of the population have excessively low concentrations, making MBL deficiency by far the commonest primary immune deficiency (at least 20 times more prevalent than IgA deficiency). This proportion (5–10%) determined by a quantitative ELISA is similar to the proportion deemed deficient by the original yeast opsonisation functional assay, but is much lower than the proportion deemed abnormal by genotyping for structural gene mutations which are found in up to a third of the European population. However deficiency be defined, evidence is steadily accumulating that deficiency correlates with an increased susceptibility to infectious diseases.

The first prospective study defined MBL deficiency by a functional yeast opsonisation assay [56]. In a small group (26 subjects plus an equal number of controls) of neonates, the incidence of infectious and atopic disorders generally during the first year of life was greater in the MBL-deficient infant group than for its controls. Later, both the opsonic defect and the presence of mutant MBL genes were shown to correlate with severe recurrent infections in three family trees [57]. The most convincing evidence, however, came from a large consecutive series of children attending a hospital paediatric department: variant MBL alleles ensuring low circulating MBL concentrations were twice as common in children with infections than in infection-free patients, constituting a highly significant difference [58]. In another substantial paediatric series, the prevalence of MBL deficiency did not seem to be greater in children with recurrent infections, but 7/9 patients with undetectable MBL had repeated infections or a single episode of severe infection sometimes in association with an IgG subclass deficiency [59]. Collectively these reports constitute a strong case for a relationship between low MBL concentrations and a general susceptibility to infections in childhood; the presenting pathology was heterogeneous, but otitis media and septicaemia were frequently found.

A rather anecdotal report involving five patients was the first indication that MBL deficiency might also predispose to infections in adults [60]. This possibility was supported by a large series of patients suspected of various non-HIV-related immune deficiencies; the frequency of homozygotes for variant alleles was very significantly increased, and several of the 20 patients concerned were older children while one was an adult [61]. An independent study provided compatible results: low MBL concentration was found to be a risk factor of comparable magnitude in two groups of adults and one group of paediatric patients with recurrent infections [62]. However, low MBL concentrations were not over-represented in a smaller group of pyrexial patients with suspected or documented infections of various kinds [63].

MBL deficiency, defined at the genetic or protein level, has also been positively associated with susceptibility to several specific infectious agents. Allelic variants of the MBL gene were strongly associated with susceptibility to symptomatic *Neisseria meningitidis* infection, with homozygotes conferring greater susceptibility than heterozygotes [64]; however, other studies have produced conflicting implications concerning the role of MBL in meningococcal disease [7,65–67]. Low MBL concentrations, and corresponding mutations in the MBL gene, were associated with increased severity of *Plasmodium falciparum* malaria in 200 Gabonese children [68]. One investigation identified the mutation in codon 52 of the MBL gene as a risk factor for chronic hepatitis B viral infection in Europeans, but the other low MBL concentration-conferring mutations were not over-represented, and nor was the codon 52 mutation present in Asians with hepatitis B [69]. This association with hepatitis B could not be confirmed by an independent and larger study [70], and yet another group of researchers found a significant association with the codon 54 mutation in symptomatic Chinese patients with cirrhosis or spontaneous bacterial peritonitis [71]. Haplotypes conferring low levels of MBL were associated with resistance to interferon-α therapy in Japanese patients with hepatitis C virus infection [72].

In addition, the relationship between MBL deficiency and HIV disease has received considerable attention, although the results obtained have not been entirely consistent with each other. Nielsen *et al.* [73] found very low MBL concentrations in a significantly higher proportion of Danish (mostly homosexual) HIV-positive subjects compared to healthy population controls, but found no association between MBL levels and disease progression or survival of the patients. This association with HIV disease susceptibility was confirmed at the genetic level by two other Scandinavian groups [74,75], and was also found, but only regarding symptomatic subjects, by a Hungarian study [76]. In HIV-infected patients with AIDS (i.e. those who are experiencing infections), the MBL level is raised [76–78], presumably as an acute-phase response, and this effect may contribute to the inconsistencies relating to MBL concentration and prognosis [73–78]. One study reported a very high risk of cryptosporidiosis (*Cryptosporidium parvum* diarrhoea) for Zambian HIV-infected patients homozygous for MBL mutant alleles [79]. In contrast to all the findings listed above,

Senaldi and co-workers [78] found no difference in the prevalence of MBL deficiency between HIV seropositive and seronegative subjects, while Maas and co-workers actually reported a weak protective influence of MBL mutant alleles on progression to AIDS and death [80]. In the special instance of babies born to HIV-infected mothers, the commonest (codon 54) mutant allele was over-represented in the children subsequently found to be HIV infected, and was strongly and significantly associated with rapid disease progression [81].

Associations with recurrent miscarriage [41,82,83] and ischaemic heart disease [84] may be secondary to a protective role against infection. Regarding the latter, evidence is accumulating that chronic inflammation associated with *Chlamydia pneumoniae* contributes to the pathogenesis of coronary heart disease [85] and MBL is known to antagonise *C. pneumoniae* attachment to cells at least *in vitro* [86]. Low concentrations of parental MBL have been found to be a risk factor for spontaneous abortion in both Scottish [41,83] and Danish [82] populations. This relationship was evident in both male and female partners with the Scottish data, but was not apparent in the small group of Danish male partners available. The most likely reason for the association would seem to be an increased susceptibility to infections within the fetoplacental unit, but other explanations, including a linked lethal gene, are possible.

The significant difference between lung function impairment/survival curves of cystic fibrosis patients with low and high MBL levels is also likely to be related to effects on infectious agents [87,88].

The relationship of MBL concentration to autoimmunity has received considerable attention. A modest but probably real association between relative MBL deficiency and systemic lupus erythematosus (SLE) has been noted in several independent investigations [89–94; see also Table 3.6]; most notably, homozygosity for MBL variant alleles was much more strongly associated with serious infections (particularly pneumonia) in SLE patients than in the general population [95]. The risk of autoimmune diseases (discoid lupus and SLE-like disorders) was also associated with variant alleles of MBL in the context of chronic granulomatous disease [96]. The dubious possibility that MBL deficiency is associated with rheumatoid arthritis [97] is discussed below.

A single family study implicating MBL deficiency with recurrent skin abscesses, prurigo and atopic dermatitis [98] is of interest in view of the very early association of the MBL-dependent opsonic defect with atopic disease [56].

The relatively high frequency of the heterozygous state in all populations studied implies some selective advantage for the mutant alleles, perhaps by limiting complement-mediated inflammatory damage to the host, or by limiting uptake of pathogens that use C3 opsonisation and uptake by complement receptors [16]. In this context, the finding that MBL exists in synovial fluid and can bind to agalactosyl (but not native) IgG and can thereby induce complement activation leads to the prediction that MBL deficiency should protect against rheumatoid arthritis [99]. However, one study actually found a positive association between MBL deficiency and rheumatoid arthritis [97], while three others found no association whatsoever [55,90,100]. On the other hand, there is some experimental support for the speculation that variant MBL alleles/low

MBL concentrations may afford some protection against mycobacterial infections [101–103]. While MBL concentration may not affect susceptibility to rheumatoid arthritis, impressive data from a cohort of 140 patients followed prospectively from 18 to 32 years after onset indicate that MBL insufficiency is a highly significant prognostic factor for severe disease [104]. The authors suggest that MBL may be useful therapeutically or prophylactically in this condition.

MBL replacement therapy was first successfully achieved in 1976 with fresh frozen plasma; clinical improvement correlated with correction of the *in vitro* opsonisation defect [105,106]. More recently, substitution therapy has been carried out on two subjects using affinity purified MBL [107]. No adverse effects were apparent, no antibody response to MBL could be detected, normalisation of MBL concentration/complement-mediated opsonisation was achieved, and a dramatic improvement was observed in a two-year-old recipient with a history of viral and other infections from birth [107].

The biological half-life of MBL was estimated to be five to seven days [107]. After death, MBL in the peripheral blood is stable for several days at least [108].

Rabbit MBL can bind IgM antibodies from several species, and this property has been exploited to isolate mouse monoclonal IgM from ascites fluid by a single affinity chromatographic step to >95% purity [109]. Bovine and human IgM can also be isolated on immobilised rabbit MBL, although the capacity of the matrix and the purity of the product were lower [109].

Rabbit serum MBL recognises a unique CD45RO glycoform found on murine thymocytes [110]. The strongest binding was to immature thymocytes with a $CD4^{+}CD8^{+}CD3^{low}$ phenotype [110].

MBL has been detected in the glomerular mesangial area (co-localised with IgA) of kidneys of some patients with IgA nephropathy and was associated with complement deposition and poorer renal function. MBL may induce glomerular injury by binding to partially deglycosylated IgA [111].

An up-to-date list of species from which the primary structure of MBL has been determined, with references to the original literature, is given in a recent review of collectin structure by Hansen and Holmskov [112].

REFERENCES

1. Wild, J., Robinson, D. & Winchester, B. (1983). Isolation of mannose-binding proteins from human and rat liver. *Biochemical Journal* **210**, 167–174.
2. Kawasaki, N., Kawasaki, T. & Yamashina, I. (1983). Isolation and characterisation of mannan-binding protein from human serum. *Journal of Biochemistry* **94,** 937–947.
3. Taylor, M.E. & Summerfield, J.A. (1987). Carbohydrate-binding proteins of human serum: isolation of two mannose/fucose-specific lectins. *Biochimica et Biophysica Acta* **915,** 60–67.
4. Terai, I., Kobayashi, K., Fujita, T. & Hagiwara, K. (1993). Human serum mannose binding protein (MBP): development of an enzyme-linked immunosorbent assay (ELISA) and determination of levels in serum from 1085 normal Japanese and in some body fluids. *Biochemical Medicine and Metabolic Biology* **50**, 111–119.
5. Ezekowitz, R.A.B., Kuhlman, M., Groopman, J.E. & Byrn, R.A. (1989). A human serum mannose-binding protein inhibits in vitro infection by the human immunodeficiency virus. *Journal of Experimental Medicine* **169**, 185–196.

6. Kilpatrick, D.C. (1997). Isolation of human mannan binding lectin, serum amyloid P component and related factors from Cohn fraction III. *Transfusion Medicine* **7**, 289–294.
7. Jack, D.L., Dodds, A.W., Anwar, N., Ison, C.A., Law, A., Frosch, M., Turner, M.W. & Klein, N.J. (1998). Activation of complement by mannose-binding lectin on isogenic mutants of *Neisseria meningitidis* serogroup B. *Journal of Immunology* **160**, 1346–1353.
8. Tan, S.M., Chung, M.C.M., Kon, O.L., Thiel, S., Lee, S.H. & Lu, J. (1996). Improvements on the purification of mannan-binding lectin and demonstration of its Ca^{2+}-independent association with a C1s-like serine protease. *Biochemical Journal* **319**, 329–332.
9. Sastry, K., Herman, G.A., Day, L., Deignan, E., Bruns, G., Morton, C.C. & Ezekowitz, R.A.B. (1989). The human mannose binding protein gene. *Journal of Experimental Medicine* **170**, 1175–1189.
10. Super, M., Gillies, S.D., Foley, S., Sastry, K., Schweinle, J.-E., Siverman, V.J. & Ezekowitz, R.A.B. (1992). Distinct and overlapping functions of allelic forms of human mannose binding protein. *Nature Genetics* **2**, 50–55.
11. Ma, Y., Shida, H. & Kawasaki, T. (1997). Functional expression of human mannan binding proteins (MBPs) in human hepatoma cell lines infected by recombinant vaccinia virus: post-translational modification, molecular assembly, and differentiation of serum and liver MBP. *Journal of Biochemistry* **122**, 810–818.
12. Ohtani, K., Suzuki, Y., Eda, S., Kawai, T., Kase, T., Keshi, H., Sakai, Y., Yamamoto, S., Sakamoto, T. & Wakamiya, N. (1999). High-level and effective production of human mannan-binding lectin (MBL) in Chinese hamster ovary (CHO) cells. *Journal of Immunological Methods* **222**, 135–144.
13. Drickamer, K., Dordal, M.S. & Reynolds, L. (1986). Mannose-binding proteins isolated from rat liver contain carbohydrate recognition domains linked to collagenous tails. *Journal of Biological Chemistry* **261**, 6878–6887.
14. Kilpatrick, D.C. (1998). Mystery plasma lectin was a contaminant. *Transfusion Medicine* **8**, 239.
15. Yakota, Y., Arai, T. & Kawasaki, T. (1995). Oligomeric structures required for complement activation of serum mannan-binding proteins. *Journal of Biochemistry* **117**, 414–419.
16. Turner, M.W. (1996). Mannose-binding lectin: the pluripotent molecule of the innate immune system. *Immunology Today* **17**, 532–540.
17. Lipscombe, R.J., Sumiya, M., Summerfield, J.A. & Turner, M.W. (1995). Distinct physiochemical characteristics of human mannose binding protein expressed by individuals of differing genotype. *Immunology* **85**, 660–667.
18. Lu, J., Thiel, S., Wiederman, H., Timpl, R. & Reid, K.B.M. (1990). Binding of the pentamer/hexamer forms of mannan-binding protein to zymosan activates the proenzyme $C1r_2$ $C1s_2$ complex, of the classical pathway of complement, without involvement of C1q. *Journal of Immunology* **144**, 2287–2294.
19. Kurata, H., Sannoh, T., Kozutsumi, Y., Yokota, Y. & Kawasaki, T. (1994). Structure and function of mannan-binding proteins isolated from human liver and serum. *Journal of Biochemistry* **115**, 1148–1154.
20. Weis, W.I., Drickamer, K. & Hendrickson, W.A. (1992). Structure of a C-type mannose-binding protein complexed with an oligosaccahride. *Nature* **360**, 127–134.
21. Drickamer, K. (1992). Engineering galactose-binding activity into a C-type mannose binding protein. *Nature* **360**, 183–186.
22. Super, M., Lu, J., Thiel, S., Levinsky, R.J. & Turner, M.W. (1989). Association of low levels of mannan-binding protein with a common defect of opsonisation. *Lancet* **2**, 1236–1239.
23. Turner, M.W. (1998). Mannose-binding lectin (MBL) in health and disease. *Immunobiology* **199**, 327–339.
24. Kuhlman, M., Joiner, K. & Ezekowitz, R.A.B. (1989). The human mannose-binding protein functions as an opsonin. *Journal of Experimental Medicine* **169**, 1733–1745.

25. Schweinle, J.E., Ezekowitz, R.A.B., Tenner, A.J., Kuhlman, M. & Joiner, K.A. (1989). Human mannose-binding protein activates the alternative complement pathway and enhances serum bactericidal activity on a mannose-rich isolate of Salmonella. *Journal of Clinical Investigation* **84**, 1821–1829.
26. Ikeda, K., Sannoh, T., Kawasaki, N., Kawasaki, T. & Yamashina, I. (1987). Serum lectin with known structure activates complement through the classical pathway. *Journal of Biological Chemistry* **262**, 7451–7454.
27. Matsushita, M. & Fujita, Y. (1995). Cleavage of the third component of complement (C3) by mannose-binding protein-associated serine protease (MASP) with subsequent complement activation. *Immunobiology* **194**, 443–448.
28. Thiel, S., Vorup-Jensen, T., Stover, C.M., Schwaeble, W., Laursen, S.B., Poulsen, K., Willis, A.C., Eggleton, P., Hansen, S., Holskov, U., Reid, K.B.M. & Jensenius, J.C. (1997). A second serine protease associated with mannan-binding lectin that activates complement. *Nature* **386**, 506–509.
29. Tenner, A.J., Robinson, S.L. & Ezekowotz, R.A.B. (1995). Mannose-binding protein (MBP) enhances mononuclear phagocyte function via a receptor that contains the 126 000 Mr component of the C1q receptor. *Immunity* **3**, 485–493.
30. Stuart, G.R., Lynch, N.J., Day, A.J., Schwaeble, W.J. & Sim, R.B. (1997). The C1q and collectin binding site within C1q receptor (cell surface calreticulin). *Immunopharmacology* **38**, 73–80.
31. Kase, T., Suzuki, Y., Kawai, T., Sakamoto, T., Ohtani, K., Eda, S., Maeda, A., Okuno, Y., Kurimura, T. & Wakamiya, N. (1999). Human mannan-binding lectin inhibits the infection of influenza A virus without complement. *Immunology* **97**, 385–392.
32. Taylor, M.E. & Summerfield, J.A. (1987). Carbohydrate-binding proteins of human serum: isolation of two mannose/fucose specific lectins. *Biochimica et Biophysica Acta* **915**, 60–67.
33. Holmskov, U., Malhotra, R., Sim, R.B. & Jensenius, J.C. (1994). Collectins: collagenous C-type lectins of the innate immune defense system. *Immunology Today* **15**, 67–74.
34. Haurum, J.S., Thiel, S., Haagsman, H.P., Laursen, S.B., Larsen, B. & Jensenius, J.C. (1993). Studies on the carbohydrate-binding characteristics of human pulmonary surfactant-associated protein A and comparison with two other collectins: mannan-binding protein and conglutinin. *Biochemical Journal* **293**, 873–878.
35. Kuroki, Y., Honma, T., Chiba, H., Sano, H., Saitoh, M., Ogasawara, Y., Sohma, H. & Akino, T. (1997). A novel type of binding specificity to phospholipids for rat mannose-binding proteins isolated from serum and liver. *FEBS Letters* **414**, 387–392.
36. Kilpatrick, D.C. (1998). Phospholipid-binding activity of human mannan-binding lectin. *Immunology Letters* **61**, 191–195.
37. Muto, S., Sakuma, K., Taniguichi, A. & Matsumoto, K. (1999). Human mannose-binding lectin preferentially binds to human colon adenocarcinoma cell lines expressing high amounts of Lewis A and Lewis B antigens. *Biological and Pharmaceutical Bulletin* **22**, 347–352.
38. Mam, Y., Uemura, K., Oka, S., Kozutsumi, Y., Kawasaki, N. & Kawasaki, T. (1999). Antitumor activity of mannan-binding protein in vivo as revealed by a virus expression system: mannan-binding protein-dependent cell-mediated cytotoxicity. *Proceedings of the National Academy of Sciences USA* **96**, 371–375.
39. Malhotra, R., Willis, A.C., Lopez Bernal, A., Thiel, S. & Sim, R.B. (1994). Mannan binding protein levels in human amniotic fluid during gestation and its interaction with collectin receptor from amnion cells. *Immunology* **82**, 439–444.
40. Malhotra, R., Wormald, M.R., Rudd, P.M., Fischer, P.B., Dwek, R.A. & Sim, R.B. (1995). Glycosylation changes of IgG associated with rheumatoid arthritis can activate complement via the mannose-binding protein. *Nature Medicine* **1**, 237–243.
41. Kilpatrick, D.C., Bevan, B.H. & Liston, W.A. (1995). Association between mannan binding protein deficiency and recurrent miscarriage. *Molecular Human Reproduction* **1** *Human Reproduction* **10**, 2501–2505.
42. Tabona, P., Mellor, A. & Summerfield, J.A. (1995). Mannose binding protein is involved in first-line host defence: evidence from transgenic mice. *Immunology* **85**, 153–159.

43. Nielsen, O.L., Jorgensen, P.H., Hedemand, J., Jensenius, J.C., Koch, C. & Laursen, S.B. (1998). Immunohistochemical investigation of the tissue distribution of mannan-binding lectin in non-infected and virus-infected chickens. *Immunology* **94**, 122–128.
44. Laursen, S.B., Hedemand, J.E., Thiel, S., Willis, A.C., Skriver, E., Madsen, P.S. & Jensenius, J.C. (1995). Collectin in a non-mammalian species: isolation and characterization of mannan-binding protein (MBP) from chicken serum. *Glycobiology* **5**, 553–561.
45. Thiel, S., Holmskov, U., Hviid, L., Laursen, S.B. & Jensenius, J.C. (1992). The concentration of the C-type lectin, mannan-binding protein, in human plasma increases during an acute phase response. *Clinical and Experimental Immunology* **90**, 31–35.
46. Madsen, H.O., Garred, P., Thiel, S., Kurtzhals, J.A.L., Lamm, L.U., Ryder, L.P. & Svejgaard, A. (1995). Interplay between promotor and structural gene variants control basal serum levels of mannan-binding protein. *Journal of Immunology* **155**, 3013–3020.
47. Madsen, H.O., Garred, P., Kurtzhals, J.A.L., Lamm, L.U., Ryder, L.P., Thiel, S. & Svejgaard, A. (1994). A new frequent allele is the missing link in the structural polymorphism of the human mannan-binding gene. *Immunogenetics* **40**, 37–44.
48. Madsen, H.O., Satz, M.L., Hogh, B., Svejgaard, A. & Garred, P. (1998). Different molecular events result in low protein levels of mannan binding lectin in populations from Southeast Africa and South America. *Journal of Immunology* **161**, 3169–3175.
49. Thiel, S., Bjerke, T., Hansen, D., Poulsen, L.K., Schiøtz, P.O. & Jensenius, J.C. (1995). Ontogeny of human mannan-binding protein, a lectin of the innate immune system. *Pediatric Allergy and Immunology* **6**, 20–23.
50. Lau, Y.L., Chan, S.Y., Turner, M.W., Fong, J. & Karlberg, J. (1995). Mannose-binding protein in preterm infants: developmental profile and clinical significance. *Clinical and Experimental Immunology* **102**, 649–654.
51. Kilpatrick, D.C., Liston, W.A. & Midgley, P.C. (1997). Mannan binding protein in human umbilical cord blood. *Natural Immunity* **15**, 234–240.
52. Terai, I. & Kobayashi, K. (1993). Perinatal changes in serum mannose-binding protein (MBP) levels. *Immunology Letters* **38**, 185–187.
53. Kilpatrick, D.C. (1996). Unpublished data presented at the Scottish Clinical Immunology Group Meeting, Perth, 1996.
54. Aittoniemi, J., Miettinen, A., Laippala, P., Isolauri, E., Viikari, J., Ruuska, T. & Soppi, E. (1996). Age-dependent variation in the serum concentration of mannan-binding protein. *Acta Paediatrica* **85**, 906–909.
55. Kilpatrick, D.C. (1997). Mannan binding protein in sera positive for rheumatoid factor. *British Journal of Rheumatology* **36**, 207–209.
56. Richardson, V.F., Larcher, V.F. & Price, J.F. (1983). A common congenital immmunodeficiency predisposing to infection and atopy in infancy. *Archives of Disease in Childhood* **58**, 799–802.
57. Sumiya, M., Super, M., Tabona, P., Levinsky, R.J., Arai, T., Turner, M.W. & Summerfield, J.A. (1991). Molecular basis of opsonic defect in immunodeficient children. *Lancet* **337**, 1569–1570.
58. Summerfield, J.A., Sumiya, M., Levin, M. & Turner, M.W. (1997). Associations of mutations in mannose binding protein gene with childhood infection in consecutive hospital series. *British Medical Journal* **314**, 1229–1232.
59. Aittoniemi, J., Baer, M., Soppi, E., Vesikari, T. & Miettinen, A. (1998) Mannan binding lectin deficiency and concomitant immunodefects. *Archives of Disease in Childhood* **78**, 245–248.
60. Summerfield, J.A., Ryder, S., Sumiya, M., Thursz, M., Gorchein, A., Monteil, M.A. & Turner, M.W. (1995). Mannose binding protein gene mutations associated with unusual and severe infection in adults. *Lancet* **345**, 886–889.
61. Garred, P., Madsen, H.O., Hofmann, B. & Svejgaard, A. (1995). Increased frequency of homozygosity of abnormal mannan-binding-protein alleles in patients with suspected immunodeficiency. *Lancet* **346**, 941–943.

62. Kakkanaiah, V.N., Shen, G.Q., Ojo-Amaize, E.A. & Peter, J.B. (1998). Association of low concentrations of serum mannose-binding protein with recurrent infections in adults. *Clinical and Diagnostic Laboratory Immunology* **5**, 319–321.
63. Aittoniemi, J., Rintala, E., Miettinen, A. & Soppi, E. (1997). Serum mannan-binding lectin (MBL) in patients with infection: clinical and laboratory correlates. *APMIS* **105**, 617–622.
64. Hibbert, M.L., Sumiya, M., Summerfield, J.A., Booy, R., Levin, M. & the Meningococcal Research Group (1999). Association of variants of the gene for mannose-binding lectin with susceptibility to meningococcal disease. *Lancet* **353**, 1049–1053.
65. Garred, P., Michaelsen, T.E., Bjune, G., Thiel, S. & Svejgaard, A. (1993). A low serum concentration of mannan-binding protein is not associated with serogroup B or C meningococcal disease. *Scandinavian Journal of Immunology* **37**, 468–470.
66. Von Emmerik, L.C., Kuijper, E.J., Fijen, C.A.P., Dankert, J. & Thiel, S. (1994). Binding of mannan-binding protein to various bacterial pathogens of meningitis. *Clinical and Experimental Immunology* **97**, 411–416.
67. Drogari-Apiranthiton, M., Fijen, C.A.P., Thiel, S., Platonov, A., Jensen, L., Dankert, J. & Kuijper, E.J. (1997). The effect of mannan-binding lectin on opsonophagocytosis of *Neisseria meningitidis*. *Immunopharmacology* **38**, 93–99.
68. Luty, A.J.F., Kun, J.F.J. & Kremsner, P.G. (1998). Mannose-binding lectin plasma levels and gene polymorphisms in *Plasmodium falciparum* malaria. *Journal of Infectious Diseases* **178**, 1221–1224.
69. Thomas, H.C., Foster, G.R., Sumiya, M., McIntosh, D., Jack, D.L., Turner, M.W. & Summerfield, J.A. (1996). Mutation of gene for mannose-binding protein associated with chronic hepatitis B viral infection. *Lancet* **348**, 1417–1419.
70. Hahler, T., Wunschlel, M., Gerken, G., Schneider, P.M., zum Buschenfelde, K.H.M. & Rittner, C. (1999). No association between mannose-binding lectin alleles and susceptibility to chronic hepatitis B virus infection in German patients. *Experimental and Clinical Immunogenetics* **15**, 130–133.
71. Yuen, M.F., Lau, C.S., Lau, Y.L., Wong, W.M., Cheng, C.C. & Lai, C.L. (1999). Mannose binding lectin gene mutations are associated with progression of liver disease in chronic hepatitis B infection. *Hepatology* **29**, 1248–1251.
72. Matsushita, M., Hijikata, M., Matsushita, M., Ohta, Y. & Mishiro, S. (1998). Association of mannose-binding lectin gene haplotype LXPA and LYPB with interferon-resistant hepatitis C virus infection. *Journal of Hepatology* **29**, 695–700.
73. Nielsen, S.L., Andersen, P.L., Koch, C., Jensenius, J.C. & Thiel, S. (1995). The level of the serum opsonin, mannan-binding protein in HIV-1 antibody-positive patients. *Clinical and Experimental Immunology* **100**, 219–222.
74. Garred, P., Madsen, H.O., Balslev, U., Hofmann, B., Pedersen, C., Gerstoft, J. & Svejgaard, A. (1997). Susceptibility to HIV infection and progression of AIDS in relation to variant alleles of mannose-binding lectin. *Lancet* **349**, 236–240.
75. Pastinin, T., Liitsola, K., Niini, P., Salminen, M. & Syvänen, A.-C. (1998). Contribution of the CCR5 and MBL genes to susceptibility to HIV type 1 infection in the Finnish population. *AIDS Research and Human Retroviruses* **14**, 695–698.
76. Prohaszka, Z., Thiel, S., Ujhelyi, E., Szlavik, J., Banhegyi, D. & Fust, G. (1997). Mannan-binding lectin serum concentrations in HIV-infected patients are influenced by the stage of disease. *Immunology Letters* **58**, 171–175.
77. Ezekowitz, R.A., Kuhlman, M., Groopman, J.E. & Byron, J. (1989). A human serum mannose-binding protein inhibits in vitro infection by the human immunodeficiency virus. *Journal of Experimental Medicine* **169**, 185–196.
78. Senaldi, G., Davies, E.T., Mahalingam, J., Lu, A., Pozniak, M., Peakman, M., Reid, K.B.M. & Vergani, D. (1995). Circulating levels of mannose binding protein in human immunodeficiency virus infection. *Journal of Infection* **31**, 145–148.
79. Jack, D.L., Kelly, P., Naeem, A., Turner, M.W. & Farthing, M.J.G. (1998). Involvement of mannose-binding lectin in *Cryptosporidium parvum* infection. *Immunology* **95** (suppl. 1), 24.

80. Maas, J., Husman, A.-M., Brouwer, M., Krol, A., Coutinho, R., Kreet, I., van Leeuwen, R. & Schuitemaker, H. (1998). Presence of the variant mannose-binding lectin alleles associated with slower progression to AIDS. *AIDS* **12**, 2275–2280.
81. Amoroso, A., Berrino, M., Boniotto, M., Crovella, S., Palomba, E., Scarlatti, G., Serra, C., Tovo, P.-A. & Vatta, S. (1999). Polymorphism at codon 54 of mannose-binding protein gene influences AIDS progression but not HIV infection in exposed children. *AIDS* **13**, 863–864.
82. Christiansen, O.B., Kilpatrick, D.C., Souter, V., Varming, K., Thiel, S. & Jensenius, J.C. (1999). Mannan-binding lectin deficiency is associated with unexplained recurrent miscarriage. *Scandinavian Journal of Immunology* **49**, 193–196.
83. Kilpatrick, D.C., Starrs, L., Moore, S., Souter, V. & Liston, W.A. (1999). Mannan binding lectin concentration and risk of miscarriage. *Human Reproduction* **14**, 2379–2380.
84. Madsen, H.O., Videm, V., Svejgaard, A., Svennevig, J.L. & Garred, P. (1998). Association of mannose-binding lectin deficiency with severe atherosclerosis. *Lancet* **352**, 959–960.
85. Danesh, J., Collins, R. & Peto, R. (1997). Chronic infections and coronary heart disease: is there a link? *Lancet* **350**, 430–436.
86. Swanson, A., Ezekowitz, R.A.B., Lee, A. & Kuo, C.-C. (1998). Human mannose-binding protein inhibits infection of HeLa cells by *Chlamydia trachomatis*. *Infection and Immunity* **66**, 1607–1612.
87. Garred, P., Pressler, T., Madsen, H.O., Fredericksen, B., Svejgaard, A., Høily, N., Schwartz, M. & Koch, C. (1999). Association of mannose-binding lectin gene heterogeneity with severity of lung disease and survival in cystic fibrosis. *Journal of Clinical Investigation* **104**, 431–437.
88. Gabolde, M., Guilloud-Bataille, M., Feingold, J. & Besmond, C. (1999). Association of variant alleles of mannose binding lectin with severity of pulmonary disease in cystic fibrosis: cohort study. *British Medical Journal* **319**, 1166–1167.
89. Davies, E.J., Snowden, N., Hillarby, M.C., Carthy, D., Grennan, D.M., Thomson, W. & Ollier, W.E.R. (1995). Mannose-binding protein gene polymorphism in systemic lupus erythematosus. *Arthritis and Rheumatism* **38**, 110–114.
90. Senaldi, G., Davies, E.T., Peakman, M., Vergani, D., Lu, J. & Reid, K.B.M. (1995). Frequency of mannose binding protein deficiency in patients with systemic lupus erythematosus. *Arthritis and Rheumatism* **38**, 1713–1714.
91. Lau, Y.L., Lau, C.S., Chan, S.Y. & Karlberg, J. (1996). Mannose-binding protein in Chinese patients with systemic lupus erythematosus. *Arthritis and Rheumatism* **39**, 706–708.
92. Sullivan, K.E., Wooten, C., Goldman, D. & Petri, M. (1996). Mannose-binding protein genetic polymorphisms in black patients with systemic lupus erythematosus. *Arthritis and Rheumatism* **39**, 2046–2051.
93. Davies, E.J., Teh, L.-S., Ordi-Ros, J., Snowden, N., Hillarby, M.C., Hajeer, A., Donn, R., Perez-Pemen, P., Vilardell-Tarres, M. & Ollier, W.E.R. (1997). A dysfunctional allele of the mannose binding protein gene associates with systemic lupus erythematosus in a Spanish population. *Journal of Rheumatology* **24**, 485–488.
94. Ip, W.K., Chan, S.Y., Lau, C.S. & Lau, Y.L. (1998). Association of systemic lupus erythematosus with promotor polymorphisms of the mannose-binding lectin gene. *Arthritis and Rheumatism* **41**, 1663–1668.
95. Garred, P., Madsen, H.O., Halberg, P., Petersen, J., Kronborg, G., Svejgaard, A., Andersen, V. & Jacobsen, S. (1999). Mannose-binding lectin polymorphisms and susceptibility to infection in systemic lupus erythematosus. *Arthritis and Rheumatism* **42**, 2145–2152.
96. Foster, C.B., Lehrnbecher, T., Mol, F., Steinberg, S.M., Venzon, D.J., Walsh, T.J., Noack, D., Rae, J., Winkelstein, J.A., Curnutte, J.T. & Chanock, S.J. (1998). Host defense molecule polymorphisms influence the risk for immune-mediated compications in chronic granulomatous disease. *Journal of Clinical Investigation* **102**, 2146–2155.
97. Graudal, N.A., Homann, C., Madsen, H.O., Svejaard, A., Jurik, A.G., Graudal, H.K. & Garred, P. (1998). Mannan binding lectin in rheumatoid arthritis, a longitudinal study. *Journal of Rheumatology* **25**, 629–635.

98. Brandrup, F., Homburg, K.M., Wang, P., Garred, P. & Madsen, H.O. (1999). Mannan-binding lectin deficiency associated with recurrent cutaneous abscesses, prurigo and possibly atopic dermatitis. A family study. *British Journal of Dermatology* **140**, 168–192.
99. Malhotra, R., Wormald, M.R., Rudd, P.M., Fischer, P.B., Dwek, R.A. & Sim, R.B. (1995). Glycosylation changes of IgG associated with rheumatoid arthritis can activate complement via the mannose-binding protein. *Nature Medicine* **1**, 237–243.
100. Stanworth, S.J., Donn, R.P., Hassall, A., Dawes, P., Ollier, W. & Snowden, N. (1998) Absence of an association between mannose-binding lectin polymorphism and rheumatoid arthritis. *British Journal of Rheumatology* **37**, 186–188.
101. Garred, P., Richter, C., Andersen, Å.B., Madsen, H.O., Mtoni, I., Svejgaard, A. & Shao, J. (1997). Mannan-binding lectin in the sub-Saharan HIV and tuberculosis epidemics. *Scandinavian Journal of Immunology* **46**, 204–208.
102. Bellamy, R., Ruwende, C., McAdam, K.P.W.J., Thursz, M., Sumiya, M., Summerfield, J., Gilbert, S.C., Corrah, T., Kwiatkowski, D., Whittle, H.C. & Hill, A.V.S. (1998). Mannose binding protein deficiency is not associated with malaria, hepatitis B carriage nor tuberculosis in Africans. *Quarterly Journal of Medicine* **91**, 13–18.
103. Hoalvan-Helden, E.G., Epstein, J., Victor, T.C., Hon, D.N., Lewis, L.A., Beyers, N., Zurakowski, D., Ezekowitz, R.A.B. & von Helden, P.D. (1999). Mannose-binding protein B allele confers protection against tuberculous meningitis. *Pediatric Research* **45**, 459–464.
104. Graudal, N.A., Madsen, H.O., Tarp, U., Svejgaard, A., Jurik, A.G., Graudal, H.K. & Garred, P. (2000). The association of variant mannose-binding lectin genotypes with radiographic outcome in rheumatoid arthritis. *Arthritis and Rheumatism* **43**, 515–521.
105. Soothill, J.F. & Harvey, B.A.M. (1976). Defective opsonisation: a common immunity deficiency. *Archives of Disease in Childhood* **51**, 91–99.
106. Candy, D.C.A., Larcher, V.F., Tripp, J.H., Harries, J.T., Harvey, B.A.M. & Soothill, J.F. (1980). Yeast opsonisation in children with chronic diarrhoeal states. *Archives of Disease in Childhood* **55**, 189–193.
107. Valdimarsson, H., Stefansson, M., Vikingsdottir, T., Arason, G.J., Koch, C., Thiel, S. & Jensenius, J.-C. (1998). Reconstitution of opsonizing activity by infusion of mannan-binding lectin (MBL) to MBL-deficient humans. *Scandinavian Journal of Immunology* **48**, 116–123.
108. Kilpatrick, D.C., James, V.S., Blackwell, C.C., Weir, D.M., Hallam, N.F. & Busuttil, A. (1998). Mannan binding lectin and the sudden infant death syndrome. *Forensic Science International* **97**, 135–138.
109. Nevens, J.R., Mallia, A.K., Wendt, M.W. & Smith, P.K. (1992). Affinity chomatographic purification of immunoglobulin M antibodies utilizing immobilized mannan binding protein. *Journal of Chromatography* **597**, 247–256.
110. Uemura, K., Yokota, Y., Kozutsumi, Y. & Kawasaki, T. (1996). A unique CD45 glycoform recognised by the serum mannan-binding protein in immature thymocytes. *Journal of Biological Chemistry* **271**, 4581–4584.
111. Matsuda, M., Shikata, K., Wada, J., Sugimoto, H., Yasushi, S., Kawasaki, T. & Makino, H. (1998). Deposition of mannan binding protein and mannan binding protein-mediated complement activation in the glomeruli of patients with IgA nephropathy. *Nephron* **80**, 408–413.
112. Hansen, S. & Holmskov, U. (1998). Structural aspects of collectins and receptors for collectins. *Immunobiology* **199**, 165–189.

Mast cell function-associated antigen

Other names: MAFA; 2F1-antigen
Mol. wt: 70 000 (native); 35 000 (subunit)
Specificity: Mannose
Source: Mammalian leukocytes

ISOLATION

MAFA was first purified from rat basophilic leukaemia cells using an immobilised murine monoclonal antibody, a disulphide-linked homodimer of 28 000–40 000 being obtained [1]. The corresponding cDNA was later cloned and expressed in transfected COS cells [2]. Subsequently, human [3] and murine [4] homologues were cloned.

STRUCTURE

The original rat MAFA is a glycoprotein that exists as a monomer and homodimer [1,2]. The protein component of each subunit is a 188 amino acid polypeptide of Mr 20 000 [2]. The human homologue is 54% identical in primary sequence to rat MAFA [3]; the murine homologue has 80% sequence identity to that of the rat [4].

All have a similar structure typical of the group V C-type lectin transmembrane receptors. The N-terminal cytoplasmic domain harbours a sequence motif similar to the functional ITIM in the tail of NK cell inhibitory receptors, and a phosphorylation site. The C-type lectin domain on the extracellular (C-terminal) side of the transmembrane domain shows most homology with those of the group V C-type lectins [2–4]. In particular, MAFA possesses an 'NK cell domain' lacking residues critical for calcium-dependent carbohydrate-binding.

BIOLOGICAL ACTIVITIES

MAFA was discovered and defined by its ability to inhibit the activation of mast cells induced by aggregation of the high affinity IgE receptor, FcεRI [1].

Recombinant MAFA could bind to immobilised neoglycoproteins in a Ca^{2+}-dependent manner [5]. Binding to BSA-mannose was stronger than to any of the five others tested. MAFA also bound to amylase, soybean agglutinin, E CorL and carboxypeptidase, but not to transferrin. It was concluded that MAFA recognised terminal mannose residues [5].

TISSUE AND/OR SUBCELLULAR DISTRIBUTION

Although MAFA was originally described as a mast cell-specific antigen in rats [1,2], the human homologue is found on peripheral blood NK cells as well [3], while the murine homologue is expressed on NK cells but not on mast cells [4].

POSSIBLE FUNCTION(S)

Essentially nothing is known about the function(s) of this antigen, and any ideas would have to take account of the species variation in cellular expression.

COMMENTS

In all species investigated, MAFA/2F1 antigen appears to be encoded by a single, non-polymorphic gene located within the NK gene region [4].

REFERENCES

1. Soto, E.O. & Pecht, I. (1988). A monoclonal antibody that inhibits secretion from rat basophilic leukaemia cells and binds to a novel membrane component. *Journal of Immunology* **141**, 4324–4332.
2. Guthmann, M.D., Tal, M. & Pecht, I. (1995). A secretory inhibitory signal transduction molecule on mast cells is another C-type lectin. *Proceedings of the National Academy of Sciences USA* **92**, 9397–9401.
3. Butcher, S., Arney, K.L. & Cook, G.P. (1998). MAFA-L, an ITIM-containing receptor encoded by the human NK cell gene complex and expressed by basophils and NK cells. *European Journal of Immunology* **28**, 3755–3762.
4. Hanke, T., Corral, L., Vance, R.E. & Raulet, D.H. (1998). 2F1 antigen, the mouse homologue of the rat 'mast cell function associated antigen', is a lectin-like type II transmembrane receptor expressed by natural killer cells. *European Journal of Immunology* **28**, 4409–4417.
5. Binsack, R. & Pecht, I. (1997). The mast cell function-associated antigen exhibits saccharide binding capacity. *European Journal of Immunology* **27**, 2557–2561.

Melanoplus sanguinipes lectin

Other name: Grasshopper lectin
Mol. wt: 700 000 (native); 40 000 and 28 000 (subunit)
Specificity: Rhamnose, fucose, galactose
Source: Haemolymph of the American grasshopper *Melanoplus sanguinipes*

ISOLATION

The lectin was purified from clarified haemolymph in one step employing adsorption to galactose–Sepharose and elution with D-galactose [1].

STRUCTURE

The grasshopper lectin is a glycoprotein of unremarkable amino acid composition rich in acidic amino acids. In native form, it is a large molecule with apparent molecular weight estimations ranging from 590 000 (PAGE) to 700 000 (gel filtration). It is an aggregate of 70 000 subunits, each a disulphide-linked heterodimer of polypeptides of apparent molecular weights 40 000 and 28 000 determined by SDS-PAGE [1].

BIOLOGICAL ACTIVITIES

The grasshopper lectin agglutinates porcine erythrocytes and (less readily) feline and bovine red blood cells. Untreated human, ovine and chicken erythrocytes are not agglutinated, but asialo-human erythrocytes give a titre similar to that of porcine cells. Haemagglutinating activity required divalent cations and was labile to heating at 56 °C for one minute, although stable at 37 °C for at least seven hours [1].

Haemagglutination was inhibited by a wide variety of small saccharides and sugar derivatives containing galactose or glucose. Of the limited range of monosaccharides tested, rhamnose was the most potent inhibitor followed by fucose and galactose; free glucose was not inhibitory [1].

TISSUE AND/OR SUBCELLULAR DISTRIBUTION

The lectin has been reported only in haemolymph where it existed as a minor component, accounting for <0.5% of the total soluble protein [1].

POSSIBLE FUNCTION(S)

A role in defence is assumed, but no evidence has been presented [1].

COMMENTS

A single lectin appears to account for all of the haemagglutinating activity present in the haemolymph of adult grasshoppers, but anomalous results with lapine erythrocytes might be an indication to the contrary [1].

The related species, *M. differentialis* possesses a very similar lectin, the characteristics of which are indistinguishable from those of the *M. sanguinipes* lectin [1].

REFERENCE

1. Stebbins, M.R. & Hapner, K.D. (1985). Preparation and properties of haemagglutinin from haemolymph of Acrididae (grasshoppers). *Insect Biochemistry* **15**, 451–462.

Modiolus modiolus lectins E and H

Other name:	Modiolin E/modiolin H
Mol. wt:	Multiple <1 300 000 (native); 14 000, 17 500 and 20 000 (subunit)
Specificity:	Sialic acid
Source:	Haemolymph of the horse mussel *Modiolus modiolus*

ISOLATION

Soluble components of haemolymph were applied to a column of bovine submaxilliary mucin–Sepharose. After extensive washing, lectin (haemagglutinating) activity was recovered using an NaCl gradient [1]. The resulting preparation appeared heterogeneous and may have consisted of aggregates with at least two separate lectin activities: a major activity towards equine red blood cells (modiolin E) and a minor activity towards human erythrocytes (modiolin H). Modiolin E/modiolin H could not be readily separated but only modiolin E activity could be precipitated by phosphorylcholine and may represent the horse mussel equivalent of C-reactive protein [1].

STRUCTURE

The modiolin E/modiolin H complex is a glycoprotein containing approximately 19% carbohydrate. Gel filtration and ultracentrifugation indicate that the bulk of the lectin activity exists as high molecular weight material (approximately 1×10^6 kDa), but some lectin activity is found below 400 000 by gel filtration. Gradient gel electrophoresis produced three major bands with apparent molecular weights of 1 300 000, 232 000 and 105 000. With SDS-PAGE under reducing conditions, three bands corresponding to subunit molecular weights of 14 000, 17 500 and 20 000 were obtained [1]. Using two-dimensional electrophoresis with immobilised pH gradients, 4 subunits were detected: Mr 14 000, pI 5.1; Mr 14 000, pI 5.5; Mr 17 500, pI 5.5; and Mr 20 000, pI 4.9 [2]. It appears that the native lectin consists of a series of aggregates made up of combinations of these subunits in a manner yet to be fully established.

BIOLOGICAL ACTIVITIES

The *M. modiolus* lectins agglutinated erythrocytes from a variety of species, including mice, dogs, humans and sheep. The highest titres were obtained with equine erythrocytes, and absorption experiments indicated the existence of at least two separate activities as noted before. Neuraminidase-treated red blood cells (of whatever species) were not agglutinated and *N*-acetylneuraminic acid was a potent inhibitor of agglutination, unlike all other monosaccharides tested. Fetuin and bovine submaxilliary mucin were also effective inhibitors [1].

Modiolin E/modiolin H agglutinated numerous bacteria; heat-killed bacterial cells absorbed out most of the erythro-agglutinating activity, and purified lipopolysaccharide from marine vibrios inhibited erythro-agglutination [1]. Modiolin E/modiolin H also exhibited strong growth inhibitory activity towards *Vibrio anguillarum*, *V. viscosus* and *V. wodanis* [2].

Modiolin E activity was Ca^{2+}-dependent; modiolin H activity was Ca^{2+}-independent. Modiolin E (but not modiolin H) precipitated with phosphorylcholine and bound to a column of *p*-amino-aminophenyl-phosphorylcholine-agarose, from which it could be eluted with EDTA [1].

TISSUE AND/OR SUBCELLULAR DISTRIBUTION

Labelling with rabbit anti-lectin antibody and protein A/gold specifically demonstrated antigen in cytoplasmic granules of haemocytes [2]. The lectin found in the haemolymph serum may originate in the haemocytes.

POSSIBLE FUNCTION(S)

A defence function is presumed since modiolin E/modiolin H agglutinated and inhibited the growth of several potentially pathogenic bacteria [2].

COMMENTS

Although the biochemical basis of the lectin activities are far from clear, it seems likely that the *M. modiolus* lectins exist as a mixture of aggregates. The generation of these aggregates from the subunits available may come about in a random manner to achieve diversity [1].

REFERENCES

1. Tunkijjanukij, S., Mokkelsen, H.V. & Olafsen, J.A. (1997). A heterogeneous sialic-binding lectin with affinity for bacterial LPS from horse mussel (*Modiolus modiolus*) hemolymph. *Comparative Biochemistry and Physiology* **117**B, 273–286.
2. Tunkijjanukij, S. & Olafsen, J.A. (1998). Sialic acid-binding lectin with antibacterial activity from the horse mussel: further characterisation and immunolocalisation. *Developmental and Comparative Immunology* **22**, 139–150.

Myelin-associated glycoprotein

Other names: MAG; siglec-4
Mol. wt: 100 000
Specificity: NeuAc-α2,3-Gal-R
Source: Mammalian brain

ISOLATION

MAG was first purified from rat brain by biochemical means involving lithium diiodosalicylate-phenol extraction and gel filtration [1].

Isolations of cDNA clones for rat [2], human [3] and murine [4] myelin-associated glycoproteins were subsequently achieved.

STRUCTURE

MAG is a glycoprotein (30% carbohydrate) with an apparent molecular weight of 100 000 determined by SDS-PAGE [1]. The deduced amino acid sequences from rats [2], humans [3] and mice [4] have been published in full, and exhibit a very high degree (>95%) of homology. In rodents, two forms of MAG RNA are produced from a single gene by alternative splicing, resulting (for rats) in a 610 amino acid mature protein of 68 000 Da (deduced) and a slightly smaller isoform (deduced Mr 63 000) differing mainly in the carboxyl-terminal cytoplasmic region [2,4]. The extracellular side of a transmembrane region contains five immunoglobulin-like domains, and shows strong overall sequence resemblance to the neural cell adhesion molecule, N-CAM [2]. It is therefore a type I integral membrane protein and a member of the immunoglobulin superfamily.

BIOLOGICAL ACTIVITIES

MAG participates in neuron–oligodendrocyte and oligodendrocyte–oligodendrocyte adhesion and can bind extracellular matrix components like collagen [5]. MAG has been shown to have lectin activity on the basis of experiments with derivatised erythrocytes and immobilised glycolipids, and inhibition of such activity with sialylated glycoconjugates [6]. A detailed study with synthetic oligosaccharides indicated recognition of terminal α2,3-linked sialic acid, but interactions were optimal with pentasaccharides [7]. MAG bound 2-keto-3-deoxy-D-glycero-D-galacto-nononic acid better than *N*-acetylneuraminic acid [7].

TISSUE AND/OR SUBCELLULAR DISTRIBUTION

MAG is expressed on myelinating oligodendrocytes and Schwann cells in the central and peripheral nervous systems [8].

POSSIBLE FUNCTION(S)

It has long been considered that MAG functions in myelination, axonal growth regulation and signal transduction, but knock-out mice have provided rather limited evidence for these assumptions. Nevertheless, abnormalities are apparent in gene-targeted mice consistent with a role for MAG in myelin/axon organisation [6].

COMMENTS

MAG is a member of the 'siglec' group of I-type lectins, along with sialoadhesin (qv), CD22 (qv) and CD33 (qv). It is also closely related to the avian Schwann cell myelin protein [9], and less closely to the Po glycoprotein [10].

MAG is notable for its developmental regulation. In rats, the larger isoform is upregulated after birth and predominates in the early post-natal period, while the smaller isoform is expressed later and persists in adults [2].

MAG is deficient in the plaques of patients with multiple sclerosis and autoantibodies to MAG are found in some patients with this disorder (and indeed in patients with several other demyelinating diseases) [3].

REFERENCES

1. Quarles, R.H., Barbarash, G.R., Figlewicz, D.A. & McIntyre, L.J. (1983). Purification and partial characterization of the myelin-associated glycoprotein from adult brain. *Biochimica et Biophysica Acta* **757**, 140–143.
2. Lai, C., Brow, M.A., Nave, K.A., Noronha, A.B., Quarles, R.A., Bloom, F.E., Milner, R.J. & Sutcliffe, J.G. (1987). Two forms of 1B236/myelin associated glycoprotein, a cell adhesion molecule for postnatal neural development, are produced by alternative splicing. *Proceedings of the National Academy of Sciences USA* **84**, 4337–4341.
3. Sato, S., Fujita, N., Kurihara, T., Kuwano, R., Sakimura, K., Takahashi, Y. & Miyatake, T. (1989). cDNA cloning and amino acid sequence for human myelin-associated glycoprotein. *Biochemical and Biophysical Research Communications* **163**, 1473–1480.
4. Fujita, N., Sato, S., Kurihara, T., Kuwano, R., Sakimura, K., Inuzuka, T., Takahashi, Y. & Miyatake, T. (1989). cDNA cloning of mouse myelin-associated glycoprotein: a novel alternative splicing pattern. *Biochemical and Biophysical Research Communications* **165**, 1162–1169.
5. Probstmeier, R., Fahrig, T., Spiess, E. & Schachner, M. (1992). Interactions of the neural cell adhesion molecule and the myelin-associated glycoprotein with collagen type I: involvement in fibrillogenesis. *Journal of Cell Biology* **116**, 1063–1070.
6. Kelm, S., Schauer, R. & Crocker, P.R. (1996). The sialoadhesions – a family of sialic acid-dependent cellular recognition molecules within the immunoglobulin superfamily. *Glycoconjugate Journal* **13**, 913–926.
7. Strenge, K., Schauer, R., Bovin, N., Hasegawa, A., Ishida, H., Kiso, M. & Kelm, S. (1998). Glycan specificity of myelin-asociated glycoprotein and sialoadhesion deduced from interactions with synthetic oligosaccharides. *European Journal of Biochemistry* **258**, 677–685.
8. Trapp, B.D. (1990). Myelin-associated glycoprotein. Location and potential functions. *Annals of the New York Academy of Sciences* **605**, 29–43.
9. Dulac, C., Tropak, M.B., Cameron-Curry, P., Rossier, J., Marshak, D.R., Roder, J. & Le Douarin, N.M. (1992). Molecular characterization of the Schwann cell myelin protein, SMP: structural similarities within the immunoglobulin superfamily. *Neuron* **8**, 323–334.
10. Lemke, G. (1988). Unwrapping the genes of myelin. *Neuron* **1**, 535–543.

N-CAM

Other names: Neural cell adhesion molecule; CD56; NKH-1
Mol. wt: 170 000–250 000
Specificity: Heparin; oligomannose
Sources: Avian and mammalian neural tissues; NK cells

ISOLATION

N-CAM was first identified and purified from retina and brain by immunoaffinity chromatography [1].

STRUCTURE

N-CAM is a glycoprotein with a high and variable sialic acid content [1,2]. Differential sialylation accounts for the large difference in apparent molecular weights determined by SDS-PAGE for N-CAM isolated from different sources; the polypeptide has an Mr of 120 000 and the fully desialated polypeptide an Mr of 140 000, but the native, sialylated form is typically around 200 000 [1,2].

N-CAM is an integral membrane protein with a short C-terminal cytoplasmic region, a transmembrane domain and a large extracellular region [1,2]. The latter consists of two fibronectin type-III domains and five (C2-set) immunoglobulin-like domains with homologies to other I-type lectins [3].

BIOLOGICAL ACTIVITIES

N-CAM mediates homotypic neural cell–cell adhesion in a calcium- and carbohydrate-independent manner [4]. N-CAM also promotes cell to substratum adhesion and this activity is mediated by a separate heparin-binding site [5]. N-CAM also binds to another cell-surface neural adhesion molecule (L1) to form a cis-complex in a lectin-like manner via oligomannose structures on L1. This Ca^{2+}-independent interaction is mediated by the fourth immunoglobulin-like domain [6].

TISSUE AND/OR SUBCELLULAR DISTRIBUTION

N-CAM is found in the central and peripheral nervous systems and in developing skeletal muscle. Transient N-CAM expression has been noted at sites associated with reorganisation of tissue structure [1].

A variant of N-CAM (CD56, NKH-1) is present on human NK cells and some activated T lymphocytes [7].

POSSIBLE FUNCTION(S)

N-CAM appears to function principally in neural cell adhesion and differentiation, and in neuron–muscle adhesion [1,2].

COMMENTS

N-CAM is an abundant adhesion molecule, comprising approximately 1% of the total cell-surface protein [1]. The single gene (located on human chromosome 11) produces a wide variety of alternative forms, mainly by gene splicing but also by post-translational modifications. One of these isoforms is the NK cell marker, CD56 [7].

REFERENCES

1. Rutishauser, U. (1984). Developmental biology of a neural cell adhesion molecule. *Nature* **310**, 549–554.
2. Edelman, G.M. (1983). Cell adhesion molecules. *Science* **219**, 450–457.
3. Powell, L.D. & Vorki, A. (1995). I-type lectins. *Journal of Biological Chemistry* **270**, 14243–14246.
4. Rutishauer, U., Hoffman, S. & Edelman, G.M. (1982). Binding properties of a cell adhesion molecule from neural tissue. *Proceedings of the National Academy of Sciences USA* **79**, 685–689.
5. Cole, G.J. & Glaser, L. (1986). A heparin-binding domain from N-CAM is involved in neural cell–substratum adhesion. *Journal of Cell Biology* **102**, 403–412.
6. Horstkorte, R., Schachner, M., Magyar, J.P., Vorherr, T. & Schmitz, B. (1993). The fourth immunoglobulin-like domain of NCAM contains a carbohydrate recognition domain for oligomannosidic glycans implicated in association with L1 and neurite outgrowth. *Journal of Cell Biology* **121**, 1409–1421.
7. Pigott, R. & Power, C. (1994). *The Adhesion Molecule Facts Book*, pp. 115–119. Academic Press, London.

Neurocan

Other name: 1D1 proteoglycan
Mol. wt: 136 000 (for core protein)
Specificity: Sulphated glycolipids
Source: Rodent brains

ISOLATION

Neurocan was first isolated from rat brain by immunoaffinity chromatography using a monoclonal antibody (1D1) from a panel of antibodies prepared against chondroitin sulphate proteoglycans [1]. It was later cloned using mixed oligonucleotide-primed PCR amplification of cDNA [2].

STRUCTURE

Neurocan is the third in size of the four hyalectans, with a much smaller central region than versican (qv) or aggrecan (qv). The cloned gene encodes a sequence of 1257 amino acids with a molecular mass of 136 kDa in the adult and/or early post-natal brain proteoglycans. The N-terminal (hyaluronic-acid-binding) region with its immunoglobulin and link protein domains is typical of the hyalectan family. The relatively small central region contains seven or more potential binding sites for glycosaminoglycans and has no significant homology to any other protein. The C-terminal region is typical of the family (about 60% identity with the corresponding region of other hyalectans) with two epidermal growth factor-like domains and a complement regulatory protein domain on either side of the C-type lectin domain [2].

BIOLOGICAL ACTIVITIES

The recombinant C-type lectin domain of neurocan can bind specifically to sulphated glycolipids in the presence of Ca^{2+}, thus permitting its inclusion as a lectin [3]. Neurocan binds strongly to neural cell adhesion molecules (Ng-CAM and N-CAM) [4], and weakly to tenascin-R [5], but these interactions may not be carbohydrate mediated.

Neurocan can inhibit neuronal adhesion and neurite outgrowth [4].

Like other hyalectans, neurocan can bind hyaluronic acid [1,2].

TISSUE AND/OR SUBCELLULAR DISTRIBUTION

Neurocan has been found only in brain, and its mRNA was not detectable in several other tissues [1,2]. Neurocan is present at highest concentration in the molecular layer and fibre tracts of developing cerebellum [1,4].

The location was predominantly extracellular shortly after birth but mainly cytoplasmic in adult cerebellum [1].

POSSIBLE FUNCTION(S)

It has been suggested that neurocan may modulate neuronal adhesion and neurite growth during development [4].

COMMENTS

Neurocan is developmentally regulated with respect to molecular size, carbohydrate composition and sulphation as well as concentration [1]. The concentration decreases from 20% of the total soluble chondroitin sulphate proteoglycan protein at 7 days post-natal to 6% in adult brain. The 136 000 core protein (sometimes designated neurocan-C), which becomes the major species within the first month of life, corresponds to the C-terminal half of a larger species present at birth [1,2].

REFERENCES

1. Rauch, U., Gao, P., Janetzko, A., Flaccus, A., Hilgenberg, L., Tekotte, H., Margolis, R.K. & Margolis, R.U. (1991). Isolation and characterization of developmentally regulated chondroitin sulphate and chondroitin/keratin sulphate proteoglycans of brain identified with monoclonal antibodies. *Journal of Biological Chemistry* **266**, 14785–14801.
2. Rauch, U., Karthikeyan, L., Mairel, P., Margolis, R.U. & Margolis, R.K. (1992). Cloning and primary structure of neurocan, a developmentally regulated, aggregating chondroitin sulphate proteoglycan of brain. *Journal of Biological Chemistry* **267**, 19536–19547.
3. Miura, R., Aspberg, A., Ethell, I.M., Hagihara, K., Schnaar, R.L., Ruoslahti, E. & Yamaguchi, Y. (1999). The proteoglycan lectin domain binds sulfated cell surface glycolipids and promotes cell adhesion. *Journal of Biological Chemistry* **274**, 11431–11438.
4. Friedlander, D.R., Milev, P., Karthikeyan, L., Margolis, R.K., Margolis, R.U. & Grumet, M. (1994). The neuronal chrondroitin sulphate proteoglycan neurocan binds to the neural cell adhesion molecules Ng-CAM/LI/NILE and N-CAM, and inhibits neuronal adhesion and neurite outgrowth. *Journal of Cell Biology* **125**, 669–680.
5. Aspberg, A., Miura, R., Bourdoulous, S., Shimonaka, M., Heinegård, D., Schachner, M., Ruoslahti, E. & Yamaguchi, Y. (1997). The C-type lectin domains of lecticans, a family of aggregating chondroitin sulphate proteoglycans, bind tenascin-R by protein–protein interactions independent of carbohydrate moiety. *Proceedings of the National Academy of Sciences USA* **94**, 10116–10121.

NKCL

Mol. wt: 26 000
Specificity: Mannose
Sources: Murine bone marrow and spleen cells

ISOLATION

The NKCL gene was identified by its overexpression in a murine model of chronic myeloid leukaemia blast crisis; the cDNA was cloned and used to transfect COS cells [1].

STRUCTURE

The cDNA and deduced amino acid sequences have been published in full [1]. NKCL is a 210 amino acid polypeptide (predicted Mr 23 000) with a type-II transmembrane receptor structure. It has a cytoplasmic tail of 17 amino acid residues, a transmembrane region (32 residues), an extracellular neck region (29 amino acids) and a C-terminal domain (132 amino acids) homologous to the carbohydrate-recognition domain of typical C-type lectins. It is glycosylated at the single N-linked glycosylation site [1]. The carbohydrate-recognition domain resembles those of the C-type group II transmembrane receptors more than those of the group V (NK type) receptors, since the residues forming the Ca^{2+}-binding site-2 are completely conserved [1].

BIOLOGICAL ACTIVITIES

NKCL bound to mannose–Sepharose, but not galactose–Sepharose, in a Ca^{2+}-dependent manner [1].

TISSUE AND/OR SUBCELLULAR DISTRIBUTION

Northern blotting revealed that the *Nkcl* gene is expressed in spleen, thymus, lung and bone marrow (but not liver, heart, brain or kidney) of adult mice. Among haematopoietic cells, it is expressed in pluripotent myeloid precursor cells, precursor and mature monocytes and mature neutrophils. Within the cell, it was localised to the cell surface of transfected COS cells by immunocytochemistry [1].

POSSIBLE FUNCTION(S)

It has been suggested that NKCL may be involved in the abnormal trafficking of myeloid cells in chronic myeloid leukaemia [1].

COMMENTS

The NKCL gene maps just proximal to the region on mouse chromosome 6 that encodes the NK gene cluster. Although NKCL resembles most of the NK (group V) lectin-like receptors in having a short neck region connecting the carbohydrate-recognition domain to the cell surface and in lacking a Ca^{2+}-binding site-1, it has a conserved Ca^{2+}-binding site-2 and its cytoplasmic and transmembrane domains are more similar to those of the chicken hepatic lectin (an asialoglycoprotein receptor, qv), a group II C-type lectin. The strongest similarity may be to *Mpcl*, a gene cloned from murine macrophages [2].

REFERENCES

1. Fernandes, M.J., Finnegan, A.A., Siracusa, L.D., Brenner, C., Iscove, N.N. & Calabretta, B. (1999). Characterization of a novel receptor that maps near the natural killer gene complex: demonstration of carbohydrate-binding and expression in hematopoietic cells. *Cancer Research* **59**, 2709–2717.
2. Balch, S.G., McKnight, A.J., Seldrin, M.F. & Gordon, S. (1998). Cloning of a novel C-type lectin expressed by murine macrophages. *Journal of Biological Chemistry* **273**, 18656–18664.

NKR-P1 family

Other names: 3.2.3 antigen; NK1.1 antigen
Mol. wt: 60 000 (native); 30 000 (subunit)
Specificity: GalNAc, GlcNAc
Source: Mammalian NK cells

ISOLATION

NKR-P1 was first identified on activated rat NK cells by a murine monoclonal antibody, 3.2.3 [1]. The corresponding cDNA was later cloned from a library created from mRNA derived from lymphokine-activated NK cells [2]. Related cDNAs encoding the murine equivalent, NK 1.1, were isolated and expressed in eukaryotic cells [3,4]. The human homologue has also been cloned [5].

STRUCTURE

The NKR-P1 family consists of homodimeric glycoproteins with the overall structure of type II transmembrane receptors. The rat molecule has 223 amino acid residues with a calculated Mr of 24 551; glycosylation accounts for the difference from the apparent Mr values of 30 000 and 60 000 determined by SDS-PAGE in the presence and absence, respectively, of reducing agent [1,2].

As well as a short cytoplasmic N-terminal region and hydrophobic transmembrane domain, the larger extracellular part of the molecule consists of a C-type carbohydrate-recognition domain homologous to those of other group V C-type lectins, but with an unusually high lysine content [2].

BIOLOGICAL ACTIVITIES

Recombinant rat NKR-P1 bound to neoglycoconjugates bearing β-D-GalNAc, β-D-GlcNAc or α-L-fucose, but not other monosaccharides. The most potent inhibitors were GalNAc and GlcNAc, although fucose >> galactose > mannose were also inhibitory. Ca^{2+} was very tightly associated, but calcium dependence of sugar binding could be demonstrated [6]. NKR-P1 may contribute to NK-cell-mediated cytotoxicity [1,7].

TISSUE AND/OR SUBCELLULAR DISTRIBUTION

NKR-P1 family members are found predominantly on NK cells [1–5], but are also expressed on a small proportion of T lymphocytes and immature thymocytes [8].

POSSIBLE FUNCTION(S)

The NKR-P1 family is associated with NK cell activation and cytotoxicity, indicating a role in innate immunity and immune surveillance [9].

Involvement in the transendothelial migration of $CD4^+$ T lymphocytes has also been suggested [8].

COMMENTS

Both in rodents and humans, the NKR-P1 family proteins are encoded in the genetic region known as the NK gene complex, which also encodes Ly-49 (qv) and CD69 (qv). Three related genes (A, B and C) have been identified in rodents, but so far only a single human homologue has been found [9].

NKR-P1 is expressed on all rodent NK cells but only a very small proportion of rodent T cells, and therefore can act as an NK cell marker. In humans, however, NKR-P1 is not present on all NK cells, and is found on a higher proportion of T lymphocytes [5].

Attempts to reproduce the recombinant form of NKR-P1A used to demonstrate lectin activity were unsuccessful, and a new recombinant carbohydrate-binding domain appears to lack carbohydrate-binding activity [10]. There is some doubt, therefore, whether this molecule is a true lectin.

REFERENCES

1. Chambers, W.H., Vujanovic, N.L., De Leo, A., Olszowy, M.W., Herbermann, R.B. & Hiserodt, J.C. (1989). Monoclonal antibody to a triggering structure expressed on rat natural killer cells and adherent lymphokine-activated killer cells. *Journal of Experimental Medicine* **169**, 1373–1389.
2. Giorda, R., Rudent, W.A., Vavassori, C., Chambers, W.H., Hiserodt, J.C. & Trucco, M. (1990). NKR-P1, a signal transduction molecule on natural killer cells. *Science* **249**, 1298–1300.
3. Yokoyama, W.M., Jacobs, L.B., Kanagawa, O., Shevach, E.M. & Cohen, D.I. (1989). A murine T lymphocyte antigen belongs to a supergene family of type II integral membrane proteins. *Journal of Immunology* **143**, 1379–1386.
4. Ryan, J.C., Turck, J., Niemi, E.C., Yokoyama, W.M. & Seaman, W.E. (1992). Molecular cloning of the NK 1.1 antigen, a member of the NKR-P1 family of natural killer cell activation molecules. *Journal of Immunology* **149**, 1631–1635.
5. Lanier, L.L., Chang, C. & Phillips, J.H. (1994). Human NKR-P1A. *Journal of Immunology* **153**, 2417.
6. Bezouska, K., Vlahas, G., Horvath, O., Jinochova, G., Fiserova, A., Giorda, R., Chambers, N.H., Feizi, T. & Pospíšil, M. (1994). Rat natural killer cell antigen, NKR-P1, related to C-type animal lectins is a carbohydrate-binding protein. *Journal of Biological Chemistry* **269**, 16945–16952.
7. Ryan, J.C., Niemi, E.C., Nakamura, M.C. & Seaman, W.E. (1995). NKR-PIA is a target-specific receptor that activates natural killer cell cytotoxicity. *Journal of Experimental Medicine* **181**, 1911–1915.
8. Poggi, A., Costa, P., Zocchi, M.R. & Moretta, L. (1997). NKRPIA molecule is involved in transendothelial migration of $CD4^+$ human T lymphocytes. *Immunology Letters* **57**, 121–123.
9. Brown, M.G., Scalzo, A.A., Matsumoto, K. & Yokoyama, W.M. (1997). The natural killer gene complex: a genetic basis for understanding natural killer cell function and innate immunity. *Immunological Reviews* **155**, 53–65.
10. Kogelberg, H., Montero, E., Bay, S., Lawson, A.M. & Feizi, T. (1999). Re-evaluation of monosaccharide binding property of recombinant soluble carbohydrate recognition domain of the natural killer cell receptor NKP-P1A. *Journal of Biological Chemistry* **274**, 30335–30336.

Non-integrin elastin/laminin-binding protein

Other names: 67 kDa Elastin-binding protein (EBP); 67 kDa laminin-binding protein
Mol. wt: 67 000
Specificity: Lactose
Source: Mammalian tissues

ISOLATION

Two independently discovered proteins, a high affinity laminin-binding protein and an elastin receptor, were subsequently found to be the same molecule [1,2]. It was first isolated by affinity chromatography on laminin–Sepharose [3], and subsequently on columns of immobilised elastin peptides [4].

STRUCTURE

EBP is a protein of Mr 67 000 determined by SDS-PAGE, which exhibits immunological cross-reactivity with galectin-1 [1,4].

EBP possesses sequence homology to the enzymatically-inactive, alternatively spliced form of human β-galactosidase [2,5]. This variant retains the galactosugar-binding site of β-galactoside while losing the enzymatic and lysosomal targeting domains and acquiring a new 32 amino acid residue, frame-shift encoded sequence in the central region of the molecule.

BIOLOGICAL ACTIVITIES

EBP binds to laminin and elastin via hydrophobic, protein–protein interactions [1,2]. It also has a galactoside-binding site [6]. These two sites are not independent: elastin is bound with high affinity in the absence of galactose or lactose, but such sugars appear to induce a conformation change, leading to a dramatic drop in affinity for elastin. Lactose was much more efficient than galactose, while glucose and fucose had no effect.

TISSUE AND/OR SUBCELLULAR DISTRIBUTION

This lectin was originally detected in tumour cells, but occurs very widely in normal cells and tissues, including muscle and nervous tissue, epithelia and various leukocytes. It is present in the cytoplasm, on the cell surface and in the extracellular space [1,2,7].

POSSIBLE FUNCTION(S)

EBP appears to mediate interaction between cells and the extracellular matrix, and has long been suspected of involvement in the spread of cancer cells. As a peripheral membrane protein, it may mediate the chemotactic response of several cell types to soluble peptides and be involved in the generation of signal transmission [2].

The main function of EBP could be to assist the secretion and assembly of tropoelastin by acting as a recycling shuttle and chaperone [2,7].

COMMENTS

This protein is of interest as an excellent example of how protein–carbohydrate interaction can regulate protein–protein interaction. Elastin is not a glycoprotein, and the EBP–elastin interaction can be inhibited by the elastin peptide sequence, Val-Gly-Val-Ala-Pro-Gly. EBP can also be eluted from elastin columns with lactose, strongly suggesting the lectin activity can regulate the non-lectin activity. This phenomenon appears to be involved in the macromolecular assembly of elastin into elastic fibres [6].

REFERENCES

1. Mecham, R.P. (1991). Receptors for laminin on mammalian cells. *FASEB Journal* **5**, 2538–2546.
2. Hinek, A. (1996). Biological roles of the non-integrin elastin/laminin receptor. *Biological Chemistry Hoppe-Seyler* **377**, 471–480.
3. Rao, N.C., Barsky, S.H., Terranova, V.P. & Liotta, L.A. (1983). Isolation of a tumor cell laminin receptor. *Biochemical and Biophysical Research Communications* **111**, 804–808.
4. Mecham, R.P., Hinek, A., Griffin, G.L., Senior, R.M. & Liotta, L.A. (1989). The elastin receptor shows structural and functional similarities to the 67-kDa tumour cell laminin receptor. *Journal of Biological Chemistry* **264**, 16652–16657.
5. Hinek, A., Rabinovitch, M., Keeley, F., Okamura-Oho, Y. & Callahan, J. (1993). The 67-kD elastin/laminin-binding protein is related to an enzymatically inactive, alternatively spliced form of β-galactosidase. *Journal of Clinical Investigation* **91**, 1198–1205.
6. Hinek, A., Wrenn, D.S., Mecham, R.P. & Barondes, S.H. (1988). The elastin receptor is a galactoside-binding protein. *Science* **239**, 1539–1541.
7. Hinek, A. & Rabinovitch, M. (1994). 67 kD Elastin-binding protein is a protective 'companion' of extracellular insoluble elastin and intracellular tropoelastin. *Journal of Cell Biology* **126**, 563–574.

P35

Other names: L-ficolin; ficolin-like protein
Mol. wt: 320 000 (native); 35 000 (subunit)
Specificity: GlcNAc
Sources: Human blood and liver

ISOLATION

P35 was first purified from recalcified human plasma by affinity chromatography on mannan–Sepharose; after elution of mannan-binding lectin (qv) with mannose, P35 was eluted with GlcNAc [1]. An apparently identical protein (L-ficolin) was obtained from human plasma by affinity chromatography on GlcNAc–Sepharose as the key step [2]. Proteins isolated by affinity chromatography on α-elastin–Sepharose (EBP-37: [3]) and on an immobilised steroid derivative (hucolin: [4]) also appear to be the same as P35.

P35 cDNA has been cloned from a human liver cDNA library [1].

STRUCTURE

The key structural feature of P35 is the presence of both collagen-like and fibrinogen-like domains [1]. The former aspect resembles collectins, but the latter, which has two potential Ca^{2+}-binding sites, appears to possess the carbohydrate-binding activity [5]. The amino acid sequence of 313 amino acid residues is homologous to porcine ficolins α and β [1,2], and to another human protein designated 'human ficolin' or 'ficolin M' because it is synthesised mainly by monocytes [6]. P35 has >75% sequence identity to the porcine ficolins, and >80% identity to human ficolin M.

Members of this family, like the collectins, have an overall structure resembling C1q, and consist of disulphide-linked oligomers of a basic tri-polypeptide structural unit [2]. The amino acid sequence [1] forms three domains: an N-terminal non-collagenous region of 13 amino acids, a collagenous region of 57 amino acids and a 218 amino acid fibrinogen-like C-terminal region. There is evidence for multimer formation [1–4]. The major form has an apparent Mr of 320 000 determined by SDS-PAGE under non-reducing conditions, suggesting a 9-mer (3×3), but there is also evidence for the formation of four trimers to form a dodecamer [7]. The predicted Mr (31 400) implies that at least one of the two potential sites is glycosylated.

BIOLOGICAL ACTIVITIES

P35 was reported as having calcium-dependent [1] and L-ficolin as having calcium-independent [2] lectin activity, although otherwise they appear identical. The elastin-binding protein named EBP-37 [3] and the steroid-binding protein named hucolin [4] may also be the same as P35. P35 bound to GlcNAc–BSA (but not to neoglycoproteins made from coupling other monosaccharides to BSA or to any of a variety of natural glycoproteins tested) and to complex GlcNAc-containing neoglycolipids [1]. It bound to neoglycolipids containing fully galactosylated bi-, tri- and tetra-antennary complex-type oligosaccharides and to neoglycolipids containing non-galactosylated bi-antennary complex-type oligosaccharides of IgG, but not to those containing high mannose-type or hybrid-type oligosaccharides. P35 bound to mannan–Sepharose, GlcNAc–agarose and asialofetuin–Sepharose, and could be eluted from all three with GlcNAc [1], but did not bind to Emphaze–mannan [8]. These observations are consistent with the unusual property described of L-ficolin that it can bind to CNBr-activated, but underivatised, Sepharose [2].

P35 binds to *Salmonella typhimurium* (Ra strain) and enhances phagocytosis by human neutrophils [1]. (Human ficolin/ficolin-M was shown to bind *Escherichia coli* [6].)

TISSUE AND/OR SUBCELLULAR DISTRIBUTION

P35 is synthesised mainly or exclusively in the liver and secreted into the bloodstream [8]. (In contrast, M-ficolin is synthesised mainly in lung, placenta and peripheral blood leukocytes, including monocytes [6,9,10].)

POSSIBLE FUNCTION(S)

Since P35 and ficolin-M can promote phagocytosis of bacteria, it is thought plasma ficolins function in innate immunity as opsonins.

COMMENTS

P35-related proteins have been found in mice as well as pigs [7]. In both these species there are two ficolin genes that code for proteins homologous to human P35 and M-ficolin; one gene is expressed in liver and is the major source of plasma ficolin. Humans have an additional, less closely related gene, which encodes the Hakata antigen (qv) [7].

P35 is found in normal adult human serum within the concentration range 1.12–12.8 μg/ml (median 3.8), but at a significantly lower level (0.2–4.7 μg/ml; median 2.5) in umbilical cord blood. Unusually low concentrations of P35 were found in a small minority of women experiencing recurrent miscarriage [11].

The genes for P35 and ficolin-M are located on human chromosome 9 [9].

REFERENCES

1. Matsushita, M., Endo, Y., Taira, S., Sato, Y., Fujita, T., Ichikawa, N., Nakata, M. & Mizuochi, T. (1996). A novel human serum lectin with collagen- and fibrinogen-like domains that functions as an opsonin. *Journal of Biological Chemistry* **271**, 2448–2454.

2. Le, Y., Tan, S.M., Lee, S.H., Kon, O.L. & Lu, J. (1997). Purification and binding properties of a human ficolin-like protein. *Journal of Immunological Methods* **204**, 43–49.
3. Harumiya, S., Omori, A., Sugiura, T., Fukumoto, Y., Tachikawa, H. & Fujimoto, D. (1995). EBP-37, a new elastin-binding protein in human plasma: structural similarity to ficolins, transforming growth factor β1-binding proteins. *Journal of Biochemistry* **117**, 1029–1035.
4. Edgar, P.F. (1995). Hucolin, a new corticosteroid-binding protein from human plasma with structural similarities to ficolins, transforming growth factor β1-binding proteins. *FEBS Letters* **375**, 159–161.
5. Le, Y., Lee, S.H., Kon, O.L. & Lu, J. (1998). Human L-ficolin: plasma levels, sugar specificity and assignment of its lectin activity to the fibrinogen-like (FBG) domain. *FEBS Letters* **425**, 367–370.
6. Lu, J., Le, O., Kon, O.L., Chan, J. & Lee, S.H. (1996). Biosynthesis of human ficolin, an *Escherichia coli*-binding protein, by monocytes: comparison with the synthesis of two macrophage-specific proteins, C1q and the mannose receptor. *Immunology* **89**, 289–294.
7. Ohashi, T. & Erickson, H.P. (1998). Oligomeric structure and tissue distribution of ficolins from mouse, pig and human. *Archives of Biochemistry and Biophysics* **360**, 223–232.
8. Kilpatrick, D.C. (1997). Isolation of human mannan binding lectin, serum amyloid P component and related factors from Cohn fraction III. *Transfusion Medicine* **7**, 289–294.
9. Endo, Y., Sato, Y., Matsushita, M. & Fujita, T. (1996). Cloning and characterization of the human lectin P35 gene and its related gene. *Genomics* **36**, 515–521.
10. Lu, J., Tay, P.N., Kon, O.L. & Reid, K.B.M. (1996). Human ficolin: cDNA cloning, demonstration of peripheral blood leukocytes as the major site of synthesis and assignment of the gene to chromosome 9. *Biochemical Journal* **313**, 473–478.
11. Kilpatrick, D.C., Fujita, T. & Matsushita, M. (1999). P35, an opsonic lectin of the ficolin family, in human blood from neonates, normal adults, and recurrent miscarriage patients. *Immunology Letters* **67**, 109–112.

Pacifasticus leniusculus β-1,3-glucan-binding protein

Other name: Crayfish laminarin-binding protein
Mol. wt: 100 000
Specificity: β-1,3-Glucans
Source: Haemolymph of the freshwater crayfish *Pacifasticus leniusculus*

ISOLATION

Haemolymph plasma was first ultracentrifuged to remove haemocyanin. The lectin was then purified by immunoaffinity chromatography using rabbit antibodies to the lectin precipitated with laminarin [1].

The lectin was cloned after constructing a cDNA library and screening with the rabbit antiserum [2].

STRUCTURE

This lectin is a glycoprotein with an amino acid composition low in cysteine and generally similar to that reported for the β-1,3-glucan-binding protein of *Bombyx mori* [1,2]. It is a monomer of apparent Mr 100 000 determined by SDS-PAGE with or without reducing agent [1]. The calculated Mr from the deduced amino acid sequence (which contained no typical signal sequence) was almost 152 000, perhaps an indication that the processed protein has undergone significant post-translational processing [2].

The isoelectric point is 5.0 [1].

BIOLOGICAL ACTIVITIES

This lectin binds to and precipitates laminarin, a β-1,3-glycan, but not dextran, mannan or cellulose. It also enhances phenoloxidase (and peptidase) activity in the presence of laminarin, but not on its own or in the presence of dextran, mannan or cellulose [1].

The lectin failed to agglutinate erythrocytes of equine, lapine or chicken origin [1], but bound to haemocytes [2], and could opsonise yeast (*Saccharomyces cerevisiae)* for phagocytosis by hyaline haemocytes [2].

TISSUE AND/OR SUBCELLULAR DISTRIBUTION

The lectin is found in haemocyte plasma and in haemocyte lysates [1]. Northern blotting analysis indicated that the hepatopancreas was the major and possibly sole site of synthesis; no specific RNA could be detected in haemocytes, gills, epidermis, green gland or abdominal muscle [2].

POSSIBLE FUNCTION(S)

It is presumed to function in defence since it recognises a typically fungal carbohydrate structure, has opsonic activity *in vitro*, and can enhance phenoloxidase activity.

COMMENTS

Although this lectin has not been shown to have haemagglutinating activity, and does not exhibit primary sequence homology with any of the major groups of animal lectins [2], it is undoubtedly a true lectin with a precise specificity for β-1,3-glycans. Its overall amino acid composition was very similar to those of the lectins with equivalent specificity found in the silkworm *Bombyx mori* and (to a lesser extent) the cockroach *Blaberus craniifer* [1].

REFERENCES

1. Duvic, B. & Soderhall, K. (1990). Purification and characterization of a β-1,3-glucan binding protein from plasma of the crayfish *Pacifastacus leniusculus*. *Journal of Biological Chemistry* **265**, 9327–9332.
2. Cerenius, L., Liang, Z., Duvic, B., Keyser, P., Hellman, U., Tapio Palva, E., Iwanaga, S. & Soderhall, K. (1994). Structure and biological activity of a 1,3-β-D-glucan-binding protein in crustacean blood. *Journal of Biological Chemistry* **269**, 29462–29467.

Pallidin

Other name: *Polysphondylium pallidum* lectin
Mol. wt: 250 000 (native); 26 500, 26 000 and 25 000 (subunit)
Specificity: Lactose, D-galactose
Source: Aggregating amoebae of *Polysphondylium pallidum*

ISOLATION

Pallidin was isolated from the soluble fraction of *P. pallidum* sonicates by adsorption onto formalin-treated human (O) erythrocytes and elution with D-galactose [1,2]. An improved method involves extraction in lactose-containing buffer and affinity chromatography on acid-treated Sepharose 6B [3].

STRUCTURE

Pallidin is a protein (no attached carbohydrate) with an amino acid composition rich in lysine which generally resembles that of discoidin II (qv) more than that of discoidin I (qv) [2]. Sedimentation equilibrium analysis of native pallidin in the presence of galactose yielded an approximate molecular weight of 250 000 [2]. The native molecule has an isoelectric point around neutral pH [2], but three distinct species of pI, 6.4 (I), 7.3 (II) and 7.5 (III), can be resolved [3].

Species II and III are isoforms of the original pallidin and consist of two very similar subunits of apparent Mr 26 500 (A) and 26 000 (B) in the ratio 2A : B (II) or A : 2B (III). It tends to form aggregates of Mr 225 000 and above, but exists as a basic triplet (Mr 73 000) in the presence of 2-mercaptoethanol [3].

Species I is a separate isolectin, not appreciably recovered from the earlier purification procedures, and which possesses a 25 000 subunit (C) in the ratio 1C : 2B. It exists as trimers and hexamers (irrespective of the presence of 2-mercaptoethanol) yielding apparent Mr values of 77 000 and 160 000 by sedimentation equilibrium [3].

BIOLOGICAL ACTIVITIES

Both original pallidin and the species I isolectin agglutinate human and lapine erythrocytes [1–3]. Additionally, exogenously added pallidin promotes the agglutination of cohesive *P. pallidum* amoebae [1]. The agglutination of erythrocytes is inhibited by sugars of the galactose-type configuration; D-galactose itself was the most potent monosaccharide, but was less effective than lactose.

TISSUE AND/OR SUBCELLULAR DISTRIBUTION

Pallidin is present on the cell surface [1].

POSSIBLE FUNCTION(S)

Pallidin exhibits a very marked developmental regulation, with specific lectin activity increasing in parallel with the cohesiveness of the cells. Monosaccharides that inhibit pallidin activity also block cell cohesion, whereas lectin non-inhibitory sugars do not affect cohesion [1]. These observations and the presence of pallidin on the cell surface are consistent with a role for pallidin in the aggregation process and/or the species-specific cell sorting phenomenon observed when amoebae of different species are mixed.

COMMENTS

The pallidin isolectins have much in common with the discoidins and related lectins from *Dictyostelium* species. A major difference, however, is the triplet of subunits as the basic unit of structure, unlike the tetrameric arrangement found in lectins from the genus, *Dictyostelium*.

REFERENCES

1. Rosen, S.D., Simpson, D.L., Rose, J.E. & Barondes, S.H. (1974). Carbohydrate-binding protein from *Polysphondylium pallidum* implicated in intracellular adhesion. *Nature* **252**, 128–151.
2. Simpson, D.L., Rosen, S.D. & Barondes, S.H. (1975). Pallidin, purification and characterization of a carbohydrate-binding protein from *Polysphondylium pallidum* implicated in intracellular adhesion. *Biochimica et Biophysica Acta* **412**, 109–119.
3. Rosen, S.D., Kaur, J., Clark, D.L., Pardos, B.T. & Frazier, W.A. (1979). Purification and characterization of multiple species (isolectins) of a slime mold lectin implicated in intercellular adhesion. *Journal of Biological Chemistry* **254**, 9408–9415.

Perch roe lectin

Other name: *Perca fluviatilis* roe lectin
Mol. wt: 13 000 (subunit)
Specificity: Mannose
Source: Roe of the perch *Perca fluviatilis*

ISOLATION

The lectin was isolated from an extract of perch roe by affinity chromatography on *O*-α-L-fucosyl-polyacrylamide and elution with L-fucose. The purified material appeared electrophoretically heterogeneous [1], but actually consists of major and minor electrophoretic forms with similar but not identical mobility [2].

STRUCTURE

The lectin is a zinc-containing protein or glycoprotein (<2% carbohydrate) with a relatively high content of aromatic amino acids. A single band corresponding to a molecular weight of 13 000 is obtained by SDS-PAGE under reducing conditions. Under non-reducing conditions, however, bands corresponding to two, three and four times that Mr are also present. Ultracentrifugal analysis revealed the presence of two major components with sedimentation coefficients of 3.3 S and 7.6 S [1]. Even larger molecular species have been detected. The lectin appears to be a complex mixture of two homomultimers and heteromultimers formed by two similar subunits. These subunits differ in amino acid composition, and while one has a blocked N-terminus the other has glutamic acid at its N-terminus [2].

BIOLOGICAL ACTIVITIES

This lectin agglutinates human erythrocytes with a distinct preference for blood group O. The minimum concentration required to agglutinate group B cells (125 μg/ml) is 30-fold higher than the minimum for group O. Haemagglutination can be abolished by EDTA; paradoxically, addition of exogenous Zn^{2+}, Mn^{2+} or Ca^{2+} was also strongly inhibitory [1].

Haemagglutination was readily inhibited (in decreasing order of potency) by D-mannose, L-fucose, D-glucose, L-arabinose, D-galactose, GalNAc and L-rhamnose. The first was 30-fold more effective than the last.

TISSUE AND/OR SUBCELLULAR DISTRIBUTION

The lectin is concentrated in the cortical vesicles of maturing oocytes, but is also found in large oocytes and follicular epithelial cells [3].

POSSIBLE FUNCTION(S)

Although originally thought of as defensive ('protectin') molecules, additional functions have been suggested for fish egg lectins including the prevention of polyspermy [3].

COMMENTS

This lectin is rather unusual. Unlike most fish egg lectins, e.g. the roach roe lectin (qv), the powan roe lectin (qv) and *Plecoglossus altivelis* egg lectin (qv), which are all blood group B specific to some extent and are most readily inhibited by L-rhamnose, the perch lectin is relatively blood group O specific and is more sensitive to inhibition by mannose and fucose. It also has an unusually high zinc content.

REFERENCES

1. Krajhanzl, A., Hořejší, V. & Kocourek, J. (1978). Studies on lectins. XLII. Isolation, partial characterization and comparison of lectins from the roe of five fish species. *Biochimica et Biophysica Acta* **532**, 215–224.
2. Bazil, V. & Entlicher, G. (1999). Complexity of lectins from the hard roe of perch (*Perca fluviatilis* L.). *International Journal of Biochemistry and Cell Biology* **31**, 431–442.
3. Nosek, J., Krajhanzl, A. & Kocourek, J. (1983). Studies on lectins. LVII. Immunofluorescence localization of lectins present in fish ovaries. *Histochemistry* **79**, 131–139.

Periplaneta americana (original) lectin

Other name: American cockroach lectin
Mol. wt: 1 500 000 (native); 30 000 (subunit)
Specificity: GalNAc, D-fucose
Source: Haemolymph of the American cockroach *Periplaneta americana*

ISOLATION

The lectin was isolated from haemolymph supernatant in two steps. The first step consisted of affinity chromatography on bovine submaxilliary gland mucin conjugated to BioGel P-150. The bound lectin was washed with a Ca^{2+}-containing buffer then eluted with EDTA-containing buffer. The second step, involving desorption from a carboxymethyl–Sepharose column with an NaCl gradient, resulted in a virtually pure protein with a 13 000-fold increase in specific activity [1].

STRUCTURE

This lectin has a large aggregate structure apparent on electron microscopy as rods of approximately 50 × 10 nm, consisting of tandemly aligned units whether in the presence of Ca^{2+} or EDTA [1].

The native molecular weight was in excess of 10^6 determined by sucrose density gradient centrifugation. A single subunit of apparent Mr 30 000 was detected by SDS-PAGE under reducing conditions. Thus the *P. americana* lectin seems to have an unusual polymeric structure made up of approximately 50 subunits [1].

BIOLOGICAL ACTIVITIES

This cockroach lectin agglutinates both ovine and human erythrocytes and requires Ca^{2+} (neither Mg^{2+} nor Mn^{2+} will substitute) to do so. With human cells, it is blood group A specific; the relative activity towards A, B and O cells was 64 : 4 : 1, respectively [1].

Erythrocyte agglutination was readily inhibited by GalNAc and D-fucose, but other monosaccharides required at least a ten-fold increase in concentration to achieve the same effect. The lectin was also inhibited to the same degree by porcine gastric mucin at a ten-fold lower concentration than that required by bovine submaxilliary mucin [1].

TISSUE AND/OR SUBCELLULAR DISTRIBUTION

The lectin was sought only in the haemolymph [1].

POSSIBLE FUNCTION(S)

The lectin is presumed to function in defence.

COMMENTS

The purified lectin is blood group A specific, in contrast to the unfractionated haemolymph, suggesting additional lectin(s) in the haemolymph. Indeed, a bacteria-agglutinating, lipopolysaccharide-binding lectin has also been isolated [2], as indeed have two other lectins [3,4], from the same species. All four may be considered to belong to a C-type lectin family called the *Periplaneta* lectin-related protein family.

REFERENCES

1. Kubo, T. & Natori, S. (1987). Purification and some properties of a lectin from the hemolymph of *Periplaneta americana* (American cockroach). *European Journal of Biochemistry* **168**, 75–82.
2. Jomori, T., Kubo, T. & Natori, S. (1990). Purification and characterization of lipopolysaccharide-binding protein from hemolymph of American cockroach *Periplaneta americana*. *European Journal of Biochemistry* **190**, 201–206.
3. Kubo, T., Kawasaki, K. & Natori, S. (1990). Sucrose-binding lectin in regenerating cockroach (*Periplaneta americana*) legs: its purification from adult haemolymph. *Insect Biochemistry* **20**, 585–591.
4. Kubo, T., Kawasaki, K. & Natori, S. (1993). Transient appearance and localization of a 26 kDa lectin, a novel member of the *Periplaneta* lectin family, in regenerating cockroach leg. *Developmental Biology* **156**, 381–390.

Periplaneta americana lipopolysaccharide-binding protein

Other name: Cockroach LPS-binding lectin
Mol. wt: 200 000 to 800 000 (native); 28 000 (subunit)
Specificity: Fucose, galactose
Source: Haemolymph of the American cockroach *Periplaneta americana*

ISOLATION

Haemolymph was first adsorbed with ovine erythrocytes to remove the original *Periplaneta americana* lectin (qv). The treated haemolymph was then applied to a column of DEAE–Sepharose and washed; LPS-binding activity was eluted with a gradient of NaCl. Finally the LPS-binding lectin was isolated in homogeneous form after affinity chromatography on mannose–agarose and elution with mannose [1].

The cDNA for this lectin has been cloned and characterised [2].

STRUCTURE

Periplaneta LPS-binding protein is a glycine-rich glycoprotein with an average molecular weight of 450 000, consisting of aggregates of a single 28 kDa subunit [1]. The deduced amino acid sequence of 223 residues yields a calculated Mr of 24 667; this is consistent with the determination by SDS-PAGE, which shifts from 28 000 to 24 500 after enzymatic deglycosylation [2].

The deduced amino acid sequence showed significant homology to the carbohydrate-recognition domains of several C-type lectins, from mammals as well as other invertebrates, including an overall 20% sequence similarity to the human asialoglycoprotein receptor [2].

BIOLOGICAL ACTIVITIES

The lectin agglutinated LPS-coated, but not untreated, ovine erythrocytes. Ca^{2+}, but not other divalent cations, was required for activity. Inhibition was most readily achieved with D- or L-fucose, followed by galactose and glucose. Mannose was a moderately effective inhibitor, but GalNAc and GlcNAc were only effective at >100 mM. Ra lipopolysaccharides prepared from some *Escherichia coli* isolates were potent inhibitors, while *E. coli* LPS lacking the hexose core was ineffective. The lectin had no affinity for Ra LPS from *Salmonella minnesota* R60 [1].

The LPS-binding protein has opsonic activity. It enhanced the clearance of *E. coli* injected intra-abdominally, and stimulated ingestion of *E. coli* by haemocytes *in vitro* [3].

TISSUE AND/OR SUBCELLULAR DISTRIBUTION

The lectin is synthesised in the fat body and secreted into the haemolymph [2].

POSSIBLE FUNCTION(S)

The lectin is thought to function in defence as an opsonin [3]. It behaves as an acute-phase response component [2] and stimulates uptake of *E. coli* by haemocytes [3].

COMMENTS

This LPS-binding protein is one of four lectins isolated from the same species [4]. It is a true lectin, binding the carbohydrate moiety of LPS and not binding the lipid A region of the molecule. It is curious, however, that it recognises *E. coli* K12 but not *Salmonella minnesota* R60, despite the similarity in the hexose core region between those bacteria [1].

REFERENCES

1. Jomori, T., Kubo, T. & Natori, S. (1990). Purification and characterization of lipopolysaccharide-binding protein from hemolymph of American cockroach *Periplaneta americana*. *European Journal of Biochemistry* **190**, 201–206.
2. Jomori, T. & Natori, S. (1991). Molecular cloning of cDNA for lipopolysaccaride-binding protein from the haemolymph of the American cockroach, *Periplaneta americana*. *Journal of Biological Chemistry* **266**, 13318–13323.
3. Jomori, T. & Natori, S. (1992). Function of the lipopolysaccharide-binding protein of *Periplaneta americana* as an opsonin. *FEBS Letters* **296**, 283–286.
4. Arai, T., Kawasaki, K., Kubo, T. & Natori, S. (1998). Cloning of cDNA for regeneration, a humoral C-type lectin of *Periplaneta americana*, and expression of the regenectin gene during leg regeneration. *Insect Biochemistry and Molecular Biology* **28**, 987–994.

Periplaneta americana 26 kDa lectin

Mol. wt: 2.6×10^6 (native); 26 000 (subunit)
Specificity: L-Rhamnose
Source: Haemolymph of the American cockroach *Periplaneta americana*

ISOLATION

The 26 kDa lectin was obtained from haemolymph serum after separation from regenectin (qv) on a column of sucrose–Sepharose. After the regenectin had been eluted with sucrose, the 26 kDa lectin was eluted with 1 mM EDTA in a calcium-free buffer. The 26 kDa lectin was then obtained in a purified form by application to a column of DEAE–Toyopearl and elution with an NaCl gradient [1].

STRUCTURE

The 26 kDa lectin is a protein or glycoprotein with a very high native molecular weight. It was found to sediment faster than any of the molecular weight markers used on a glycerol density gradient, and by extrapolation an approximate Mr value of 2 600 000 was calculated. On SDS-PAGE under reducing conditions, it appears as a single polypeptide of apparent Mr 26 000 [1].

The 26 kDa lectin shows immunological cross-reactivity with the original *Periplaneta americana* lectin (qv) and with regenectin, and, like them, appears as heterogeneous rod-shaped structures by electron microscopy [1].

BIOLOGICAL ACTIVITIES

The 26 kDa lectin agglutinates lapine erythrocytes. This activity is Ca^{2+}-dependent and neither Mg^{2+} nor Mn^{2+} can substitute for Ca^{2+} [1].

Haemagglutination was most readily inhibited by L-rhamnose, but D-arabinose, D-fucose, L-fucose, D-mannose and L-mannose were also potent inhibitors. None of the other monosaccharides tested was inhibitory [1].

TISSUE AND/OR SUBCELLULAR DISTRIBUTION

The 26 kDa lectin is found in haemolymph, and appears transiently in leg homogenates during regeneration after amputation. Immunofluorescent studies on sections from regenerating metathoracic legs indicate a localisation outside the epidermal layers [1].

POSSIBLE FUNCTION(S)

It has been suggested that the 26 kDa lectin contributes to the organisation or stabilisation of the epidermis [1].

COMMENTS

This 26 kDa lectin appears to belong to the *Periplaneta* lectin family, a sub-group of C-type lectins. The other members of the family also exist as large aggregates ($0.8–2.0 \times 10^6$ kDa) made up of single polypeptides of apparent Mr 28 000–37 000. Each member of the family has a distinct sugar specificity.

Unlike the *Periplaneta* (original) lectin or the *Periplaneta* LPS-binding protein (qv), which appear to function in defence, the 26 kDa lectin resembles regenectin in appearing transiently during late regeneration and is presumed to function in that process. It is noteworthy, however, that the 26 kDa lectin appears first after leg amputation (and can be detected in homogenates from days 9 to 20) whereas regenectin appears at a later stage (days 14 to 29). Moreover, the localisation of the 26 kDa lectin around the outside of the newly formed epidermal cells is different from that of regenectin. It seems plausible that these two lectins have distinct functions during tissue regeneration.

REFERENCE

1. Kubo, T., Kawasaki, K. & Natori, S. (1993). Transient appearance and localization of a 26-kDa lectin, a novel member of the *Periplaneta* lectin family, in regenerating cockroach leg. *Developmental Biology* **156**, 381–390.

Phalera flavescens lectin

Other name: *Phalera flavescens* agglutinin (PFA)
Mol. wt: 74 000 (native); 17 000 and 18 000 (subunit)
Specificity: *N*-acetyllactosamine
Source: Larvae of the moth *Phalera flavescens*

ISOLATION

An ammonium sulphate fraction was obtained from haemolymph plasma and applied to a column of DEAE–Toyopearl, which was subsequently eluted with a linear NaCl gradient. Lectin-containing fractions were again concentrated with ammonium sulphate before application to a column of Cellulofine GCL-1000-sf. The purified lectin eluted as a single peak of electrophoretically homogeneous protein [1].

STRUCTURE

The lectin is a glycoprotein (1% carbohydrate) with a native Mr estimated by gel filtration of 74 000. Two subunits (17 000 and 18 000) were detected by SDS-PAGE in the presence or absence of reducing agent. The subunits had a virtually identical amino acid composition rich in alanine, arginine, leucine and threonine residues, as well as glutamic and aspartic acids. N-terminal analyses of the first 26 residues revealed sequence identity between the subunits; however, only the 18 000 subunit was glycosylated. It seems likely that the only difference between the subunits is the attached carbohydrate, and that the lectin is a tetramer of glycosylated and non-glycosylated subunits held together by non-covalent interactions [1].

BIOLOGICAL ACTIVITIES

PFA agglutinates human and lapine erythrocytes. The titre against cells of both species was enhanced by neuraminidase treatment, and was not divalent cation dependent. The activity towards human cells was ABO non-specific, but umbilical cord blood cells were agglutinated at higher titre than adult cells [1].

Haemagglutination was not inhibited by any monosaccharide tested, but *N*-acetyllactosamine was an effective inhibitor. Among glycoproteins tested, only laminin inhibited. Adsorption experiments with immobilised saccharide structures indicated that PFA binds only to Gal-β1,4-GlcNAc-containing glycoconjugates [1].

PFA was found to be non-mitogenic for native or neuraminidase-treated murine lymphocytes [1].

TISSUE AND/OR SUBCELLULAR DISTRIBUTION

Nothing was reported beyond its occurrence in haemolymph plasma.

POSSIBLE FUNCTION(S)

Nothing is known concerning possible functions.

COMMENTS

PFA was the first lectin purified from a species of the insect family Notodontidae. It was found to be a new type of *N*-acetyllactosamine-specific lectin which differed in detailed specificity from previously reported agglutinins. Since then, a lectin with apparently similar specificity, HOL-II, has been isolated from the toxic marine sponge *Halichondria okadai* (qv) [2].

REFERENCES

1. Umetsu, K., Yamashita, K., Suzuki, J., Yamashita, T. & Suzuki, T. (1993). Purification and characterization of an *N*-acetyllactosamine-specific lectin from larvae of a moth, *Phalera flavescens*. *Archives of Biochemistry and Biophysics* **301**, 200–205.
2. Kawagishi, H., Yamawaki, M., Isobe, S., Usui, T., Kimura, A. & Chiba, S. (1994). Two lectins from the marine sponge, *Halichondria okadai*. *Journal of Biological Chemistry* **269**, 1375–1379.

Phosphomannosyl receptor I

Other names: Cation-independent mannose-6-phosphate receptor; insulin-like growth factor II receptor
Mol. wt: Approx. 300 000
Specificity: Mannose-6-phosphate
Source: Vertebrate cells generally

ISOLATION

The receptor was first isolated from a detergent extract of membranes from bovine liver using a strategy based on its affinity for β-galactosidase. Corresponding proteins were isolated from human skin fibroblasts, Chinese hamster ovary cells and rat hepatocytes in a similar way [1].

cDNA clones encoding part [2] and all [3] of the bovine receptor sequence have been isolated. The human [4] and rat [5] homologues have also been cloned.

STRUCTURE

The receptor is a large glycosylated protein with an apparent Mr of 215 000 determined by SDS-PAGE [1]. The deduced primary sequence of the bovine receptor indicates a mature protein of 2455 amino acid residues and a calculated Mr of 270 000. After glycosylation, the true Mr could be as high as 300 000, significantly greater than the value derived from SDS-PAGE. The C-terminal cytoplasmic region has 163 residues, while 23 residues are found in the transmembrane domain. The extracellular region has 2269 residues forming 15 repeating domains of about 147 residues and a distinctive pattern of eight cysteine residues. One domain has an additional inserted sequence similar to the type II repeat of fibronectin. Homology with the phosphomannosyl receptor II (qv) was noted [3].

The very similar (80% identity) human homologue has a sequence identical to that of the human insulin-like growth factor II receptor [4]. The rat insulin-like growth factor II receptor was found to be the rat homologue of the phosphomannosyl receptor at about the same time [5]. The receptor has two sugar phosphate binding sites distinct from the insulin-like growth factor II binding site, but it is not clear whether the receptor exists as a monomer or as an oligomer or how the quaternary structure might influence binding affinities [6].

BIOLOGICAL ACTIVITIES

The phosphomannosyl receptor I binds lysosomal enzyme glycoproteins in a divalent cation-independent but highly pH-dependent manner. This binding is specifically inhibited by mannose-6-phosphate and fructose-1-phosphate; various other phosphorylated sugars are non-inhibitory. Maximal binding occured at pH 6.0 and above, and there was a precipitous drop in binding as the pH dropped below 5.7 [1].

The receptor also binds insulin-like growth factor II with high affinity, although homologues found in non-mammals lack this high-affinity binding site [7]. As this ligand is a non-glycosylated polypeptide, the binding must involve protein–protein interaction.

TISSUE AND/OR SUBCELLULAR DISTRIBUTION

This receptor has been found on many cell types, where it is concentrated in the Golgi region [6–8]. It is cycled constitutively among the Golgi apparatus, endosomes and plasma membrane, with most being present in endosomes and little if any in lysosomes [7].

POSSIBLE FUNCTION(S)

A major function of this lectin is to direct newly synthesised lysosomal enzymes into lysosomes and away from the secretory pathway [6]. In the Golgi, phosphomannosyl residues act as ligands and the resulting lectin–lysosomal enzyme complex is dissociated after transport to the acidic endosomal compartment.

Other functions include the capture of extracellular lysosomal enzymes by endocytosis, a role in extracellular matrix degradation, and participation in signal transduction [7].

Studies of knock-out mice have confirmed that fibroblasts lacking the receptor secrete lysosomal enzymes excessively and have a diminished content of certain hydrolases within their lysosomes. Fibroblasts lacking both phosphomannosyl receptors I and II possess only 20% of the normal level of lysosomal hydrolases and accumulate undigested material (resembling mucolipodoses II and III). These gene-targeted embryos are not viable and die *in utero* [9,10].

COMMENTS

This receptor and its analogue, the phosphomannosyl receptor II, are clearly of major physiological significance. They have also found application in the detection and isolation of phosphorylated glycoproteins [11,12]. Screening and diagnosis of mucolipidoses II and III may be possible using an iodinated soluble form of the phosphomannosyl receptor I with samples immobilised on membranes [11].

REFERENCES

1. Sahagian, G.G., Distler, J. & Jourdian, G.W. (1981). Characterization of a membrane-associated receptor from bovine liver that binds phosphomannosyl residues of bovine testicular β-galactosidase. *Proceedings of the National Academy of Sciences USA* **78**, 4289–4293.
2. Lobel, P., Dahms, N.M., Breitmeyer, J., Chirgwin, J.M. & Kornfeld, S. (1987). Cloning of the bovine 215-kDa cation-independent mannose-6-phosphate receptor. *Proceedings of the National Academy of Sciences USA* **84**, 2233–2237.
3. Lobel, P., Dahms, N.M. & Kornfeld, S. (1988). Cloning and sequence analysis of the cation-independent mannose-6-phosphate receptor. *Journal of Biological Chemistry* **263**, 2563–2570.
4. Oshima, A., Nolan, C.M., Kyle, J.W., Grubb, J.H. & Sly, W.S. (1988). The human cation-independent mannose 6-phosphate receptor. *Journal of Biological Chemistry* **263**, 2553–2562.
5. MacDonald, R.G., Pfeffer, S.R., Coussens, L., Tepper, M.A., Brocklebank, C.M., Mole, J.E., Anderson, J.K., Chen, E., Czech, M.P. & Ullrich, A. (1988). A single receptor binds both insulin-like growth factor II and mannose-6-phosphate. *Science* **239**, 1134–1137.
6. Dahms, N.M. (1996). Insulin-like growth factor II/cation-independent mannose 6-phosphate receptor and lysosomal enzyme recognition. *Biochemical Society Transactions* **24**, 136–141.
7. Kornfeld, S. (1992). Structure and function of the mannose 6-phosphate insulinlike growth factor II receptors. *Annual Review of Biochemistry* **61**, 307–330.
8. Brown, W.J. & Farquhar, M.G. (1984). The mannose-6-phosphate receptor for lysosomal enzymes is concentrated in *Cis* Golgi cisternae. *Cell* **36**, 295–307.
9. Ludwig, T., Le Borgne, R. & Hoflock, B. (1995). Roles for mannose-6-phosphate receptors in lysosomal enzyme sorting, IGF-II binding and clathrin-coat assembly. *Trends in Cell Biology* **5**, 202–206.
10. Wang, Z.-Q., Fung, M.R., Barlow, D.P. & Wagner, E.F. (1994). Regulation of embryonic growth and lysosomal targeting by the imprinted Igf2/Mpr gene. *Nature* **372**, 464–467.
11. Valenzano, K.J., Kallay, L.M. & Lobel, P. (1993). An assay to detect glycoproteins that contain mannose-6-phosphate. *Analytical Biochemistry* **209**, 156–162.
12. Cummings, R.D. (1997). Lectins as tools for glycoconjugate purification and characterization. In: *Glycosciences: Status and Perspectives* (eds H.-J. Gabius & S. Gabius), pp. 191–199. Chapman & Hall, Weinheim.

Phosphomannosyl receptor II

Other name: Cation-dependent mannose-6-phosphate receptor
Mol. wt: 90 000 (native); 46 000 (subunit)
Specificity: Mannose-6-phosphate
Source: Vertebrate cells generally

ISOLATION

A second phosphomannosyl receptor was first isolated from murine macrophages and bovine liver by procedures involving affinity chromatography with immobilised lysosomal enzymes [1]. The corresponding receptors from human and rat sources were subsequently purified [2].

cDNAs have been cloned from human [3], bovine [4] and murine [5,6] sources. Human and murine cDNAs have been expressed in COS cells, and the bovine cDNA in *Xenopus* oocytes [4,5].

STRUCTURE

This receptor is a heavily glycosylated protein rich in hydrophobic amino acids [1–6]. The deduced amino acid sequence for the mature protein has just over 250 residues and a calculated Mr of about 30 000 whatever the species [3–6]. Homology between the species is very high, and differences in apparent subunit Mr depend as much on the tissue of origin as on the species [2].

The primary structure consists of a C-terminal cytoplasmic domain, a short transmembrane domain, and an extracellular N-terminal domain of about 164 amino acids. The extracellular region has more than twice as many residues as the cytoplasmic domain, and possesses significant homology to the phosphomannosyl receptor I (qv), especially with regard to the short sequences bordered by cysteines [3–6].

The apparent Mr of the subunit determined by SDS-PAGE is about 46 000. Oligomerisation can undoubtedly occur, and there is evidence that the receptor exists as a dimer in membranes [2].

BIOLOGICAL ACTIVITIES

The phosphomannosyl receptor II binds lysosomal enzymes, but notably not those from *Dictyostelium* [1,2]. It binds to phosphorylated saccharides including phosphomannan and neoglycoproteins containing mannose-6-phosphate [1,2,4,5,7]. Such binding typically requires the presence of divalent cations, particularly Mn^{2+}, but the human form seems to be divalent cation independent [7]. Binding can be inhibited by mannose-6-phosphate, glucose-6-phosphate and related phosphorylated saccharides [1,8,9]. It appears each dimeric receptor has two binding sites, both of which can be occupied by a single oligosaccharide containing two mannose-6-phosphate residues [8,9].

TISSUE AND/OR SUBCELLULAR DISTRIBUTION

This receptor seems to have a similar tissue and subcellular distribution to that of the phosphomannosyl receptor I, and is concentrated in the Golgi apparatus of many cell types [10].

POSSIBLE FUNCTION(S)

The major function of this receptor is to direct newly synthesised lysosomal enzymes into the endosomal compartment and away from the secretory pathway. The two phosphomannosyl receptors appear to fulfil similar and complementary, but not identical, functions [10]. Consequently, knock-out mice lacking both receptors exhibit a major missorting of multiple lysosomal enzymes and have a phenotype resembling mucolipodosis II; mutants lacking only one receptor have fibroblasts with lysosomal enzymes abnormalities, but a very much milder phenotype [11].

COMMENTS

Phosphomannosyl receptors I and II are homologous glycoproteins that constitute their own small family, structurally unrelated to those of the other animal lectins. Their function is perhaps the best defined of all the animal lectin family groups with the possible exception of the selectins. Although encoded by genes on separate chromosomes (human 6 and 12), it is likely the two receptors resulted from gene duplication. The structure and function of these two lectins were reviewed and compared by Kornfeld [12].

REFERENCES

1. Hoflack, B. & Kornfeld, S. (1985). Purification and characterization of a cation-dependent mannose 6-phosphate receptor from murine P388D$_1$ macrophages and bovine liver. *Journal of Biological Chemistry* **260**, 12008–12014.
2. Stein, M., Meyer, H.E., Hasilik, A. & Von Figura, K. (1987). 46-kDa mannose 6-phosphate-specific receptor: purification, subunit composition, chemical modification. *Biological Chemistry Hoppe-Seyler* **368**, 927–936.

3. Pohlmann, R., Nagel, G., Schmidt, B., Stein, M., Lorkowski, G., Krentler, C., Cully, J., Meyer, H.E., Grzeschik, K.-H., Mersmann, G., Hasilik, A. & Von Figura, K. (1987). Cloning of a cDNA encoding the human cation-dependent mannose 6-phosphate receptor. *Proceedings of the National Academy of Sciences USA* **84**, 5575–5579.
4. Dahms, N.M., Lobel, P., Breitmeyer, J., Chirgwin, J.M. & Kornfeld, S. (1987). 46 kD mannose 6-phosphate receptor: cloning, expression and homology to the 215 kD mannose 6-phosphate receptors. *Cell* **50**, 181–192.
5. Ma, Z., Grubb, J.H. & Sly, W.S. (1991). Cloning, sequencing and functional characterization of the murine 46-kDA mannose 6-phosphate receptor. *Journal of Biological Chemistry* **266**, 10589–10595.
6. Köster, A., Nagel, G., Von Figura, K. & Pohlmann, R. (1991). Molecular cloning of the mouse 46-kDa mannose 6-phosphate receptor (MPR46). *Biological Chemistry Hoppe-Seyler* **372**, 297–300.
7. Junghuns, U., Waheed, A. & Von Figura, K. (1988). The 'cation-dependent' mannose 6-phosphate receptor binds ligands in the absence of divalent cations. *FEBS Letters* **237**, 81–84.
8. Tong, P.Y. & Kornfeld, S. (1989). Ligand interactions of the cation-dependent mannose 6-phosphate receptor. *Journal of Biological Chemistry* **264**, 7970–7975.
9. Distler, J.J., Guo, J., Jourdain, G.W., Srivastava, O.P. & Hindsgaul, O. (1991). The binding specificity of high and low molecular weight phosphomannosyl receptors from bovine testes. *Journal of Biological Chemistry* **266**, 21687–21692.
10. Ludwig, T., Le Borgne, R. & Hoflack, B. (1995). Roles for mannose-6-phosphate receptors in lysosomal enzyme sorting, IGF-II binding and clathrin-coat assembly. *Trends in Cell Biology* **5**, 202–206.
11. Ludwig, T., Munier-Lehmann, H., Bauer, U., Hollinshead, M., Ovitt, C., Lobel, P. & Hoflack, B. (1994). Differential sorting of lysosomal enzymes in mannose 6-phosphate receptor-deficient fibroblasts. *EMBO Journal* **13**, 3430–3437.
12. Kornfeld, S. (1992). Structure and function of the mannose 6-phosphate/insulinlike growth factor II receptors. *Annual Review of Biochemistry* **61**, 307–330.

Physarumin

Other name: *Physarum polycephalum* lectin
Mol. wt: 21 500 (native); 10 000 (subunit)
Specificity: Complex
Source: Plasmodia of *Physarum polycephalum*

ISOLATION

Physarumin was purified from plasmodial extracts by ion-exchange chromatography followed by hydroxyapatite chromatography [1]. A similar lectin was obtained by other workers, who used affinity chromatography on thyroglobulin–Toyopearl as the first major fractionation step [2].

STRUCTURE

The apparent molecular weight of the native molecule has been estimated by gel filtration to be 22 000 [1] or 21 000 [2]. SDS-PAGE under reducing or non-reducing conditions indicated a subunit molecular mass of 8.7 kDa [1] to 11 kDa [2]. Physarumin is therefore likely to be a dimer. Its isoelectric point is 4.5 and its amino acid composition has been determined [1].

BIOLOGICAL ACTIVITIES

Physarumin agglutinates red blood cells from several species, but is most active towards lapine erythrocytes; its weaker activity towards human erythrocytes is ABO non-specific [1]. It can also agglutinate several murine tumour cells, and induce the apoptotic cell death of murine tumour cell lines [3].

Agglutinating activity was calcium-dependent [1,3]. Agglutinating activity was not inhibited by any of the simple sugars tested, but several glycoproteins, notably asialofetuin and thyroglobulin, were effective inhibitors [1–3].

TISSUE AND/OR SUBCELLULAR DISTRIBUTION

Physarumin is present in the slime layer coating the plasmodial surface [1,2].

POSSIBLE FUNCTION(S)

It has been conjectured that physarumin might influence the fusion of plasmodia [2]. Alternatively, it may have a role in interactions with bacteria [4].

COMMENTS

Physarumin is not the only agglutinin present in *P. polycephalum* plasmodial extracts [2,3]. The other lectin(s) shows some ability to agglutinate ovine erythrocytes (unlike physarumin), but is inactive towards human red blood cells [2,3]. This activity is at least partly represented by a molecule of 12 kDa made up of 6 kDa subunits [2]. The two *Physarum* lectins have similar amino acid sequences around the binding site, including a trytophan residue essential for carbohydrate-binding [5].

REFERENCES

1. Yokota, M. & Nitta, K. (1996). Purification and some properties of hemagglutinin from the myxomycete, *Physarum polycephalum*. *Experientia* **52**, 544–548.
2. Morita, M. & Nishi, A. (1992). Isolation and partial characterization of hemagglutinin from plasmodia of *Physarum polycephalum*. *Journal of General Microbiology* **138**, 619–625.
3. Numata, C., Yokota, M., Hosono, M., Ogawa, Y., Suzuki, T., Takayanagi, Y. & Nitta, K. (1998). Apoptotic cell death induced by physarumin (hemagglutinin from myxomycete, *Physarum polycephalum*). *Biological and Pharmacological Bulletin* **21**, 214–218.
4. Morita, M., Minami, E., Zheng, L. & Nishi, A. (1995). *Physarum polycephalum* haemagglutinins: effect of nutrition on synthesis, and their possible role in nature. *Microbiology-UK* **141,** 2315–2322.
5. Morita, M., Iwado, Y. & Okamura, S. (1998). Identification of a carbohydrate-binding site in *Physarum* haemagglutinin I. *Biochemistry and Molecular Biology International* **46,** 233–240.

Pieris brassicae lectin

Other names: PBA; white cabbage butterfly lectin
Mol. wt: 40 000 (native); 23 000 (subunit)
Specificity: GlcNAc
Source: Wings from larvae of the butterfly *Pieris brassicae*

ISOLATION

The starting material was a microsomal membrane fraction prepared from homogenate of wings of pupae or an acetone powder obtained from dissected wings. Both starting materials were solubilised in a buffer containing Triton X-100 before application to a column of immobilised GlcNAc. After washing, elution with GlcNAc resulted in a 640-fold increase in specific activity with 50% recovery of an electrophoretically pure protein [1].

STRUCTURE

PBA appears to be a homodimer since SDS-PAGE under reducing conditions yields a single band of apparent Mr 23 000 while gel filtration on Sephadex G-100 indicates a molecular size of around 40 000 [1].

BIOLOGICAL ACTIVITIES

PBA agglutinates fixed trypsin-treated lapine erythrocytes and chitin (scorpion cuticle) particles with a similar titre. These activities are readily inhibited by GlcNAc, but not by other monosaccharides. Oligomers of GlcNAc (chitobiose and chitotriose) were also effective inhibitors [1].

TISSUE AND/OR SUBCELLULAR DISTRIBUTION

PBA was extracted from epidermal cell membranes. Notably, no similar lectin activity could be detected in the haemolymph [1].

POSSIBLE FUNCTION(S)

It has been suggested that PBA may play some role in the metabolism of the cuticle [1].

COMMENTS

Lectins that bind both GlcNAc and oligomers of GlcNAc are uncommon, even in the plant kingdom, where only a few such lectins have been found, most of them from monocot sources [2].

REFERENCES

1. Mauchamp, B. (1982). Purification of an *N*-acetyl-D-glucosamine specific lectin (PBA) from epidermal cell membranes of *Pieris brassica* L. *Biochimie* **64**, 1001–1008.
2. Van Damme, E.J.M., Peumans, W.J., Pusztai, A. & Bardocz, S. (1998). *Handbook of Plant Lectins: Properties and Biomedical Applications*. John Wiley & Sons, Chichester.

Pinctada fucata lectin

Other name: Pearl oyster agglutinin
Mol. wt: 440 000 (native); 20 000 (subunit)
Specificity: D-Galactose, GalNAc
Source: Haemolymph of the pearl oyster *Pinctada fucata martensii*

ISOLATION

Haemolymph was dialysed against phosphate buffer and the precipitated material redissolved in a calcium-containing buffer and subjected to affinity chromatography on bovine submaxillary mucin–Sepharose (elution with 3 M $MgCl_2$); this was followed by anion exchange chromatography on a Mono Q column and finally gel filtration on Superose 6HR [1].

STRUCTURE

The apparent Mr of the native molecule estimated by gel filtration was 440 000. Gradient PAGE in the presence of SDS (but no reducing agent) yields a single band of Mr 20 000 [1]. The molecule therefore appears to be an aggregate of perhaps 20 or more non-covalently attached subunits.

Antiserum prepared in rabbits against the purified lectin cross-reacted with haemolymph components from several other bivalves, but not from gastropod sources [1].

BIOLOGICAL ACTIVITIES

The lectin agglutinates equine erythrocytes, and also those of sheep, rabbits and chickens in a Ca^{2+}-dependent manner. This activity was most readily inhibited by D-galactose and GalNAc, and much less readily so by *N*-acetylneuraminic acid and *N*-acetylmannosamine. D-Glucose and D-mannose were quite without effect [1].

TISSUE AND/OR SUBCELLULAR DISTRIBUTION

The lectin has only been reported in the haemolymph [1].

POSSIBLE FUNCTION(S)

The possible role of this lectin in cell recognition and defence is discussed by Suzuki and Mori [2].

COMMENTS

The lectin described appears to be the sole or major lectin of pearl oyster haemolymph, as anti-lectin antibodies completely inhibited the agglutination of erythrocytes from several different species [1].

Lectins have also been described to some extent from two other oyster species, *Crassostrea virginica* [3] and *C. gigans* [4,5]. The former, a major component of haemolymph, is also a large (33 S), Ca^{2+}-dependent molecule consisting of identical non-covalently attached subunits of Mr 20 000 [3]. The latter possesses at least two lectins, one preferring equine erythrocytes, the other a sialic acid-binding protein preferring human red blood cells [4]. Both 'Gigalin E' and 'Gigalin H' agglutinated bacteria and were induced by exposure to bacteria [5].

REFERENCES

1. Suzuki, T. & Mori, K. (1989). A galactose-specific lectin from the hemolymph of the pearl oyster, *Pinctada fucata martensii*. *Comparative Biochemistry and Physiology* **92B**, 455–462.
2. Suzuki, T. & Mori, K. (1990). Hemolymph lectin of the pearl oyster, *Pinctada fucata martensii*: a possible non-self recognition system. *Developmental and Comparative Immunology* **14**, 161–173.
3. Acton, R.T., Bennet, J.C., Evans, E.E. & Schrohenloher, R.E. (1969). Physical and chemical characterization of an oyster hemagglutinin. *Journal of Biological Chemistry* **244**, 4128–4135.
4. Hardy, S.W., Grant, P.T. & Fletcher, T.C. (1977). A haemagglutinin in the tissue fluid of the pacific oyster, *Crassostrea gigans*, with specificity for sialic acid residues in glycoproteins. *Experientia* **33**, 767–768.
5. Olafsen, J.A., Fletcher, T.C. & Grant, P.T. (1992). Agglutinin activity in pacific oyster (*Crassostrea gigans*) haemolymph following in vivo *Vibrio anguillarum* challenge. *Developmental and Comparative Immunology* **16**, 123–138.

Placental Gp120 receptor

Mol. wt: 46 000
Specificity: Mannan; L-fucose, D-mannose
Source: Human placenta

ISOLATION

A human placental cDNA library was screened by expression cloning and a cDNA encoding an HIV-Gp120-binding protein unrelated to CD4 was isolated. Stable expression of the lectin was obtained in transfected HeLa cells [1].

STRUCTURE

The cDNA sequence predicts a protein of 404 amino acids with a calculated Mr of 45 775; this is similar to the size (46 000) estimated from immunoprecipitation of the lectin expressed on transfected HeLa cells. The predicted structure consists of an N-terminal cytoplasmic region (40 residues) connected to a 21-residue hydrophobic transmembrane domain; a series of 8 tandem-repeats; and a C-type lectin domain of about 150 amino acid residues at the C-terminal end. This carbohydrate-recognition domain is most closely related in sequence to members of the asialoglycoprotein receptor (qv) family. The sequence, published in full, has a single N-linked glycosylation site [1].

BIOLOGICAL ACTIVITIES

The Gp120 receptor binds natural and recombinant Gp120 and whole human immunodeficiency virus. Binding to Gp120 is Ca^{2+}-dependent, of high affinity and inhibited by mannan, L-fucose, α-methylmannoside, mannose and GlcNAc (in that order). Galactose, GalNAc, sialic acid and mannose-6-phosphate were not inhibitory [1].

Gp120 bound by cells transfected with the Gp120 receptor was rapidly internalised. This process was unaffected by alloantibodies that blocked Gp120–CD4 interaction [1].

TISSUE AND/OR SUBCELLULAR DISTRIBUTION

A placental cDNA library was chosen because placental membranes were known to bind Gp120 in a CD4-independent manner. No information has been reported concerning expression of the lectin in other tissues.

POSSIBLE FUNCTION(S)

A role in first-line host defence seems plausible, but no direct evidence has been obtained relevant to function.

COMMENTS

This lectin is clearly a type II transmembrane receptor protein, belonging to the group II C-type lectins typified by the asialoglycoprotein receptor family. This lectin may mediate the CD4-independent uptake of human immunodeficiency virus by the placenta, but it is not known if this would result in viral inactivation in the endosomal compartment or (perhaps more likely) productive infection by the virus.

REFERENCE

1. Curtis, B.M., Scharnowske, S. & Watson, A.J. (1992). Sequence and expression of a membrane-associated C-type lectin that exhibits CD4-independent binding of human immunodeficiency virus envelope glycoprotein gp120. *Proceedings of the National Academy of Sciences USA* **89**, 8356–8360.

Plecoglossus altivelis egg lectin

Other name: Ayu fish egg lectin
Mol. wt: 15 000
Specificity: L-Rhamnose, L-mannose, L-lyxose
Source: Eggs of the Ayu fish *Plecoglossus altivelis*

ISOLATION

Fish eggs were extracted as an acetone powder, and the major lectin isolated by sequential ion-exchange chromatography on DEAE–cellulose and CM–cellulose followed by affinity chromatography on L-rhamnose–Sepharose [1].

STRUCTURE

This lectin has little or no attached carbohydrate and an amino acid composition rich in aspartic and glutamic acids, but poor in histidine and proline. Methionine was not detected. The molecular weight determined by SDS-PAGE, with or without reducing agent, was 14 800 [1].

BIOLOGICAL ACTIVITIES

The Ayu fish lectin agglutinated human blood group B erythrocytes and Ehrlich tumour cells. Human A and O cells were not agglutinated even at a 500-fold greater concentration [1].

Haemagglutination was extremely sensitive to L-rhamnose, L-mannose and L-lyxose, more than to D-galactose, D-fucose or L-arabinose, although the latter group also constituted potent inhibitors. Glucose (D- or L-), D-lyxose, D-mannose and D-arabinose (among others tested) were not inhibitory. Both α- and β-galactosides were potent inhibitors, but a clear preference was shown for the α anomer [1].

TISSUE AND/OR SUBCELLULAR DISTRIBUTION

Tissues other than ova do not seem to have been investigated [1].

POSSIBLE FUNCTION(S)

No suggestions regarding function were made [1].

COMMENTS

This is one of several rhamnose-specific fish egg lectins with specificity for human blood group B (see Chapter 4).

REFERENCE

1. Sakakibara, F., Kawauchi, H. & Takayanagi, G. (1985). Blood group B-specific lectin of *Plecoglossus altivelis* (Ayu fish) eggs. *Biochimica et Biophysica Acta* **841**, 103–111.

Polyandrocarpa misakiensis lectin

Other name: TC-14
Mol. wt: 14 000
Specificity: D-Galactose
Source: The budding tunicate *Polyandrocarpa misakiensis*

ISOLATION

An extract was obtained by homogenising the tunicates, followed by centrifugation, heat clarification and ammonium sulphate salting out. This material was then subjected to gel filtration on Ultrogel AcA44 and ion-exchange chromatography on TSK Gel DEAE-5PN. A homogeneous protein was obtained from the latter using an NaCl gradient [1].

STRUCTURE

This lectin appears to be a non-glycosylated polypeptide, the complete amino acid sequence of which has been determined. It has 125 residues, including 4 half-cysteines, and is high in aspartic acid and low in basic amino acids, especially histidine. It exhibits substantial (20–30% identity) homology with C-type lectins from various vertebrate and invertebrate sources. Additionally, the N-terminal 35 residues show marked homology with an Ig κ-chain variable region sequence [1].

The calculated molecular weight of 14 034 is consistent with estimated values of 15 000 (gel filtration) and 17 000 (SDS-PAGE), suggesting that it may normally exist as a monomer [1]. Multimeric forms, however, can also exist [2], and analytical ultracentrifugation revealed that TC-14 behaves as a dimer in solution [3]. The X-ray crystal structure of the lectin complexed with D-galactose at 0.22 nm has been determined [3].

BIOLOGICAL ACTIVITIES

The lectin bound to immobilised D-galactose, but not to immobilised GalNAc or D-mannose, in the presence of Ca^{2+}. It could be eluted with EDTA [1].

The lectin also has 'a strong anti-bacterial factor activity' [1].

TISSUE AND/OR SUBCELLULAR DISTRIBUTION

TC-14 lectin activity is detectable only during budding [1]. The lectin is detectable in granules of inner epithelial cells at the primordial bud stage, and subsequently is secreted into a developing extracellular matrix. The lectin can also be detected in granular leukocytes, and lectin receptors were present on the surface of haemoblast precursor cells [2].

POSSIBLE FUNCTION(S)

A defence function has been conjectured, as has a role in asexual reproduction [1]. Based on the experimental treatment of developing buds with galactose and anti-TC-14 antibodies, both of which prevent organogenesis, it was concluded that TC-14 plays a crucial role in morphogenesis, perhaps as an extracellular matrix component directing stem cells (haemoblasts) to the epithelium [2].

COMMENTS

Several kinds of calcium-dependent, galactose-binding tunicate lectins have been found to facilitate cell aggregation and migration *in vivo* and *in vitro*; their role in cell adhesion during budding has been reviewed by Kawamura and Sugino [4].

Affinity chromatography with immobilised *P. misakiensis* lectin has been used to isolate the endogenous lectin-binding factors which promote the maintenance of *Polyandrocarpa* mesochymal cells *in vitro* and have potent anti-bacterial activity [5].

REFERENCES

1. Suzuki, T., Takagi, T., Furukohri, T., Kawamura, K. & Nakauchi, M. (1990). A calcium-dependent galactose-binding lectin from the tunicate *Polyandrocarpa misakiensis*. *Journal of Biological Chemistry* **265**, 1274–1281.
2. Kawamura, K., Fujiwara, S. & Sugino, Y.M. (1991). Budding-specific lectin induced in epithelial cells is an extracellular matrix component for stem cell aggregation in tunicates. *Development* **113**, 995–1005.
3. Poget, S.F., Legge, G.B., Proctor, M.R., Butler, P.J.G., Bycroft, M. & Williams, R.L. (1999). The structure of a tunicate C-type lectin from *Polyandrocarpa misakiensis* complexed with D-galactose. *Journal of Molecular Biology* **290**, 867–879.
4. Kawamura, K. & Sugino, Y.M. (1999). Cell adhesion in the process of asexual reproduction of tunicates. *Microscopy Research and Technique* **44**, 269–278.
5. Kawamura, K. & Fujiwara, S. (1995). Establishment of cell lines from multipotent epithelial sheet in the budding tunicate, *Polyandrocarpa misakiensis*. *Cell Structure and Function* **20**, 97–106.

Powan roe lectin

Other name: *Coregonus lavaretus* lectin
Mol. wt: 27 000
Specificity: L-rhamnose
Source: Roe of the powan fish *Coregonus lavaretus maraena*

ISOLATION

This lectin was isolated from an extract of roe by affinity chromatography on *O*-α-D-galactosyl-polyacrylamide, eluting with galactose [1].

STRUCTURE

This lectin is a glycoprotein (10% carbohydrate) with a sedimentation coefficient of 2.9 S. Its amino acid composition is rich in aspartic and glutamic acids, but has little or no histidine, and a rather low content of aromatic amino acids. [1].

The molecular size determined by SDS-PAGE was 27 000, whether reducing agent was present or not. A similar value was obtained by thin-layer gel filtration [1].

BIOLOGICAL ACTIVITIES

The lectin agglutinated human blood group B erythrocytes; cells of groups A_1, A_2 or O were not agglutinated even at 300-fold greater concentration. A modest increase in titre was observed with the addition of exogenous $ZnCl_2$ or $MgCl_2$, while the presence of EDTA was associated with 75% inhibition of activity [1].

Agglutination was inhibited by D-galactose, L-arabinose, GalNAc and particularly L-rhamnose amongst monosaccharides. Methyl-α-D-galactoside was considerably more effective, and methyl-β-D-galactoside slightly less effective, than D-galactose.

TISSUE AND/OR SUBCELLULAR DISTRIBUTION

Only the roe was investigated for lectin activity [1].

POSSIBLE FUNCTION(S)

No information concerning function was obtained [1].

COMMENTS

Although blood group B-specific lectins are common in the roe of species of the Salmonidae, powan roe has a much higher lectin content than those of the commoner species [1].

REFERENCE

1. Krajhanzl, A., Hořejší, V. & Kocourek, J. (1978). Studies on lectins. XLI. Isolation and characterization of a blood group B specific lectin from the roe of the powan (*Coregonus lavaretus maraena*). *Biochimica et Biophysica Acta* **532**, 209–214.

P-selectin

Other names: GMP-140; CD62P; PADGEM; LECAM-3
Mol. wt: 126 000
Specificity: Sialyl-Lewis^X
Sources: Mammalian platelets and endothelium

ISOLATION

P-selectin was first isolated from detergent extracts of human platelets by immunoaffinity chromatography, making use of murine monoclonal antibodies specific for activated platelets [1,2]. The corresponding cDNA was later cloned from a human umbilical vein endothelial cell cDNA library [3]. Murine P-selectin has also been cloned [4].

STRUCTURE

P-selectin is a glycoprotein of Mr 126 000 containing approximately 30% carbohydrate [5], despite an apparent Mr of about 140 000 determined by SDS-PAGE [1,2]. The human cDNA encodes a signal sequence plus a mature protein of 789 residues, predicting a polypeptide molecular mass of 86 000. The structure is typical of the selectin family: an N-terminal C-type lectin domain, followed by an epidermal growth factor-like domain, a series of complement regulatory protein-like domains, a short transmembrane domain and a slightly longer domain forming the C-terminal cytoplasmic tail region [3]. The primary sequence is highly homologous to that of L-selectin (qv) and E-selectin (qv), with about 50% identity in the C-type lectin and epidermal growth factor-like domains [3,6]. P-selectin differs most obviously from the other selectins in having nine short consensus repeats, making it the largest member of the family.

BIOLOGICAL ACTIVITIES

P-selectin is a Ca^{2+}-dependent adhesion molecule. Like the other selectins, P-selectin makes a contribution to spontaneous leukocyte rolling [7,8] and can cooperate with L- and P-selectins in the process. P-selectin collaborates with E-selectin in mediating leukocyte–endothelial cell interactions at sites of inflammation [9]. P-selectin is the only selectin on platelets, and can mediate adhesion of activated platelets to neutrophils [10,11], monocytes [11], NK cells and T cell subsets [12]. There is evidence for a P-selectin-dependent homing of lymphocytes to Peyer's patches, perhaps involving effector cells [13].

In addition to simple adhesion, P-selectin may have a signalling role in some circumstances, which may involve the translocation of the transcription factor NF-κB from cytoplasm to nucleus [14–16].

The best characterised endogenous ligand for P-selectin is a glycoprotein with extensive fucose and sialic acid-containing polylactosamine side chains called

P-selectin glycoprotein ligand-1 (PSGL-1), which may be a physiological counter-receptor for all three selectins [17].

The interactions of P-selectin can be inhibited by sialyl-Lewis^X (see Figure 3.4) and related structures [18–20], and by sulphated glycans [21] or sulphated galactoceramides [22].

P-selectin can also bind to heparin in a Ca^{2+}-independent manner [23].

TISSUE AND/OR SUBCELLULAR DISTRIBUTION

P-selectin is found in megakaryocytes, platelets and endothelial cells; it is stored in α-granules (platelets) or Weibel–Palade bodies (endothelial cells) and is translocated to the cell surface upon activation [3,24].

POSSIBLE FUNCTION(S)

P-selectin undoubtedly functions in cellular adhesion, mediating the attachment of leukocytes to activated platelets and endothelial cells, a role held in common with other selectins [6]. The overlapping role of the three members of the family makes the distinct contribution of any one selectin hard to unravel, even with a set of knock-out mice with different combinations of selectin deficiencies [25]. While such studies show most clearly that selectins operate in an overlapping and synergistic manner, P-selectin appears to be the most versatile in mediating moderate neutrophil recruitment in several models of inflammation [25]. P-selectin may differ from other family members in having a role in haemostasis; P-selectin-deficient mice showed a 40% prolongation of bleeding time after tail cutting [26].

COMMENTS

Activated T-helper cell subsets can be distinguished on the basis of selectin receptors: murine Th-1 cells, but not Th-2 cells, are able to bind P-selectin (and E-selectin) [27].

P-selectin, both on the platelet surface and in soluble form in the circulation, signifies platelet activation and may be used as a disease marker. Elevated blood P-selectin levels are found in cardiovascular diseases [28], insulin-dependent diabetes [29], rheumatoid arthritis [30] and pre-eclampsia/eclampsia [31]. In rheumatoid arthritis, soluble E- and L-selectins were slightly increased, but only P-selectin correlated with disease activity [30].

Both plasma membrane-bound [32] and soluble P-selectin [33] are elevated in smokers compared to non-smokers.

P-selectin can promote tumour growth and metastases, while, conversely, P-selectin-deficient mice were relatively protected from metastases in an experimental colon cancer model [34].

In animal models of complement-mediated lung injury [35] and ischaemia-induced reperfusion injury [36], blocking of P-selectin resulted in diminished neutrophil adherence to endothelium and protection against tissue damage.

Curiously, experimental collagen-induced arthritis was more severe in P-selectin-deficient mice compared to wild-type mice, and the onset of arthritis was accelerated [37].

The gene for P-selectin has been localised to human and murine chromosomes 1 [38].

REFERENCES

1. Hsu-Lin, S.-C., Berman, C.L., Furie, B.C., August, D. & Furie, B. (1989). A platelet membrane protein expressed during platelet activation and secretion. Studies using a monoclonal antibody specific for thrombin-activated platelets. *Journal of Biological Chemistry* **259**, 9121–9126.
2. McEver, R.P. & Martin, M.N. (1984). A monoclonal antibody to a membrane glycoprotein binds only to activated platelets. *Journal of Biological Chemistry* **259**, 9799–9804.
3. Johnstone, G.I., Cook, R.G. & McEver, R.P. (1989). Cloning of GMP-140, a granule membrane protein of platelets and endothelium: sequence similarity to proteins involved in cell adhesion and inflammation. *Cell* **56**, 1033–1044.
4. Weller, A., Isenmann, S. & Vestweber, D. (1992). Cloning of the mouse endothelial selectins. Expression of both E- and P-selectin is inducible by tumour necrosis factor. *Journal of Biological Chemistry* **267**, 15176–15183.
5. Johnstone, G.I., Kurosky, A. & McEver, R.P. (1989). Structural and biosynthetic studies of the granule membrane protein, GMP-140, from human platelets and endothelial cells. *Journal of Biological Chemistry* **264**, 1816–1823.
6. Kansas, G.S. (1996). Selectins and their ligands: current concepts and controversies. *Blood* **88**, 3259–3287.
7. Doré, M., Korthuis, R.J., Granger, D.N., Entman, M.L. & Smith, C.W. (1993). P-selectin mediates spontaneous leukocyte rolling in vivo. *Blood* **82**, 1308–1316.
8. Ley, K., Bullard, S., Arbones, M.L., Bosse, R., Vestweber, D., Tedder, T.F. & Beaudet, A.L. (1995). Sequential contribution of L- and P-selectin to leukocyte rolling in vivo. *Journal of Experimental Medicine* **181**, 669–675.
9. Catabina, M.D., Estess, P. & Siegelman, M.H. (1999). Selective requirements for leukocyte adhesion molecules in models of acute and chronic cutaneous inflammation: participation of E- and P- but not L-selectin. *Blood* **93**, 580–589.
10. Hamburger, S.A. & McEver, R.P. (1990). GMP-140 mediates adhesion of stimulated platelets to neutrophils. *Blood* **75**, 550–554.
11. Larsen, E., Celi, A., Gilbert, G.E., Furie, B.C., Erban, J.K., Bonfanti, R., Wagner, D.D. & Furie, B. (1989). PADGEM protein: a receptor that mediates the interaction of activated platelets with neutrophils and monocytes. *Cell* **59**, 305–312.
12. Moore, K.L. & Thompson, L.F. (1992). P-selectin (CD62) binds to subpopulations of human memory T lymphocytes and natural killer cells. *Biochemical and Biophysical Research Communications* **186**, 173–181.
13. Kunkel, E.J., Ramos, C.L., Steeber, D.A., Muller, N., Wagner, N., Tedder, T.F. & Ley, K. (1998). The roles of L-selectin, $\beta 7$ integrins, and P-selectin in leukocyte rolling and adhesion in high endothelial venules of Peyer's patches. *Journal of Immunology* **161**, 2449–2456.
14. Cooper, D., Butcher, C.M., Berndt, M.C. & Vodas, M.A. (1994). P-selectin interacts with a β_2-integrin to enhance phagocytosis. *Journal of Immunology* **153**, 3199–3209.
15. Weyrich, A.S., McIntyre, T.M., McEver, R.P., Prescott, S.M. & Zimmermann, G.A. (1995). Monocyte tethering by P-selectin regulates monocyte chemotactic protein-1 and tumour necrosis factor-α secretion. *Journal of Clinical Investigation* **95**, 2297–2303.
16. Elstad, M.R., La Pine, T.R., Cowley, F.S., McEver, R.P., McIntyre, T.M., Prescott, S.M. & Zimmermann, G.A. (1995). P-selectin regulates platelet-activating factor synthesis and phagocytosis by monocytes. *Journal of Immunology* **155**, 2109–2122.
17. Yang, J., Furie, B.C. & Furie, B. (1999). The biology of P-selectin glycoprotein ligand-1: its role as a selectin counterreceptor in leukocyte–endothelial and leukocyte–platelet interaction. *Thrombosis and Haemostasis* **81**, 1–7.
18. Larsen, E., Palabrica, T., Sajer, S., Gilbert, G.E., Wagner, D.D., Furie, B.C. & Furie, B. (1990). PADGEM-dependent adhesion of platelets to monocytes and neutrophils is mediated by a lineage-specific carbohydrate LNF III (CD15). *Cell* **63**, 467–474.
19. Zhou, Q., Moore, K.L., Smith, D.F., Varki, A., McEver, R.P. & Cummings, R.D. (1991). The selectin GMP-140 binds to sialylated, fucosylated lactosaminoglycans on both myeloid and nonmyeloid cells. *Journal of Cell Biology* **115**, 557–564.

20. Polley, M.J., Phillips, M.L., Wayner, E., Nudelman, E., Singhal, A.K., Hakomori, S. & Paulson, J.C. (1991). CD62 and endothelial cell–leukocyte adhesion molecule 1 (ELAM-1) recognise the same carbohydrate ligand, sialyl-Lewis X. *Proceedings of the National Academy of Sciences USA* **88**, 6224–6228.
21. Skinner, M.P., Lucas, C.M., Burns, G.F., Chesterman, C.N. & Berndt, M.C. (1991). GMP-140 binding to neutrophils is inhibited by sulphated glycans. *Journal of Biological Chemistry* **266**, 5371–5374.
22. Todderud, G., Alford, J., Millsap, K.A., Aruffo, A. & Tramposch, K.M. (1992). PMN binding to P-selectin is inhibited by sulfatide. *Journal of Leukocyte Biology* **52**, 85–88.
23. Skinner, M.P., Fournier, D.J., Andrews, R.K., Gorman, J.J., Chesterman, C.N. & Berndt, M.C. (1989). Characterization of human platelet GMP-140 as a heparin-binding protein. *Biochemical and Biophysical Research Communications* **164**, 1373–1379.
24. Beckstead, J.H., Stenberg, P.E., McEver, R.P., Shuman, M.A. & Bainton, D.F. (1986). Immunohistochemical localization of membrane and α-granule proteins in humam megakaryocytes: application to plastic-embedded bone marrow biopsy specimens. *Blood* **67**, 285–293.
25. Jung, U. & Ley, K. (1999). Mice lacking two or all three selectins demonstrate overlapping and distinct functions for each selectin. *Journal of Immunology* **162**, 6755–6762.
26. Subramanian, M., Frenette, P.S., Saffaripour, S., Johnson, R.C., Hynes, R.O. & Wagner, D.D. (1996). Defects in hemostasis in P-selectin deficient mice. *Blood* **87**, 1238–1242.
27. Austrup, F., Vestweber, D., Borges, E., Lohning, M., Brauer, R., Herz, U., Renz, H., Hallermann, R., Scheffold, A., Radbruch, A. & Hamann, A. (1997). P- and E-selectin mediate recruitment of T-helper-1 but not T-helper-2 cells into inflamed tissues. *Nature* **385**, 81–83.
28. Blann, A.D. & McCollum, C.N. (1998). Increased soluble P-selectin in peripheral artery disease: a new marker for the progression of atherosclerosis. *Thrombosis and Haemostasis* **80**, 1031–1032.
29. Jilma, B., Fasching, P., Ruthner, C., Rumplmayr, A., Ruzicka, S., Kapiotis, S., Wagner, O.F. & Eichler, H.-G. (1996). Elevated circulating P-selectin in insulin dependent diabetes mellitis. *Thrombosis and Haemostasis* **76**, 328–332.
30. Littler, A.J., Buckley, C.D., Wordsworth, P., Collins, I., Martinson, J. & Simmons, J.L. (1997). A distinct profile of six soluble adhesion molecules (ICAM-1, ICAM-3, VCAM-1, E-selectin, L-selectin and P-selectin) in rheumatoid arthritis. *British Journal of Rheumatology* **36**, 164–169.
31. Halim, A., Kanayama, N., Maradny, E.E., Nakashima, A., Bhuiyan, A.B., Khatun, S. & Terao, T. (1996). Plasma P selectin (GMP-140) and glycocalicin are elevated in preeclampsia and eclampsia: their significancies. *American Journal of Obstetrics and Gynecology* **174**, 272–277.
32. Pernerstorfer, T., Stohlawetz, P., Stummvoll, G., Kapiotis, S., Szekeres, T., Eichler, H.-G. & Jilma, B. (1998). Low dose aspirin does not lower in vivo platelet activation in healthy smokers. *British Journal of Haematology* **102**, 1229–1231.
33. Blann, A.D., Steele, C. & McCollum, C.N. (1997). The influence of smoking on soluble adhesion molecules and endothelial cell markers. *Thrombosis Research* **85**, 433–438.
34. Kim, Y.J., Borsig, L., Varki, N.M. & Varki, A. (1998). P-selectin deficiency attenuates tumor growth and metatasis. *Proceedings of the National Academy of Sciences USA* **95**, 9325–9330.
35. Mulligan, M.S., Polley, M.J., Bayer, R.J., Nunn, M.F., Paulson, J.C. & Ward, P.A. (1992). Neutrophil-dependent acute lung injury. Requirement for P-selectin (GMP-140). *Journal of Clinical Investigation* **90**, 1600–1607.
36. Weyrich, A.S., Ma, X., Lefer, D.J., Albertine, K.H. & Lefer, A.M. (1993). In vivo neutralization of P-selectin protects feline heart and endothelium in myocardial ischaemia and reperfusion injury. *Journal of Clinical Investigation* **91**, 2620–2629.
37. Bullard, D.C., Mobley, J.M., Justen, J.M., Sly, L.M., Chosay, J.G., Dunn, C.J., Lindsey, J.R., Beaudet, A.L. & Staite, N.D. (1999). Acceleration and increased severity of collagen-induced arthritis in P-selectin mutant mice. *Journal of Immunology* **163**, 2844–2849.
38. Watson, M.L., Kingsmore, S.F., Johnstone, G.I., Siegelman, M.H., Le Beau, M.M., Lemons, R.S., Bora, N.S., Howard, T.A., Weissman, I.L. & McEver, P.R. (1990). Genomic organisation of the selectin family of leukocyte adhesion molecules on human and mouse chromosome 1. *Journal of Experimental Medicine* **172**, 263–272.

Purpurin

Other name: *Dictyostelium purpureum* lectin
Mol. wt: 90 000 (native); 22 000 and 23 000 (subunit)
Specificity: Lactose, GalNAc
Source: Aggregating amoebae of *Dictyostelium purpureum*

ISOLATION

A mixture of isolectins collectively known as purpurin was isolated by affinity chromatography on Sepharose 4B, eluting with galactose [1]. Two of the isolectins can be separated from the others on lactose–agarose (with partial separation of five others) using a step-wise gradient of galactose [2]. Affinity purified purpurin can be resolved into two major and five less abundant electrophoretic forms on 3.5% polyacrylamide at pH 8.9 [2].

STRUCTURE

Purpurin is a protein without detectable methionine. The apparent molecular weight was estimated at 140 000 by sucrose density gradient centrifugation, but was noted to be similar to discoidin I (qv), which has a true molecular weight nearer 100 000 [1]. Analytical ultracentrifugation indicated a value of less than 90 000 [2].

The apparent subunit Mr is around 23 000 [1,2]. The different electrophoretic forms are due to combinations of four closely related, but distinct, subunits [2]. Three isolectins are composed of heterotetramers of the commonest subunits (I_1II_3, I_2II_2, I_3II_1); two consist of homotetramers of the same subunits (I_4 and II_4); and two consist of homotetramers of the other subunits (III_4 and IV_4).

BIOLOGICAL ACTIVITIES

Purpurin agglutinates formalin-treated lapine erythrocytes, and agglutination is inhibited most potently by lactose and GalNAc. D-galactose and L-fucose (surprisingly) were ineffective as inhibitors as were (as expected) glucose and mannose.

Purpurin is weakly mitogenic towards murine splenic mononuclear cells; non-T cells seem to be the main responders [3].

TISSUE AND/OR SUBCELLULAR DISTRIBUTION

Purpurin is found in aggregating (cohesive) amoebae, but not appreciably in vegetative cells, and is located partly on the cell surface but mainly intracellularly. There is some evidence that purpurin can be induced to translocate to the cell surface by divalent immunoglobulins (and, by implication, endogenous ligands) [4].

POSSIBLE FUNCTION(S)

By analogy with discoidins and other cellular slime mould lectins, it is assumed that purpurin is involved in cell–cell cohesion and/or species-specific cell segregation.

COMMENTS

The family of purpurin isolectins has much in common with the discoidins, including overall similarities in molecular size, amino acid composition and sugar specificity. Nevertheless, these characteristics differ in fine detail between the species. Another cellular slime mould, *Dictyostelium mucoroides*, also has a lectin or lectins belonging to this family [1]. The lectins from all three species serologically cross-react, although they are all antigenically distinct. Interestingly, the *D. mucoroides* lectin activity has a sugar inhibition profile resembling the discoidins much more than the purpurin isolectins [1].

REFERENCES

1. Barondes, S.H. & Haywood, P.L. (1979). Comparison of developmentally regulated lectins from three species of slime mold. *Biochimica et Biophysica Acta* **550**, 297–308.
2. Cooper, D.N. & Barondes, S.H. (1981). Isolectins from *Dictyostelium purpureum*. *Journal of Biological Chemistry* **256**, 5046–5051.
3. Lipsick, J.S., Beyer, E.C., Barondes, S.H. & Kaplan, N.O. (1980). Lectins from chicken tissues are mitogenic for Thy-1 negative murine spleen cells. *Biochemical and Biophysical Research Communications* **97**, 56–61.
4. Springer, W.R., Haywood, P.L. & Barondes, S.H. (1980). Endogenous cell surface lectin in *Dictyostelium*: quantitation, elution by sugar, and elicitation by divalent immunoglobulin. *Journal of Cell Biology* **87**, 682–690.

Rana catesbeiana egg lectin

Other name: Frog egg lectin
Mol. wt: 210 000
Specificity: (GalNAcα1,3[4]Galβ1,4GlcNAc)
Source: Eggs of the frog *Rana catesbeiana*

ISOLATION

The lectin was purified from an extract of frog eggs by gel filtration followed by preparative PAGE at pH 4.3. The isolated lectin appeared homogeneous by electrophoresis under various conditions [1].

STRUCTURE

The frog egg lectin is a basic glycoprotein (3% carbohydrate) with an apparent Mr of 210 000 determined by SDS-PAGE under reducing conditions [1].

BIOLOGICAL ACTIVITIES

The lectin agglutinates human erythrocytes but is highly specific for blood group A. (Groups B and O erythrocytes can be agglutinated after neuraminidase treatment.) The lectin also caused precipitation of a blood group A glycolipid fraction and group A glycoprotein from ovarian cyst [1].

Lectin activity is most readily inhibited by *N*-acetyllactosamine, lacto-*N*-tetraose and lactose. Galactose and GalNAc, but not other monosaccharides, were also inhibitory [1].

TISSUE AND/OR SUBCELLULAR DISTRIBUTION

Although known to be present in eggs, information about its presence or absence in other tissues is lacking.

POSSIBLE FUNCTION(S)

It was originally assumed that such lectins participate in anti-bacterial defence ('protectins') but no function has been clearly demonstrated.

COMMENTS

Extracts of eggs from *Rana* species and other frogs were surveyed for lectin activity, but only *R. catesbeiana* and *R. nigromaculata* had human blood group A specificity. The *R. catesbeiana* lectin was the first amphibian lectin with blood group specificity to be recognised and purified. From its reactivity profile and the profile of inhibitors, it appears to recognise not just the terminal GalNAc but the sub-terminal (Gal-β1,4-GlcNAc) as well [1].

REFERENCE

1. Sakakibara, F., Takayanagi, G., Kawauchi, H., Wantanabe, K. & Hakomori, S. (1976). An anti-A-like lectin of *Rana catesbiana* eggs showing unusual reactivity. *Biochimica et Biophysica Acta* **444**, 386–395.

Rana sialic acid-binding lectin

Other name: Frog egg sialic acid-binding lectin
Mol. wt: 12 500
Specificity: Sialic acid; heparin
Source: Eggs from species of the frog genus *Rana*

ISOLATION

A tumour cell-agglutinating lectin was isolated from eggs of *Rana japonica* by gel filtration and ion-exchange chromatography [1]. A very similar lectin was isolated in a broadly similar way from the related species *R. catesbeiana* (which also has a separate blood group A-specific lectin) [2].

STRUCTURE

The lectin is a basic (pI ~ 12) polypeptide (no attached carbohydrate) rich in threonine with an apparent Mr of 13 000 determined by SDS-PAGE [1,2].

The complete amino acid sequences from *R. japonica* [3] and *R. catesbeiana* [4] have been reported. Both have 111 amino acid residues, a γ-pyroglutamyl *N*-terminal residue, and a calculated Mr of approximately 12 500 [3,4]. No sequence homology with other lectins (animal or plant) was found, but there was 80% identity between the corresponding lectins from the two species. Some homology was also found with human angiogenin and with a family of pancreatic ribonucleases. Notably, all have a similar disulphide loop structure [3].

BIOLOGICAL ACTIVITIES

The *R. japonica* lectin agglutinates transformed cells *in vitro* and *in vivo*, including Ehrlich ascites tumour cells, but does not agglutinate normal fibroblasts or erythrocytes [1,5]. The *R. catesbeiana* lectin, however, weakly agglutinates some normal fibroblasts, lymphocytes and erythrocytes in addition to readily agglutinating tumour cells [2]. Monosaccharides do not inhibit agglutination but agglutination can be inhibited by sialylated glycoproteins or glycolipids; neuraminidase treatment of the glycoconjugate inhibitors abolishes or greatly reduces their inhibitory potency [1,2]. Agglutination was enhanced by the presence of Ca^{2+} [2].

The homologues from both species possess ribonuclease activity [3,6], and could act as inhibitors of protein synthesis [7]. The anti-proliferative activity of *Rana* lectins towards P388 leukaemia cells was inhibited by heparin, as was its agglutination of tumour cells [7]. RNase activity was also inhibited by heparin [7].

TISSUE AND/OR SUBCELLULAR DISTRIBUTION

These lectins were extracted from eggs; other tissues do not seem to have been investigated [1,2].

POSSIBLE FUNCTION(S)

The *Rana* sialic acid-binding lectin is primarily a pyrimidine base-specific RNase, and chemical modifications that destroy enzymatic activity also destroy lectin/ anti-tumour activities [8]. It may function as a cytostatic factor, inducing apoptotic cell death [8]. This may depend on the protein's bi-functional nature, requiring (i) binding to the cell surface receptor and (ii) internalisation and degradation of RNA.

COMMENTS

The *Rana* sialic acid-binding lectin belongs to a family of angiogenin/turtle/ chicken/frog type RNAses, reviewed by Irie *et al.* [8].

Administration of the lectin *in vivo* led to a significantly increased survival rate for tumour-bearing mice [7], and it has been suggested that frog RNAses might have a future application in cancer therapy [8].

REFERENCES

1. Sakakibara, F., Kawauchi, H., Takayanagi, G. & Ise, H. (1979). Egg lectin of *Rana japonica* and its receptor glycoprotein of Ehrlich tumour cells. *Cancer Research* **39**, 1347–1352.
2. Nitta, K., Takayanagi, G., Kawauchi, H. & Hakomori, S. (1987). Isolation and characterization of *Rana catesbeiana* lectin and demonstration of the lectin-binding glycoprotein of rodent and human tumor cell membrane. *Cancer Research* **47**, 4877–4883.
3. Kamiya, Y., Oyama, F., Oyama, R., Sakkakibara, F., Nitta, K., Kawauchi, H., Takayanagi, Y. & Titani, K. (1990). Amino acid sequence of a lectin from Japanese frog (*Rana japonica*) eggs. *Journal of Biochemistry* **108**, 139–143.
4. Titani, K., Takio, K., Kuwada, M., Nitta, K., Sakakibara, F., Kawauchi, H., Takayanagi, G. & Hakomori, S. (1987). Amino acid sequence of sialic acid binding lectin from frog (*Rana catesbeiana*) eggs. *Biochemistry* **26**, 2189–2194.
5. Kawauchi, H., Sakakibara, F. & Watanabe, K. (1975). Agglutinins of frog eggs: a new class of proteins causing preferential agglutination of tumor cells. *Experientia* **31**, 364–365.
6. Nitta, K., Oyama, F., Oyama, R., Sekiguchi, K., Kawauchi, H. & Takayanagi, Y. (1991). Ribonuclease activity of a sialic acid binding lectin from *Rana catesbeiana*. *Glycobiology* **3**, 37–45.
7. Nitta, K., Ozaki, K., Ishikawa, M., Furusawa, S., Hosono, M., Kawauchi, H., Sasaki, K., Takayanagi, Y., Tsuiki, S. & Hakomori, S. (1994). Inhibition of cell proliferation by *Rana catesbeiana* and *Rana japonica* lectins belonging to the ribonuclease superfamily. *Cancer Research* **54**, 920–927.
8. Irie, M., Nitta, K. & Nonaka, T. (1998). Biochemistry of frog ribonucleases. *Cellular and Molecular Life Sciences* **54**, 775-784.

Regenectin

Other name: *Periplaneta americana* sucrose-binding lectin
Mol. wt: 2×10^6 (native); 37 000 (subunit)
Specificity: Sucrose
Source: Haemolymph of the American cockroach *Periplaneta americana*

ISOLATION

Regenectin was purified from haemolymph serum by a single step of affinity chromatography on sucrose–Sepharose, eluting with sucrose [1].

A cDNA for regenectin has been isolated [2].

STRUCTURE

Regenectin is a glycoprotein which consists of huge aggregates with an average molecular weight of 2×10^6 [1]. The single subunit from which these aggregates are formed has an apparent Mr of 37 000 estimated by SDS-PAGE [1]. The deduced amino acid sequence of 238 residues yields a calculated Mr of 25 000, close to the value of 29 000 obtained by SDS-PAGE of deglycosylated regenectin [2].

Regenectin immunologically cross-reacts with the original *Periplaneta americana* lectin (qv), but the tryptic peptide maps of the two lectins are dissimilar [1]. The amino acid sequence derived from the cDNA nucleotide sequence exhibited 52% identity with that of the *Periplaneta* lipopolysaccharide-binding protein (qv) and about 40% identity with the other *Periplaneta* lectins [2]. The sequence included a C-type carbohydrate-recognition domain homologue near the C-terminus.

BIOLOGICAL ACTIVITIES

Regenectin agglutinates lapine erythrocytes at high titre, but not ovine erythrocytes. Human cells occupy an intermediate position, with group O cells slightly less readily agglutinated than those of groups A or B. Haemagglutinating activity was Ca^{2+}-dependent; Mg^{2+}, but not Mn^{2+} or Ba^{2+}, was partially effective at replacing Ca^{2+} [1].

Haemagglutination was not readily inhibited by any monosaccharide tested, although 50% inhibition was achieved with D-galactose, D-fucose and rhamnose at 100 mM concentrations. The same effect, however, was obtained with sucrose at 10 mM concentration [1].

TISSUE AND/OR SUBCELLULAR DISTRIBUTION

Regenectin is found in haemolymph, but also appears in regenerating tissue transiently between days 17 and 26 after leg amputation [1]. By Northern blotting and *in situ* hybridisation, regenectin was found to be synthesised *de novo* by epidermal cells. It has been suggested that the lectin is secreted into the regenerating leg saccule and assembles around myoblasts [2].

POSSIBLE FUNCTION(S)

Regenectin synthesis is upregulated *de novo* in regenerating leg tissue, and is not affected by bacterial infection [2]. The clear implication is that regenectin is not a defence molecule, but may have a role in muscle differentiation during leg regeneration.

COMMENTS

Regenectin belongs to the superfamily of C-type animal lectins and more specifically to the *Periplaneta* lectin-related protein family, of which it is one of four that have been purified, characterised and entered into this Directory.

Although regenectin resembles the LPS-binding protein most in primary structure, functionally it seems to be more closely related to the 26 kDa lectin. Both have been implicated in tissue regeneration, but the latter appears earlier [3].

REFERENCES

1. Kubo, T., Kawasaki, K. & Natori, S. (1990). Sucrose-binding lectin in regenerating cockroach (*Periplaneta americana*) legs: its purification from adult hemolymph. *Insect Biochemistry* **20**, 585–591.
2. Arai, T., Kawasaki, K., Kubo, T. & Natori, S. (1998). Cloning of cDNA for regenectin, a humoral C-type lectin of *Periplaneta americana*, and expression of the regenectin gene during leg regeneration. *Insect Biochemistry and Molecular Biology* **28**, 987–994.
3. Kubo, T., Kawasaki, K. & Natori, S. (1993). Transient appearance and localisation of a 26-kDa lectin, a novel member of the *Periplaneta* lectin family in regenerating cockroach leg. *Developmental Biology* **156**, 381–390.

Roach roe lectin

Other name: *Rutilus rutilus* roe lectin
Mol. wt: 25 000
Specificity: L-Rhamnose
Source: Roe of the roach *Rutilus rutilus*

ISOLATION

The lectin was isolated from an extract of roach roe by affinity chromatography on *O*-α-D-galactosyl-polyacrylamide and elution with D-galactose. The purified material possessed one major electrophoretic form [1].

STRUCTURE

The lectin is a glycoprotein (2.6% carbohydrate) with an unremarkable amino acid composition. The major form has an apparent Mr of about 25 000 determined by gel filtration and ultracentrifugation (2.1 S), but has a tendency towards pH- and ionic strength-dependent aggregation. The subunit size determined by SDS-PAGE under non-reducing conditions was also 25 000; however, under reducing conditions, two bands corresponding to Mr values of 25 000 and 32 000 were observed. The lectin contains zinc.

BIOLOGICAL ACTIVITIES

The roach roe lectin agglutinates human erythrocytes with a slight preference for blood group B. Agglutination was partially inhibited by EDTA, but the addition of exogenous divalent cations (Ca^{2+}, Mn^{2+}, Mg^{2+} or Zn^{2+}) had no effect. Haemagglutination was extremely sensitive to L-rhamnose, and D-galactose and L-arabinose were also potent inhibitors (although much less so). Other common monosaccharides were not inhibitory [1].

In immunodiffusion, *O*-glycosyl polyacrylamide copolymers containing bound α-L-rhamnosyl, α-D-galactosyl or L-arabinosyl residues were precipitated by the lectin [1].

TISSUE AND/OR SUBCELLULAR DISTRIBUTION

The lectin was found to be concentrated in cortical vesicles adjacent to the plasma membrane of vitellogenic oocytes [2].

POSSIBLE FUNCTION(S)

Although originally thought of as defensive ('protectin') molecules, additional functions, including the prevention of polyspermy, have been suggested for fish ovary lectins based on localisation findings [2].

COMMENTS

The roach lectin is very similar to lectins purified from the roe of related species from the Cyprinidae family: *Vimba vimba* and *Scardinus erythrophtalmus* [1].

REFERENCES

1. Krajhanzl, A., Hořejší, V. & Kocourek, J. (1978). Studies on Lectins. XLII. Isolation, partial characterization and comparison of lectins from the roe of five fish species. *Biochimica et Biophysica Acta* **532**, 215–224.
2. Nosek, J., Krajhanzl, A. & Kocourek, J. (1983). Studies on lectins. LVII. Immunofluorescence localization of lectins present in fish ovaries. *Histochemistry* **79**, 131–139.

Salmo salar mannose-binding lectin

Other name: Atlantic salmon serum lectin
Mol. wt: 17 000
Specificity: Mannose
Source: The Atlantic salmon *Salmo salar*

ISOLATION

The lectin was purified from diluted blood serum by affinity chromatography on mannose–agarose followed by passage over unsubstituted agarose to remove contaminating proteins [1].

STRUCTURE

The lectin is a protein with an amino acid composition broadly similar to those of human and chicken mannan-binding lectins (qv). The subunit Mr determined by SDS-PAGE was 17 000 under reducing conditions; under non-reducing conditions, a vast series of bands was obtained, forming a ladder corresponding to molecular sizes from 46 000 to hundreds of thousands separated by Mr intervals of about 17 000 [1].

An N-terminal sequence of 28 amino acid residues shared no definite similarity to known protein sequences, but indicated possible homology to a short sequence available from the 'ladder lectin' of the rainbow trout (*Oncorhynchus mykiss*) [1,2].

BIOLOGICAL ACTIVITIES

The lectin bound to the surface of two bacterial species, *Vibrio anguillarum* and *Aeromonas salmonicida*, and could be eluted with mannose. The lectin binds Ca^{2+} which is essential for binding to immobilised mannose during purification [1].

The lectin promoted the phagocytosis of *A. salmonicida* by *S. salar* macrophages [3].

TISSUE AND/OR SUBCELLULAR DISTRIBUTION

The lectin has been described only in serum.

POSSIBLE FUNCTION(S)

In view of its opsonic activity towards a pathogenic bacterium, it is assumed that the lectin is part of a defence mechanism against bacteria [1,3].

COMMENTS

This lectin has a high glycine content and other characteristics that suggest it may be a piscine homologue of the mammalian mannan-binding lectin. However, no support for that possibility was obtained from the limited sequence data available, and as yet it cannot be classified into an established animal lectin family.

There are some grounds, however, for supposing it to be a relative of the 'ladder lectin' of the rainbow trout [2]. Although lectins from the salmon family have been known for decades, including certain human group-B-specific lectins in roe [4], it is only very recently that lectins from this family have been isolated and studied in purified form. As well as the ladder lectin, the rainbow trout is known to possess pentraxins and RT-L37, a calcium-dependent LPS-binding protein [5].

REFERENCES

1. Ewart, K.V., Johnson, S.C. & Ross, N.W. (1999). Identification of a pathogen-binding lectin in salmon serum. *Comparative Biochemistry and Physiology C* **123**, 9–15.
2. Jensen, L.E., Thiel, S., Peterson, T.E. & Jensenius, J.C. (1997). A rainbow trout lectin with multimeric structure. *Comparative Biochemistry and Physiology B* **116**, 385–390.
3. Ottinger, C.A., Johnson, S.C., Ewart, K.V., Brown, L.L. & Ross, N.W. (1999). Enhancement of anti-*Aeromonas salmonicida* activity in Atlantic salmon (*Salmo salar*) macrophages by a mannose-binding lectin. *Comparative Biochemistry and Physiology C* **123**, 53–59.
4. Kilpatrick, D.C. & Green, C. (1992). Lectins as blood typing reagents. In: *Advances in Lectin Research*, Volume 5 (ed. H. Franz), pp. 51–94. Ullstein Mosby, Berlin.
5. Hoover, G.J., El Mowafi, A., Simko, E., Kocal, T.E., Ferguson, H.W. & Hayes, M.A. (1998). Plasma proteins of rainbow trout (*Oncorhynchus mykiss*) isolated by binding to lipopolysaccharide from *Aeromonus salmonicida*. *Comparative Biochemistry and Physiology B* **120**, 559–569.

Sarcolectin

Mol. wt: 55 000
Specificity: D-Galacturonic acid
Sources: Normal muscle and sarcomas

ISOLATION

Sarcolectin was originally purified from pepsin-treated extracts of hamster [1] and human [2] sarcomas and normal muscle tissue by gel filtration and ion-exchange chromatography. Recently a more complex purification procedure has been devised, which appears to indicate that the earlier method resulted in the isolation of a complex of sarcolectin and albumin [3].

Sarcolectin has been cloned and a recombinant molecule expressed [4].

STRUCTURE

Sarcolectin appears to be a glycoprotein of apparent Mr 200 000 estimated by gel filtration [1], but the major form of circulating sarcolectin is a complex with albumin, while sarcolectin itself exists as a minor form in serum with an Mr of 55 000 determined by SDS-PAGE [3].

The structure of sarcolectin is related to that of cytokeratin intermediate filaments, and its carbohydrate-binding activity is located close to the N-terminal end of the molecule [4].

Sarolectin cross-reacts immunologically with galectins and has areas of primary structural homology to human galectin-1 [3,4]. Those sequences are found in the N-terminal β-sheet region, whereas the middle α-helical region of the molecule has homologies with human keratin, human vimentin and human neurofilament [4].

BIOLOGICAL ACTIVITIES

Sarcolectin antagonises the anti-viral activity of interferons α and β and can reverse previously established anti-viral protection mediated by interferon [1,2]. Sarcolectin does not interact or compete with interferons directly [1], but rather acts by stimulating DNA synthesis and promoting cell growth [4]. The major ligand of sarcolectin in human placenta is the macrophage migration inhibitory factor [5].

Sarcolectin could also agglutinate a variety of rodent cell lines, murine T lymphocytes and (at lower titre) murine B lymphocytes [1]. Both hamster and human sarcolectins were inhibited by D-galacturonic acid, but while the hamster sarcolectin showed some modest sensitivity to fucose [1], human sarcolectin was potently inhibited by *N*-acetylneuraminic acid [2].

TISSUE AND/OR SUBCELLULAR DISTRIBUTION

Sarcolectin has been found in murine costal cartilage, hamster sarcomas/normal muscle, and human sarcomas/normal muscle; it is secreted for localisation on the cell surface and in the extracellular matrix [1,2].

POSSIBLE FUNCTION(S)

The function(s) of sarcolectin(s) has not been determined, but it has been postulated that sarcolectins (and interferons) operate as part of a balanced feedback system concerned with the regulation of normal growth.

COMMENTS

Immobilised human sarcolectin has been used to isolate the lymphokine, macrophage migration inhibitory factor, by affinity chromatography [5].

Expression of binding sites for sarcolectin on tissue sections of lung carcinomas correlated positively and significantly with patient survival [6].

REFERENCES

1. Jiang, P.H., Chany-Fournier, F., Robert-Galliot, B., Sarragne, M. & Chany, C. (1983). Sarcolectin: an interferon antagonist extracted from hamster sarcomas and normal muscles. *Journal of Biological Chemistry* **258**, 12361–12367.
2. Jiang, P.H., Chany-Fournier, F., Sarragne, M., Grégoire, A. & Chany, C. (1984). Detection of an interferon antagonist, sarcolectin, in human sarcomas and muscles. *International Journal of Cancer* **34**, 625–632.
3. Jiang, P.H., Chany-Fournier, F. & Chany, C. (1999). Sarcolectin: complete purification for molecular cloning. *Biochimie* **81**, 701–707.
4. Kaba, A., Jiang, P.H., Chany-Fournier, F. & Chany, C. (1999). Sarcolectin (SCL): structure and expression of the recombinant molecule. *Biochimie* **81**, 709–715.
5. Zeng, F.-Y., Weiser, W.Y., Kratzin, H., Stahl, B., Karas, M. & Gabius, H.-J. (1993). The major binding protein of the interferon antagonist sarcolectin in human placenta is a macrophage migration inhibitory factor. *Archives of Biochemistry and Biophysics* **303**, 74–80.
6. Kayser, K., Bovin, N.V., Korchagina, E.Y., Zeilinger, C., Zeng, F.-Y. & Gabius, H.-J. (1994). Correlation of expression of binding sites for synthetic blood group A-, B- and H-trisaccharides and for sarcolectin with survival of patients with bronchial carcinoma. *European Journal of Cancer* **30A**, 653–657.

Sarcophaga peregrina humoral lectin

Other name: *Sarcophaga* lectin
Mol. wt: 190 000 (native); 32 000 and 30 000 (subunit)
Specificity: D-galactose
Source: Haemolymph of the flesh fly *Sarcophaga peregrina*

ISOLATION

Haemolymph was removed from surgically injured larvae, clarified, and applied to an affinity column of Sepharose 4B. The lectin was eluted with galactose [1].

A cDNA from *Sarcophaga* lectin mRNA was subsequently isolated, cloned and characterised [2].

STRUCTURE

The *Sarcophaga* lectin is a protein or glycoprotein with an apparent molecular weight of 190 000. The active lectin consists of four α (Mr 32 000) and two β (Mr 30 000) subunits not linked by disulphide bonds [1]. However, there is a single lectin gene which codes for the α subunit, and it appears the β subunit is derived from the former by post-translational modification [2,3].

The complete nucleotide and deduced amino acid sequences of the *Sarcophaga* lectin have been published. The gene codes for 260 amino acid residues plus an additional signal sequence [2].

BIOLOGICAL ACTIVITIES

The *Sarcophaga* lectin agglutinates ovine erythrocytes and certain murine (e.g. Ehrlich ascites) tumour cells. Agglutination requires divalent cations and was inhibited by D-galactose and lactose, but not by other monosaccharides tested [1].

The *Sarcophaga* lectin has various immunity-enhancing and anti-tumour activities. It potentiates the cytotoxic activity of murine macrophages and polymorphonuclear leukocytes towards syngeneic and allogeneic mouse tumour cells *in vitro* [4–6]; these effects may be mediated by a 45 000 protein with tumour necrosis factor-like activity [7,8]. Both this protein and the *Sarcophaga* lectin were effective *in vitro* on tumour-bearing mice [8,9]. The *Sarcophaga* lectin can also enhance the cytotoxic activity of murine bone marrow cells towards *Candida parapsilosis* [10], and can stimulate human peripheral blood lymphocytes to produce γ-interferon [11]. Most significantly, the lectin mediates the lysis of ovine erythrocytes introduced into the abdominal cavity of *Sarcophaga* larvae [12].

TISSUE AND/OR SUBCELLULAR DISTRIBUTION

Sarcophaga lectin is synthesised in the fat body (in response to injury) and then secreted into the haemolymph. Lectin was also synthesised *de novo* during pupation, but decreased rapidly before emergence. The lectin was detected on the cell surface of haemocytes as well as being free in the haemolymph plasma [1,10].

POSSIBLE FUNCTION(S)

The *Sarcophaga* lectin appears to have a significant role in host defence. The lectin is not expressed during the normal larval stage but appears in the haemolymph shortly after injury to the body wall. An approximately 50-fold increase in specific lectin activity occurs within two days [1]. Curiously, haemocytes from injured larvae bind the lectin more readily than haemocytes from untraumatised animals [10]. The immunological activities listed above provide indirect, but persuasive, circumstantial evidence that the lectin might activate haemocytes to increase their phagocytic capacity and release effector molecules. Certainly, the *Sarcophaga* lectin mediates the lysis and removal of ovine erythrocytes injected into the abdomen of *S. peregrina* larvae [12].

The developmental regulation of the lectin could also be consistent with a defensive function, since the lectin might promote the scavenging of disintegrating larval tissues by haemocytes, in a manner analogous to removal of infectious agents or exogenously introduced xenogeneic erythrocytes.

However, the observed developmental regulation is also consistent with another function, the development of imaginal discs into adult structures such as wings and legs. Imaginal discs synthesise *Sarcophaga* lectin, and when cultured with anti-*Sarcophaga* lectin antibodies or in the presence of galactose, fail to differentiate normally [13].

COMMENTS

This *Sarcophaga* lectin is perhaps the best known and most intensively studied of all the insect lectins, and it is unusual in appearing to have two independent biological functions, one in defence and another in development. The same species is also the source of a less well-known lectin, granulocytin (qv). Curiously, granulocytin is recognised by a monoclonal antibody specific for a physiological ligand of the better known *Sarcophaga* lectin [14].

REFERENCES

1. Komano, H., Mizuno, D. & Natori, S. (1980). Purification of lectin induced in the hemolymph of *Sarcophaga peregrina* larvae on injury. *Journal of Biological Chemistry* **255**, 2919–2924.

2. Takahashi, H., Komano, H., Kawaguchi, N., Kitamura, N., Nakanishi, S. & Natori, S. (1985). Cloning and sequencing of cDNA of *Sarcophaga peregrina* humoral lectin induced on injury of the body wall. *Journal of Biological Chemistry* **260**, 12228–12233.
3. Komano, H., Mizuno, D. & Natori, S. (1981). A possible mechanism of induction of insect lectin. *Journal of Biological Chemistry* **256**, 7087–7089.
4. Nakajima, H., Komano, H., Esumi-Kurisu, M., Abe, S., Yamayaki, M., Natori, S. & Mizuno, D. (1982). Induction of macrophage-mediated tumour lysis by an animal lectin, *Sarcophaga peregrina* agglutinin. *Gann* **73**, 627–632.
5. Yamazaki, M., Ikenami, M., Komano, H., Tsunawaki, S., Kamiya, H., Natori, S. & Mizuno, D. (1983). Polymorphonuclear leukocyte-mediated cytolysis induced by animal lectin. *Gann* **74**, 576–583.
6. Ohkuma, Y., Komano, H. & Natori, S. (1985). Identification of target proteins participating in a lectin-dependent macrophage-mediated cytotoxic reaction. *Cancer Research* **45**, 288–292.
7. Itoh, A., Iizuka, K. & Natori, S. (1984). Induction of TNF-like factor by murine macrophage-like cell line J774.1 on treatment with *Sarcophaga* lectin. *FEBS Letters* **175**, 59–62.
8. Itoh, A., Ohsawa, F. & Natori, S. (1986). Purification of a cytotoxic protein produced by the murine macrophage-like cell line J774.1 in response to *Sarcophaga* lectin. *Journal of Biochemistry* **99**, 9–15.
9. Itoh, A., Iizuka, K. & Natori, S. (1985). Antitumour effect of *Sarcophaga* lectin on murine transplanted tumours. *Japanese Journal of Cancer Research* **76**, 1027–1033.
10. Komano, H., Nozawa, R., Mizuro, D. & Natori, S. (1983). Measurement of *Sarcophaga peregrina* lectin under various physiological conditions by radioimmunoassay. *Journal of Biological Chemistry* **258**, 2143–2147.
11. Tamura, M. & Natori, S. (1984). Induction of human γ interferon by *Sarcophaga peregrina* humoral lectin. *FEBS Letters* **175**, 325–328.
12. Komano, H. & Natori, S. (1985). Participation of *Sarcophaga peregrina* humoral lectin in the lysis of sheep red blood cells injected into the abdominal cavity of larvae. *Developmental and Comparative Immunology* **9**, 31–40.
13. Kawaguchi, N., Komano, H. & Natori, S. (1991). Involvement of *Sarcophaga* lectin in the development of imaginal discs of *Sarcophaga peregrina* in an autocrine manner. *Developmental Biology* **144**, 86–93.
14. Fujita, Y., Kurata, S., Homma, K. & Natori, S. (1998). A novel lectin from *Sarcophaga*. *Journal of Biological Chemistry* **273**, 9667–9672.

Saxidomus purpuratus isolectins

Other names: SPA-I and SPA-III
Mol. wt : 40 000 (native); 14 000 (subunit)
Specificity: GlcNAc
Source: The shellfish *Saxidomus purpuratus*

ISOLATION

Haemagglutinating activity was purified from crude extracts by ion-exchange chromatography on DEAE–cellulose and affinity chromatography on GlcNAc–Sepharose. By these means, two major isolectins were obtained plus a minor component (SPA-II), which was not studied further [1].

STRUCTURE

Both isolectins are proteins or possibly glycoproteins (0.3% carbohydrate in the form of GlcNAc) with similar amino acid compositions. The apparent molecular weights were estimated at 40 000 by gel filtration, and single bands corresponding to 14 000 were obtained by SDS-PAGE under reducing conditions. The isolectins differed in isoelectric points: 4.4 (SPA-I) and 4.1 (SPA-III) [1].

BIOLOGICAL ACTIVITIES

SPA-I and SPA-III agglutinated human group A erythrocytes; there was no activity towards group B or group O erythrocytes even at high concentrations [1].

SPA-I and SPA-III are highly specific for GlcNAc and its methylated derivatives [1].

TISSUE AND/OR SUBCELLULAR DISTRIBUTION

The highest specific lectin activity is found in the mantle of the mollusc, followed by the foot. Lectin activity has also been detected in various other tissues but at much lower specific activities [1].

POSSIBLE FUNCTION(S)

Nothing is known concerning the possible function(s) of these isolectins.

COMMENTS

Several mollusc species are sources of blood group A specific lectins (see Chapter 4).

REFERENCE

1. Tatsumi, M., Arai, Y. & Itoh, T. (1982). Purification and characterization of a lectin from the shellfish *Saxidomus purpuratus*. *Journal of Biochemistry* **91**, 1139–1146.

Scylla serrata lectin

Mol. wt: 55 000 (native); 30 000 and 25 000 (subunit)
Specificity: Sialic (*N*-glycolylneuraminic) acid
Source: Haemolymph of the edible crab *Scylla serrata*

ISOLATION

The lectin was isolated from the soluble fraction of crab haemolymph by affinity chromatography on bovine thyroglobulin–agarose. The starting material was applied at 4 °C, and washed extensively at 4 °C then at 32 °C with Ca^{2+}-containing buffers. Finally, the lectin was eluted at 32 °C with a Ca^{2+}-free buffer containing EDTA. Over 80% of the starting activity was recovered with a 27 000-fold increase in specific activity [1].

STRUCTURE

SDS-PAGE yielded a single band of apparent molecular weight 55 000 under non-reducing conditions; under reducing conditions, two bands (Mr 30 000 and 25 000) were obtained [1].

BIOLOGICAL ACTIVITIES

The *S. serrata* lectin agglutinated murine, lapine and human (blood group B only) erythrocytes at high titre, but was much less active towards red blood cells from other species [1]. This activity was Ca^{2+}-dependent [1,2].

The lectin did not agglutinate neuraminidase-treated erythrocytes. Agglutination of untreated erythrocytes was readily inhibited by an *N*-glycolylneuraminic acid but not by other monosaccharides, including *N*-acetylneuraminic acid. Among glycoproteins, bovine and porcine thyroglobulins were the most potent inhibitors; bovine, but not porcine, submaxilliary mucin was moderately inhibitory; human glycophorin, human acid glycoprotein and bovine fetuin had little or no effect. The lectin appears to have a uniquely narrow specificity for *N*-glycolylneuraminic acid [1].

TISSUE AND/OR SUBCELLULAR DISTRIBUTION

Only haemolymph appears to have been tested for lectin activity [1,2].

POSSIBLE FUNCTION(S)

It has been suggested that the lectin might function in defence as an opsonin [1], but no direct evidence is available.

COMMENTS

This crab lectin has two very unusual qualities which are difficult to reconcile with each other. Firstly, among animal lectins, it appears to have a uniquely narrow specificity for *N*-glycolylneuraminic acid. Secondly, it appears to be highly specific for human blood group B, although the latter is determined by galactose, and *N*-glycolylneuraminic acid is not a significant component of human erythrocytes. The lectin shows relatively little activity towards human blood group O cells, and none towards group A. Although it has not been evaluated as a reagent, it appears very promising because blood group B specificity is rare among both plant and animal lectins and none has been found to be ideal for routine typing purposes [3].

N-glycolylneuraminic acid is not normally found in human tissues, but human tumour tissue (e.g. melanoma) may contain glycoconjugates with this sialic acid. The *S. serrata* lectin therefore has potential as a human cancer marker.

Erythrocyte adsorption experiments provided no evidence for more than one lectin in *S. serrata* haemolymph [1]. Therefore, the relationship with scyllin (qv), a 5000 Mr lectin with independent anti-microbial activity, is unclear [4,5].

REFERENCES

1. Mercy, P.D. & Ravindranath, M.H. (1993). Purification and characterization of *N*-glycolylneuraminic acid specific lectin from *Scylla serrata*. *European Journal of Biochemistry* **215**, 697–704.
2. Mercy, P.D. & Ravindranath, M.H. (1992). An agglutinin with a unique specificity for *N*-glycol sialic acid residues of thyroglobulin in the haemolymph of a marine crab *Scylla serrata* (Forskal). *Experientia* **48**, 498–500.
3. Kilpatrick, D.C. & Green, C. (1992). Lectins as blood typing reagents. In: *Advances in Lectin Research*, Volume 5 (ed. H. Franz), pp. 51–94. Ullstein Mosby, Berlin.
4. Chattopadhyay, T. & Chatterjee, B.P. (1993). A low molecular weight lectin from the edible crab *Scylla serrata* hemolymph; purification and partial characterization. *Biochemical Archives* **9**, 65–72.
5. Chattopadhyay, T., Guha, A.K. & Chatterjee, B.P. (1996). Novel antimicrobial activity of scyllin, a hemolymph lectin of the edible crab *Scylla serrata*. *Biomedical Letters* **53**, 29–40.

Scyllin

Other name: Low molecular weight *Scylla serrata* lectin
Mol. wt: Approx. 5000
Specificity: Complex
Source: Haemolymph of the estuarine edible crab *Scylla serrata*

ISOLATION

Scyllin was purified some 1200-fold from clarified haemolymph by adsorption to a GalNAc–Separon column and subsequent elution with a citrate-phosphate buffer of pH 3.5 [1].

STRUCTURE

Little is known of the structure, but molecular weight estimations both by gel filtration (Sephadex G-75) and by SDS-PAGE indicated a size smaller than the smallest molecular weight marker used [1]. By extrapolation, an Mr value of $\leqslant$5000 was calculated.

BIOLOGICAL ACTIVITIES

Scyllin agglutinated human (ABO non-specifically), lapine, rat, hamster and guinea-pig erythrocytes; haemagglutinating activity was independent of Ca^{2+} concentration and could not be inhibited by any monosaccharide tested. Several glycoproteins, however, including fetuin, caeruloplasmin and human glycophorin, were potent inhibitors [1].

Scyllin agglutinated various bacteria, particularly the Gram-positive species *Bacillus subtilis*, *Bacillus cereus* and several *Lactobacilli*, but also *Escherichia coli* and *Pseudomonas aeruginosa* [1,2]. Scyllin also inhibited the growth of both Gram-positive and Gram-negative bacteria and could inhibit endogenous respiration and exogenous glucose oxidation in *B. cereus* and *E. coli* [2]. Although scyllin did not agglutinate the yeast *Saccharomyces cerevisiae* [1], it was able to inhibit the growth of *S. cerevisiae* [2]. The glycoprotein inhibitors of scyllin-mediated haemagglutination, however, did not affect its anti-microbial activity [2].

TISSUE AND/OR SUBCELLULAR DISTRIBUTION

Scyllin was purified from haemolymph; its presence in other tissues does not seem to have been investigated.

POSSIBLE FUNCTION(S)

Scyllin is presumed to function in defence [1,2].

COMMENTS

Scyllin is clearly different from the calcium-dependent, sialic acid-binding lectin from the same species (qv) [3].

REFERENCES

1. Chattopadhyay, T. & Chatterjee, B.P. (1993). A low molecular weight lectin from the edible crab *Scylla serrata* hemolymph: purification and partial characterization. *Biochemical Archives* **9**, 65–72.
2. Chattopadhyay, T., Guha, A.K. & Chatterjee, B.P. (1996). Novel antimicrobial activity of scyllin, a haemolymph lectin of the edible crab *Scylla serrata*. *Biomedical Letters* **53**, 29–40.
3. Mercy, P.D. & Ravindranath, M.H. (1993). Purification and characterization of *N*-glycolylneuraminic acid specific lectin from *Scylla serrata*. *European Journal of Biochemistry* **215**, 697–704.

Serum amyloid P component

Other names: SAP; C1t; 9.5S α_1-glycoprotein
Mol. wt: 250 000 (native); 25 000 (subunit)
Specificity: Complex
Source: Blood plasma of vertebrates

ISOLATION

SAP was perhaps first identified in 1965 [1] as a component of amyloid, although it was subsequently isolated in different laboratories and given different names before their common identity as a relative of C-reactive protein (qv) was established [2]. The isolation of SAP is easily achieved by Ca^{2+}-dependent affinity chromatography on unsubstituted Sepharose followed by additional steps to remove trace impurities [3].

The isolation of cDNA clones has also been reported [4].

STRUCTURE

SAP is a glycoprotein arranged as two discs interacting face to face, each disc composed of a pentameric ring of five identical subunits [5]. Each subunit has 204 amino acids; both determined [6] and deduced [4] sequences have been published. The sequence exhibits 52–54% identity to human C-reactive protein [4,6].

The apparent Mr of the subunit has been estimated at approximately 25 000 [4–7], consistent with a molecular mass of 240 000 Da determined for the native molecule.

The three-dimensional structure has been solved at 0.2 nm resolution [8]. SAP has a 'jelly-roll' tertiary fold similar to that of the legume lectins; carboxylate and phosphate compounds bind directly to two calcium ions.

BIOLOGICAL ACTIVITIES

SAP binds to a variety of ligands in a Ca^{2+}-dependent manner. It is the major DNA-binding protein in serum [9]; SAP bound to chromatin and nucleoli (unlike CRP) in frozen sections [10], and to extracellular chromatin *in vivo* [11]. As well as binding amyloid fibres and agarose, through which properties it was first isolated [1,2], SAP binds phosphorylethanolamine (but not phosphorylcholine) [12] and various carbohydrates including snail galactans [13], zymosan, oligosaccharides with penultimate galactose residues, and glycoconjugates with terminal mannose [14]. The best characterised carbohydrate ligand is the cyclic 4,6-pyruvate acetal of galactose [15].

SAP can activate the complement system, apparently via binding to C1q [16]. SAP can also bind C4b-binding protein [12], and C5b6, the latter in a Ca^{2+}-independent manner [17]. SAP binding to model soluble immune complexes (aggregated IgG) has also been reported [18].

SAP is strongly anti-mitogenic towards human mononuclear cells stimulated with PPD, PHA or PWM [13]. As well as binding to lymphocytes and erythrocytes [13], SAP has high-affinity binding sites on neutrophils and can compete with CRP for binding to those sites [19].

TISSUE AND/OR SUBCELLULAR DISTRIBUTION

SAP is synthesised by hepatocytes in the liver and secreted into the bloodstream. It can be a normal component of extracellular matrix and is a universal constituent of primary and secondary amyloid deposits, but is present mainly in the circulation as a plasma protein [2].

POSSIBLE FUNCTION(S)

Mice with targeted deletion of the SAP gene spontaneously develop anti-nuclear autoimmunity and glomerulonephritis [20], perhaps indicating that SAP is essential for the clearance of chromatin from cell debris, and/or functions in the regulation of the immune response to self antigens.

It has also been suggested that SAP might participate in the initial stages of host defence since it recognises structures that are common in micro-organisms and can activate complement [8].

Functionally, SAP may overlap with CRP to some extent, and the two are likely to complement each other.

COMMENTS

SAP is present in normal plasma at a concentration of about 30 μg/ml, and shows little or no increase during the acute phase of inflammation [21]. SAP is present on all amyloid deposits and may act to stabilise these deleterious fibrils. In a mouse model, inhibition of SAP binding to amyloid fibrils *in vivo* retarded amyloidosis [22]; potential application of inhibitors of SAP as drugs might therefore be valid, and the availability of the high-resolution structure of SAP including the binding site would permit direct modelling of competitive inhibitors and facilitate the production of binding site homologues [8].

Radiolabelled SAP can be used to locate amyloid deposits, and scintigraphy after intravenous ^{123}I-SAP injection is of major value in diagnosing and locating amyloidosis, and for monitoring changes with time or treatment [23].

REFERENCES

1. Cathcart, E.S., Comerford, F.R. & Cohen, A.S. (1965). Immunological studies on a protein extracted from human secondary amyloid. *New England Journal of Medicine* **273**, 143–146.
2. Pepys, M.B. & Baltz, M.L. (1983). Acute phase proteins with special reference to C-reactive protein and related proteins (pentraxins) and serum amyloid A protein. *Advances in Immunology* **34**, 141–212.
3. De Beer, F.C. & Pepys, M.B. (1982). Isolation of human C-reactive protein and serum amyloid P component. *Journal of Immunological Methods* **50**, 17–31.
4. Mantzouranis, E.C., Dowton, S.B., Whitehead, A.S., Edge, M.D., Burns, G.A.P. & Colten, H.R. (1985). Human serum amyloid P component. cDNA isolation, complete sequence of pre-serum amyloid P component, and localization of the gene to chromosome 1. *Journal of Biological Chemistry* **260**, 7752–7756.
5. Osmand, A.P., Friederson, B., Gewurz, H., Painter, R.H., Hofmann, T. & Shelton, E. (1977). Characterization of C-reactive protein and the complement subcomponent C1t as homologous proteins displaying cyclic pentameric symmetry (pentraxins). *Proceedings of the National Academy of Sciences USA* **74**, 739–743.
6. Prelli, F., Pras, M. & Frangione, B. (1985). The primary structure of human tissue amyloid P component from a patient with primary idiopathic amyloidosis. *Journal of Biological Chemistry* **260**, 12895–12898.
7. Perkins, S.J. & Pepys, M.B. (1986). X-ray and neutron scattering studies of CRP and SAP. *Protides of the Biological Fluids* **34**, 323–326.
8. Emsley, J., White, H.E., O'Hara, B.P., Oliva, G., Srinivasan, N., Tickle, I.J., Blundell, T.L., Pepys, M.B. & Wood, S.P. (1994). Structure of pentameric human serum amyloid P component. *Nature* **367**, 338–345.
9. Pepys, M.B. & Butler, P.J.G. (1987). Serum amyloid P component is the major calcium-dependent specific DNA binding protein of the serum. *Biochemical and Biophysical Research Communications* **148**, 308–313.
10. Pepys, M.B., Booth, S.E., Tennent, G.A., Butler, P.J.G. & Williams, D.G. (1994). Binding of pentraxins to different nuclear structures: C-reactive protein binds to small nuclear ribonucleoprotein particles, serum amyloid P component binds to chromatin and nucleoli. *Clinical and Experimental Immunology* **97**, 152–157.
11. Breathnach, S.M., Kofler, H., Sepp, N., Ashworth, J., Woodrow, D., Pepys, M.B. & Hintner, H. (1989). Serum amyloid P component binds to cell nuclei in vitro and to in vivo deposits of extracellular chromatin in systemic lupus erythematosus. *Journal of Experimental Medicine* **170**, 1433–1438.
12. Schwalbe, R.A., Dahlbäck, B., Coe, J.E. & Nelsestuen, G.L. (1992). Pentraxin family of proteins interact specifically with phosphorylcholine and/or phosphorylethanolamine. *Biochemistry* **31**, 4907–4915.
13. Li, J.J., Pereira, M.E.A., De Lellis, R.A. & McAdam, P.W.J. (1984). Human amyloid P component: a circulating lectin that modulates immunological responses. *Scandinavian Journal of Immunology* **19**, 227–236.
14. Kubak, B.M., Potempa, L.A., Anderson, B., Mahklouf, S., Venegas, M., Gewurz, H. & Gewurz, A.T. (1988). Evidence that serum amyloid P component binds to mannose-terminated sequences of polysaccharides and glycoproteins. *Molecular Immunology* **25**, 851–858.
15. Hind, C.R.K., Collins, P.M., Renn, D., Cook, R.B., Caspi, D., Baltz, M.L. & Pepys, M.B. (1984). Binding specificity of serum amyloid P component for the pyruvate acetal of galactose. *Journal of Experimental Medicine* **159**, 1058–1069.
16. Ying, S.-C., Gewurz, A.T., Jiang, H. & Gewurz, H. (1993). Human serum amyloid P component oligomers bind and activate the classical complement pathway via residues 14–26 and 76–92 of the A chain collagen-like region of C1q. *Journal of Immunology* **150**, 169–176.

17. Barbashov, S.F., Wang, C. & Nicholson-Weller, A. (1997). Serum amyloid P component forms a stable complex with human C5b6. *Journal of Immunology* **158**, 3830–3835.
18. Brown, M.R. & Anderson, B.E. (1993). Receptor–ligand interactions between serum amyloid P component and model soluble immune complexes. *Journal of Immunology* **151**, 2087–2095.
19. Landsmann, P., Rosen, O., Pontet, M., Pras, M., Levartowsky, D., Shephard, E.G. & Fridkin, M. (1994). Binding of human serum amyloid P component (hSAP) to human neutrophils. *European Journal of Biochemistry* **223**, 805–811.
20. Bickerstaff, M.C.M., Botto, M., Hutchinson, W.L., Herbert, J., Tennent, G.A., Bybee, A., Mitchell, D.A., Cook, H.T., Butler, P.J.G., Walpont, M.J. & Pepys, M.B. (1999). Serum amyloid P component controls chromatin degradation and prevents antinuclear autoimmunity. *Nature Medicine* **5**, 694–697.
21. Steele, D.M. & Whitehead, A.S. (1994). The major acute phase reactants: C-reactive protein, serum amyloid P component and serum amyloid A protein. *Immunology Today* **15**, 81–88.
22. Botto, M., Hawkins, P.N., Bickerstaff, M.C.M., Herbert, J., Bygrave, A.E., McBride, A., Hutchinson, W.L., Tennent, G.A., Walport, M.J. & Pepys, M.B. (1997). Amyloid deposition is delayed in mice with targeted deletion of the serum amyloid P component gene. *Nature Medicine* **3**, 855–859.
23. Hawkins, P.N., Lavender, J.P. & Pepys, M.B. (1990). Evaluation of systemic amyloidosis by scintigraphy with ^{123}I-labelled serum amyloid P component. *New England Journal of Medicine* **323**, 508–513.

Sialoadhesin

Other names: Siglec-1; Sn
Mol. wt: 185 000
Specificity: NeuAc-α2,3-Gal-R
Source: Murine tissue macrophages

ISOLATION

Sialoadhesin was first identified as a lectin-like macrophage receptor for sheep erythrocytes [1,2], and purified from murine splenic lysate by immunoaffinity chromatography [3].

A cDNA was cloned and expressed in COS cells [4].

STRUCTURE

Sialoadhesin is a glycoprotein with an apparent Mr of 185 000 determined by SDS-PAGE under reducing conditions [2]. It is a type I transmembrane protein of 1675 amino acid residues, with a predicted protein molecular mass of 181 000 Da, and extensive homology to CD22, CD33, myelin-associated glycoprotein and CD54. Like other members of the I-type lectin family, it has a C-terminal cytoplasmic region, a transmembrane domain and a large extracellular region. It differs from the others, however, in the vastly greater size of its extracellular region, consisting of 17 (16 × C2-sets plus one V-set) immunoglobulin-like domains [4].

BIOLOGICAL ACTIVITIES

Sialoadhesin can agglutinate ovine and human erythrocytes in a sialic acid-dependent manner [3]. It can also bind to murine neutrophils > bone marrow cells > lymphocytes > thymocytes, and selectively bound myeloid cells in complex cell mixtures [5].

All cellular interactions are dependent on cell surface sialic acid, and can be inhibited by (NeuAc-α2,3-Galβ1,3-GalNAc) and related structures [3]. The interaction is mainly with the terminal sialic acid but contacts with subterminal galactose and GlcNAc/GalNAc contribute additional binding strength [6]. Unlike myelin-associated glycoprotein, sialoadhesin prefers *N*-acetylneuraminic acid to 2-keto-3-deoxy-D-glycero-D-galacto-nononic acid [6].

TISSUE AND/OR SUBCELLULAR DISTRIBUTION

Sialoadhesin is found on tissue macrophages in bone marrow, spleen and lymph nodes [1,2]. Within clusters of bone marrow macrophages and other cells, sialoadhesin was highly concentrated at contacts with developing myelomonocytic precursor cells (but not erythroblasts) visualised by immunofluorescence or by electron microscopy [7].

POSSIBLE FUNCTION(S)

Obvious possible functions include a role in the differentiation of myeloid cells in bone marrow, and in trafficking of leukocytes in lymphoid tissues, but the function of sialoadhesin has not been defined.

COMMENTS

Desialylation of endogenous glycoconjugates enhances sialoadhesin-mediated lectin activity, and sialylation of sialoadhesin itself inhibits its lectin activity [8].

Although sialoadhesin was discovered and intensively studied in the mouse, rat [9] and human [10] homologues are known to exist.

REFERENCES

1. Crocker, P.R. & Gordon, S. (1986). Properties and distribution of a lectin-like hemagglutinin differentially expressed by murine stromal tissue macrophages. *Journal of Experimental Medicine* **164**, 1862–1875.
2. Crocker, P.R. & Gordon, S. (1989). Mouse macrophage hemagglutinin (sheep erythrocyte receptor) with specificity for sialylated glycoconjugates characterised by a monoclonal antibody. *Journal of Experimental Medicine* **169**, 1333–1346.
3. Crocker, P.R., Kelm, S., Dubois, C., Martin, B., McWilliam, A.S., Shotton, D.M., Paulson, J.C. & Gordon, S. (1991). Purification and properties of sialoadhesion, a sialic acid-binding receptor of murine tissue macrophages. *EMBO Journal* **10**, 1661–1669.
4. Crocker, P.R., Mucklow, S., Bouckson, V., McWilliam, A., Willis, A.C., Gordon, S., Milon, G., Kelm, S. & Bradfield, P. (1994). Sialoadhesion, a macrophage sialic acid binding receptor for haemopoietic cells with 17 immunoglobulin-like domains. *EMBO Journal* **13**, 4490–4503.
5. Crocker, P.R., Freeman, S., Gordon, S. & Kelm, S. (1995). Sialadhesion binds preferentially to cells of the granulocytic lineage. *Journal of Clinical Investigation* **95**, 635–643.
6. Strenge, K., Schauer, R., Bovin, N., Hasegawa, A., Ishida, H., Kiso, M. & Kelm, S. (1998). Glycan specificity of myelin-associated glycoprotein and sialoadhesion deduced from interactions with synthetic oligosaccharides. *European Journal of Biochemistry* **258**, 677–685.
7. Crocker, P.R., Werb, Z., Gordon, S. & Bainton, D.F. (1990). Ultrastructural localization of a macrophage-restricted sialic acid binding hemagglutinin, SER, in macrophage-hematopoietic cell clusters. *Blood* **76**, 1131–1138.
8. Barnes, Y.C., Skelton, T.P., Stamenkovic, I. & Sgroi, D.C. (1999). Sialylation of the sialic acid binding lectin sialoadhesion regulates its ability to mediate cell adhesion. *Blood* **93**, 1245–1252.
9. Van Den Berg, T.K., Brevé, J.J.P., Damoiseaux, J.G.M.C., Dopp, E.A., Kelm, S., Crocker, P.R., Dijkstra, C.D. & Kraal, G. (1992). Sialoadhesion on macrophages: its identification as a lymphocyte adhesion molecule. *Journal of Experimental Medicine* **176**, 647–655.
10. Mucklow, S., Hartnell, A., Mattei, M.-G., Gordon, S. & Crocker, P.R. (1995). Sialoadhesin (Sn) maps to mouse chromosome 2 and human chromosome 20 and is not linked to the other members of the sialoadhesin family, CD22, MAG, and CD33. *Genomics* **28**, 344–346.

Spodoptera exigua lectin

Other name: Beet armyworm lectin
Mol. wt: 700 000 (native); 33 000 (subunit)
Specificity: D-Galactose
Source: Haemolymph of the insect *Spodoptera exigua*

ISOLATION

The lectin can be isolated from haemolymph plasma by affinity chromatography, either by application to Affi-Gel-ovalbumin and elution with EDTA [1], or by application to galactose–agarose and elution with galactose [2,3].

STRUCTURE

The *S. exigua* lectin is a glycoprotein [1,3]. The native molecule exists as aggregates ranging in mass from about 100 000 to 700 000 Da. After SDS-PAGE, it appears as a doublet of apparent Mr 33 200–34 400, of which only the latter is glycosylated. The amino acid compositions of the two bands were very similar and N-terminal analyses revealed an identical first 32 amino acid sequence, suggesting that the two SDS-PAGE bands differ only in glycosylation. No homology was detected between the N-terminal sequence and those of several other invertebrate galactose-binding lectins [3].

BIOLOGICAL ACTIVITIES

The lectin agglutinates trypsin-treated lapine and human erythrocytes in the presence of Ca^{2+}. Haemagglutination was most readily inhibited by D-galactose and lactose [1].

The lectin also binds to blastospores of hyphomycetes (like *Paecilomycetes farinosus*) [1] and opsonises them [2]; both binding and opsonisation are inhibited by galactose. *In vivo*, *P. farinosus* cells are rapidly cleared from haemolymph, while *Nomuraea rileyi* blastospores (which lack galactosidic residues) are not cleared [2].

TISSUE AND/OR SUBCELLULAR DISTRIBUTION

The lectin is found in haemolymph plasma, haemocytes and the fat body. The larval fat body is the major site of synthesis. Haemocyte lectin is found only in the granules of granulocytes [3].

POSSIBLE FUNCTION(S)

The lectin is presumed to function in defence as an opsonin [3].

COMMENTS

There is no evidence that the production of the *S. exigua* lectin is induced by injury or injection of fungal cells [1]. Nevertheless, the lectin binds to fungal cell surfaces with galactosyl residues and promotes uptake of fungi by haemocytes. It has been suggested that on binding to foreign galactosyl residues, a conformational change is induced in the lectin that enables it to attach to haemocyte surfaces to which it is normally inert [3].

REFERENCES

1. Pendland, J.C. & Boucias, D.G. (1986). Characteristics of a galactose-binding hemagglutinin (lectin) from hemolymph of *Spodoptera exigua* larvae. *Developmental and Comparative Immunology* **10**, 477–487.
2. Pendland, J.C., Heath, M.A. & Boucias, D.G. (1988). Function of a galactoside-binding lectin from blastospores from entomogenous hyphomycetes. *Journal of Insect Physiology* **34**, 533–540.
3. Boucias, D.G. & Pendland, J.C. (1993). The galactose binding lectin from the beet armyworm, *Spodoptera exigua*: distribution and site of synthesis. *Insect Biochemistry and Molecular Biology* **23**, 233–242.

Stichopus japonicus coelomic plasma lectin-1

Other name: SPL-1
Mol. wt: 400 000 (native); 17 000 (subunit)
Specificity: Uronic acids
Source: Coelomic plasma of the sea cucumber *Stichopus japonicus*

ISOLATION

A solution derived from coelomic plasma was applied to a column of porcine gastric mucin immobilised on agarose, which, after washing, was eluted with an EDTA-containing buffer. The resulting lectin mixture was resolved into four peaks by gel filtration on Superose 6; the second peak (the first with high haemagglutinating titre) was designated SPL-1 and trace contaminants removed by ion-exchange chromatography [1].

STRUCTURE

SPL-1, subjected to SDS-PAGE, migrated with apparent molecular weights of 17 000 and 16 000, respectively, under reducing and non-reducing conditions. The native Mr was estimated as 400 000 by gel filtration, suggesting an aggregate of perhaps 24 non-covalently linked subunits [1].

The amino terminal sequence of SPL-1 exhibited limited homology to *S. japonicus* coelomic plasma lectin-2 (SPL-2) (qv) and *S. japonicus* lectin-I (SJL-I) (qv), as well as several other C-type lectins from both invertebrates and vertebrates [1].

BIOLOGICAL ACTIVITIES

SPL-1 agglutinated glutaraldehyde-fixed trypsin-treated lapine erythrocytes in a Ca^{2+}-dependent manner. Haemagglutination was readily inhibited by galacturonic and glucuronic acids, but not by chondroitin sulphate or heparin. Lactose was weakly inhibitory, but other galactosides and simple sugars were without effect. SPL-1 did not bind to a lactose-conjugated affinity column [1].

SPL-1 bound to coelomocyte clots, but only in the presence of calcium and not in the presence of galacturonic acid. However, porcine gastric mucin did not interfere with the clotting process [1].

TISSUE AND/OR SUBCELLULAR DISTRIBUTION

SPL-1 was found in coelomic fluid; its distribution elsewhere was not reported [1].

POSSIBLE FUNCTION(S)

No information indicative of function was reported [1].

COMMENTS

The coelomic plasma of *S. japonicus* contains at least four or five lectins, of which SPL-1 and SPL-2 show the most activity towards fixed, trypsin-treated lapine erythrocytes. The coelomic plasma also contains activity towards fresh human erythrocytes and fixed human platelets, although it is not clear how much of those activities are associated with either SPL-1 or SPL-2 [1].

Aggregation of coelomocytes in response to wounding (clotting) is necessary to prevent loss of haemolymph. Although SPL-1 can bind to aggregated coelomocytes in a lectin-like manner, this is not essential to the clotting process, which might instead involve the RGD-integrin recognition system.

REFERENCE

1. Matsui, T., Ozeki, Y., Suzuki, M., Hino, A. & Titani, K. (1994). Purification and characterization of two Ca^{2+}-dependent lectins from coelomic plasma of sea cucumber, *Stichopus japonicus*. *Journal of Biochemistry* **116**, 1127–1133.

Stichopus japonicus coelomic plasma lectin-2

Other name: SPL-2
Mol. wt: 60 000 (native); 17 000 (subunit)
Specificity: GalNAc > Gal
Source: Coelomic plasma of the sea cucumber *Stichopus japonicus*

ISOLATION

A mixture of calcium-dependent lectins was obtained from coelomic plasma by affinity chromatography on porcine gastric mucin–agarose, eluting with EDTA. Five peaks were resolved by gel filtration on Superose 6; the third peak was designated SPL-2. Most (80%) of the activity towards the target cells used was present in peaks 2 (SPL-1) and 3 (SPL-2). Trace contaminants were removed from SPL-2 by ion-exchange chromatography [1].

STRUCTURE

SPL-2 eluted from the Superose column in a volume corresponding to an apparent Mr of 60 000. An identical Mr value was obtained by SDS-PAGE under non-reducing conditions, although a single 17 000 band was obtained in the presence of reducing agent. It seems likely therefore that SPL-2 is a homotetramer linked by disulphide bonds.

The amino terminal sequence of SPL-2 exhibited limited homology to *S. japonicus* coelomic plasma lectin-1 (SPL-1) (qv), and *S. japonicus* lectin-I (SJL-I) (qv) as well as to several other C-type lectins [1].

BIOLOGICAL ACTIVITIES

SPL-2 agglutinated glutaraldehyde-fixed, trypsin-treated lapine erythrocytes in a calcium-dependent manner. Haemagglutination was readily inhibited by D-galactose and galactosides, but the most potent inhibitor was GalNAc. Asialofetuin was as good an inhibitor as porcine gastric mucin, but untreated fetuin was 60-times less effective [1].

SPL-2 bound to coelomocyte clots in the presence of calcium and this adhesion could be inhibited by GalNAc. The presence of GalNAc, however, did not prevent the clotting process [1].

TISSUE AND/OR SUBCELLULAR DISTRIBUTION

SPL-2 was purified from coelomic plasma; its presence in other tissues was not investigated [1].

POSSIBLE FUNCTION(S)

The function of SPL-2 is unknown.

COMMENTS

SPL-2 is one of several lectins present in coelomic fluid of *S. japonica* and/or extracted from homogenates of the whole animal [1,2]. SPL-2 resembles SJL-I (qv) with respect to saccharide specificity, and can bind to an immobilised lactose column similar to that used to purify SJL-I [1,2]. These two lectins have a different primary structure, however, and it is perhaps more likely that SPL-2 is identical to the Mr 68 000 protein eluted from lactosyl–Sepharose by Hatakeyama and co-workers but not investigated further [2].

S. japonica coelomocytes have saccharide ligands for SPL-2 and SPL-1, but this lectin–receptor interaction is not essential for clot formation (aggregation), which instead may be integrin mediated [1].

REFERENCES

1. Matsui, T., Ozeki, Y., Suzuki, M., Hino, A. & Titani, K. (1994) Purification and characterization of two Ca^{2+}-dependent lectins from coelomic plasma of sea cucumber, *Stichopus japonicus*. *Journal of Biochemistry* **116**, 1127–1133.
2. Hatakeyama, T., Himeshima, T., Komatsu, A. & Yamasaki, N. (1993). Purification and characterization of two lectins from the sea cucumber, *Stichopus japonicus*. *Bioscience, Biotechnology and Biochemistry* **57**, 1736–1739.

Stichopus japonicus lectin I

Other name: SJL-I
Mol. wt: 15 837
Specificity: GalNAc > Gal
Source: The sea cucumber *Stichopus japonicus*

ISOLATION

Several Ca^{2+}-dependent lectins were obtained from a homogenate of *S. japonicus* by affinity chromatography on lactosyl–Sepharose, eluting with EDTA. SJL-I was separated from the others by ion-exchange chromatography on Q-Sepharose [1].

STRUCTURE

SJL-I is a protein (no attached carbohydrate) with a relatively high content of glycine, aspartic acid, glutamic acid and serine, which has been completely sequenced [2]. The calculated Mr from its 143 amino acid residues is 15 837 [2], slightly higher than the 13 000 estimated by SDS-PAGE [1]. No information was reported concerning its native molecular size; its isoelectric point is 5.2 [2].

SJL-I possesses significant sequence homology to several invertebrate C-type lectins, including that of the sea urchin *Anthocidaris crassispina* (echinoidin, qv). The N-terminal sequence is similar but nonetheless clearly distinct from those of the *S. japonicus* coelomic plasma lectins, SPL-1 (qv) and SPL-2 (qv) [3].

BIOLOGICAL ACTIVITIES

SJL-I agglutinated lapine and human (with a preference for blood group A) erythrocytes, but not the corresponding cells from equine, ovine or chicken sources. Compared to SJL-II, much higher concentrations of both lectin protein and Ca^{2+} were required to achieve the same titre against lapine erythrocytes. Haemagglutination was readily inhibited by galactose-related sugars, of which GalNAc was the most effective. D-Galactose was a more effective inhibitor than D-fucose, but various α and β galactosides, including melibiose, raffinose and lactose, were of similar potency to galactose [1].

TISSUE AND/OR SUBCELLULAR DISTRIBUTION

Nothing was reported concerning the distribution of this lectin [1].

POSSIBLE FUNCTION(S)

The function of this lectin is not known [1].

COMMENTS

There is evidence for at least five lectins present in *S. japonicus* [1,3]. SJL-I has a distinct primary structure from the two haemagglutinins isolated from coelomic plasma, but, in the absence of primary structural data, its relationship to SJL-II cannot be described with confidence. Nevertheless, SJL-I and SJL-II are clearly distinct lectins with different erythrocyte and saccharide specificities, and quantitatively different requirements for Ca^{2+} [1].

REFERENCES

1. Hatakeyama, T., Himeshima, T., Komatsu, A. & Yamasaki, N. (1993). Purification and characterization of two lectins from the sea cucumber, *Stichopus japonicus*. *Bioscience, Biotechnology and Biochemistry* **57**, 1736–1739.
2. Himeshima, T., Hatakeyama, T. & Yamasaki, N. (1994). Amino acid sequence of a lectin from the sea cucumber, *Stichopus japonicus*, and its structural relationship to the C-type animal lectin family. *Journal of Biochemistry* **115**, 689–692.
3. Matsui, T., Ozeki, Y., Suzuki, M., Hino, A. & Titani, K. (1994). Purification and characterisation of two Ca^{2+}-dependent lectins from coelomic plasma of sea cucumber, *Stichopus japonicus*. *Journal of Biochemistry* **116**, 1127–1133.

Stichopus japonicus lectin II

Other name: SJL-II
Mol. wt: 15 000
Specificity: Complex (galactosides)
Source: The sea cucumber *Stichopus japonicus*

ISOLATION

A homogenate of *S. japonicus* was fractionated by application to a column of lactosyl–Sepharose and elution with EDTA. SJL-II was then separated from the other lectins by ion-exchange chromatography on Q-Sepharose [1].

STRUCTURE

SJL-II is a protein or glycoprotein with an amino acid composition broadly similar to that of SJL-I. The Mr estimated by SDS-PAGE was 15 000, but the migration was slower than SJL-I, which has a true Mr of almost 16 000. The native Mr was not determined [1].

BIOLOGICAL ACTIVITIES

SJL-II agglutinated lapine but not human, equine, ovine or chicken erythrocytes. The titre with lapine cells, on a gravimetric basis, was 90-times higher than that of SJL-I, and there was a much lower Ca^{2+} requirement (optimal activity at 0.1 mM compared to 10 mM for SJL-I). Haemagglutination was not inhibited by any of the monosaccharides tested; some weak inhibition was observed, however, with lactose, melibiose and raffinose [1].

TISSUE AND/OR SUBCELLULAR DISTRIBUTION

No information concerning distribution was reported [1].

POSSIBLE FUNCTION(S)

The function of this lectin is not known.

COMMENTS

Although SJL-II has not been fully characterised, its saccharide specificity and calcium dependence indicate that it must represent a separate gene product from SJL-I. These properties also exclude any possibility of its identity to either of the lectins isolated from *S. japonicus* coelomic plasma [2].

REFERENCES

1. Hatakeyama, T., Himeshima, T., Komatsu, A. & Yamasaki, N. (1993). Purification and characterisation of two lectins from the sea cucumber *Stichopus japonicus*. *Bioscience, Biotechnology and Biochemistry* **57**, 1736–1739.
2. Matsui, T., Ozeki, Y., Suzuki, M., Hino, A. & Titani, K. (1994). Purification and characterization of two Ca^{2+}-dependent lectins from coelomic fluid of sea cucumber, *Stichopus japonicus*. *Journal of Biochemistry* **116**, 1127–1133.

Styela clava opsonin

Other name: *Styela clava* lectin
Mol. Wt: 18 000
Specificity: GalNAc, galactose
Source: Haemolymph of the solitary tunicate *Styela clava*

ISOLATION

The opsonin was isolated in a single step by gel filtration of concentrated haemolymph plasma on Sephadex G-50. Affinity adsorption to yeast particles (and elution with EDTA) also resulted in the isolation of an apparently homogeneous protein [1].

STRUCTURE

The purified protein or glycoprotein migrated as a single band on SDS-PAGE corresponding to an apparent Mr of 17 500 under reducing or non-reducing conditions. A similar molecular size (around 19 000) was estimated by gel filtration, indicating a monomeric structure. A single isoelectric point of 7.0 was determined by isoelectric focusing [1].

BIOLOGICAL ACTIVITIES

The lectin opsonised yeast (*Saccharomyces cerevisiae*) for *S. clava* haemocytes at an optimal concentration of 1 μg/ml [1,2]. This activity required divalent cations, and could be inhibited by GalNAc and galactose but not by GlcNAc or glucose. Mannan was also inhibitory, but sucrose and lactose were not [2].

Although the *S. clava* opsonin could opsonise yeasts for haemocytes from the congeneric species *S. plicata*, some degree of species specificity was evident, presumably reflecting a species difference in cellular receptors [2].

TISSUE AND/OR SUBCELLULAR DISTRIBUTION

No information is available.

POSSIBLE FUNCTION(S)

A defensive function is presumed [1,2].

COMMENTS

This lectin is perhaps the best described representative of a group of tunicate lectins that have been isolated but not studied in great detail, including the serum and cellular lectins of *Phallusia mamillata* [3] and the lectins from *Ascidia malaca* [4], which are thought to form part of an innate immune system.

REFERENCES

1. Kelly, K.L., Cooper, E.L. & Raftos, D.A. (1992). Purification and characterization of a humoral opsonin from the solitary urochordate *Styela clava*. *Comparative Biochemistry and Physiology* **103B**, 749–753.
2. Kelly, K.L., Cooper, E.L. & Raftos, D.A. (1993). A humoral opsonin from the solitary urochordate *Styela clava*. *Developmental and Comparative Immunology* **17**, 29–39.
3. Parrinello, N. & Arizza, V. (1989). Sugar specific cellular lectins of *Phallusia mamillata* hemocytes: purification, characterization and evidence for cell surface localization. *Developmental and Comparative Immunology* **13**, 113–121.
4. Parrinello, N. & Arizza, V. (1988). D-Galactoside binding lectins from the tunicate *Ascidia malaca*: subunit characterization and hemocyte surface distribution. *Developmental and Comparative Immunology* **12**, 495–507.

Surfactant protein A

Other names: SP-A; SP-35
Mol. wt: Approx. 700 000 (native); $\geqslant$28 000 (subunit)
Specificity: *N*-acetylmannosamine
Sources: Mammalian lung tissue and lavage

ISOLATION

SP-A was initially purified from canine lung lavage fluid by a procedure involving a series of butanol extractions [1]. SP-A from various species have been isolated following butanol precipitation, but extraction with *n*-octyl-*β*-D-glucopyranoside and subsequent isolation by cation-exchange chromatography gives a product retaining the ability to interact with macrophages [2].

The cDNAs for canine [3], human [4] and rat [5] SP-A have been isolated and cloned.

STRUCTURE

SP-A is a glycoprotein with a typical collectin structure. The full amino acid sequences from canine [3], human [4] and rat [5] SP-As have been published. The overall sequence identity between the species is 64% [5].

The apparent Mr of the basic chain of approximately 32 000 varies according to the degree of glycosylation [3]. These chains form triplet subunits, each chain with an N-terminal cysteine-rich domain, a relatively short collagen-like domain, a linking (neck) domain and a carboxyl-terminal C-type lectin domain. In humans, the two gene products available can form homotrimers or heterotrimers [6]. The trimeric subunits undergo oligomerisation and are predominantly assembled as hexamers (6 × 3) with an apparent Mr of approximately 700 000 [7,8].

The structure visualised by electron microscopy resembled C1q of the complement system [8]. SP-A in bronchoalveolar lavage fluid exists as a mixture of oligomeric forms, and there was a shift towards lower (smaller) oligomeric forms in patients with pollen allergy [9]. SP-A exists as acidic isoforms of pI 4.2–5.0.

BIOLOGICAL ACTIVITIES

Calcium-dependent carbohydrate-binding activity is manifested by binding to agarose-immobilised monosaccharides [10], neoglycoconjugates [11] and solid-phase mannan [12]. Some (Ba^{2+}, Sr^{2+} and Mn^{2+}) but not all (e.g. Mg^{2+}) divalent cations could substitute to some extent for Ca^{2+} in the monosaccharide affinity chromatography system [10]. Although there were some inconsistencies (particularly in relation to galactose), L-fucose, D-mannose and D-glucose appear to be among the best monosaccharide inhibitors. With binding to solid-phase mannan, *N*-acetylmannosamine was the most potent inhibitor found [12].

SP-A can bind to a variety of lipids, including dipalmitoylphosphatidylcholine [13]. Binding both to glycolipids and non-glycosylated ceramides is Ca^{2+} dependent, and involves the carbohydrate-recognition domain [11,13,14]. However, the binding of SP-A to the lipid A moiety of rough LPS was found to be calcium-independent [15] and insensitive to carbohydrate inhibitors [16].

SP-A has opsonic activity, enhancing the phagocytosis of bacteria, fungi and viruses by alveolar macrophages [15–18]. SP-A is also a chemoattractant for neutrophils and monocytes [19]. Binding to macrophages may involve attachment to C1q (collectin) receptors [20] and other receptors shared with type II epithelial cells [21]. SP-A has been shown to stimulate the respiratory burst of alveolar macrophages [2].

SP-A is anti-mitogenic for human T lymphocytes [22], although it can act as a co-mitogen with concanavalin A [22,23]. SP-A has multiple effects on the cytokine profile of human mononuclear cells in culture [24].

TISSUE AND/OR SUBCELLULAR DISTRIBUTION

Surfactant is synthesised by type II alveolar epithelial cells and SP-A has been localised to secretory granules (lamellar bodies); SP-A has also been found in non-ciliated bronchiolar (Clara) cells of rat and canine, but not human, lung [25]. SP-A has also been localised to tracheal glands, bronchial glands and bronchiolalveolar portals [26]. There is evidence for SP-A expression in rat (but not human) intestine [27].

POSSIBLE FUNCTION(S)

SP-A gene knock-out mice have been generated and observed [28]. There are no apparent abnormalities in respiratory function, seemingly refuting the early concept that SP-A was essential for the production of normal surfactant. On the other hand, these animals clear group B streptococci and *Pseudomonas aeruginosa* less efficiently than wild-type mice after experimental infection, and impaired clearance correlated with decreased binding and uptake by alveolar macrophages.

While the principal function of SP-A may be to protect against microbial pathogens, there is circumstantial evidence for its protecting against allergens. SP-A can bind pollen [29] and house dust mite [30] extracts. The immunoregulatory properties of SP-A, including the inhibition of allergen-induced histamine release, might function to limit the damaging effects of allergen-mediated respiratory inflammation [30].

COMMENTS

SP-A is encoded by two genes on human chromosome 10 with several alleles at each locus [4,31], but most mammals have only one SP-A gene. In both humans and rodents, expression is developmentally regulated. There is an increase in late gestation associated with SP-1 gene activity, and a further increase in adult tissue associated with SP-2 gene expression [32]. SP-A may behave as an acute-phase reactant, responding to lung injury or to the influence of glucocorticoids [32,33].

The restricted expression of SP-A within the lung permits its application as a disease marker in various clinical settings [34,35]. SP-A in amniotic fluid (antenatally) and in tracheal aspirates (infants) accurately predicts lung immaturity and its measurement may be useful in relation to respiratory distress syndrome. Adult respiratory distress syndrome is associated with decreased levels of SP-A in lung lavage fluid, but increased levels have been reported in tracheal aspirates and sputum. SP-A (and SP-A/phospholipid ratio) in lung lavage is also decreased in pulmonary fibrosis patients, but increased in sarcoidosis, hypersensitivity pneumonitis and (especially) pulmonary alveolar proteinosis. The very high level of SP-A in the sputum of alveolar proteinosis patients is diagnostically useful and its measurement much less invasive than alternative procedures.

Elevated SP-A can be detected in the circulation of patients with certain lung diseases. Serum SP-A is often elevated in alveolar proteinosis, idiopathic pulmonary fibrosis, interstitial pneumonia with collagen vascular disease and adult respiratory distress syndrome.

Finally, SP-A detected by immunohistochemical or molecular biological techniques can serve as a marker for lung adenocarcinomas.

REFERENCES

1. King, R.J. & MacBeth, M.C. (1979). Physiochemical properties of dipalmitoyl phosphatidylcholine after interaction with an apolipoprotein of pulmonary surfactant. *Biochimica et Biophysica Acta* **557**, 86–101.
2. Van Iwaarden, J.F., Teding Van Berkhout, F., Whitsett, J.A., Oosting, R.S. & Van Golde, L.M.G. (1995). A novel procedure for the rapid isolation of surfactant protein A with retention of its alveolar-macrophage-stimulating properties. *Biochemical Journal* **309**, 551–555.
3. Benson, B., Hawgood, S., Schilling, J., Clements, J., Damm, D., Cordell, B. & White, R.T. (1985). Structure of canine pulmonary surfactant apoprotein: cDNA and complete amino acid sequence. *Proceedings of the National Academy of Sciences USA* **82**, 6379–6383.
4. Whyte, R.T., Damm, D., Miller, J., Spratt, K., Schilling, J., Hawgood, S., Benson, B. & Cordell, B. (1985). Isolation and characterization of the human pulmonary surfactant apoprotein gene. *Nature* **317**, 361–363.
5. Sano, K., Fisher, J., Mason, R.J., Kuroki, Y., Schilling, J., Benson, B. & Voelker, D. (1987). Isolation and sequence of a cDNA clone for the rat pulmonary surfactant-associated protein (PSP-A). *Biochemical and Biophysical Research Communications* **144**, 367–374.
6. Voss, T., Melchers, G., Scheirle, G. & Schafer, K.P. (1991). Structural comparison of recombinant pulmonary surfactant protein SP-A derived from two human coding sequences: implications for the chain composition of natural human SP-A. *American Journal of Respiratory Cellular and Molecular Biology* **4**, 88–94.

7. King, R.J., Simon, D. & Horowitz, P.M. (1989). Aspects of secondary and quaternary structure of surfactant protein A from canine lung. *Biochimica et Biophysica Acta* **1001**, 294–301.
8. Voss, T., Eistetter, H. & Schafer, K.P. (1988). Macromolecular organisation of natural and recombinant lung surfactant protein SP 28–36. *Journal of Molecular Biology* **201**, 219–227.
9. Hickling, T.P., Malhatra, R. & Sim, R.B. (1998). Human lung surfactant protein A exists in several different oligomeric states: oligomer size distribution varies between patient groups. *Molecular Medicine* **4**, 266–275.
10. Haagsman, H.P., Hawgood, S., Sargeant, T., Buckley, D., White, R.T., Drickamer, K. & Benson, B.J. (1987). The major lung surfactant protein, SP28–36 is a calcium dependent, carbohydrate-binding protein. *Journal of Biological Chemistry* **262**, 13877–13880.
11. Childs, R.A., Wright, J.R., Ross, G.F., Yuen, C.T., Lawson, A.M., Chai, W., Drickamer, K. & Feizi, T. (1992). Specificity of lung surfactant protein SP-A for both the carbohydrate and the lipid moieties of certain neutral glycolipids. *Journal of Biological Chemistry* **267**, 9972–9979.
12. Haurum, J.S., Thiel, S., Haagsman, H.P., Laursen, S.B., Larsen, B. & Jensenius, J.C. (1993). Studies on the carbohydrate-binding characteristics of human pulmonary surfactant-associated protein A and comparison with two other collectins: mannan-binding protein and conglutinin. *Biochemical Journal* **293**, 873–878.
13. Kuroki, Y. & Akino, T. (1991). Pulmonary surfactant protein A (SP-A) specifically binds dipalmitoylphosphatidylcholine. *Journal of Biological Chemistry* **266**, 3068–3073.
14. Ogasawara, Y., McCormack, F.X., Mason, R.J. & Voelker, D.R. (1994). Chimeras of surfactant proteins A and D identify the carbohydrate recognition domains as essential for phospholipid interaction. *Journal of Biological Chemistry* **269**, 29785–29792.
15. Kalina, M., Blau, H., Riklis, S. & Kravtsov, V. (1995). Interaction of surfactant protein A with bacterial lipopolysaccharide may affect some biological functions. *American Journal of Physiology* **268**, L144–L151.
16. Van Iwaarden, J.F., Pikaar, J.C., Storm, J., Brouwer, E., Verhoef, J., Oosting, R.S., Van Golde, L.M.G. & Van Strijp, J.A.G. (1994). Binding of surfactant protein A to the lipid moiety of bacterial lipopolysaccharides. *Biochemical Journal* **303**, 407–411.
17. Pikaar, J.C., Voorhout, W.F., Van Golde, L.M.G., Verhoef, J., Van Strijp, J.A.G. & Van Iwaarden, J.F. (1995). Opsonic activities of surfactant proteins A and D in phagocytosis of gram-negative bacteria by alveolar macrophages. *Journal of Infectious Diseases* **172**, 481–489.
18. Haagsman, H.P. (1998). Interactions of surfactant protein A with pathogens. *Biochimica et Biophysica Acta* **1408**, 264–277.
19. Madan, T., Eggleton, P., Kishore, U., Strong, P., Aggrawal, S.S., Sarma, P.U. & Reid, K.B.M. (1997). Binding of pulmonary surfactant proteins A and D to *Aspergillus fumigatus conidia* enhances phagocytosis and killing by human neutrophils and alveolar macrophages. *Infection and Immunity* **65**, 3171–3179.
20. Malhotra, R., Haurum, J., Thiel, S. & Sim, R.B. (1992). Interaction of C1q receptor with lung surfactant protein A. *European Journal of Immunology* **22**, 1437–1445.
21. Chroneos, Z.C., Abdolrasulnia, R., Whitsett, J.A., Rice, W.R. & Shepherd, V.L. (1996). Purification of a cell-surface receptor for surfactant protein A. *Journal of Biological Chemistry* **271**, 16375–16383.
22. Borron, P., Veldhuigen, R.A.W., Lewis, J.F., Possmayer, F., Caveney, A., Inchley, K., McFadden, R.G. & Fraher, L.J. (1996). Surfactant associated protein A inhibits human lymphocyte proliferation and IL-2 production. *American Journal of Respiratory Cell and Molecular Biology* **15**, 115–121.
23. Kremlev, S.G., Umstead, T.M. & Phelps, D.S. (1994). Effects of surfactant protein A and surfactant lipids on lymphocyte proliferation in vitro. *American Journal of Physiology* **267**, L357–L364.

24. Kremler, S.G. & Phelps, D.S. (1994). Surfactant protein A stimulation of inflammatory cytokine and immunoglobulin production. *American Journal of Physiology* **267**, L712–L719.
25. Weaver, T.E. & Whitsett, J.A. (1991). Function and regulation of expression of pulmonary surfactant-associated proteins. *Biochemical Journal* **273**, 249–264.
26. Khoor, A., Gray, M.E., Hall, W.M., Whitsett, J.A. & Stahlmann, M.T. (1993). Developmental expression of SP-A and SP-A mRNA in the proximal and distal respiratory epithelium in the human fetus and newborn. *Journal of Histochemistry and Cytochemistry* **41**, 1311–1319.
27. Rubio, S., Lacaze-Masmonteil, B., Chailley-Heu, B., Kahn, A., Bourbon, J.R. & Ducroc, R. (1995). Pulmonary surfactant protein A (SP-A) is expressed by epithelial cells of small and large intestine. *Journal of Biological Chemistry* **270**, 12162–12169.
28. Korfhagen, T.R., Le Vine, A.M. & Whitsett, J.A. (1998). Surfactant protein A (SP-A) gene targeted mice. *Biochimica et Biophysica Acta* **1408**, 296–302.
29. Malhotra, R., Haurum, J., Thiel, S., Jensenius, J.C. & Sim, R.B. (1993). Pollen grains bind to lung alveolar type II cells (A459) via lung surfactant protein A (SP-A). *Bioscience Reports* **13**, 79–90.
30. Wang, J.-Y., Shieh, C.-C., You, P.-F., Lei, H.-Y. & Reid, K.B.M. (1998). Inhibitory effect of pulmonary surfactant proteins A and D on allergen-induced lymphocyte proliferation and histamine release in children with asthma. *American Journal of Critical Care Medicine* **158**, 510–518.
31. Karinch, A.M. & Floros, J. (1995). 5′ Splicing and allelic variants of the human pulmonary surfactant protein A genes. *American Journal of Respiratory Cell and Molecular Biology* **12**, 77–88.
32. McCormick, S.M. & Mendelson, C.R. (1994). Human SP-A1 and SP-A2 genes are differentially regulated during development and by cAMP and glucocorticoids. *American Journal of Physiology* **266**, L367–L374.
33. Crouch, E.C. (1998). Collectins and pulmonary host defence. *American Journal of Respiratory Cell and Molecular Biology* **19**, 177–201.
34. Yuroki, Y., Takahashi, H., Chiba, H. & Akino, T. (1998). Surfactant proteins A and D: disease markers. *Biochimica et Biophysica Acta* **1408**, 334–345.
35. Mason, R.J., Greene, K. & Voelker, D.R. (1998). Surfactant protein A and surfactant protein D in health and disease. *American Journal of Physiology* **275**, L1–L13.

Surfactant protein D

Other names: SP-D; CP4
Mol. wt: 620 000 (native); 43 000 (subunit)
Specificity: Maltose
Sources: Mammalian lung surfactant and amniotic fluid

ISOLATION

SP-D was first identified in the medium of cultured rat alveolar type II cells and in EDTA extracts of rat surfactant obtained from lung lavage [1]. It was purified from these sources by reverse-phase HPLC or ion-exchange chromatography [1,2]. Human SP-D has been isolated from the soluble component of amniotic fluid by affinity chromatography on maltose–Sepharose, as has SP-D from bovine lung lavage [3].

SP-D has been cloned from human [3,4] and rat [5] lung cDNA libraries.

STRUCTURE

SP-D is a glycoprotein with a typical collectin triplet chain structure. Each polypeptide chain has an N-terminal cysteine-rich domain, a collagen-like domain, a link (neck) region and a C-terminal carbohydrate-recognition domain [3–5]. The collagen-like region differs from that of SP-A in being significantly longer and by containing hydroxylysine and hydroxylysyl-glycosides [2]. SP-D most closely resembles bovine conglutinin, which has 66% primary structural identity to both human [3] and rat [5] SP-Ds; this compares with approximately 40% identity between SP-D and the other collectins [5].

The apparent Mr of each chain is approximately 43 000 [1–3]; this compares with a deduced Mr of 35 482 for the polypeptide component [5]. SP-D is predominantly assembled as dodecamers (4×3), but other forms as small as trimers and as large as multimers of eight or more triplets occur [3,6,7].

SP-D is a more basic protein than SP-A; the predicted pI of rat SP-D (7.7) is consistent with the observed isoforms of pI 6–8 [2].

BIOLOGICAL ACTIVITIES

SP-D exhibits calcium-dependent binding to albumin-based neoglycoproteins and to saccharide-substituted affinity supports [8]. The most efficient inhibitors of these interactions were α-glucosyl-containing saccharides, particularly maltose. Divalent cations other than Ca^{2+} (Ba^{2+}, Sr^{2+}, Mn^{2+} and Mg^{2+}) were inhibitory [8].

SP-D has lipid-binding activity distinct from that of SP-A. SP-D binds avidly to phosphatidylinositol [9] and to glucosylceramide [10], but not to other glycolipids. This activity is Ca^{2+}-dependent and can be inhibited by appropriate sugars [9,10].

SP-D can bind to certain viruses (e.g. influenza A), Gram-negative bacteria (specific strains of *Pseudomonas aeruginosa, Haemophilus influenzae, Escherichia coli* etc.) and fungal pathogens in a Ca^{2+}-dependent and sugar-inhibitable manner [11–13].

SP-D can bind to (non-lectin) receptor(s) on alveolar macrophages, but these are distinct from the C1q (collectin) receptors [14]. Rat SP-D is an extremely potent chemoattractant for human monocytes and neutrophils, and the carbohydrate-binding site seems to be involved [15].

Rat SP-D is anti-mitogenic towards human mononuclear leukocytes, but this activity is independent of carbohydrate binding ability [16].

TISSUE AND/OR SUBCELLULAR LOCATION

SP-D is predominantly found in lung tissue, where it is mainly synthesised by type II alveolar epithelial cells and Clara cells and stored in secretory granules [17]. However, Northern blotting and RT-PCR analyses provide evidence for non-pulmonary synthesis in gastric mucosa and other sites [18]. Immunochemical evidence supports the production of SP-D in human salivary, sweat, mammary and lacrymal glands [19].

POSSIBLE FUNCTION(S)

It is believed that SP-D, like other collections, has a defensive function. It is likely to complement SP-A in lung surfactant (and perhaps acts without SP-A elsewhere) in protecting mucosal surfaces from pathogens, and possibly also by regulating the inflammatory response to pathogens or allergens [20]. It is interesting that the increased susceptibility of diabetic mice and humans to influenza A infection could be explained by glucose-mediated inhibition of SP-D [21].

SP-D undoubtedly has a separate role in surfactant homeostasis, since SP-D-deficient knock-out mice accumulated surfactant lipids in the alveolar space and also accumulated multinucleated, foamy macrophages [22,23].

COMMENTS

SP-D is the product of a single gene; in humans, the gene is located on chromosome 10, and no variant alleles have yet been characterised [24]. Like SP-A, the expression of SP-D is developmentally regulated, in a manner similar to, but nonetheless distinct from, SP-A, with a rise throughout prenatal and postnatal life to reach maximum levels in adult lung [25].

Like SP-A, SP-D can be considered a marker of lung disorders [12,26]. SP-D is exceptionally high in lung lavage of patients with pulmonary alveolar proteinosis (like SP-A), and somewhat reduced in patients with idiopathic pulmonary fibrosis. SP-D can also be detected in serum, and is detected at elevated levels in some patients with certain lung diseases (notably alveolar proteinosis and adult respiratory distress syndrome), and may reflect disease activity. SP-D (like SP-A) in amniotic fluid reflects lung maturity and is therefore relevant to neonatal respiratory distress syndrome.

REFERENCES

1. Persson, A., Rust, K., Chang, D., Moxley, M., Longmore, W. & Crouch, E. (1988). CP4: a pneumocyte-derived collagenous surfactant-associated protein. Evidence for heterogeneity of collagenous surfactant proteins. *Biochemistry* **27**, 8576–8584.
2. Persson, A., Chang, D., Rust, K., Moxley, M., Longmore, W. & Crouch, E. (1989). Purification and biochemical characterization of CP4 (SP-D), a collagenous surfactant-associated protein. *Biochemistry* **28**, 6361–6367.
3. Lu, J., Willis, A.C. & Reid, K.B.M. (1992). Purification, characterization and cDNA cloning of human lung surfactant protein D. *Biochemical Journal* **284**, 795–802.
4. Rust, K., Grosso, L., Zhang, V., Chang, D., Persson, A., Longmore, W., Cai, G.-Z. & Crouch, E. (1991). Human surfactant protein D: SP-D contains a C-type lectin carbohydrate recognition domain. *Archives of Biochemistry and Biophysics* **290**, 116–126.
5. Shimizu, H., Fisher, J.H., Papst, P., Benson, B., Lau, K., Mason, R.J. & Voelker, D.R. (1992). Primary structure of rat pulmonary surfactant protein D. *Journal of Biological Chemistry* **267**, 1853–1857.
6. Crouch, E., Persson, A., Chang, D. & Heuser, J. (1994). Molecular structure of pulmonary surfactant protein D (SP-D). *Journal of Biological Chemistry* **269**, 17311–17319.
7. Hartshorn, K., Chang, D., Rust, K., White, M., Heuser, J. & Crouch, E. (1996). Interactions of recombinant human pulmonary surfactant protein D and SP-D multimers with influenza A. *American Journal of Physiology* **271**, L753–762.
8. Persson, A., Chang, D. & Crouch, E. (1990). Surfactant protein D is a divalent cation-dependent carbohydrate-binding protein. *Journal of Biological Chemistry* **265**, 5755–5760.
9. Persson, A.V., Gibbons, B.J., Shoemaker, J.D., Moxley, M.A. & Longmore, W.J. (1992). The major glycolipid recognised by SP-D in surfactant is phosphatidyl-inositol. *Biochemistry* **31**, 12183–12189.
10. Kuroki, Y., Gasa, S., Ogasawara, Y., Shiratori, M., Makita, A. & Akino, T. (1992). Binding specificity of lung surfactant SP-D for glucosylceramide. *Biochemical and Biophysical Research Communications* **187**, 963–969.
11. Crouch, E.C. (1998). Structure, biological properties and expression of surfactant protein D (SP-D). *Biochimica et Biophysica Acta* **1408**, 278–289.
12. Mason, R.J., Green, K. & Voelker, D.R. (1998). Surfactant protein A and surfactant protein D in health and disease. *American Journal of Physiology* **275**, L1–L13.
13. Reid, K.B.M. (1998). Interactions of surfactant protein D with pathogens, allergens and phagocytes. *Biochimica et Biophysica Acta* **1408**, 290–295.

14. Miyamura, K., Leigh, L.E.A., Lu, J., Hopkin, J., López-Bernal, A. & Reid, K.B.M. (1994). Surfactant protein D binding to alveolar macrophages. *Biochemical Journal* **300**, 237–242.
15. Crouch, E.C., Persson, A., Griffin, G.L., Chang, D. & Senior, R.M. (1995). Interactions of pulmonary surfactant protein D (SP-D) with human blood leukocytes. *American Journal of Respiratory Cell and Molecular Biology* **12**, 410–415.
16. Borron, P.J., Crouch, E.C., Lewis, J.F., Wright, J.R., Possmayer, F. & Fraher, L.J. (1998). Recombinant rat surfactant-associated protein D inhibits human T lymphocyte proliferation and IL-2 production. *Journal of Immunology* **161**, 4599–4603.
17. Crouch, E., Parghi, D., Kuan, S.F. & Persson, N. (1992). Surfactant protein D: subcellular localisation in nonciliated bronchiolar epithelial cells. *American Journal of Physiology* **263**, L60–L66.
18. Motwani, M., White, R.A., Guo, N., Dowler, L.L., Tauber, A.I. & Sastry, K.N. (1995). Mouse surfactant protein D. cDNA cloning, characterization and gene localization to chromosome 14. *Journal of Immunology* **155**, 5671–5677.
19. Hansen, S. & Holmskov, U. (1998). Structural aspects of collectins and receptors for collectins. *Immunobiology* **199**, 180–189.
20. Wang, J.-Y., Shieh, C.-C., You, P.-F., Lei, H.-Y. & Reid, K.B.M. (1998). Inhibitory effect of pulmonary surfactant proteins A and D on allergen-induced lymphocyte proliferation and histamine release in children with asthma. *American Journal of Respiratory and Critical Care Medicine* **158**, 510–518.
21. Reading, P.C., Allison, J., Crouch, E.C. & Anders, E.M. (1998). Increased susceptibility of diabetic mice to influenza virus infection: compromise of collectin-mediated host defense of the lung by glucose? *Journal of Virology* **72**, 6884–6887.
22. Botas, C., Poulain, F., Akiyama, J., Brown, C., Allen, L., Goerke, J., Clements, J., Carlson, E., Gillespie, A.M., Epstein, C. & Hawgood, S. (1998). Altered surfactant homeostasis and alveolar type II cell morphology in mice lacking surfactant protein D. *Proceedings of the National Academy of Sciences* **95**, 11869–11874.
23. Korfhagen, T.R., Sheftelyevich, V., Burhans, M.S., Bruno, M.D., Ross, G.F., Wert, S.E., Stahlman, M.T., Jobe, A.H., Ikegami, M., Whitsett, J.A. & Fisher, J.H. (1998). Surfactant protein D regulates surfactant phospholipid homeostasis in vivo. *Journal of Biological Chemistry* **273**, 28438–28443.
24. Floros, J. & Hoover, R.R. (1998). Genetics of the hydrophilic surfactant proteins A and D. *Biochimica et Biophysica Acta* **1408**, 312–322.
25. Ogasawara, Y., Kuroki, Y., Shiratori, M., Shimizu, H., Miyamura, K. & Akino, T. (1991). Ontogeny of surfactant apoprotein D, SP-D, in the rat lung. *Biochimica et Biophysica Acta* **1083**, 252–256.
26. Yuroki, Y., Takahashi, H., Chiba, H. & Akino, T. (1998). Surfactant proteins A and D: disease markers. *Biochimica et Biophysica Acta* **1408**, 334–345.

Tachylectin-1

Other name: Limulus lectin L-6
Mol. wt: 24 400
Specificity: Complex
Source: The Japanese horseshoe crab *Tachypleus tridentatus*

ISOLATION

Tachylectin-1 was purified from a Lubrol extract of homogenised haemocyte debris by affinity chromatography on either LPS–agarose or Sepharose CL-6B and elution with EDTA. Tachylectin-1 could also be recovered by elution with high concentrations (0.5–1.0 M) of galactose, glucose or mannose.

STRUCTURE

The amino acid sequence has been determined from CNBr-generated peptides. Tachylectin-1 is a non-glycosylated polypeptide consisting of 221 amino acid residues composed of six tandem repeats (33–38 residues each). It contains three intrachain disulphide bonds and one SH–cysteine. Inductively coupled plasma spectrometry indicated the presence of Zn^{2+} at approximately one cation per molecule of protein, but no other divalent cations were detected. Overall, no sequence similarity between tachylectin-1 and other proteins was noted originally (but see **Comments**).

The deduced molecular weight (24 383) is lower than the apparent molecular weight (27 000) estimated by SDS-PAGE. The isoelectric point was calculated to be 9.7.

BIOLOGICAL ACTIVITIES

Tachylectin-1 can bind to LPS purified from Gram-negative bacteria. It can agglutinate both Gram-positive and Gram-negative bacteria, the latter more readily. It inhibits the growth of Gram-negative, but not Gram-positive, bacteria. Tachylectin-1 failed to agglutinate human, ovine or lapine erythrocytes [1].

TISSUE AND/OR SUBCELLULAR DISTRIBUTION

Tachylectin-1 is present in haemocytes (amoebocytes) and is localised to the large secretory granules [1]. It is not normally present as a soluble protein in the haemolymph plasma, but can be released by exocytosis in response to external stimuli [2]. Tachylectin-1 is also expressed in skeletal muscle, heart and stomach [2].

POSSIBLE FUNCTION(S)

Tachylectin-1 is presumed to function in defence. It is thought that upon stimulation with LPS from invading microbes, tachylectin-1 is secreted into the extracellular fluid where it contributes its anti-bacterial activity in concert with other defence molecules [1].

COMMENTS

Tachylectin-1 appears to be identical to a protein (L6) isolated from large granules of haemocytes [3]. For a summary of the similarities to, and differences from, other tachylectins, see Table 2.4.

Tachylectin-1 is virtually identical in primary structure to a new lectin designated 'tachylectin-P', found in the perivitelline fluid of the horseshoe crab embryo [4]. Despite sharing 218/221 amino acid residues, tachylectin-1 and tachylectin-P have astonishing differences in biochemical characteristics. The latter is a haemagglutinin that can be inhibited by *N*-acetylated sugars, and has no significant affinity for bacterial lipopolysaccharides nor anti-bacterial growth activity. It may be relevant that tachylectin-P appears to form a dimer in solution, whereas tachylectin-1 is present as a monomer. These isolectins are now known to have considerable sequence similarity to tectonins from *Physarum polycephalum*, but do not seem to be structurally related to any group of true lectins [4].

REFERENCES

1. Saito, T., Kawabata, S., Hirata, M. & Iwanaga, S. (1995). A novel type of limulus lectin-L6. *Journal of Biological Chemistry* **270**, 14493–14499.
2. Inamori, K., Saito, T., Iwaki, D., Nagira, T., Iwanaga, S., Arisaki, F. & Kawabata, S. (1999). A newly identified horseshoe crab lectin with specificity for blood group A antigen recognises specific O-antigens of bacterial lipopolysaccharides. *Journal of Biological Chemistry* **274**, 3272–3278.
3. Shigenaga, T., Takayenoki, Y., Kawasaki, S., Seki, N., Muta, T., Toh, Y., Ito, A. & Iwanaga, S. (1993). Separation of large and small granules from horseshoe crab (*Tachypleus tridentatus*) hemocytes and characterization of their components. *Journal of Biochemistry* **114**, 307–316.
4. Nagai, T., Kawabata, S., Shishikura, F. & Sugita, H. (1999). Purification, characterization, and amino acid sequence of an embryonic lectin in perivitelline fluid of the horseshoe crab. *Journal of Biological Chemistry* **274**, 37673–37678.

Tachylectin-2 isolectins

Other name: L10 isolectins
Mol. wt: 27 000
Specificity: GlcNAc, GalNAc
Source: The Japanese horseshoe crab *Tachypleus tridentatus*

ISOLATION

Haemocyte lysate was applied to a column of dextran sulphate–Sepharose CL-6B, and the flow-through material containing the isolectins was then applied sequentially to columns of CM-Sepharose CL-6B, Sephacryl S-200 and Mono S. This resulted in the isolation of three similar isolectins, termed L10a, L10b and L10c [1].

A cDNA coding for the L10c isoform of tachylectin-2 was isolated from a haemocyte cDNA library [1].

STRUCTURE

Tachylectin-2 is a mixture of three very similar single-chain proteins [1]. The amino acid compositions are characterised by high proportions of glycine, aspartic acid, phenylalanine and glutamic acid and an absence of cysteine. SDS-PAGE, under reducing or non-reducing conditions, revealed single bands corresponding to an Mr of 27 000; this is in perfect agreement with an Mr of 26 757 calculated from the 236 residues deduced from the cDNA of L10c. The amino acid sequence exhibited no significant sequence homology to other proteins. Analytical ultracentrifugation of L10b was consistent with the presence of a single monomer of 27 000 [1].

The structure has been resolved by X-ray crystallography and found to display a five-bladed β-propeller structure, exhibiting five almost identical binding sites within the single polypeptide chain [2].

BIOLOGICAL ACTIVITIES

Tachylectin-2 agglutinates human blood group A erythrocytes in a Ca^{2+}-independent manner. Neither human B or O erythrocytes, nor those of rabbits, sheep, cows or horses are agglutinated. The minimum concentrations required for haemagglutination differed slightly between the isolectins, but for all three the titre was increased by treating the cells with trypsin. Haemagglutination was potently inhibited by GlcNAc and (to a lesser extent) GalNAc, but not by other monosaccharides. *N*-acetylallolactosamine (Gal-β1,6-GlcNAc) and oligomers of GlcNAc were also potent inhibitors. Notably, lactose and *N*-acetyllactosamine were not inhibitory [1].

L-10b agglutinated *Staphylococcus saprophyticus* but not various other bacteria tested; this activity was also specifically inhibited by GlcNAc [1].

TISSUE AND/OR SUBCELLULAR DISTRIBUTION

In the haemolymph, tachylectin-2 is found in the large granules of haemocytes [1], and can be released by exocytosis in response to external stimuli [3]. Tachylectin-2 has also been in detected in skeletal muscle, heart, stomach and hepatopancreas [3].

POSSIBLE FUNCTION(S)

Tachylectin-2 is presumed to have a defensive role as a bacterial agglutinin. As a granule protein, it would be released into the haemocyte plasma in response to lipopolysaccharide [1,3].

COMMENTS

The tachylectin-2 isoforms constitute a major component of the large granules of *T. tridentatus* haemocytes, and are identical to the L-granule protein previously designated L10 [4]. The three isolectins separated by cation-exchange chromatography are distinct gene products despite very similar amino acid sequences. An unusual feature of those sequences is an internal homology consisting of five tandem repeats of 47 amino acid residues [1].

For a comparison of the properties of tachylectin-2 with other tachylectins, see Table 2.4.

REFERENCES

1. Okino, N., Kawabata, S., Saito, T., Hirata, M., Takagi, T. & Iwanaga, S. (1995). Purification, characterisation and cDNA cloning of a 27-kDa lectin (L10) from horseshoe crab hemocytes. *Journal of Biological Chemistry* **270**, 31008–31015.
2. Beisel, H.G., Kawabata, S., Iwanaga, S., Huber, P. & Bode, W. (1999). Tachylectin-2: crystal structure of a specific GlcNAc/GalNAc-binding lectin involved in the innate immunity host defence of the Japanese horseshoe crab, *Tachypleus tridentatus*. *EMBO Journal* **18**, 2313–2322.
3. Inamori, K., Saito, T., Iwaki, D., Nagira, T., Iwanaga, S., Arisaka, F. & Kawabata, S. (1999). A newly identified horseshoe crab lectin with specificity for blood group A antigen recognises specific O-antigens of bacterial lipopolysaccharides. *Journal of Biological Chemistry* **274**, 3272–3278.
4. Shigenaga, T., Takayenoki, Y., Kawasaki, S., Seki, N., Muta, T., Toh, Y., Ito, A. & Iwanaga, S. (1993). Separation of large and small granules from horseshoe crab (*Tachypleus tridentatus*) hemocytes and characterization of their components. *Journal of Biochemistry* **114**, 307–316.

Tachylectin-3

Mol. wt: 28 000 (native); 14 000 (subunit)
Specificity: Complex
Source: The Japanese horseshoe crab *Tachypleus tridentatus*

ISOLATION

Haemocyte lysate in high ionic strength buffer was first applied to a Sepharose CL-6B column to remove the tachylectins-2 and -4 in the flow-through. The slightly retarded material with haemagglutinating activity was then applied to a CM-Toyopearl column and eluted with an NaCl gradient to obtain a homogeneous protein [1].

A cDNA coding for tachylectin-3 was isolated from a haemocyte cDNA library [1].

STRUCTURE

Tachylectin-3 is a protein of apparent Mr 28 850 estimated by sedimentation equilibrium. It appears to exist as a homodimer of the single band of apparent Mr 14 000 detected by SDS-PAGE under reducing or non-reducing conditions [1].

The nucleotide and deduced amino acid sequences have been published in full. The cDNA encodes for a 24-residue signal sequence and a mature protein of 123 amino acids. The calculated Mr is 13 723 and the calculated pI is 9.1. The sequence exhibits some homology to that of the haemagglutinin-neuraminidase of rinderpest virus [1].

BIOLOGICAL ACTIVITIES

Tachylectin-3 agglutinates human, ovine, equine, bovine and lapine erythrocytes in a divalent cation-independent manner. Only human blood group A cells, not groups B or O, were agglutinated, and the titre was enhanced by pre-treating the cells with trypsin [1].

Haemagglutination could not be inhibited by monosaccharides, but was readily inhibited by a blood group A pentasaccharide (but not a similar trisaccharide). Haemagglutination was potently inhibited by S-type lipopolysaccharides from various Gram-negative bacteria and (to a lesser extent) lipoteichoic acid from *Staphylococcus aureus* and other staphylococci. It was inferred that tachylectin-3 recognises an extended structure on the O-antigen involving GalNAc/galactose disaccharide units [1].

TISSUE AND/OR SUBCELLULAR DISTRIBUTION

Tachylectin-3 has been localised to the large granules of haemocytes, from which it can be exocytosed. Tachylectin-3 is also found in muscular tissue including the heart [1].

POSSIBLE FUNCTION(S)

Tachylectin-3 is presumed to function in defence, forming part of a group of antibacterial lectins released from haemocytes in response to recognition of non-self [1].

COMMENTS

All four haemocyte tachylectins are presumably released in response to invading microbes to mediate in concert (and perhaps synergistically) a defensive action. For a comparative summary of their properties, see Table 2.4.

REFERENCE

1. Inamori, K., Saito, T., Iwaki, D., Nagira, T., Iwanaga, S., Arisaka, F. & Kawabata, S. (1999). A newly identified horseshoe crab lectin with specificity for blood group A antigen recognises specific O-antigens of bacterial lipopolysaccharides. *Journal of Biological Chemistry* **274**, 3272–3278.

Tachylectin-4

Mol. wt: 470 000 (native); 30 000 and 31 500 (subunit)
Specificity: Complex
Source: The Japanese horseshoe crab *Tachypleus tridentatus*

ISOLATION

Components of haemocyte lysate that bound to a column of dextran sulphate-Sepharose CL-6B were eluted with an NaCl gradient. The agglutinin was further purified by two affinity chromatography steps on concanavalin A-Sepharose and agarose–fucose [1].

A cDNA coding for tachylectin-4 was isolated from a haemocyte cDNA library [1].

STRUCTURE

Tachylectin-4 is a glycoprotein with an apparent Mr of 470 000 estimated by gel filtration. On SDS-PAGE, it appears as a dimer of 30 000 and 31 500 bands, but the larger subunit appears to be a glycosylated form of the smaller subunit. In the absence of reducing conditions, the apparent molecular weights determined by SDS-PAGE were somewhat smaller at 23 500 and 26 000 [1].

A tachylectin-4 cDNA encoded a signal sequence of 23 amino acid residues and a mature protein of 232 residues. The calculated Mr was 26 625 and the calculated pI was 6.1 [1].

A homology search indicated limited sequence similarity to a pentraxin from *Xenopus laevis* [1].

BIOLOGICAL ACTIVITIES

Tachylectin-4 is an extremely potent agglutinin for human blood group A erythrocytes (active above 14 ng/ml); human B and O cells were also agglutinated but 120- and 60-fold higher concentrations, respectively, were required. Equine, bovine and lapine erythrocytes were not agglutinated. Haemagglutination required Ca^{2+}; other divalent cations could not substitute [1].

Haemagglutination was not inhibited by monosaccharides other than a weak effect of fucose and *N*-acetylneuraminic acid at 100 mM. However, bacterial lipopolysaccharides from various sources were potent inhibitors, as were, to a lesser extent, lipoteichoic acids. It was inferred that colitose (3-deoxy-L-fucose) in the O-antigen of *Escherichia coli* is critical for recognition by tachylectin-4 [1].

TISSUE AND/OR SUBCELLULAR DISTRIBUTION

Tachylectin-4 has been reported only in haemocytes [1].

POSSIBLE FUNCTION(S)

A defence function, in concert with other lectins, is presumed from its activity towards bacterial structures [1].

COMMENTS

Tachylectin-4 shares with tachylectins-1 and -3 the property of being an LPS-binding protein and with tachylectins-2 and -3 its blood group A specificity (see Table 2.4). However, in terms of primary sequence it has little in common with the other tachylectins or other LPS-binding proteins, and instead exhibits some homology to a *Xenopus* pentraxin.

REFERENCE

1. Saito, T., Hatada, M., Iwanaga, S. & Kawabata, S. (1997). A newly identified horseshoe crab lectin with binding specificity to O-antigen of bacterial lipopolysaccharides. *Journal of Biological Chemistry* **272**, 30703–30708.

Tachylectins-5A and -5B

Other names: TL-5A and TL-5B
Mol. wt: 210 000 (A) and 160 000 (B) (native); 41 000 (subunit)
Specificity: Sialic acid, GlcNAc, GalNAc
Source: The Japanese horseshoe crab *Tachypleus tridentatus*

ISOLATION

Haemolymph plasma was applied to a vinyl polymer (Toyopearl) substituted with acetyl groups; TL-5A was eluted with 25 mM GlcNAc followed by TL-5B with 250 mM GlcNAc [1].

Both lectins were cloned: TL-5A by screening a heart cDNA library, TL-5B by screening a haemocyte cDNA library [1].

STRUCTURE

The tachylectins-5 are glycoproteins with apparent molecular weights of 210 000 (A) and 160 000 (B) estimated by gel filtration. SDS-PAGE analysis yielded a diffuse (A) or sharp (B) band corresponding to 41 000 under reducing conditions, and a smear (A) or a sharp band (B) corresponding to 72 000 under non-reducing conditions. Electron microscopy revealed two- to four-bladed propeller-like structures, consistent with each blade representing a disulphide-linked homodimer. TL-5A would seem to exist as oligomers formed from three or four such dimers, TL-5B from two.

The predicted primary structures of the mature protein subunits have 269 (A) and 289 (B) amino acid residues and share 45% identity. Both sequences were highly homologous to those of vertebrate fibrinogen β- and γ-chains. A slightly higher homology (approximately 50% identity) with the fibrinogen-like domain of mammalian ficolins was noted, but the tachylectins-5 do not have a collagenous region [1].

BIOLOGICAL ACTIVITIES

The tachylectins-5 potently agglutinated human erythrocytes in a Ca^{2+}-dependent manner. TL-5A was a much stronger agglutinin than TL-5B, and both showed a slight preference for blood groups A and O over group B [1].

Haemagglutination was readily inhibited by sugars containing an acetyl group, including *N*-acetylneuraminic acid, GlcNAc and GalNAc, in that order of potency. Other acetylated compounds like aspirin, acetyl coenzyme A and even sodium acetate were also very inhibitory, whereas methylated sugars were not. The acetyl group is therefore required and sufficient for recognition [1].

Both tachylectins-5 also agglutinated Gram-negative (e.g. *Escherichia coli*) and Gram-positive (e.g. *Staphylococcus aureus*) bacteria and enhanced the antimicrobial activity of big defensin towards Gram-positive (but not Gram-negative) organisms [1].

TISSUE AND/OR SUBCELLULAR DISTRIBUTION

TL-5A is expressed in heart, intestine and (to a limited extent) hepatopancreas, but notably not in haemocytes. In contrast, TL-5B was detectable only in haemocytes. No expression of either lectin was detectable in nervous or skeletal muscular tissue. Both lectins are present in haemolymph plasma [1].

POSSIBLE FUNCTION(S)

It is presumed these lectins function to discriminate self from non-self, and indeed it has been suggested they are the primary lectins recognising invading microbes in first-line defence of the horseshoe crab [1].

COMMENTS

These lectins are clearly distinct from the *Tachypleus tridendatus* sialic acid-binding lectin (qv), and are components of a mixture of lectins in the haemolymph of this species, the complexity of which has been obvious from the earliest reports (see Chapter 2). The tachylectins-5 are not true sialic acid-specific lectins, but acetyl group recognition proteins; as such, they resemble the *Diogenes affinis* agglutinin (qv).

It is noteworthy that ficolins like P35 (qv) and the Hakata antigen (qv) share with these tachylectins a fibrinogen-type domain conferring lectin activity; the ficolins, however, also have a collagenous region like the collectins, whereas these tachylectins do not.

REFERENCE

1. Gokudan, S., Muta, T., Tsuda, R., Koori, K., Kawahara, T., Seki, N., Mizunoe, Y., Wai, S.N., Iwanaga, S. & Kawabata, S. (1999). Horseshoe crab acetyl group-recognizing lectins involved in innate immunity are structurally related to fibrinogen. *Proceedings of the National Academy of Sciences USA* **96**, 10086–10091.

Tachypleus tridentatus sialic acid-binding lectin

Mol. wt: 530 000 (native); 30 000 and 32 000 (subunit)
Specificity: Sialic acid
Source: The Japanese horseshoe crab *Tachypleus tridentatus*

ISOLATION

The lectin was isolated from haemolymph plasma by affinity chromatography on immobilised equine glycophorin (eluting with a Ca^{2+}-free buffer) followed by gel filtration on Sephacryl S-300 [1].

STRUCTURE

This lectin is a protein (no attached carbohydrate) characterised by a high content of glycine and serine. The molecular weight, estimated by HPLC using an Asahipak GS-510 column, is approximately 530 000, and the molecule is composed of two non-identical subunits of apparent Mr 30 000 and 32 000 determined by SDS-PAGE [1].

This lectin exhibits immunological cross-reactivity with one of the haemolymph lectins from *Limulus polyphemus* [1].

BIOLOGICAL ACTIVITIES

This lectin agglutinates equine erythrocytes at high titre and human erythrocytes at low titre, but has no effect on neuraminidase-treated cells from either species. Haemagglutinating activity is Ca^{2+}-dependent and neither Mg^{2+} nor Mn^{2+} could substitute for Ca^{2+}[1].

Haemagglutination could not be inhibited by monosaccharides, except for sialic acids at very high concentrations; however, human and equine glycophorins were strong inhibitors, while bovine submaxilliary mucin and fetuin were moderately effective. The asialo equivalents of all those glycoproteins were not inhibitory [1].

TISSUE AND/OR SUBCELLULAR DISTRIBUTION

This lectin has only been reported in haemolymph [1].

POSSIBLE FUNCTIONS

A role in defence may be presumed, but no direct evidence regarding function has been described [1].

COMMENTS

Shimizu *et al.* [2] and Shishikura & Sekiguchi [3] both described agglutinins for equine erythrocytes (called M4 and TTA-III, respectively) in *T. tridentatus* haemolymph, but it is unclear how they relate to each other or to this lectin.

A summary comparing the properties of this lectin and the four 'tachylectins' obtained from haemocyte lysates of the same species is given in Table 2.4.

REFERENCES

1. Tsuboi, I., Matsukawa, M., Sato, N. & Kimura, S. (1993). Isolation and characterization of a sialic acid-specific binding lectin from the hemolymph of Asian horseshoe crab, *Tachypleus tridentatus*. *Biochimica et Biophysica Acta* **1156**, 255–262.
2. Shimizu, S., Ito, M. & Niwa, M. (1977). Lectins in the hemolymph of the Japanese horseshoe crab, *Tachypleus tridentatus*. *Biochimica et Biophysica Acta* **500**, 71–79.
3. Shishikura, F. & Sekiguchi, K. (1983). Agglutinins in the horseshoe crab hemolymph: purification of a potent agglutinin of horse erythrocytes from the hemolymph of *Tachypleus tridentatus*, the Japanese horseshoe crab. *Journal of Biochemistry* **93**, 1539–1546.

Taglin

Other name:	Trypsin activated *Giardia* lectin
Mol. wt:	28 000 and 30 000 (subunit)
Specificity:	Mannose-6-phosphate
Source:	The protozoan *Giardia lamblia*

ISOLATION

A lectin activity similar to that of taglin was first partially purified by extraction of *G. lamblia* trophozoites [1], but this material was probably heterogeneous. A monoclonal antibody that inhibits the lectin activity of taglin was used to identify a doublet of 28/30 kDa by Western blotting [2]. Confirmation that this doublet was indeed taglin was obtained by the demonstration of direct erythrocyte binding to nitrocellulose membranes.

STRUCTURE

Little is known about the structure of taglin, which is made up of two subunits in the presence of SDS. Each subunit appears to possess lectin activity. The possibility that the native lectin exists as a heterodimer is supported by rather weak evidence for 57 000 [1] and 56 000 [3] molecular species in partially purified preparations. It is clear, however, that its activation by trypsin does not bring about an appreciable alteration in molecular size [2].

BIOLOGICAL ACTIVITIES

Taglin can agglutinate lapine erythrocytes [2] and also murine enterocytes [3], but not ovine, murine or human erythrocytes [3]. It requires divalent cations (Ca^{2+} or Mn^{2+}) for its agglutinating activity, which is confined to a very narrow pH range with an optimum at pH 6.5 [2].

Haemagglutinating activity is most readily inhibited by mannose-6-phosphate; mannose-1-phosphate and other phosphorylated sugars are ineffective, while mannose itself is a weak inhibitor [2].

TISSUE AND/OR SUBCELLULAR DISTRIBUTION

There is consistent evidence from different approaches that at least some of the lectin is expressed on the cell surface [1–3].

POSSIBLE FUNCTION(S)

There is abundant circumstantial evidence that taglin plays a role in host–parasite interaction [4]. Most noteworthy is the consideration that trypsin is secreted into the proximal small intestine at the exact location of colonisation by *Giardia* and where the pH of the luminal fluid coincides with the narrow pH optimum of the lectin.

COMMENTS

A most remarkable and unusual feature of taglin is its conversion by limited proteolysis from prolectin to an active form [2–4]. This can be achieved by trypsin, pronase or human duodenal fluid containing trypsin [4]. The mechanism involved could in principle be the digestion of an endogenous ligand exposing a cryptic site.

An earlier report by Farthing and co-workers [1] describes a lectin activity similar to, but with some differences from, taglin. Those workers may have been describing a related lectin or a mixture of taglin with another *Giardia* lectin.

REFERENCES

1. Farthing, M.J.G., Pereira, M.E.A. & Keusch, G.T. (1986). Description and characterization of a surface lectin from *Giardia lamblia*. *Infection and Immunity* **51**, 661–667.
2. Ward, H.D., Lev, B.I., Kane, A.V., Keusch, G.T. & Pereira, M.E.A. (1987). Identification and characterisation of taglin, a mannose 6-phosphate binding, trypsin-activated lectin from *Giardia lamblia*. *Biochemistry* **26**, 8669–8675.
3. Lev, B., Ward, H., Kreusch, G.T. & Pereira, M.E.A. (1986). Lectin activation in *Giardia lamblia* by host protease: a novel host–parasite interaction. *Science* **232**, 71–73.
4. Ward, H.D., Kreusch, G.T. & Pereira, M.E.A. (1990). Induction of a phosphomannosyl binding lectin activity in *Giardia*. *BioEssays* **12**, 211–215.

Teleogryllus commodus lectin

Other names: Cricket lectin; Walker insect lectin
Mol. wt: >10^6 (native); 31 000 and 53 000 (subunit)
Specificity: Sialic acid, GlcNAc
Source: Haemolymph of the adult cricket *Teleogryllus commodus* (Walker)

ISOLATION

The lectin was purified from haemolymph supernatant by affinity chromatography on fetuin–Sepharose and elution with *N*-acetylneuraminic acid [1].

STRUCTURE

The Walker insect lectin is a glycoprotein consisting of large aggregates (Mr estimated at >10^6 by gel filtration). SDS-PAGE in the presence and absence of mercaptoethanol indicates a dimeric structure consisting, respectively, of subunits of apparent Mr 31 000 and 53 000 joined by disulphide bonds [1].

BIOLOGICAL ACTIVITIES

The lectin agglutinated pigeon, chicken, rat and human (ABO non-specifically) erythrocytes at relatively high titres, and feline and ovine cells at low titres. Equine and lapine erythrocytes were not agglutinated [1].

Haemagglutination was inhibited by EDTA, but the activity could not be restored by the addition of a molar excess of Ca^{2+}, Mg^{2+} or Mn^{2+} [1].

Haemagglutination was inhibited most readily by *N*-acetylneuraminic acid and GlcNAc; GalNAc required five-times the concentration to achieve the same effect, while other monosaccharides tested were ineffective [1].

TISSUE AND/OR SUBCELLULAR DISTRIBUTION

Lectin activity was only sought in the haemolymph [1].

POSSIBLE FUNCTION(S)

A defence function was assumed, but no direct evidence was provided [1].

COMMENTS

The lectin is a minor haemolymph component and is significantly more labile to heat, lyophilisation, ultrafiltration, dialysis or freezing after purification [1].

REFERENCE

1. Hapner, K.D. & Jermyn, M.A. (1981). Haemagglutinin activity in the haemolymph of *Teleogryllus commodus* (Walker). *Insect Biochemistry* **11**, 287–295.

Tetranectin

Other names: TN; plasminogen kringle 4-binding protein
Mol. wt: 60 000 (native); 20 000 (subunit)
Specificity: Sulphated polysaccharides
Sources: Vertebrate tissues

ISOLATION

Tetranectin was accidentally discovered after its co-purification with α_2-plasmin inhibitor on plasminogen–Sepharose. It was then isolated from human plasma by a complex method in which the key steps were the removal of plasminogen by lysine–Sepharose and positive affinity chromatography using immobilised kringle 4 structure of plasminogen and elution with tranexamic acid [1]. It has also been isolated by immunoaffinity chromatography [2].

The human tetranectin gene [3] and cDNA [4] have been isolated from placental libraries by screening with degenerate oligonucleotide probes. Murine tetranectin cDNA has also been cloned [5].

STRUCTURE

Human tetranectin is a glycoprotein composed of three identical chains [6], with an apparent molecular weight of around 60 000 estimated by gel filtration [1,2]. Each chain consists of 181 amino acid residues with a molecular weight of 20 100 [2–4]. The sequence exhibits significant homology to the carbohydrate-recognition domains of C-type lectins including the asialoglycoprotein receptors [2–4]. The murine homologue is about 80% identical in primary structure to the human molecule [5].

The crystal structure of human tetranectin has been determined at 0.28 nM resolution [6]. Each monomer consists of a C-terminal carbohydrate-recognition domain connected to a long α-helix providing trimerisation. Two calcium ions have been located at the same positions as calcium-binding sites-1 and -2 in mannan-binding lectin (qv); calcium-binding site-1 is highly conserved, but the coordination geometry at calcium-binding site-2 is less similar to that of MBL. Overall, there is much structural similarity between tetranectin and the collectins [6].

BIOLOGICAL ACTIVITIES

Tetranectin binds plasminogen by a highly specific and lysine-sensitive interaction with the kringle 4 domain, and this binding enhanced plasminogen activation via the tissue-type plasminogen activator [1]. Binding to plasminogen is calcium-sensitive and involves the putative carbohydrate-binding site within the C-type lectin domain [7].

Tetranectin binds fucoidin and to some extent interacts with other sulphated polysaccharides like heparin sulphate and chondroitin sulphate [8]. Any divalent cation dependence was not investigated [8], but in an earlier report, tetranectin was described as binding heparin–Sepharose Ca^{2+}-independently [1]. Evidence that this tetranectin–carbohydrate interaction is calcium-independent and mediated by residues close to the N-terminus has been cited as unpublished work [7].

TISSUE AND/OR SUBCELLULAR DISTRIBUTION

Tetranectin is synthesised by a wide range of tissues and organs including lung, spleen, heart, skeletal muscle, liver and brain [3,4]. It has been found in the cytoplasm of many cell types including lymphocytes, granulocytes, macrophages, fibroblasts, smooth muscle cells, striated muscle cells and Schwann cells, as well as most epithelial and endothelial cells [9]. The immunohistochemical staining was often granular [9]. Tetranectin has also been found in the extracellular matrix of various carcinomas, but not in the corresponding locations of normal tissues [4,10].

POSSIBLE FUNCTION(S)

It has been suggested that tetranectin might target plasminogen to specific carbohydrate ligands on cell surfaces or in the extracellular matrix, or to fibrin [6]. Other possibilities include a role in the clearance of toxic glycosaminoglycan derivatives, or in bone mineralisation.

COMMENTS

This lectin was first thought to be a tetramer [1], so *tetra*nectin is a misnomer; the name has persisted although 'trinectin' would be more logical.

Tetranectin is present in healthy human blood plasma or serum within a concentration range of approximately 8–17 (median 13) μg/ml [1,11]. Significantly lower levels have been reported in association with various malignancies, including myeloma and breast, ovarian and colonic carcinomas [11,12]. Tetranectin seems a promising analyte for the detection of residual disease and as a prognostic marker. More specifically, measurement of tetranectin in combination with CA125 may be useful for ovarian cancer screening, and is becoming established as an independent prognostic variable in ovarian, colorectal and metastatic breast cancers [12].

Although TN concentration was slightly decreased in patients with acute pelvic inflammatory disease, TN is not a negative acute-phase reactant but appears to be a negative cancer marker [12].

As well as homologues of the human form, other proteins forming a tetranectin-related subfamily are a stem cell growth factor [13] and a cartilage protein [14,15].

Heterogeneity in plasma tetranectin is due to a sequence dichotomy at position 85, and the presence of a partially sialylated oligosaccharide structure attached to a threonine at position 4 [16].

REFERENCES

1. Clemmensen, I., Petersen, L.C. & Kluft, C. (1986). Purification and characterization of a novel, oligomeric, plasminogen kringle 4 binding protein from human plasma: tetranectin. *European Journal of Biochemistry* **156**, 327–333.
2. Fuhlendorff, J., Clemmensen, I. & Magnusson, S. (1987). Primary structure of tetranectin, a plasminogen Kringle 4-binding plasma protein: homology with asialoglycoprotein receptors and cartilage proteoglycan core protein. *Biochemistry* **26**, 6757–6764.
3. Berglund, L. & Petersen, T.E. (1992). The gene structure of tetranectin, a plasminogen binding protein. *FEBS Letters* **309**, 15–19.
4. Wewer, U.M. & Albrechtsen, R. (1992). Tetranectin, a plasminogen kringle 4-binding protein. Cloning and gene expression pattern in human colon cancer. *Laboratory Investigation* **67**, 253–262.
5. Sorensen, C.B., Berglund, L. & Petersen, T.E. (1995). Cloning of a cDNA encoding murine tetranectin. *Gene* **152**, 243–245.
6. Nielsen, B.B., Kastrup, J.S., Rasmussen, H., Holtet, T.L., Graversen, J.H., Etzerodt, M., Thøgersen, H.C. & Larsen, I.K. (1997). Crystal structure of tetranectin, a trimeric plasminogen-binding protein with an α-helical coiled coil. *FEBS Letters* **412**, 388–396.
7. Graversen, J.H., Lorentsen, R.H., Jacobsen, C., Moestrup, S.K., Sigurskjold, B.W., Thøgersen, H.C. & Etzerodt, M. (1998). The plasminogen binding site of the C-type lectin tetranectin is located in the carbohydrate recognition domain, and binding is sensitive to both calcium and lysine. *Journal of Biological Chemistry* **273**, 29241–29246.
8. Clemmensen, I. (1989). Interaction of tetranectin with sulphated polysaccharides and trypan blue. *Scandinavian Journal of Clinical and Laboratory Investigation* **49**, 719–725.
9. Christensen, L. & Clemmensen, I. (1989) Tetranectin immunoreactivity in normal human tissues. *Histochemistry* **92**, 29–35.
10. Christensen, L. & Clemmensen, I. (1991). Differences in tetranectin immunoreactivity between benign and malignant breast tissue. *Histochemistry* **95**, 427–433.
11. Thougaard, A.V., Høgdall, C.K., Kjaer, S.K., Blaakaer, J., Jaliashvili, I. & Christiansen, M. (1998). Determination of serum tetranectin: technical and clinical evaluation of three sandwich immunoassays. *Clinica Chimica Acta* **276**, 19–34.
12. Høgdall, C.K. (1998). Human tetranectin: methodological and clinical studies. *APMIS* **106** (Suppl. 86), 7–31.
13. Mio, H., Kagami, N., Yokokawa, S., Kawai, H., Nakagawa, S., Takeuchi, K., Sekine, S. & Hiraoka, A. (1998). Isolation and characterization of a cDNA for human, mouse, and rat full-length stem cell growth factor, a new member of C-type lectin superfamily. *Biochemical and Biophysical Research Communications* **249**, 124–130.
14. Neame, P.J., Young, C.N. & Treep, J.T. (1992). Primary structure of a protein isolated from reef shark (*Carcharhinus springeri*) cartilage that is similar to the mammalian C-type lectin homologue, tetranectin. *Protein Science* **1**, 161–168.
15. Neame, P.J., Tapp, H. & Grimm, D.R. (1999). The cartilage-derived C-type lectin (CLECSF1): structure of the gene and chromosomal location. *Biochimica et Biophysica Acta* **1446**, 193–202.
16. Jaquinod, M., Holtet, T.L., Etzerodt, M., Clemmensen, I., Thogersen, H.C. & Roepstorff, P. (1999). Mass spectrometric characterisation of post-translational modification and genetic variation in human tetranectin. *Biological Chemistry* **380**, 1307–1314.

Thrombolectin

Mol. wt: 28 000 (native); 15 000 (subunit)
Specificity: Lactose
Source: Venom of the fer-de-lance *Bothrops atrox*

ISOLATION

Thrombolectin was purified from venom by affinity chromatography on asialofetuin–Sepharose 4B, eluting with lactose [1].

STRUCTURE

Thrombolectin gave a single band on SDS-PAGE; the apparent Mr was 15 000 in the presence of reducing agent, and 28 000 under non-reducing conditions. The bands were PAS-negative, indicating an absence of glycosylation [1]. Thrombolectin was heterogeneous on isoelectric focusing: a sharp band at pH 4 was observed plus a triplet of bands at around pH 9.6 [2].

BIOLOGICAL ACTIVITIES

Thrombolectin agglutinated formaldehyde-fixed, bovine erythrocytes in a divalent cation-dependent manner. The abolition of activity with EDTA could be overcome by adding excess calcium. Dithiothreitol was also a potent inhibitor of agglutination, as were various sugars. Thiodigalactoside was the most potent saccharide inhibitor, followed by lactose and D-galactose. Methyl-β-galactoside was possibly more effective than D-galactose, methyl-α-galactoside somewhat less so [1,2].

Thrombolectin, unlike several closely related lectins, was non-mitogenic for bovine lymph node mononuclear leukocytes [3].

Thrombolectin did not reproducibly induce the aggregation of human platelets, but did so occasionally [4].

TISSUE AND/OR SUBCELLULAR DISTRIBUTION

Thrombolectin has only been reported from venom.

POSSIBLE FUNCTION(S)

Since the venom gland of crotalid snakes contains a mucin to which the lectin can bind, it has been suggested that the lectin might facilitate formation and attachment of mucin to the epithelial cells of the gland and thereby protect the gland from proteolytic damage [2].

COMMENTS

Although lectin activities were first discovered in snake venoms during the 19th century, thrombolectin was the first to be purified to homogeneity from such a source. Similar lectins were later isolated from other species, including *Agkistrodon piscivorus* (qv), *Crotalus atrox* (qv), *Lachesis muta* (qv) and *Trimeresurus stejnegeri* (qv).

These true lectins are related to, but distinct from, the botrocetin group of C-type lectin-like molecules which appear to lack carbohydrate-binding activity (see Chapter 4).

REFERENCES

1. Gartner, T.K., Stocker, K. & Wlliams, D.C. (1980). Thrombolectin: a lectin isolated from *Bothrops atrox* venom. *FEBS Letters* **117**, 13–16.
2. Gartner, T.K. & Ogilvie, M.L. (1984). Isolation and characterization of three Ca^{2+}-dependent β-galactoside-specific lectins from snake venoms. *Biochemical Journal* **224**, 301–307.
3. Mastro, A.M., Hurley, D.J., Winning, R.K., Filipowski, R., Ogilvie, M.L. & Gartner, T.K. (1986). Mitogenic activity of snake venom lectins. *Cell and Tissue Kinetics* **19**, 557–566.
4. Ogilvie, M.L., Wilson-Byl, J. & Gartner, T.K. (1989). Platelet aggregation is stimulated by lactose-inhibitable snake venom lectins. *Thrombosis and Haemostasis* **62**, 704–707.

Thrombospondin

Other names: TSP; thrombin-sensitive protein; endogenous platelet lectin
Mol. wt: 400 000 (native); 133 000 (subunit)
Specificity: Amino sugars; heparin
Sources: Human platelets etc.

ISOLATION

Thrombospondin can be obtained in a reasonably pure form simply by gel filtration of the supernatant from thrombin-treated platelets, although additional purification can be achieved by binding to immobilised heparin and eluting with an NaCl gradient [1].

Human thrombospondin cDNA has been cloned from libraries of endothelial cells and fibroblasts [2,3].

STRUCTURE

Thrombospondin is a glycoprotein (about 4% carbohydrate) with a high cysteine content; it exists as a homotrimer of chains linked by two inter-subunit disulphide bridges near the N-terminus [1–4]. The apparent Mr values of 450 000 and 150 000 determined by SDS-PAGE under non-reducing and reducing conditions, respectively, are somewhat higher than the values of 410 000 and 142 000 estimated by sedimentation equilibrium, or that calculated from the deduced primary sequence of 1152 amino acids (127 524 plus carbohydrate).

The structure has a number of interesting features including a globular heparin-binding domain near the N-terminus, a cysteine-rich sequence homologous to the N-propeptide of collagen, cysteine-rich repeats homologous to those found in complement proteins, cysteine-rich epidermal growth factor-like repeats, and a set of homologous repeats forming a Ca^{2+}-binding region which also contains the arginine–glycine–aspartate cell adhesion sequence. This last region shows some homology to calmodulin (qv).

BIOLOGICAL ACTIVITIES

Thrombospondin promotes the autoagglutination of thrombin-activated human platelets, and the haemagglutination of trypsin-treated, formalin-fixed ovine and bovine red blood cells in the presence of calcium [1,5–7]. These activities can be inhibited by the amino acids lysine and arginine, and also by amino sugars (mannosamine, glucosamine and galactosamine) but not by GlcNAc, GalNAc or other common monosaccharides [5–7].

Thrombospondin can bind to heparin, fibrinogen, fibrin, fibronectin, plasminogen, CD36 and other macromolecules [4,8].

TISSUE AND/OR SUBCELLULAR DISTRIBUTION

In platelets, thrombospondin is a major component of α-granules, is translocated to the cell surface upon activation, and much of it is secreted [8,9]. However, it is synthesised by many cell types other than platelets, including fibroblasts, endothelial cells and monocytes [8], and is found within the extracellular matrix of a wide variety of human tissues [10]. The level of circulating thrombospondin at around 100 ng/ml is rather low [4,8].

POSSIBLE FUNCTION(S)

Thrombospondin appears to function (in general terms) as an adhesion molecule, promoting the formation and stabilisation of platelet aggregates [4,8], modulating cell movement/cell–substratum interactions [4], and influencing cell growth and proliferation [11]. Thrombospondin may also contribute to the clearance of senescent (apoptotic) neutrophils by macrophages [12]. As an inhibitor of angiogenesis, thrombospondin limits vessel density in normal tissues and may help to curtail tumour growth [13].

COMMENTS

Although human TSP has been particularly well studied, this lectin has been found in other animals, including the mouse, in which a striking developmental regulation is evident [14].

In humans, there is a strong correlation ($r=0.97$) between platelet count and serum TSP, but no correlation ($r=-0.1$) between plasma TSP and platelet count (due to contributions from non-platelet sources) [15].

TSP is of interest in oncology, since it has been argued that it has a positive, pivotal role in the metastatic process [16]. Although there is some evidence that TSP could stimulate tumour motility and angiogenesis, anti-tumour effects and other conflicting data have also been reported [11,13].

REFERENCES

1. Lawler, J.W., Slayter, H.S. & Coligan, J.E. (1978). Isolation and characterisation of a high molecular weight glycoprotein from human blood platelets. *Journal of Biological Chemistry* **253**, 8609–8616.

2. Lawler, J. & Hynes, R.O. (1986). The structure of human thrombospondin, an adhesive glycoprotein with multiple calcium-binding sites and homologies with several different proteins. *Journal of Cell Biology* **103**, 1635–1648.
3. Hennesy, S.W., Frazier, B.A., Kim, D.D., Deckwerth, T.L., Baumgartel, D.M., Rotwein, P. & Frazier, W.A. (1989). Complete thrombospondin mRNA sequence includes potential regulatory sites in the 3′ untranslated region. *Journal of Cell Biology* **108**, 729–736.
4. Mosher, D.F. (1990). Physiology of thrombospondin. *Annual Review of Medicine* **41**, 85–97.
5. Gartner, T.K., Williams, D.C., Minion, F.C. & Phillips, D.R. (1978). Thrombin-induced platelet aggregation is mediated by a platelet plasma membrane-bound lectin. *Science* **200**, 1281–1283.
6. Gartner, T.K., Gerrard, J.M., White, J.G. & Williams, D.C. (1981). Fibronectin is the receptor for the endogenous lectin of human platelets. *Nature* **289**, 688–690.
7. Jaffe, E.A., Leung, L.L.K., Nachmann, R.L., Levin, R.I. & Mosher, D.F. (1982). Thrombospondin is the endogenous lectin of human platelets. *Nature* **295**, 246–248.
8. Silverstein, R.L., Leung, L.L.K. & Nachman, R.L. (1986). Thrombospondin: a versatile multifunctional glycoprotein. *Arteriosclerosis* **6**, 245–253.
9. Phillips, D.R., Jennings, L.K. & Prasanna, H.R. (1980). Ca^{2+}-mediated association of glycoprotein G (thrombin-sensitive protein, thrombospondin) with human platelets. *Journal of Biological Chemistry* **255**, 11629–11632.
10. Wight, T.N., Raugi, G.J., Mumby, S.M. & Bornstein, P. (1985). Light microscopic immunolocation of thrombospondin in human tissues. *Journal of Histochemistry and Cytochemistry* **33**, 295–302.
11. Yamashita, Y., Sendo, S., Hosokawa, T., Tuszynski, G., Kurohiji, T., Beppu, R., Shinohara, T., Kinugasa, T. & Shirakusa, T. (1998). Exogenous thrombospondin stimulates the proliferation of non-thrombospondin-producing cells. *International Journal of Oncology* **13**, 355–359.
12. Murphy, J.F., McGregor, J.L. & Leung, L.L.K. (1998). Senescent human neutrophil binding to thrombospondin (TSP): evidence for a TSP-independent pathway of phagocytosis by macrophages. *British Journal of Haematology* **102**, 957–964.
13. Jiménez, B., Volpert, O.V., Crawford, S.E., Febbraio, M., Silverstein, R.L. & Bouck, N. (2000). Signals leading to apoptosis-dependent inhibition of neovascularization by thrombospondin-1. *Nature Medicine* **6**, 41–48.
14. O'Shea, K.S. & Dixit, V.M. (1988). Unique distribution of the extracellular matrix component thrombospondin in the developing mouse embryo. *Journal of Cell Biology* **107**, 2737–2748.
15. Dawes, J., Pratt, D.A., Dewar, M.S. & Preston, F.E. (1988). Do extra-platelet sources contribute to the plasma levels of thrombospondin? *Thrombosis and Haemostasis* **59**, 273–276.
16. Tuszynski, G.P., Gasic, T.B., Rothman, V.L., Knudsen, K.A. & Gasic, G.J. (1987). Thrombospondin, a potentiator of tumor cell metastasis. *Cancer Research* **47**, 4130–4133.

Tridacna derasa lectin

Other name: TDL
Mol. wt: 480 000 (native); 23 000 and 46 000 (subunit)
Specificity: GalNAc
Source: Haemolymph of the bivalve *Tridacna derasa*

ISOLATION

Concentrated haemolymph supernatant was subjected to affinity chromatography on Sepharose 4B, and a homogeneous lectin was eluted with a EGTA-containing Ca^{2+}-free buffer [1].

STRUCTURE

TDL is a protein (no attached carbohydrate) with an apparent Mr of 480 000 determined by gel filtration, and consists of a single large subunit after SDS-PAGE under non-reducing conditions. However, SDS-PAGE in the presence of 2-mercaptoethanol yields two bands of apparent Mr 23 000 and 46 000. The subunits have identical N-terminal sequences and essentially identical amino acid compositions; both have isoelectric points of 6.0. The 46 000 subunit therefore appears to be a homodimer of 23 000 subunits linked via a covalent bond other than a disulphide bridge. The native TDL must consist of these two molecular species linked by disulphide bonds [1]. (The *Tridacna maxima* lectin (qv) may have a similar quaternary structure brought about by the same unusual means [2].)

The N-terminal sequence (29 residues) of TDL showed homology to some C-type lectins.

BIOLOGICAL ACTIVITIES

TDL agglutinates human erythrocytes ABO group non-specifically. This activity was pH- and Ca^{2+}-dependent and could be completely inhibited by 0.5 mM GalNAc or 50 mM D-galactose, but 200 mM GlcNAc was without effect [1].

TISSUE AND/OR SUBCELLULAR DISTRIBUTION

TDL has only been reported in haemolymph plasma [1].

POSSIBLE FUNCTION(S)

Nothing is really known concerning function. However, it is suspected that the unusual pH-dependence of its ligand-binding activity (see **Comments**) may play a physiological role and provide a clue to function [1].

COMMENTS

More than 80% of the haemolymph protein consists of this lectin. The amino terminal sequence showed some homology to C-type lectins while the unusual pH dependence of its haemagglutinating activity (optimal at and above pH 6.6; dramatically reduced below pH 6.5) is reminiscent of vertebrate asialoglycoprotein receptors [1].

REFERENCES

1. Odo, S., Kamino, K., Kanai, S., Maruyama, T. & Harayama, S. (1995). Biochemical characterisation of a Ca^{2+}-dependent lectin from the hemolymph of a photosymbiotic marine bivalve, *Tridacna derasa* (Röding). *Journal of Biochemistry* **117**, 965–973.
2. Baldo, B.A., Sawyer, W.H., Stick, R.V. & Uhlenbruck, G. (1978). Purification and characterization of a galactin-reactive agglutinin from the clam *Tridacna maxima* (Röding) and a study of its combining site. *Biochemical Journal* **175**, 467–477.

Tridacna maxima lectin

Other names: Tridacnin; TML
Mol. wt: 470 000 (native); 10 000, 20 000 and 40 000 (subunit)
Specificity: GalNAc
Source: The elongate clam *Tridacna maxima* (Röding)

ISOLATION

Tridacnin can be isolated by affinity chromatography, with polyleucyl larch galactan, D-galactosamine or acid-treated Sepharose as the solid phase, and eluting with GalNAc, lactose or galactosamine [1,2].

STRUCTURE

Tridacnin is a glycoprotein (3–7% carbohydrate [2,3]) with an amino acid composition devoid of methionine [2]. It sedimented as an asymmetrical boundary on ultracentrifugation with a sedimentation coefficient of 15 S; the native molecular weight was calculated to be 470 000. SDS-PAGE under reducing conditions yielded three bands of apparent Mr 10 000, 20 000 and 40 000. Isoelectric focusing indicated at least three species all with isoelectric points close to pH 4.0 [2].

It has been suggested that the native molecule consists of perhaps 12 units of 40 000, each divisible into two 20 000 or four 10 000 subunits [2].

BIOLOGICAL ACTIVITIES

Tridacnin agglutinates various kinds of erythrocytes including human, lapine, rat and chicken cells; it does not agglutinate equine erythrocytes, while ovine and murine cells react weakly [2]. It precipitates some galactans and some polysaccharides and glycopeptides [1,2,4]. These reactions are dependent on Ca^{2+}, and are readily inhibited by β-D-galactosides [2,4]. With monosaccharides, on a molar basis, GalNAc was about twenty times as inhibitory as galactose, and ten times as inhibitory as galactosamine or lactose [2].

In contrast to *Tridacna derasa* lectin (qv), tridacnin exhibited a broad pH optimum between pH 4 and pH 10.6 [2].

TISSUE AND/OR SUBCELLULAR DISTRIBUTION

Tridacnin is found in haemolymph, but other tissues do not seem to have been examined [1–4].

POSSIBLE FUNCTION(S)

It has been suggested that tridacnin may play a role in symbiosis by participation in the elimination and utilisation of algae which during senility harbour exposed galactan structures on their surfaces [5].

COMMENTS

Tridacnin has considerable potential as an immunochemical tool [5].

REFERENCES

1. Baldo, B.A. & Uhlenbruck, G. (1975). Purification of tridacnin, a novel anti-β-(1-6)-digalactobiose precipitin from the haemolymph of *Tridacna maxima* (Röding). *FEBS Letters* **55**, 25–29.
2. Baldo, B.A., Sawyer, W.H., Stick, R.V. & Uhlenbruck, G. (1978). Purification and characterization of a galactin-reactive agglutinin from the clam *Tridacna maxima* (Röding) and a study of its combining site. *Biochemical Journal* **175**, 467–477.
3. Odo, S., Kamino, K., Kanai, S., Muruyama, T. & Harayama, S. (1995). Biochemical characterisation of a Ca^{2+}-dependent lectin from the hemolymph of a photosymbiotic marine bivalave, *Tridacna derasa* (Röding). *Journal of Biochemistry* **117**, 965–973.
4. Baldo, B.A. & Uhlenbruck, G. (1975). Anti-galactan activity in *Tridacna maxima* (Röding) haemolymph. *Immunology* **29**, 1161–1170.
5. Uhlenbruck, G. & Steinhausen, G. (1977). Tridacnins: Symbiosis-profit or defense-purpose. *Developmental and Comparative Immunology* **1**, 183–192.

Trimeresurus stejnegeri venom lectin

Other name: TSL
Mol. wt: 40 000 (native); 20 000 (subunit)
Specificity: Galactose
Source: *Trimeresurus stejnegeri* venom

ISOLATION

TSL was purified from venom by procedures based on affinity chromatography [1,2].

The cDNA has also been cloned and sequenced [2].

STRUCTURE

TSL is a glycoprotein homodimer of 135 amino acid residue-chains, each with an apparent Mr of approximately 20 000 determined by SDS-PAGE under reducing conditions. Under non-reducing conditions, several bands presumed to represent dimers, trimers and tetramers can be detected. The isoelectric point is 8.5 [1–3].

The deduced amino acid sequence shows >80% identity to the venom lectins from *Crotalus atrox* (qv) and *Lachesis muta* (qv), and substantial homology to a number of other C-type lectins, including the rat asialoglycoprotein receptor [2].

BIOLOGICAL ACTIVITIES

TSL binds asialo-α_1-acid glycoprotein and agglutinates lapine erythrocytes in a divalent cation-dependent manner. Galactose was the most effective saccharide inhibitor [1].

TISSUE AND/OR SUBCELLULAR DISTRIBUTION

TSL was obtained from venom. Its presence or absence in other tissues has not been reported.

POSSIBLE FUNCTION(S)

The function of this lectin is not known.

COMMENTS

TSL is obviously one of the thrombolectin (qv) group of closely related snake venom lectins, and was the first to be characterised at the cDNA level [2].

REFERENCES

1. Liang, X.-L. & Wang, K.-Y. (1993). Isolation and characterization of a lectin from the venom of the snake *Trimeresurus stejnegeri* Schmidt. *Acta Biochimica et Biophysica Sinica* **25**, 515–521.
2. Xu, Q., Wu, X.-F., Xia, Q.-C. & Wang, K.-Y. (1999). Cloning of a galactose-binding lectin from the venom of *Trimeresurus stejnegeri*. *Biochemical Journal* **341**, 733–737.
3. Zeng, R., Xu, Q., Shao, X.-X., Wang, K.-Y. & Xia, Q.-C. (1999). Characterization and analysis of a novel glycoprotein from snake venom using liquid chromatography-electrospray mass spectrometry and Edman degradation. *European Journal of Biochemistry* **266**, 352–358.

Tritrichomonas foetus lectin

Other name: TFL
Mol. wt: 600 000–700 000 (native); 66 000 (subunit)
Specificity: Sialic acid
Source: Culture supernatant of the protozoan *Tritrichomonas foetus*

ISOLATION

TSL was purified in two steps: affinity adsorption onto human erythrocytes followed by affinity chromatography on fetuin–agarose. Elution from both supports was with *N*-acetylneuraminic acid [1].

STRUCTURE

TFL is a glycoprotein (4% carbohydrate) with an amino acid composition rich in leucine. Although TFL shares with *Tritrichomonas mobilensis* lectin (qv) a surprisingly low content of acidic amino acids, otherwise the amino acid compositions of the two lectins have little in common. The apparent Mr values determined by SDS-PAGE were 306 000 under non-reducing and 66 000 under reducing conditions; a surprisingly high concentration of mercaptoethanol was required to dissociate the subunits completely [1]. TFL therefore appears to exist as a high molecular weight complex, the exact molecular arrangement of which is uncertain.

BIOLOGICAL ACTIVITIES

TFL agglutinates human (ABO non-specifically) and equine erythrocytes equally, but has a three-fold lower titre for rat and lapine red blood cells. Divalent cations were not required for haemagglutinating activity. Haemagglutination was inhibited by sialic acids and their derivatives, but not by other common sugars. Free *N*-acetylneuraminic acid was twice as effective as free *N*-glycolylneuraminic acid, but both were potent inhibitors, while colominic acid was weakly inhibitory. Equal inhibition was evident with α2,3- and α2,6-linked neuraminic acid derivatives [1].

TFL histochemistry indicated binding to bovine vaginal epithelium, cervical mucosa and endometrial glands [2].

TISSUE AND/OR SUBCELLULAR DISTRIBUTION

Immunocytochemical staining of *T. foetus* with specific rabbit antibodies showed fine granular cytoplasmic staining and strong cell surface staining [1,2].

POSSIBLE FUNCTION(S)

It is suspected that the lectin could mediate attachment and adherence of the parasite to host epithelium [1,2].

COMMENTS

T. foetus is a parasite of the bovine genital tract which can cause abortion and sterility.

Although TFL has a very similar (but not identical) saccharide and erythrocyte specificity to *T. mobilensis* lectin, they are clearly two separate lectins with distinct amino acid and carbohydrate compositions.

REFERENCES

1. Babál, P., Pindak, F.F., Russell, L.C. & Gardner, W.A. (1999). Sialic acid-specific lectin from *Tritrichomonas foetus*. *Biochimica et Biophysica Acta* **1428**, 106–116.
2. Babál, P. & Russell, L.C. (1999). Sialic acid-specific lectin-mediated adhesion of *Tritrichomonas foetus* and *Tritrichomonas mobilensis*. *Journal of Parasitology* **85**, 33–40.

Tritrichomonas mobilensis lectin

Other name: TML
Mol. wt: 556 000 and 491 000 (native); 66 000, 61 000 and 56 000 (subunit)
Specificity: Sialic acid
Source: Culture supernatant of the protozoan *Tritrichomonas mobilensis*

ISOLATION

TML was purified to apparent homogeneity by affinity chromatography on bovine submaxillary mucin–Sepharose [1].

STRUCTURE

TML is a glycoprotein (approximately 20% carbohydrate) with an amino acid composition rich in cysteine residues. TML forms multimeric complexes with apparent molecular weights of 556 000 and 491 000 on gel filtration. These complexes have three different subunits held together by intramolecular thiol groups which can only be fully reduced in the presence of urea. The assembly of subunits is to some extent concentration dependent [1].

BIOLOGICAL ACTIVITIES

TML readily agglutinates simian, murine and equine erythrocytes, and agglutinates but does not discriminate between human ABO groups [1]. It has a strict specificity for sialic acids, displaying no inhibition of agglutination in the presence of other monosaccharides at up to 200 mM concentrations. TML does not differentiate between α2,3- and α2,6-linked terminal neuraminic acid, both of which are more potent inhibitors than *N*-glycolylneuraminic acid. Agglutinating activity does not require divalent cations [1].

TML mediates the adherance of *T. mobilensis* to Chinese hamster ovary cells, bovine cervical mucus and simian colonic mucus [2].

TISSUE AND/OR SUBCELLULAR DISTRIBUTION

T. mobilensis is a unicellular protozoan which synthesises TML and secretes it into the culture medium. Within the protozoan, the lectin is localised to cytoplasmic granules and the surface membrane [2].

POSSIBLE FUNCTION(S)

T. mobilensis is a parasite of squirrel monkeys, and may use its lectin to colonise its hosts via adhesion to mucosal surfaces [2].

COMMENTS

TML has been used as a reagent in histochemistry [3].

A lectin of similar specificity has been found in the related protozoan *T. foetus* (qv) [2,4].

REFERENCES

1. Babál, P., Pindak, F.F., Wells, D.J. & Gardner, W.A. (1994). Purification and characterization of a sialic acid-specific lectin from *Tritrichomonas mobilensis*. *Biochemical Journal* **299**, 341–346.
2. Babál, P. & Russell, L.C. (1999). Sialic acid-specific lectin-mediated adhesion of *Tritrichomonas foetus* and *Tritrichomonas mobilensis*. *Journal of Parasitology* **85**, 33–40.
3. Babál, P., Gardner, W.A., Pindak, F.F. & Wells, D.J. (1992). Histochemistry of a novel lectin. *Modern Pathology* **5**, 117A.
4. Babál, P., Pindak, F.F., Russell, L.C. & Gardner, W.A. (1999). Sialic acid-specific lectin from *Tritrichomonas foetus*. *Biochimica et Biophysica Acta* **1428**, 106–116.

Trout egg lectins

Other names: STL-1 and STL-2 etc.
Mol. wt: 64 000 (native); 21 000 (subunit) (STL-2)
Specificity: L-Rhamnose
Sources: Ova from fish of the genus *Oncorhynchus*

ISOLATION

The best-characterised trout eggs lectins have been isolated from acetone extracts of *Oncorhynchus mykiss* (steelhead trout or rainbow trout) by affinity chromatography on L-rhamnose–Sepharose, and separation into isolectins (STL-1 and STL-2) by ion-exchange chromatography [1]. An apparently rather similar lectin has been isolated from the Chinook salmon, *Oncorhynchus tshawytscha*, by affinity chromatography on agarose [2]. The lectin(s) from the brown trout (*Salmo trutta*) has only been studied in crude extracts [3].

STRUCTURE

The Chinook salmon lectin has an apparent Mr of 122 000 estimated by gel filtration, and an isoelectric point of 4.5 [2]; STL-1 and STL-2 have molecular weights of 84 000 and 68 000 estimated by gel filtration and pI values of 6.8 and 4.7, respectively [1].

STL-2 has been fully sequenced. It is a non-glycosylated trimer of chains, each of which has 195 amino acid residues. The size of each chain was determined to be 22 000 or 28 000 by SDS-PAGE in the absence or presence, respectively, of reducing agent; the former value is close to the value determined by mass spectrometry (21 500) and to the calculated Mr of 21 349. The primary sequence is composed of two tandemly repeated homologous (52% identity) domains, and shows extensive homology to the sea urchin egg lectin from *Anthocidaris crassispina* (about 40% identity) and with the ligand-binding domains of mammalian lipoprotein receptors [1].

STL-1 is also a non-covalently linked trimer, but of rather larger subunits (31 000), and a partial amino acid sequence showed some homology to STL-2.

Homogenates of ova from *O. mykiss* and *O. tshawytscha* formed a line of identity in double diffusion against rabbit antiserum to the lectin purified from the latter species [2].

BIOLOGICAL ACTIVITIES

All the purified lectins previously mentioned readily agglutinate lapine erythrocytes and this activity is sensitive to inhibition by rhamnose and (to a lesser extent) galactose. L-Arabinose, but not other monosaccharides, is also inhibitory. The saccharide specificities of STL-1 and STL-2 are very similar, but not identical, and the specific activity of the latter is 45 times higher [1,2]. Agglutination by STL-1 and STL-2 is divalent cation-independent [1].

The Chinook salmon lectin, like the brown trout extracts, agglutinates human erythrocytes with a strong preference for blood group B [2,3]. This activity was inhibited by D-galactose but not by rhamnose [2].

STL-2 also showed L-rhamnose-dependent binding to certain egg yolk proteins [1].

The Chinook salmon lectin inhibited the growth of several bacterial pathogens of fish [2].

TISSUE AND/OR SUBCELLULAR DISTRIBUTION

These lectins were extracted from ova; other tissues seem not to have been examined.

POSSIBLE FUNCTION(S)

A role in defence against virulent bacteria seems to be the most supported possible function [2], but it has also been suggested that these lectins might participate in fertilisation and/or the development of the embryo, or in decoding carbohydrate structures of glycoproteins to influence compartmentalisation [1].

COMMENTS

STL-2 is among the best characterised of the many fish egg lectins described, most of which, irrespective of family, have the unusual property of high affinity for L-rhamnose. Those of the trout/salmon family (Salmonidae) often show a strong preference for human blood group B erythrocytes (see Chapter 4).

The limited sequence data available [1] is consistent with the possibility that fish egg lectins generally may belong to a distinct structural family of animal lectins related to the low density lipoprotein receptor superfamily [1].

The species *O. mykiss* is also a source of various other lectins present in the serum, including 'ladderlectin' [4] and other LPS-binding lectins [5].

A lectin homologous to STL-2 has been purified from catfish (*Siluris asotus*) eggs and fully sequenced [6].

REFERENCES

1. Tateno, H., Saneyoshi, A., Ogawa, T., Muramoto, K., Kamiya, H. & Saneyoshi, M. (1998). Isolation and characterization of rhamnose-binding lectins from eggs of

steelhead trout (*Oncorhynchus mykiss*) homologous to low density lipoprotein receptor superfamily. *Journal of Biological Chemistry* **273**, 19190–19197.

2. Voss, E.W., Fryer, J.L. & Banowetz, G.M. (1978). Isolation, purification and partial characterization of a lectin from Chinook salmon ova. *Archives of Biochemistry and Biophysics* **186**, 25–34.
3. Rees, P., Cotton, R., Holt, P.D.J. & Anstee, D.J. (1976). Trout ova, an alternative source of anti-B. *Medical Laboratory Sciences* **33**, 13–21.
4. Jensen, L.E., Thiel, S., Petersen, T.E. & Jensenius, J.C. (1997). A rainbow trout lectin with multimeric structure. *Comparative Biochemistry and Physiology B* **116**, 385–390.
5. Hoover, G.J., El Mowafi, A., Simko, E., Kocal, T.E., Ferguson, H.W. & Hayes, M.A. (1998). Plasma proteins of rainbow trout (*Oncorhynchus mykiss*) isolated by binding to lipopolysaccharide from *Aeromonas salmonicida*. *Comparative Biochemistry and Physiology B* **120**, 559–569.
6. Hosono, M., Ishikawa, K., Mineki, R., Murayama, K., Numata, C., Ogawa, Y., Takayanagi, Y. & Nitta, K. (1999). Tandem repeat structure of rhamnose-binding lectin from catfish (*Silurus asotus*) eggs. *Biochimica et Biophysica Acta* **1472**, 668–675.

Tumour necrosis factor

Other names: TNFα; cachectin; cytotoxin
Mol. wt: 51 000 (native); 17 000 (subunit)
Specificity: Diacetylchitobiose
Sources: Macrophage/monocyte cultures

ISOLATION

TNFα was first discovered as an anti-tumour activity in mice. Human TNFα was first reported in homogeneous form as a recombinant protein expressed in *Escherichia coli* after isolation of cDNA clones [1,2], although it had previously been purified from HL-60 cell culture medium by a series of chromatographic steps as the first stage of a cloning strategy [3]. Murine TNF was also purified first from culture medium, using high performance anion-exchange and gel filtration steps [4] before being cloned [5].

STRUCTURE

Human TNFα is a homotrimeric protein with a high content of leucine as well as glutamic and aspartic acid residues [3,6]. The 155-amino acid subunit sequence, which has a single intra-molecular disulphide bond and a calculated Mr of around 17 000, shows approximately 30% sequence identity to lymphotoxin (TNFβ) [1,2]. Molecular weight determination by SDS-PAGE gave values of 17 000 irrespective of the presence of reducing agent, but the apparent molecular weight of the native trimer was estimated at 45 000 [3]. The isoelectric point lies between 5 and 6 [2,3]. The tertiary structure is formed by edge-to-face packing of β-sheets and resembles the 'jelly-roll' motif [6].

Murine TNFα possesses about 80% primary sequence identity to the human form and is generally similar. Murine TNFα binds to both human TNF receptors, although the human TNFα only binds one of the two murine receptors. Murine TNF differs most obviously from the human in being glycosylated and by associating into higher molecular weight oligomers [5,7].

BIOLOGICAL ACTIVITIES

TNFα binds to two receptors (expressed on many cell types and which it shares with lymphotoxin, TNFβ) by protein–protein interaction. As a cytokine, TNF mediates a wide variety of biological effects, often in concert with other cytokines. These include anti-tumour and anti-microbial activities, regulation of cell growth and differentiation, mediation of fever and septic shock, and such immunopathological manifestations as cachexia and inflammatory tissue destruction [7–9].

TNFα also binds Gram-negative bacteria like *Shigella flexneri* [10] and trypanosomes [11] independently of TNF receptors. The trypanolytic interaction of murine or human TNFα with *Trypanosoma brucei* involves a separate (non-cytokine) binding site and can be inhibited by *N,N'*-diacetylchitobiose but not by cellobiose [11]. The physiological significance of the earlier observation that recombinant murine TNFα had lectin activity towards Tamm–Horsfall glycoprotein [12] was called into question by the failure of other workers to demonstrate specific binding at neutral pH [13,14].

TISSUE AND/OR SUBCELLULAR DISTRIBUTION

While monocytes–macrophages have always been regarded as the principal source of TNF, it is also synthesised and secreted by many other cell types [7,9]. It is synthesised in precursor form and can be detected at the cell surface as a type II membrane protein [7].

POSSIBLE FUNCTION(S)

TNFα is a paracrine and endocrine mediator of pro-inflammatory immune functions, often interacting with other cytokines to promote host defence and to inhibit neoplastic growth [7–9]. Knock-out mice have provided confirmation of a role in host defence (in relation to candidiasis) [15]; surprisingly, knock-out mice have also provided evidence that TNFα can promote the early stages of carcinogen-induced malignant skin tumours [16].

COMMENTS

TNFα has necessary and beneficial roles in the immune response, but is also commonly associated with immunopathology; the net biological effect(s) of this cytokine may be desirable or injurious [7–9]. Many attempts have been made to exploit the desirable (especially anti-tumour) activities therapeutically, and likewise to prevent the injurious consequences of TNFα production. In general terms, supplementation of endogenous TNF, which is limited by toxicity, has enjoyed little success; however, strategies to limit the consequences of TNF production have seemed more promising. For example, anti-TNF antibody infusion has been used to achieve significant short-term benefit in rheumatoid arthritis [17,18].

The physiological significance of the lectin activity of TNFα has received relatively little attention, but this activity may be amenable to therapeutic manipulation.

REFERENCES

1. Pennica, D., Nedwin, G.E., Hayflick, J.S., Seeburg, P.H., Derynck, R., Palladino, M.A., Kohr, W.J., Aggarwal, B.B. & Goeddel, D.V. (1984). Human tumour necrosis factor: precursor structure, expression and homology to lymphotoxin. *Nature* **312**, 724–729.
2. Shirai, T., Yamaguchi, H., Ito, H., Todd, C.W. & Wallace, R.B. (1985). Cloning and expression in *Escherichia coli* of the gene for human tumour necrosis factor. *Nature* **313**, 803–806.
3. Aggarwal, B.B., Kohr, W.J., Hass, P.E., Moffat, B., Spencer, S.A., Henzel, W.J., Bringman, T.S., Nedwin, G.E., Goeddel, D.V. & Harkins, R.N. (1985). Human tumour necrosis factor. Production, purification and characterization. *Journal of Biological Chemistry* **260**, 2345–2354.
4. Beutler, B., Greenwald, D., Hulmes, J.D., Chang, M., Pan, Y.-C. E., Mathison, J., Ulevitch, R. & Cerami, A. (1985). Identity of tumour necrosis factor and the macrophage-secreted factor cachectin. *Nature* **316**, 552–554.
5. Pennica, D., Hayflick, J.S., Bringman, T.S., Palladino, M.A. & Goeddel, D.V. (1985). Cloning and expression in *Escherichia coli* of the cDNA for murine tumor necrosis factor. *Proceedings of the National Academy of Sciences USA* **82**, 6060–6064.
6. Jones, E.Y., Stuart, D.I. & Walker, N.P.C. (1989). Structure of tumour necrosis factor. *Nature* **338**, 225–228.
7. Callard, R. & Gearing, A. (1994). *The Cytokine Facts Book*, pp. 241–246. Academic Press, London.
8. Beutler, B. & Cerami, A. (1986). Cachectin and tumour necrosis factor as two sides of the same biological coin. *Nature* **320**, 584–588.
9. Tracey, K.J., Vlassara, H. & Cerami, A. (1989). Cachectin/tumour necrosis factor. *Lancet* **1**, 1122–1126.
10. Luo, G., Niesel, D.W., Shaban, R.A., Grimm, E.A. & Klimpel, G.R. (1993). Tumour necrosis factor alpha binding to bacteria: evidence for a high affinity receptor and alteration of bacterial virulence properties. *Infection and Immunity* **61**, 830–835.
11. Lucas, R., Magez, S., De Leys, R., Fransen, L., Scheerlinck, J.-P., Rampelberg, M., Sablon, E. & De Baetselier, P. (1994). Mapping the lectin-like activity of tumour necrosis factor. *Science* **263**, 814–817.
12. Sherblom, A.P., Decker, J.M. & Muchmore, A.V. (1988). The lectin-like interaction between recombinant necrosis factor and uromodulin. *Journal of Biochemistry* **263**, 5418–5424.
13. Moonen, P., Gaffner, R. & Wingfield, P. (1988). Native cytokines do not bind to uromodulin (Tamm Horsfall protein). *FEBS Letters* **226**, 314–318.
14. Lambert, C., Brealey, R., Steele, J. & Rook, G.A.W. (1993). The interaction of Tamm–Horsfall protein with the extracellular matrix. *Immunology* **79**, 203–210.
15. Netea, M.G., Van Tits, L.J.H., Curfs, J.H.A.J., Amiot, F., Meis, J.F.G.M., Van Der Meer, J.W.M. & Kullberg, B.J. (1999). Increased susceptibility to TNF-α lymphotoxin-α double knockout mice to systemic candidiasis through impaired recruitment of neutrophils and phagocytosis of *Candida albicans*. *Journal of Immunology* **163**, 1498–1505.
16. Moore, R.J., Owens, D.M., Stamp, G., Arnott, C., Burke, F., East, N., Holdsworth, H., Turner, L., Rollins, B., Pasparakis, M., Kollias, G. & Balkwill, F. (1999). Mice deficient in tumour necrosis factor-α are resistant to skin carcinogenesis. *Nature Medicine* **5**, 828–831.
17. Elliott, M.J., Maini, R.N., Feldmann, M., Kalden, J.R., Antoni, C., Smolen, J.S., Leeb, B., Breedveld, F.C., Macfarlane, J.D., Bijl, H. & Woody, J.N. (1994). Randomised double-blind comparison of chimeric monoclonal antibody to tumour necrosis factor α (cA2) versus placebo in rheumatoid arthritis. *Lancet* **344**, 1105–1110.
18. Firestein, G.S. & Zvaifler, N.J. (1997). Anticytokine therapy in rheumatoid arthritis. *New England Journal of Medicine* **337**, 195–197.

Urechis unicinctus isolectins

Mol. wt: 35 000 (native); 31 000 or 34 000 (subunit)
Specificity: Lactose
Source: Coelomocytes of the echiuran *Urechis unicinctus*

ISOLATION

A crude extract was prepared from cells in the coelomic fluid of echiurnoid worms by sonication in a buffer containing lactose, Triton X-100, mercaptoethanol and EDTA. The soluble fraction of this extract was fractionated with ammonium sulphate (50–90% cut) and applied to a column of lactose–Sepharose. The lectins were eluted with lactose [1].

STRUCTURE

Apparent Mr values ranging between 26 000 and 35 000 were estimated by gel filtration in the presence of lactose, and a similar value (31 000) was obtained by SDS-PAGE under non-reducing conditions. In the presence of 2-mercaptoethanol, two bands were detected after SDS-PAGE, with mobilities corresponding to 31 000 and 34 000. Each was isolated by preparative SDS-PAGE and found to retain haemagglutinating activity susceptible to inhibition by galactosides. The affinity-purified material appears to be a mixture of two similar monomeric isolectins, and this mixture was used for general characterisation. These isolectins were focused on ampholytes to a single peak of activity with a pI of 5–5.5. The isolectin mixture contained 8.4% carbohydrate [1].

BIOLOGICAL ACTIVITIES

The *Urechis* isolectins agglutinated lapine erythrocytes at high titre and human (ABO non-specific), bovine and porcine cells at low titre. Haemagglutinating activity did not require divalent cations, and was more stable in the presence of 2-mercaptoethanol [1].

Haemagglutinating activity was potently inhibited by D-galactose and various galactosides, of which thiodigalactoside was the most potent. D-Fucose (but not L-fucose), L-arabinose and GalNAc were also potent inhibitors, but less so than galactose. Asialofetuin, but not native fetuin, was a powerful inhibitor [1].

TISSUE AND/OR SUBCELLULAR DISTRIBUTION

In coelomic fluid, the isolectins were detected in the coelomocytes but not in the plasma. Other tissues were not investigated [1].

POSSIBLE FUNCTION(S)

These isolectins were sought as possible participants in a defence mechanism; however, they do not appear to promote the clumping of coelomocytes. No physiological role could therefore be suggested [1].

COMMENTS

Although not as well characterised as some lectins, these are included here as the sole representatives of the phylum Echiurida.

REFERENCE

1. Matsui, T. (1984). D-Galactoside specific lectins from coelomocytes of the echiuran, *Urechis unicinctus*. *Biological Bulletin* **166**, 178–188.

Versican

Other name: PG-M
Mol. wt: 265 000–388 000 (core protein)
Specificity: Fucose, GlcNAc
Sources: Various vertebrate tissues

ISOLATION

The human fibroblast proteoglycan (versican) gene was identified from overlapping cDNA clones obtained by screening a placental library [1]. A homologous chondroitin sulphate proteoglycan (PG-M) gene expressed in developing chick limb buds was also identified from cDNA clones [2]. The C-terminal (lectin) region of PG-M has been expressed as a glutathione S-transferase fusion protein in bacteria [3], and a corresponding human recombinant region has been expressed in mammalian (CHO) cells [4].

STRUCTURE

Versican is the largest of the hyalectan family, the deduced sequences of 2409 and 3562 residues, respectively, providing calculated molecular weights of 265 000 (human) to 388 000 (avian) [1,2].

The N-terminal hyaluronic acid-binding region is somewhat smaller than the corresponding region from aggrecan (qv). The central region, which shows the least inter-species similarity, consists of two large subregions containing up to 30 glycosaminoglycan attachment sites for chondroitin sulphate and dermatan sulphate. The C-terminal region, like that of aggrecan, contains two epidermal growth factor-like repeats, the C-type lectin domain and a terminal complement regulatory protein-like module.

BIOLOGICAL ACTIVITIES

The recombinant human versican C-type lectin domain bound to agarose-immobilised fucose and GlcNAc (but not to immobilised GalNAc, mannose or glucose) [4], while the recombinant C-terminal region of PG-M bound to immobilised mannose, galactose and heparin as well as fucose and GlcNAc [3]. The binding was calcium-dependent.

Versican can bind tenascin-R via its C-type lectin domain in a Ca^{2+}-independent manner [5], as well as hyaluronic acid via its N-terminal domain [6]. Versican, like aggrecan, but unlike neurocan (qv) or brevican (qv), combines with high affinity to the widely distributed extracellular matrix protein fibulin-1, a possible physiological ligand, in a Ca^{2+}-independent manner, but to the protein part of fibulin-1 [7].

TISSUE AND/OR SUBCELLULAR DISTRIBUTION

Versican is found in many tissues and is abundant in the brain [6,8], where it is principally associated with the granular layer of cerebellum [4]. It is expressed in many cell types, including fibroblasts, keratinocytes and smooth muscle cells [8].

POSSIBLE FUNCTION(S)

Versican presumably functions in the organisation of the extracellular matrix, possibly contributing to the regulation of cell migration and pattern formation [9].

COMMENTS

During skeletal development in the vertebrate limb, versican is a major extracellular proteoglycan of the condensation area but disappears with the formation of cartilage, a developmental regulation inversely related to that of aggrecan [2].

REFERENCES

1. Zimmermann, D.R. & Ruoslahti, E. (1989). Multiple domains of the large fibroblast proteoglycan, versican. *EMBO Journal* **8**, 2975–2981.
2. Shinomura, T., Nishida, Y., Ito, K. & Kimata, K. (1993). cDNA cloning of PG-M, a large chondroitin sulphate proteoglycan expressed during chondrogenesis in chick limb buds. *Journal of Biological Chemistry* **268**, 14461–14469.
3. Ujita, M., Shinomura, T., Ito, K., Kitagawa, Y. & Kimata, K. (1994). Expression and binding activity of the carboxyl-terminal portion of the core protein of PG-M, a large chondroitin sulphate proteoglycan. *Journal of Biological Chemistry* **269**, 27603–27609.
4. Aspberg, A., Binkert, C. & Ruoslahti, E. (1995). The versican C-type lectin domain recognises the adhesion protein tenasin-R. *Proceedings of the National Academy of Sciences USA* **92**, 10590–10594.
5. Aspberg, A., Miura, R., Bourdoulous, S., Shimonaka, M., Heinegard, D., Schachner, M., Ruoslahti, E. & Yamaguchi, Y. (1997). The C-type lectin domains of lecticans, a family of aggregating chondroitin sulphate proteoglycans, bind tenasin-R by protein–protein interactions independent of carbohydrate moiety. *Proceedings of the National Academy of Sciences USA* **94**, 10116–10121.
6. Ruoslahti, E. (1996). Brain extracellular matrix. *Glycobiology* **6**, 489–492.
7. Aspberg, A., Adam, S., Kostka, G., Timpl, R. & Heinegard, D. (1999). Fibulin-1 is a ligand for the C-type lectin domains of aggrecan and versican. *Journal of Biological Chemistry* **274**, 20444–20449.
8. Dours-Zimmermann, M.T. & Zimmermann, D.R. (1994). A novel glycosaminoglycan attachment domain identified in two alternative splice variants of human versican. *Journal of Biological Chemistry* **269**, 32992–32998.
9. Iozzo, R.V. (1998) Matrix proteoglycans: from molecular design to cellular function. *Annual Review of Biochemistry* **67**, 609–653.

VIP36

Other name: Vesicular integral membrane protein of 36 kDa
Mol. wt: 36 000
Specificity: GalNAc
Source: Mammalian tissues

ISOLATION

VIP36 was isolated from detergent-insoluble complexes from canine epithelial (MDCK) cells by preparative two-dimensional gel electrophoresis [1].

A cDNA encoding VIP36 was subsequently isolated and translated *in vitro* [1].

STRUCTURE

VIP36 is a type I transmembrane glycoprotein with an apparent molecular weight of 36 000 determined by SDS-PAGE under reducing conditions. The calculated Mr is 35 500 and the predicted and observed pI is around pH 6.5. No evidence could be found for oligomerisation [1,2].

The molecule consists of a large N-terminal domain, a short transmembrane domain and a very short C-terminal sequence containing polar or charged residues. The N-terminal, luminal domain exhibits about 30% identity to ERGIC-53 and >20% identity to several legume lectins. The primary sequence predicts an anti-parallel β-strand secondary structure and fold highly homologous to that characteristic of legume lectins [1,3].

BIOLOGICAL ACTIVITIES

VIP36 binds Ca^{2+} but not Mn^{2+}. An immunofluorescence assay was used to demonstrate cell membrane binding that was calcium-dependent and moderately inhibited by GalNAc but not by other common saccharides. Immobilised VIP36 bound galactose-containing glycopeptides which could be specifically eluted with GalNAc [2]. However, other workers have reported Ca^{2+}-independent binding to high mannose-type glycans [4].

TISSUE AND/OR SUBCELLULAR DISTRIBUTION

In the mouse, Northern blotting has revealed expression of VIP36 in many organs including kidney, liver, lungs, spleen and heart [2]. Within canine epithelial cells, it was localised mainly to the Golgi apparatus and the plasma membrane [1].

POSSIBLE FUNCTION(S)

VIP36 is believed to function as a sorting receptor in the secretory pathway, transporting glycoproteins from the Golgi to the cell surface, but its precise role remains to be defined [1,2].

COMMENTS

VIP36 appears to belong to a separate lectin family of intracellular sorting receptors which includes ERGIC-53 (qv).

REFERENCES

1. Fiedler, K., Parton, R.G., Kellner, R., Etzold, T. & Simons, K. (1994). VIP36, a novel component of glycolipid rafts and exocytic carrier vesicles in epithelial cells. *EMBO Journal* **13**, 1729–1740.
2. Fiedler, K. & Simons, K. (1996). Characterization of VIP36, an animal lectin homologous to leguminous lectins. *Journal of Cell Science* **109**, 271–276.
3. Fiedler, K. & Simons, K. (1994). A putative novel class of animal lectins in the secretory pathway homologous to leguminous lectins. *Cell* **77**, 625–626.
4. Yamashita, K., Hara-Kuge, S. & Ohkura, T. (1999). Intracellular lectins associated with N-linked glycoprotein traffic. *Biochimica et Biophysica Acta* **1473**, 147–160.

Xenopus laevis egg lectin

Other name: XL-43
Mol. wt: 480 000 (native); 43 000 and 45 000 (subunit)
Specificity: Galactose
Sources: Eggs and embryos of the clawed toad *Xenopus laevis*

ISOLATION

The lectin was isolated from extracts of homogenates of *X. laevis* oocytes or embryos by affinity chromatography on immobilised melibiose, eluting with the same sugar [1].

STRUCTURE

The lectin is a glycoprotein (20% carbohydrate) with an apparent Mr of 480 000 estimated by gel filtration. Both SDS-PAGE and isoelectric focusing indicated subunit heterogeneity, possibly due to variable glycosylation. The former resulted in two major diffuse bands with mobilities corresponding to molecular weights of 43 000 and 45 000, while the latter revealed four discrete molecular species with isoelectric points in the vicinity of 4.6 [1].

BIOLOGICAL ACTIVITIES

The lectin agglutinated trypsin-treated, glutaraldehyde-fixed lapine erythrocytes in a calcium-dependent manner. Agglutination was inhibited by galactose, but not by other monosaccharides. Thiodigalactose, melibiose and lactose were more effective inhibitors than galactose; methyl-α-D-galactose was of comparable potency to galactose, but methyl-β-galactose was considerably less potent [1].

TISSUE AND/OR SUBCELLULAR DISTRIBUTION

The lectin was found in blastula stage embryos and oocytes of mixed stages, where it comprises 1–2% of the total soluble protein, and similar activity was detected as a minor component of extracts of adult liver [1]. The lectin was localised to the yolk platelets, cortical granules and vitelline envelope of oocytes and unfertilised eggs [2]. (What may have been the same lectin was previously localised to cortical granules of unfertilised eggs [3].) In blastula-stage embryos, the lectin was found extracellularly in the cleavage furrow region [2].

POSSIBLE FUNCTION(S)

It is possible that the lectin contributes to inhibition of polyspermy [2,3], but the observed distribution of the lectin may be an indication of multiple roles during development [2].

COMMENTS

This lectin is clearly distinct from the other frog lectins, the *Rana catesbeiana* egg lectin (qv) and the *Rana* sialic acid-binding lectin (qv).

REFERENCES

1. Robertson, M.M. & Barondes, S.H. (1982). Lectin from embryos and oocytes of *Xenopus laevis*. *Journal of Biological Chemistry* **257**, 7520–7524.
2. Robertson, M.M. & Barondes, S.H. (1983). *Xenopus laevis* lectin is localised at several sites in *Xenopus* oocytes, eggs and embryos. *Journal of Cell Biology* **97**, 1875–1881.
3. Wyrick, R.E., Nishihara, T. & Hedrick, J.L. (1974). Agglutination of jelly coat and cortical granule components and the block to polyspermy in the amphibian *Xenopus laevis*. *Proceedings of the National Academy of Sciences USA* **71**, 2067–2071.

Appendix A: Classification of Lectins According to Sugar Specificity

As a generalisation, different lectins have different specificities. Even when two lectins appear to recognise the same sugar(s) on preliminary investigation, more detailed studies usually reveal differences. Nevertheless, it is often convenient and helpful to group together lectins with similar sugar-binding properties. Therefore, the entries in Chapter 5 are listed below under various headings that are not mutually exclusive and, for some lectins, the classification is fairly arbitrary. As far as possible, each lectin has been placed in a single category, but some have been appropriately placed under two headings. A distinction has been made between some nominally sugar-specific lectins into 'simple' and 'complex' specificities. The former refers to activity which is readily inhibited by a low concentration of the free sugar and presumably results mainly from recognition of a terminal monosaccharide in glycoconjugates. The latter term is used where there is evidence of involvement of saccharides but no inhibition by free sugars (or very high concentrations are required), or where there is other evidence for recognition of a more extended binding site. A miscellaneous group including lectins of unknown specificity and others known to recognise an ill-defined extended structure have been classified together as 'complex'. For an accurate understanding of the sugar specificity, the reader should consult the original literature, to which he or she will be guided by consulting the individual entries in Chapter 5.

SPECIFICITY

Acetylated sugars:
- *Diogenes affinis* agglutinin
- *Halichondria okadai* lectin I
- Tachylectins-5A and -5B
- *Teleogryllus commodus* lectin

Arabinose:
- Anadarin P

Complex (miscellaneous):
- *Charybdis japonica* blood type B lectin
- Echinoidin
- Granulocytin
- Physarumin
- Scyllin
- Serum amyloid P component

Tachylectin-1
Tachylectin-3
Tachylectin-4

Fucose:
Eel agglutinins
Hakata antigen
Periplaneta americana lipopolysaccharide-binding protein

Fucose/GalNAc:
Kupffer cell receptor
Periplaneta americana (original) lectin

Fucose/GlcNAc:
Versican

Fucose/glucose/mannose:
Surfactant protein A
Surfactant protein D

Galactose:
Agkistrodon piscivorus venom lectin
Anthocidaris crassispina egg lectin
Aplysia kurodai skin lectin
Axinella polypoides agglutinins I and II
CD23
Didemnum candidum I and II isolectin families
Drosophila melanogaster lectin
Halocynthia roretzi lectin
Pallidin
Polyandrocarpa misakiensis lectin
Sarcophaga peregrina humoral lectin
Spodoptera exigua lectin
Trimeresurus stejnegeri venom lectin

GalNAc:
Cucumaria echinata lectin I
Helix pomatia agglutinin
Heterometrus granulomanus lectin
Homarus americanus agglutinin II
Tridacna derasa lectin
Tridacna maxima lectin
VIP36

GalNAc/galactose/fucose:
Aggrecan
Blaberus discoidalis lectin-2
Discoidin I
Discoidin II

GalNAc/galactose:
Asialoglycoprotein receptors
Cucumaria echinata lectin III
Cucumaria echinata lectin IV

Entamoeba histolytica lectin
Pinctada fucata lectin
Stichopus japonicus lectin I
Stichopus japonicus coelomic plasma lectin-2
Styela clava opsonin
GalNAc/GlcNAc:
CD69
NKR-P1 family
GlcNAc:
Didemnum ternatanum lectin
P35
Pieris brassicae lectin
Saxidomus purpuratus isolectins
Tachylectin-2
GlcNAc oligomers:
Aaptos papillata agglutinins
Interleukin-2
Tumour necrosis factor
β-1,3-Glucans:
Blaberus discoidalis β-1,3-glucan-specific lectin
Bombyx mori β-1,3-glucan-binding protein
CD11b/CD18
Galleria mellonella β-1,3-glucan-binding lectin
Pacifasticus leniusculus β-1,3-glucan-binding protein
Heparin:
Amphoterin
Anadara granosa heparin-binding lectin
CD31
Factor H
N-CAM
Rana sialic acid-binding lectin
Thrombospondin
Lactose/β-galactosides:
Allomyrina dichotoma isolectins
Amphibian galectins
Botryllus schlosseri galectin-like lectin
Caenorhabditis elegans galectin
Chicken galectins
Crotalis atrox venom lectin
Electrolectin
Galectin-1
Galectin-2
Galectin-3
Galectin-4/6
Galectin-5
Galectin-7

Galectin-8
Galectin-9
Geodia cydonium galectin
Halichondria okadai lectin II
HIP/PAP
Lachesis muta venom lectin
Non-integrin elastin/laminin-binding protein
Phalera flavescens lectin
Purpurin
Rana catesbeiana egg lectin
Stichopus japonicus lectin II
Thrombolectin
Urechis unicinctus isolectins
Xenopus laevis egg lectin

Mannose (simple):
Blaberus discoidalis lectin-1
ERGIC-53
Galectin-10
Mast cell function-associated antigen
NKCL
Perch roe lectin
Salmo salar mannose-binding lectin

Mannose (complex):
Calreticulin/calnexin
Cerebellar soluble lectin(s)
Interleukin-2

Mannose/fucose/GlcNAc (broad pattern recognition):
Collectin-43
Conglutinin
Macrophage mannose receptor
Mannan-binding lectin
Placental Gp120 receptor

Phosphorylated mannose or galactose:
C-reactive protein
Phosphomannosyl receptor I
Phosphomannosyl receptor II
Taglin

Rhamnose:
Melanoplus sanguinipes lectin
Periplaneta americana 26 kDa lectin
Plecoglossus altivelis egg lectin
Powan roe lectin
Roach roe lectin
Trout egg lectins

Sialic acid (simple):
$\text{Achatinin}_\text{H}$

Cancer antennarius lectin
Carcinoscorpin
Cepaea hortensis sialic acid-binding lectin
Homarus americanus (lobster) agglutinin I
Limax flavus lectin
Limulin
Liocarcinus depurator lectin
Lophopsetta maculata agglutinin
Modiolus modiolus lectins H and E
Scylla serrata lectin
Tritrichomonas foetus lectin
Tritrichomonas mobilensis lectin

Sialic acid (complex):
Acorn barnacle isolectins
Biomphalaria glabrata lectin-I
CD22
CD33
E-selectin
Factor H
L-selectin
Myelin-associated glycoprotein
P-selectin
Rana sialic acid-binding lectin
Sialoadhesin
Tachypleus tridentatus sialic acid-binding lectin

Sucrose:
Regenectin

Sulphated sugars:
Brevican
Echinonectin
E-selectin
Interleukin-1β
L-selectin
Ly-49 family
Neurocan
P-selectin
Tetranectin

Uronic acids (miscellaneous):
Aggrecan
Brevican
CD44
CD54
Neurocan
Sarcolectin
Stichopus japonicus coelomic plasma lectin-1
Versican

Appendix B: Lectins of Human Origin by Tissue Source

Blood/bone marrow cells:
- Asialoglycoprotein receptors
- CD11b/CD18
- CD22
- CD23
- CD31
- CD69
- Galectin-5
- Galectin-10
- Interleukin-2
- Macrophage mannose receptor
- N-CAM (CD56)
- NKR-P1
- Sialoadhesin

Blood plasma/serum:
- C-reactive protein
- P35
- Mannan-binding lectin
- Serum amyloid P

Bone/cartilage:
- Aggrecan

Epithelium:
- Asialoglycoprotein receptors (liver)
- C-reactive protein (liver)
- Galectin-2 (intestine)
- Galectin-4/6 (intestine)
- Galectin-7
- HIP/PAP (intestine, pancreas)
- P35 (liver)
- Serum amyloid P component (liver)
- Surfactant protein A (lung)
- Surfactant protein D (lung)

Muscle:
- Sarcolectin

Nervous tissue:
- Brevican
- Cerebellar soluble lectins
- Myelin-associated glycoprotein
- N-CAM

Placenta:
- Placental Gp 120 receptor

Tissue non-specific:
- Calreticulin
- CD44
- CD54
- ERGIC-53
- Galectin-1
- Galectin-3
- Galectin-8
- Galectin-9
- Interleukin-1
- Phosphomannosyl receptor I
- Phosphomannosyl receptor II
- Tetranectin
- Thrombospondin
- Tumour necrosis factor
- Versican

Appendix C: Commerical Suppliers of Animal Lectins

Most animal lectins are not commercially available. However, the following lectins listed in Chapter 5 can be purchased in purified and unlabelled form from the British suppliers indicated below.

Lectin	Supplier
Allomyrina dichotoma	3
Cancer antennarius	3
C-reactive protein	2, 7
Eel (*Anguilla anguilla*)	1, 3, 7
E-selectin	2, 6
(Complement) Factor H	7
Galectin-1 (Bovine spleen 'Galaptin')	7
Helix pomatia agglutinin	3, 4, 7
Homarus americanus	3
Interleukin-1	2, 4, 5, 6, 7
Interleukin-2	2, 4, 5, 6, 7
Limax flavus (slug)	2, 3, 4
Limulin (*Limulus polyphemus*)	2, 3, 4, 7
L-selectin	2, 6
P-selectin	2, 6
Serum amyloid P component	2, 7
Tumour necrosis factor-α	2, 4, 5, 6, 7
Tritrichomonas mobilensis	2

SUPPLIERS

1. Boehringer-Mannheim UK Ltd
   Bell Lane
   Lewes
   East Sussex
   BN7 1LG
2. Calbiochem-Novabiochem (UK) Ltd
   Boulevard Industrial Park
   Padze Road
   Beeston
   Nottingham
   NG9 2JR
3. EY Laboratories
   TCS (Microbiology)
   Botolph Claydon
   Bucks
   MK18 2LR
4. ICN Pharmaceuticals Inc.
   1 Elmwood
   Chineham Business Park
   Basingstoke
   Hampshire
   RG24 8WG
5. ReproTech EC Ltd
   23 St James Square
   London
   SW1Y 4JH
6. R&D Systems Ltd
   4–10 The Quadrant
   Barton Lane
   Abingdon
   OX14 3YS
7. Sigma-Aldrich Co. Ltd
   The Old Brickyard
   New Road
   Gillingham
   Dorset
   SP8 4BR

INDEX

The entries in the Directory (Chapter 5) are arranged in alphabetical order of the names by which the lectins are best known, so Chapter 5 acts as its own index. The following therefore only applies to Chapters 1 to 4.